The changing geography of the United Kingdom

Providing an integrated survey of how the geography of the United Kingdom has changed through the 1990s, this book offers an understanding of the social, economic, political and physical forces bringing about change. It captures the changing geography of the country at a time of particular interest, with the 1990s seeing not only a change in government but also radical shifts in policies in many areas.

All aspects of the social, economic, political and physical geography of the United Kingdom are covered. The basic primary use of the resources afforded by the physical environment are assessed through chapters on water, energy and bulk resources. The book then proceeds to address the geographies of manufacturing and service sectors, before considering the social issues arising, and the political frameworks within which the social, economic and physical processes moulding the geography of the country operate. The relationships between the people of the United Kingdom and their environment are considered further in chapters addressing topics such as physical hazards, pollution, conservation and climatic change. The book focuses on changes throughout the 1990s, although the final chapters look towards possible changes in the future.

Vince Gardiner is Director of the School of Social Science, Liverpool John Moores University. **Hugh Matthews** is Professor of Geography at University College Northampton.

The changing geography of the United Kingdom

Third edition

Edited by Vince Gardiner
and Hugh Matthews

Routledge
Taylor & Francis Group
LONDON AND NEW YORK

First published 1982 by Methuen
Second edition published by Routledge 1991
Third edition 2000

2 Park Square, Milton Park, Abingdon, Oxon OX14 4RN
711 Third Avenue, New York, NY 10017, USA

Routledge is an imprint of the Taylor & Francis Group, an informa business

First issued in hardback 2016

Transferred to Digital Printing 2005

Typeset in Ehrhardt by Keystroke, Jacaranda Lodge, Wolverhampton

British Library Cataloguing in Publication Data
A catalogue record for this book is available from the British Library

Library of Congress Cataloging in Publication Data
The changing geography of the UK / [edited by] Vince Gardiner and Hugh Matthews. – 3rd ed.
Rev. ed. of: The changing geography of the United Kindom / edited for the Institute of British Geographers by R.J. Johnston and V. Gardiner. 2nd ed. 1991.
Includes bibliographical references and index.
1. Great Britain – Economic conditions – 20th century. 2. Great Britain – Population – History – 20th century. 3. Great Britain – Social conditions – 20th century. 4. Natural resources – Great Britain – History – 20th century. I. Gardiner, V. II. Matthews, M. H. (Michael Hugh). III. Changing geography of the United Kingdom.
HC256.C47 1999
914.1–dc21 99–28270

ISBN 978-0-415-17901-0 (pbk)
ISBN 978-1-138-15660-9 (hbk)

For our families

Contents

Figures

Tables

Boxes

Contributors

John Blunden is Reader in Geography at The Open University and is director of a postgraduate environment programme at Fitzwilliam College, Cambridge. He has been a visiting professor at universities in Canada and the United States and has been engaged in research for the European Commission in the field of rural development. Publications include: *An Overcrowded World* (1995), *Energy, Resources, and Environment* and *Environment, Population and Development* (both 1996).

Ian Bowler is a Reader in Human Geography at the University of Leicester. His teaching and research specialisms lie in agricultural geography and the rural geography of the European Union. His books include: *The Geography of Agriculture in Developed Market Economies* (Longman), *CAP and the Regions* (INRA), *Agricultural Change in Developed Countries* (Cambridge University Press) and *Agriculture Under the Common Agricultural Policy* (Manchester University Press).

Tony Champion is Professor of Population Geography at the University of Newcastle upon Tyne. His principal interests are the monitoring and analysis of population and social change in Britain, particularly in their regional and local dimension. Publications include *Population Matters: The Local Dimension* (1993), *The Population of Britain in the 1990s* (1996) and *Urban Exodus* (1998).

Keith Chapman is Professor of Geography at the University of Aberdeen. Interests in the oil, gas and chemical industries are reflected in various publications including *The International Petrochemical Industry: Evolution and Location* (Blackwell, 1991) and in current research on industry restructuring in the EU.

David Crouch is Professor of Cultural Geography at Anglia University and Visiting Professor of Geography and Tourism at Karlstad University, Sweden. David has researched and written widely on leisure, culture, landscape and tourism, and has a special interest in the making of lay geographical knowledge.

Daniel Dorling is a lecturer in the School of Geographical Sciences at the University of Bristol. His research interests are in the visualisation of spatial social structure, the changing economic, medical and political geographies of Britain, and, in particular, the polarisation of health and life chances. He has published a *New Social Atlas of Britain* (Wiley, 1995) and helped edit *Statistics in Society* (Arnold, 1998) among other books and papers.

Vince Gardiner is Director of the School of Social Science and Professor of Geography, Liverpool John Moores University. His work on water resources includes *Water Demand Forecasting*, with Paul Herrington (GeoBooks, 1986), and he edited the Royal Geographical Society (with IBG)'s *Geographical Journal* from 1995 to 1998.

David Herbert is Senior Pro Vice-Chancellor and Professor of Geography at University of Wales, Swansea. His publications in Urban Geography include *Cities in Space: City as Place* (3rd edition 1997, with C. J. Thomas); *Communities within Cities* (1993, with W.K.D. Davies), and *Crime, Policing and Place* (1992, edited with D.J. Evans and N.R. Fyfe). He has also published on aspects of heritage and tourism.

Jonathan Horner is a Senior Lecturer in Environmental Studies at Roehampton Institute London and was formerly a Scientific Office (Pollution Control) working for Rotherham's Environmental Health Department. He has an MSc in Environmental Technology (taken at Imperial College) which included research on heavy metal pollution monitoring, a PhD (also Imperial College) which considered the effects of acid rain and fluoride pollution on plants and the Royal Society of Health's Diploma in Air Pollution Control.

Mike Hulme is a Reader in Climatology in the Climatic Research Unit at the University of East Anglia. He has published extensively in the field of climate change, including climate scenario reports commissioned by the UK Government, the European Commission, the United Nations Intergovernmental Panel on Climate Change, the World Bank, and WWF International. He writes a monthly column for the *Guardian* newspaper on UK climate and edited *The Climates of the British Isles* published by Routledge in 1997.

Peter Jackson is Professor of Human Geography at the University of Sheffield where he teaches social and cultural geography. His current research includes studies of the production, content and readership of men's lifestyle magazines, the changing identity of Ukrainians in Bradford, and the cultural geography of transnational commodities flows. His most recent book is *Shopping, Place and Identity* (Routledge, 1998).

Ron Johnston is a Professor of Geography at the University of Bristol, having previously been Professor of Geography at the University of Sheffield (1972–92) and Vice-Chancellor of the University of Essex (1992–5). He has written widely on electoral, urban and political geography, and was a co-editor of the first two editions of this book.

David Jones is Professor of Physical Geography at the London School of Economics and Political Science and currently Convenor of the Department of Geography and Environment. His publications on environmental hazards and risk include contributions in *Risk: Analysis, Perception and Management* (Royal Society, 1992), co-editorship (with C. Hood) of *Accident and Design: Contemporary Debates in Risk Management* (UCL Press, 1996) and co-authorship (with M. Lee) of *Landsliding in Britain* (HMSO, 1994).

Hugh Matthews is Professor of Geography and Director of the Centre of Children and Youth at University College Northampton. Hugh is editor of the *Journal of Geography in Higher Education* and past Director of the Geography Discipline Network. He has written many books and journal articles including *Making Sense of Place: Children's Understanding of Large-scale Environments* (Harvester-Wheatsheaf, 1992).

Malcolm Moseley is Professor of Rural Community Development in the Countryside and Community Research Unit of Cheltenham and Gloucester College of Higher Education. He was Director of the ACRE, the national association of England's Rural Community Councils, for six years. Since 1992 he has been an adviser to the European Union's 'LEADER Observatory' in Brussels. His recent research has related to the provision of services in sparsely populated areas, to regeneration partnerships and to the involvement of local people in rural development.

Joe Painter is Lecturer in Geography at the University of Durham. His research interests lie in the relations between politics, space and place in contemporary Europe and North America. He has published widely on the local state, local governance, urban politics and regulation theory and is the author of *Politics, Geography and 'Political Geography'* (Arnold, 1995).

Chris Park is Senior Lecturer in Geography and Principal of The Graduate College at Lancaster University. A graduate of the universities of Ulster and Exeter, he has written about environmental issues for more than two decades. He is author of the Routledge textbook *The Environment: Principles and Applications.*

Charles Pattie is Senior Lecturer in Geography at the University of Sheffield, where he teaches political geography and quantitative methods. His main research is in electoral geography, and he has published widely on regional trends in voting, economic influences on the vote, and party campaigning. He is co-author of a major study of the constituency redistricting process in Britain, *The Boundary Commission* (Manchester University Press, 1999), and has recently completed a project on the 1997 Scottish devolution referendum.

Jamie Peck is Professor of Geography and Director of the International Centre for Labour Studies at the University of Manchester. With research interests in economic regulation and governance, urban political economy, welfare reform and labour-market restructuring, his publications include *Work-Place: the Social Regulation of Labour Markets* (Guilford, New York, 1996) and *Workfare States* (Guilford, New York, 1999).

David Sadler is Reader in Geography at the University of Durham. His research interests are centred on questions of uneven regional development in Europe. He is managing editor of *European Urban and Regional Studies*, co-author of *A Place Called Teesside: a Locality in a Global Economy* (1994) and *Approaching Human Geography: an Introduction to Contemporary Theoretical Debates* (1991), author of *The Global Region:*

Production, State Policies and Uneven Development (1992), and co-editor of *Europe at the Margins: New Mosaics of Inequality* (1995).

Mary Shaw has a PhD in Sociology from the University of Queensland in Australia. She is currently a Research Fellow in the School of Geographical Sciences at the University of Bristol. Her research interests include social inequalities in health, housing and homelessness. She has previously conducted research in the universities of Kent, California and the Central European University (Prague).

Adam Tickell is Professor of Geography at the University of Southampton. He has published widely in economic geography on the geography of finance, regulation theory, local governance and regional development. He has held appointments at the universities of Leeds and Manchester, and fellowships from the Economic and Social Research Council and the Canadian High Commission. He has also acted as adviser and consultant to policy-making bodies, including the European Parliament and national and local government departments within the UK.

Brian Turton is Senior Lecturer in Geography at the University of Keele. He has contributed chapters to *Modern Transport Geography* (1992 and 1998) and is co-author of *Transport Systems, Policy and Planning: a Geographical Approach*.

Preface

The first edition of this book was published in 1982, the second in 1991. The first edition was stimulated by a wish to mark the fiftieth anniversary of the foundation of the Institute of British Geographers, and it summarised what geographers had learned about the changing geography of the United Kingdom during those fifty years. The second edition brought the story up to date, with authors in that edition concentrating on changes during the 1980s. Both editions were very favourably received by fellow geographers, reviewers, and students, and were purchased in substantial numbers. Copies are still in use, but there is clearly a need for updating, for two reasons.

First, the geography of the United Kingdom continues to change, reflecting changes in both global social, economic and political conditions, and conditions in the United Kingdom itself. Both previous editions of this book were written during the long period of political domination of the United Kingdom by Conservative governments, from 1979 to 1997, and reflected the impacts of the policies of those governments on the country's geography. Thus chapters describing spatial patterns of industrial and other economic activity, land use, resource exploitation and the interrelationships between people and the physical environment all variously explored the outcomes resulting from the impacts of government policies.

The present book brings the story begun in the previous editions to an end, and potentially heralds the beginning of a new one, following the election of a New Labour government in 1997. Authors were asked to concentrate on changes during the 1990s, but, as in the previous editions, were given considerable freedom to organise and develop their material as they wished, after editorial co-ordination to ensure a minimum of overlap and a maximum of appropriate cross-referencing. Authors have developed their chapters in a variety of ways, but each is characterised by a scholarly approach, founded on a detailed and intimate knowledge of their subject areas; many chapters incorporate the results of the authors' own researches into the topics concerned. Chapters were begun in 1997 and completed in spring and early summer of 1998, and the data used in them reflects this time-scale. One major change from previous editions is that many authors have been able to refer readers to Internet web pages which contain the most up-to-date data available.

A second reason for updating this book is that geography itself has changed. Concepts such as gender and sexuality, lifestyle, commodification, consumption and globalisation now play a much more central role in the study of human geography, and within physical geography issues such as environmental risk, climatic change and sustainable development have assumed greater significance. These changes in the nature of geography are reflected within the book in both the content of the chapters, and by the addition of new ones. Whereas the second edition of this book had essentially the same authors as the first, on this occasion there are several new names in the list of authors, and some changes to the topics authors elected to cover. Many changes resulted from the inevitable progression of authors through academic life. Some volunteered to stand down as they felt that they had moved on academically away from the topics that they had covered – into new academic areas, into university administration or management, into retirement, or because they felt that they wished to make way for younger colleagues to bring in fresh perspectives. One change was sadly enforced because of the death of the author. Other changes were made in order to reflect the changing nature of the subject, with new chapters on some of the topics mentioned above. Not all authors would necessarily describe themselves as geographers, but they have in common informed perspectives on geographical issues. Continuing the tradition established by the first two editions, one editor has remained for this edition, and one has changed. However we are grateful to Ron Johnston, who co-edited both of the previous editions, for contributing to some initial discussions on the book, and for providing such a thought-provoking and stimulating introduction.

The structure of the book broadly proceeds as a progression through consideration of the resources provided by the physical environment of the UK; their development in terms of primary, secondary and tertiary activities, as tempered by social considerations; the political framework determining the activities of people; and the environmental implications of these activities within various physical, social, economic and political frameworks. Although we have largely avoided the obvious temptation to labour the 'end of the millennium/dawn of a new millennium' perspectives, the final two chapters do postulate on aspects of citizenship in Britain today and tomorrow.

Because of the changes mentioned above, this third edition is at the same time both a new book and a new edition of an old book. It does however stand alone in elucidating the changing geography of the UK during the 1990s. We hope that anyone who is interested in how and why the United Kingdom varies from place to place will gain from reading it. The target readership is students of the UK's geography, mainly at undergraduate level – although we do know that previous editions of the book have been referred to by reading lists ranging from advanced school level to postgraduate level, and across a wide range of disciplines.

We owe a major debt to all of the authors, who produced such excellent chapters, mainly on time! Thanks are also due to the anonymous secretaries and cartographers throughout the country who provided text and maps and figures for authors, and specifically cartographers Mary MacKenzie at Roehampton Institute London, and Paul Stroud at University College Northampton; our publishers at Routledge (Sarah Lloyd and Sarah Carty) for their support and assistance; the many universities which have provided support in so many ways, and over many years, for the research reported; and finally, but by no means least, those geographers and others who have contributed by their academic endeavours to the present understanding of the richly complex geography of the UK. The book is a testament to them, and to their work, without which the book would not have been possible.

Vince Gardiner, Liverpool
Hugh Matthews, Northampton

Chapter 1

Introduction

Ron Johnston

The east side of Sheffield offers one of the most stark and dramatic industrial vistas in Britain. Heavy industry and steelworks pack the valley of the Don as it flows not so quietly towards Rotherham.

Robert Waller, *The Almanac of British Politics* (1983)

The east side of Sheffield still offers one of the most stark and dramatic industrial vistas in Britain. Whereas a few years ago heavy industry and steelworks packed the valley of the Don . . . now a scene of industrial – or rather post-industrial – desolation strikes the observer. The furnaces are extinguished, most of the great buildings demolished . . . [to be replaced in part by] the giant (and highly successful) Meadowhall out-of-town hyper-shopping centre . . . just as large a source of employment . . . as the old British Steel works at Tinsley on the Rotherham border.

Robert Waller and Byron Criddle, *The Almanac of British Politics* (1996)

These two brief descriptions of Sheffield Attercliffe, one of the United Kingdom's inner-city Parliamentary constituencies in what was formerly a great centre of the country's steel industry, encapsulate many of the changes wrought on Britain's urban landscape during the last two decades. Economic restructuring has meant the death of numerous traditional heavy industrial complexes – in coal-mining, iron and steel manufacture, shipbuilding, and so forth – and widespread dereliction of their former landscapes which were already ecologically dead, or virtually so. In their place, if anything, have come the new centres of the post-modern consumption society. Elsewhere – usually a long way from the industrial wastelands – other urban scenes have been transformed by construction of the temples of the finance industries that have powered the restructuring, and where Britain's new wealth is now being created. None of that can be found in Sheffield Attercliffe, however.

The rapid elimination of the manufacturing industries on which most of urban Britain's nineteenth- and early twentieth-century prosperity was based, especially urban Britain north of the Watford Gap, was short and sharp, ending a long period of slow decline. In some cases, it was presaged by troubled times within industries that were struggling to survive and in almost continual conflict with workforces represented by militant trade unions. The precipitous run-down of the coal-mining industry, for example, followed a year-long strike which paralysed most of the coalfields during 1984–5 and was characterised by violent clashes between police and pickets: unlike most employer–union confrontations, the strike was not over wages and conditions but rather the future of the industry and job security, and it started only a few miles from Sheffield Attercliffe when British Coal proposed to close a pit (Cortonwood) in Rotherham for 'economic reasons'. Soon after the strike was ended, and the union was emasculated, the rate of pit closures gathered pace, and mining is now such a small residual of its former size that its main union, the National Union of Mineworkers, no longer qualifies for a guaranteed place on the inner councils of the Trades Union Council, which is increasingly dominated by unions representing workers in 'white-collar industries'. In somewhat similar fashion only a few years previously, the decimation of the British steel industry also followed a major strike, in which the steelworks that occupied part of the site now filled by the Meadowhall shopping centre were an important focus.

One aspect of economic change in the UK over recent decades has been the virtual disappearance of what were long considered the country's main manufacturing industries – not only the heavy trades but also other major staples, such as textiles. These were the economic cores of the country's industrial conurbations, and most of their major cities – Glasgow, Manchester and Liverpool, Leeds and Bradford, Sheffield, Belfast and Birmingham, and Cardiff and Swansea – were founded on those industries. Several million jobs, most of them filled by men, disappeared: unemployment grew exponentially and replacement jobs were extremely hard to find – certainly jobs offering the same security and incomes. This was not a temporary slump from which recovery was expected, with the factories only remaining idle until demand was rekindled and staff could be re-employed: the jobs would never return and the plant, even very new plant in which there had been substantial recent investment, was redundant. The asset-strippers and demolition firms soon moved in, and many thousands of hectares of derelict industrial wastelands were created in what were once the hearts of the British manufacturing economy.

Some new jobs appeared after a few years, resulting from efforts to regenerate the impacted areas but, as Meadowhall shows, they were in very different types of industry. Most were very different types of jobs, too: they demanded fewer well-honed skills – certainly not those involving years of apprenticeship and on-the-job training which characterised the industries they replaced; many were part-time and relatively poorly paid, with little security or career prospects; and many more than was the case with those that they replaced were taken by women. For those made redundant in the declining industries, opportunities were few, and many settled for early retirement, perhaps linked with various forms of part-time 'grey economy' employment; the new jobs catered for new entrants to the labour force.

The new shopping centres such as Meadowhall at the centre of many redevelopment and regeneration strategies are not only places for buying consumption goods but also consumption centres themselves. They are destinations for day outings, by car or by coach, leisure experiences of a sort that never characterised Saturday visits to traditional city-centre shopping areas. To cater for them, the new malls open late in the evenings and, increasingly, all through the weekends – as do the 'theme parks' which are at the centre of so many tourist visits. Meanwhile, the old city-centre shopping areas languished, as Sheffield illustrates: many of their most profitable outlets moved to the suburban malls while those that remained slowly declined, servicing a more restricted clientele based on local employees plus inner-city residents who lacked the mobility to visit the consumption cornucopia of the motorway exits and whose purchasing power was low. Many High Streets today, even those in thriving market towns such as Salisbury, seem to have more charity shops and estate agents than any other types of commercial enterprise, as the trend to the suburban and out-of-town shopping centres has accelerated. Urban authorities have tried to make their traditional centres more attractive to counter such trends – as with Eldon Square in central Newcastle and the rebuilding of Manchester's Arndale Centre after its near destruction by an IRA bomb in 1994, and many smaller 'environmental enhancement' schemes, most of which involve closing main streets to private vehicles – but their long-term prospects look at best uncertain and at worst bleak.

Countering the decline of manufacturing industry nationally was a major expansion of service industries, many involved in some aspect of finance. London was the predominant focus of activity, as one of three major 'world cities', and office development there proceeded apace – both in the City itself and in satellite districts, such as Docklands and Croydon. A wide new range of jobs was on offer, especially during the late 1980s, many of them paying very high salaries (with even higher bonuses) and attracting a new type of labour force, the so-called 'yuppies', who contributed to the life of the eating and drinking places and patronised the shops in the surrounding areas – but only from Monday to Friday! Those workers took their large salaries back to their suburban and exurban homes, and their weekend cottages, where they contributed both to a boom in housing values within London's ever-expanding commuter hinterland and to the success of a wide range of service industries, not only the traditional suppliers of food, drink, clothing, furniture and 'white goods' but also those retailing an increasing range of electronic gadgets and 'specialist interest commodities' such as antiques.

Other manufacturing industries have been established, many associated with information technology in some form or another and financed (perhaps with government subsidies and sweeteners in the case of the largest) by foreign investors. But they differ from their predecessors: most are in small plants not locationally tied to major urban agglomerations, while their workforces tend not to be unionised and are much more flexible. They are to be found throughout urban England, especially urban southern England, in small towns and large, where the attractive environments drew the initial investors and from whose successes further new firms are then 'spun off'. Competition to attract them is intense. The old industrial areas rarely succeed, their desolate landscapes, their labour histories, and their distances from London's 'bright lights' act strongly against them when competing with rapidly expanding towns such as Swindon, the focus of large recent investments by the Japanese car giant Honda.

Many of the late twentieth century's new jobs are not associated with making things, however, but rather with consumption associated with leisure. Such are the contemporary levels of labour productivity that relatively few people are needed to manufacture goods, and many of our needs can be met more cheaply by overseas producers (though such is the depression in some UK industrial areas that their prevailing wage rates were low enough to attract investment in assembly plants for corporations based in the successful 'tiger economies' of Asia seeking access to the large European Union markets, before their problems in 1997–8; parts of Britain were able to compete with the 'third world' for cheap labour!). A substantial proportion of those in work were earning incomes far in excess of what was needed for their basic daily and weekly needs purchasable from the shopping malls – despite the high levels of unemployment in some places and the low wages associated with part-time temporary, low-skilled jobs in other sectors of the economy – and their recreation and leisure demands have stimulated major growth sectors of the economy: the UK became divided between work-rich and work-poor households.

Large-scale leisure and entertainment developments are exemplified in Sheffield Attercliffe by the Don Valley Stadium and Don Valley Arena, both built for the World Student Games in 1991 and justified to the major 'investors' – Sheffield city ratepayers who face decades paying off the interest on loans of over £150 million – on the grounds that they would form major foci for redevelopment and regeneration (of which there is as yet little sign). A range of national and international events is held there, including the Labour Party's premature celebration rally before the 1992 General Election. Other cities have similar strategies for generating economic growth: Birmingham, for example, has a major node at the National Exhibition Centre on the city's eastern edge, close to its international airport; Edinburgh boasts that its 'Hogmanay street party' is the biggest New Year's Eve celebration in the world; Manchester bid twice for the Olympic Games as a focus for its regeneration strategy, and will host the 2002 Commonwealth Games; and the Millennium Exhibition at Greenwich in 2000 has been presented not only as a centrepoint of the country's celebrations then but also as a permanent exhibition site on what was a wasteland following the decline of the Thames-side industries.

These major developments, and the host of smaller ones throughout the country, are the centrepieces of local redevelopment strategies, catering not only for the leisure interests of local residents but also for large projected numbers of tourists, both home and overseas. Sheffield's Lower Don Valley is just one of many; it includes a redevelopment of the canal that runs through the area, with conversions of the warehouses at the terminus close to the city centre. Almost every town has been developing what it identifies as its unique attractions, with an ever-wider range of facilities presented to the would-be visitor: Britain's traditional manufacturing industries are now heritage, not producers of wealth, with industrial museums and districts 'selling' the 'industrial experience' in a wide variety of ways – as at Sheffield's Kelham Island and County Durham's Beamish Outdoor Museum. (Local government financial problems are impacting on these facilities, however: Sheffield's long-established Abbeydale Industrial Hamlet, in the west of the city and based on a pre-Industrial Revolution settlement, was 'mothballed' in 1997.)

This discussion of economic restructuring and associated cultural changes has focused on urban Britain, using one of the country's archetypal industrial cities, Sheffield, to illustrate much of what has been happening. But change is also occurring elsewhere – just as fast and just as obviously to the observer – in the countryside and small towns. New agricultural practices produce major landscape changes: crops such as oilseed rape and linseed, for example, mean that parts of England's green and pleasant land are at certain times of the year England's yellow and pale blue land. Crises of over-production, associated with removal of subsidies and occasional health scares about food products (eggs and beef, for example), have led to some marginal farms being abandoned, while many more farmers (or their successors, the agri-business managers) take advantage of subsidies to create set-aside land or to move into potentially more profitable activities such as golf courses and horse-riding; and the pressures to diversify sources of income have stimulated the conversion of many agricultural buildings to either permanent or visitor occupation whereas others are transformed into 'business units'.

The villages and small towns serving agricultural areas have experienced the consequences of these changes. As farm employment has declined, job opportunities have been reduced and the traditional rural communities eroded. In their place have come the commuters, the retirees, and the visitors, many of whom form their own communities separate from their long-established neighbours'. They gentrify the villages and small towns with their barn conversions and bijou upgraded cottages, but conduct their consumer behaviour in the larger towns: village and small-town post offices and shops are no longer viable, and many pubs are unable to survive – especially those which fail to diversify successfully into 'rural restaurants': in 1998, the monthly journal *Wiltshire Life* was so concerned about these trends that it launched a 'Save Our Villages' campaign a few months after the well-supported 'Countryside March' to London in October 1997 which gained massive media coverage of a range of concerns about threats to the 'rural way of life'.

These economic changes are linked to parallel cultural and social changes. Culturally, as we have seen, economic restructuring is partly founded on new forms of consumption catering for rapidly altering niche markets, based on 'position goods' which adorn the home and the body until they fall out of fashion, along with commercially produced entertainment of an ever-multiplying diversity of forms. New lifestyles are being created and encouraged by the media and other trendsetters.

These new lifestyles are enjoyed by only a proportion of the population, however – mainly those with marketable skills (most, but not all, based on many years of education and life-long learning, hence the expansion not only of the traditional universities but also a range of public and private sector further and higher education institutions, not least the business schools and their ubiquitous MBAs – full-time and part-time, executive and distance learning, modular and prior-learning accredited). These bring high incomes and relative job security, on which the ever-available credit for 'good risks' can be based. For the remainder of the population life is a continual struggle, with always-impending threats of redundancy consequent on restructuring, possibly following a take-over, down-sizing and/or de-layering, while pressures for greater productivity lead to people working longer and more unsociable hours, with consequent stresses for family lives and social relationships. Thus alongside the wealth of 'booming Britain' we find the hardship and poverty of 'struggling Britain' and an increasing proportion of the population so marginalised from the country's mainstream that there are increasing concerns about social exclusion of whole sections of society, with some alienated from it and others unable to find a niche within it.

This contrast between affluence and poverty is producing a polarised society of haves and have-nots, not only in one generation but for a sequence of generations. They tend to be spatially segregated, at a variety of scales, regional and local. Market researchers and others use a range of data on the characteristics and buying habits of these different communities to segregate them into distinct market sectors – such as 'Clever capitalists', 'Rising materialists', 'Corporate careerists', 'Co-op club and colliery', 'Smokestack shiftwork', 'Bijou homemakers' and 'Sweatshop sharers': some are the targets of aggressive marketing strategies; others are ignored. Geographical polarisation is exacerbated.

With relative poverty come other symptoms of a divided society. Health differentials are widening – between groups and between places: certain types of crime are increasing and more people (though not only the excluded) are seeking relief through addictive drugs (another source of economic enterprise – albeit illegal, except in the cases of nicotine and alcohol). Poverty, unemployment, addiction and crime interact and generate spirals of decay. They cause stresses within households and initiate the break-up of many partnerships and families. Sheffield illustrates this for us through the film *The Full Monty*, whose underlying theme is not just of men made unemployed by the collapse of the steel industry but also of men without prospects, even hope; their roles in life are increasingly taken over by women who succeed in the new service economy and who make men feel not only redundant but also obsolescent. In some places, it is not just individual households that have fractured under such strains (with consequent implications for housing and child-rearing) but entire communities. The pit villages of South Yorkshire illustrate this, and another film – *Brassed Off* – delivers the message superbly. The miners' strike empowered women there for the first time, both economically and politically, and when the strike was over many were unwilling to return to their former domestic roles as expected by the men.

Social and cultural changes have been manifold within the culture of contentment in the more affluent sectors of the economy too. Changing attitudes to childbearing and child-rearing accompanied the increased participation of women in the labour force throughout their lives, with a consequent expansion in the number of dual-career families dependent on a range of formal and informal ways of providing childcare. Many of those families, especially in the professions, involve one or both of the partners in long-distance commuting: substantial numbers operate two homes during the working week. Commuting patterns are increasingly complex and cross-cutting: many depend on private transport and contribute to the build-up of traffic and its associated problems in rush-hours which begin on Sunday evenings and end on Saturday mornings.

Social, cultural and economic changes are intertwined and inscribed in the country's landscapes. They interact with the political processes, and these too have been very substantially modified in recent years. The post-war settlement which lasted from 1945 until the late 1970s was based on the goals of full employment and the reduction of inequalities, enshrined in the five pillars of the welfare state – combating want, disease, ignorance, squalor and idleness. This came under increasing attack from economists and right-wing politicians as both unsustainable and a major constraint on the individual and corporate enterprise necessary for successful economic restructuring in a globalising world: the corporate-welfare state's high and spiralling costs meant that taxation rates were too high to encourage risk-taking while the necessary government borrowing both led to high interest rates (another brake on investment and enterprise) and stimulated inflation. By the mid-1990s, the case for reforming the welfare state was accepted by all political parties.

The critique of the welfare state was part of a broader attack on the state's role in contemporary society, accompanied by a rhetoric that it encouraged dependency and discouraged self-help. The state was implicated in too many aspects of people's lives, it was argued, and its role should be reduced: people should take more control of their present and future, and more of their necessities should be provided through much more efficient and effective market mechanisms. At the same time, those services which remained with the state – such as health-care and education – should be subject to the rigours of the marketplace: many others, such as the public utilities, should be returned to that marketplace, with the state, if necessary, acting only as a regulator to safeguard society's interests.

Much of this rhetoric was implemented during the 1980s and 1990s, creating a new political map of the United Kingdom. One of the main mechanisms chosen was to place a wide range of public services – such as health-care trusts and many educational institutions – under the management of non-elected boards of directors and governors which were accountable to central government for their expenditure and the achievement of specified targets, but not subject to direct democratic control. These quangos, as they became known, were dominated by political appointees who accepted the pervasive rhetoric. Complementing this trend was a decline in the role of the democratic state: central government oversight of the detailed implementation of policies was reduced while local government powers were eroded and in some cases eliminated.

Once again, Sheffield Attercliffe illustrates what has happened. Economic and social regeneration was vested not in the democratically elected and accountable Sheffield City Council (South Yorkshire County Council having already been eliminated as redundant in 1987) but in a quango – the Lower Don Development Corporation, whose membership was dominated by government appointees (most of them businessmen) and who were responsible for developing public–private partnerships to redevelop the valley. The City Council's responsibilities for land-use planning were substantially reduced and control of what went on in a substantial component of their city was largely taken away from Sheffield residents. At the same time, local government powers in other areas were reduced – they were required to sell their council housing to sitting tenants at cut-price rates, for example, and prevented from using the receipts to build more homes. Eventually, even those local councils dominated by large left-wing Labour majorities were forced to accept the new ideology and methodologies – radical Sheffield, in Patrick Seyd's telling phrase, shifted its emphasis from socialism to entrepreneurialism, replacing its role as a focus for anti-Thatcherism opposition by joining the competition for inward investment in an 'enterprise-favourable environment', while some of its leaders abandoned their socialist credentials and became avid supporters of New Labour.

Local government was restructured too, by legislative fiat in Scotland, Wales and Northern Ireland, and by a complex system of pseudo-consultation in England that largely failed, producing a complicated system of authorities. The goal was to make local government more efficient and effective, reaping the economies of scale and scope – though the scope could be that of the purchaser of services from outside contractors rather than the traditional role of provider.

Government became both more centralised and less open to democratic accountability: even the functions remaining with the local state apparatus were under strict central control, notably (but not only) on the amount raised through local taxation. There was a growing democratic deficit. Much of this occurred when the country was very divided politically: the northern areas which suffered most were predominantly represented at Westminster and governed locally by the Labour Party, which was in opposition nationally from 1979 until 1997, whereas almost all of the booming southern regions' MPs were Conservatives, as were a majority of councillors in most of their local governments until the Tory fall from grace after 1992. The national political map changed in 1997, not because the northern regions were becoming much more prosperous again relative to their southern counterparts but because the electorate in the latter were increasingly disillusioned with the government, with its divisions (especially over the major issue of British participation in the European Union), sleaze, lack of leadership, and failed macro-economic policies (as exemplified by Black Wednesday in September 1992 when sterling was forced out of the Exchange Rate Mechanism, the precursor of European Monetary Union). New Labour replaced it with a government having the largest House of Commons majority for sixty years: it won middle-class support and seats in the south, not because it offered an alternative macro-economic policy – indeed, it pledged that it would continue many of its opponents' policies – but because it looked better able to continue providing much of what had been delivered by the Conservatives in the 1980s.

The democratic deficit has stimulated calls for constitutional reform, though much of this has been elite-driven rather than populist. The main exception has been in Scotland, where pressure for a devolved Parliament with tax-raising powers was substantially confirmed in a 1997 referendum and the first elections were held in 1999. The Welsh also voted in 1997 (by a slim majority) for a weaker form of devolution – a Welsh Assembly, also elected in 1999; following the 1998 'Good Friday Agreement' a devolved Assembly was rapidly created in Northern Ireland, as part of a programme to end thirty years of bloody inter-community strife there, with its first elections only three months later; and Londoners voted in May 1998 for their own reconstituted metropolitan government led by a separately elected 'strong mayor'. There is considerable pressure for devolution to some English regions – notably the North East and the South West – too. The political map is being redrawn yet again, with potential implications for economic and social directions.

All of this change is taking place on a landscape which is a palimpsest of many thousands of years of human occupance, and which itself presents major constraints and opportunities. It is experiencing growing demands to provide the fundamentals needed for affluent living, not only agricultural products but also the raw materials for construction projects; it is under pressure to contain those lifestyles, not least the further 4 million homes which the government estimates are required over the next two decades; it is required to absorb an increasing volume of waste materials; and it is affected by various global environmental changes. Conservation and preservation of physical landscapes and the human imprint upon them is in increasing conflict with those demands, and makes sustainable living for future generations increasingly difficult to ensure.

All of the change described here has taken place in the last twenty years: it represents another epoch in the remaking of the UK landscape. Each edition of this book has appeared at what seemed to be a critical moment in that continual process: in 1983, at the end of a long period of reconstruction after the major slump of the early 1930s, punctuated by a world war; in 1991, after a decade of change as the welfare state came under fire and the assumed superiority of the market was being put to the test; and now in 1999 in the early stages of a new approach to that experiment. The basic themes identified here are developed in the various chapters that follow; who can guess how Robert Waller will represent their playing out in Sheffield Attercliffe before the general election of 2007?

Chapter 2

Water resources

Vince Gardiner

In concluding the chapter on water resources in the previous edition of this book, Park said: 'The water industry in the United Kingdom during the 1990s will look and act very differently from how it did in the 1980s' (Park 1991: 169). This has undoubtedly proved to be the case, with the privatisation of the water industry, legislation concerning water quality, and changes in patterns of both demand and supply all interacting to ensure rising public awareness of, interest in and controversy about water resource issues. In this Chapter the basic geography of use and supply of water is outlined, and then developments and events during the 1990s are discussed. The major theme of this discussion is the changing structure of the water industry, especially in England and Wales.

The basic geography of water use and supply in the United Kingdom

Although it is not always possible to differentiate clearly between supply of and demand for water (see Park's discussion, 1991), it is possible to think in at least conceptual terms of the ways in which water is used, and the sources of supply of that water. In the United Kingdom water is used for a great variety of purposes, which may be broadly categorised as domestic, industrial and agricultural. Total use is discussed first, before individual uses are examined.

Total abstraction of water from all sources in England and Wales is shown in Table 2.1. Although there is significant year-to-year variation, there is little evidence of a systematic trend of increasing abstraction. In 1995, the total abstraction in England and Wales was divided as in Table 2.2. About a third was used for public water supply, by the water companies; this includes both water used in homes and water supplied to a wide variety of commercial, business and public buildings. Figures for water put into public supply over the last twenty years show a 15 per cent increase from 1974/5 to 1994/5, but only a 3 per cent increase within the last decade. Although OFWAT (see p. 16) suggested in

TABLE 2.1 Estimated actual abstractions from all surface and groundwater sources in England and Wales, 1985–95 (for limitations of these data, see source)

Year	*Quantity (ml/day)*
1985	51,206
1986	50,883
1987	52,122
1988	55,855
1989	61,885
1990	60,208
1991	59,196
1992	67,909
1993	53,742
1994	53,640
1995	55,970

Source: Abstracted from data in the Department of the Environment, Transport and the Regions (DETR) World Wide Web pages.

TABLE 2.2 Total abstraction of water in England and Wales, 1995

Use	*Per cent*
Public water supply	31.0
Spray irrigation	0.6
Other agriculture	0.2
Electricity supply industry	46.1
Other industry	13.4
Mineral washing	0.5
Fish farming, cress growing, amenity ponds	7.63
Private water supply	0.2
Other	0.4

Source: Abstracted from data in the DETR World Wide Web pages.

1994 that the demand for public water supply was virtually static, and would remain so until 2014–15, there is some suggestion (Figure 2.1) that periodic and almost cyclic fluctuations in demand are superimposed on a generally rising trend. Troughs in total demand occurred in 1988 and the early 1990s, but overall demand continues to rise slowly, for both the UK as a whole and its constituent countries. A more detailed analysis of the uses of water put into public water supply in particular regions is shown in Figure 2.2 for the end of the period considered above.

Although the demand for water supply fell until 1993, there has since been an upturn in demand for both metered and unmetered supplies. The growth in demand for metered supplies varies from company to company, being least in the areas served by the South West and Yorkshire companies, where rates are high and there have been strong public campaigns against charges. For unmetered supplies again variation exists, with North West, Southern, South West and Yorkshire not indicating consistent growth, unlike in the other areas.

Domestic use accounts for about 40 per cent of all water abstracted for the public supply, and is affected by increasing affluence and higher standards of living. The average household uses about 380 litres per day, or about 160 litres for every person, one-third of which is used for each of toilet flushing and miscellaneous uses. Only around 5 litres per day is used for drinking and cooking. The main increases in household water use are for personal hygiene and for watering of gardens, both of which are most significant in summer, when supplies are under most stress. The accuracy of forecasts of future demand will remain heavily dependent on predictions of economic, lifestyle and cultural changes within the communities of consumers, and are likely to be further influenced by climatic change.

About half of the water abstracted in 1995 in England and Wales went to the electricity supply industry, mainly for hydropower. Other industry accounted for about an eighth of the total. Only a very small proportion was used for irrigation in agriculture (less than 1 per cent), although there is an underlying trend for this to increase. Although spray irrigation forms only a very small proportion of the total demand for water, it can be very significant because the water is immediately lost as a resource, and it occurs in a concentrated period of the year, usually in the driest areas, and in driest years (Figure 2.3). Over the last forty years changes in agricultural practices have moved the emphasis in irrigation away from increasing yields and towards securing better quality and a more reliable supply. Irrigation rates have doubled from approximately 1,300 litres per day per hectare in 1982 to 2,500 litres per day per hectare in 1992, although these rates depend greatly on weather conditions. Potatoes account for most of the water used (59 per cent in 1992), closely followed by other vegetables (see also Chapters 5 and 19).

There has been a general reduction in the amount of water abstracted for industry and general agricultural purposes over the last decade, attributed partly to increased efficiency in use, and partly to the shift in the national economy away from manufacturing and towards service industries using less water (see Chapter 7). This decline has however been accompanied by a general increase in the demand for water for irrigation, fish farming and associated uses, and hydropower.

Since 1989, the vast majority of people in the UK have received their water and sewerage services from companies appointed under licence and regulated by official bodies (see pp. 16–17). However, such services can be provided by others. These supplies are not regulated, but need to comply with recognised water quality standards, and need to be licensed by the Environment Agency for abstraction and discharge of water. An unregulated supply could be a well supplying the owner's premises and perhaps several neighbouring sites, or could be a larger supply serving a village through a pipe network. There are around 50,000 self-contained water supplies in England and Wales which supply more than one property, and many industrial customers have their own sewage treatment works; 4 per cent of domestic properties in England and Wales are not connected to main sewers, having either individual septic tanks and cesspools, or private treatment works.

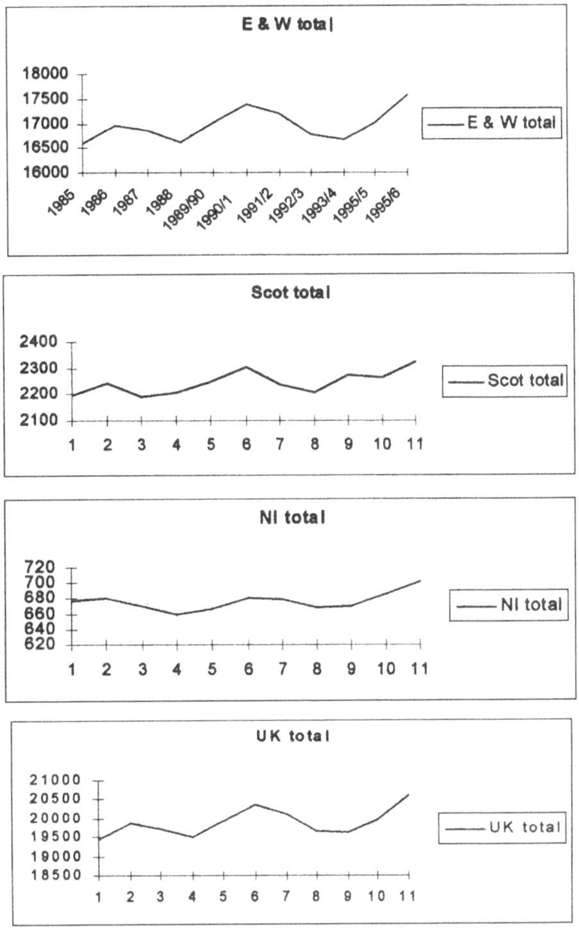

FIGURE 2.1 Trends in public water supply in the United Kingdom, 1985–96

Source: From figures in the DETR World Wide Web pages.

Note: For all figures and data in this chapter derived from Environment Agency, OFWAT and DETR data, there are considerable caveats and qualifications to be attached to the data, and the original source should be consulted.

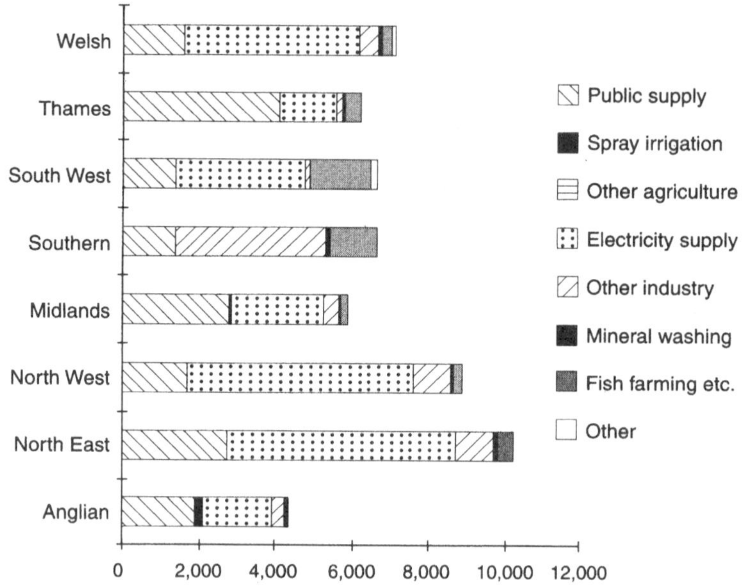

FIGURE 2.2 Variations in use of water abstracted from surface and groundwater for Environment Agency Regions, 1995
Source: From figures in the DETR World Wide Web pages.

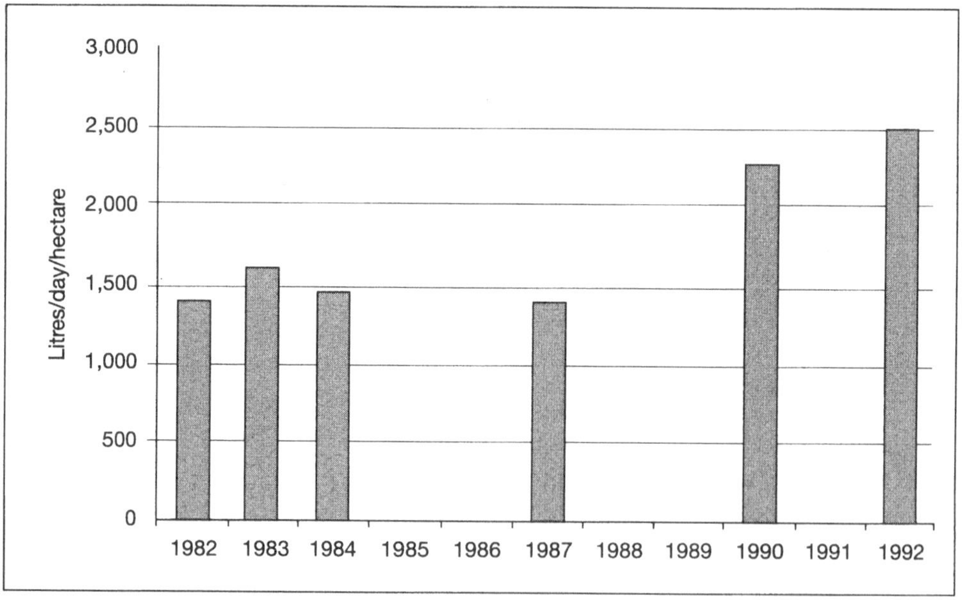

FIGURE 2.3 Irrigation water used per hectare of irrigated land, England and Wales (litres per day)
Source: Derived from figure in the DETR World Wide Web pages.

Sources of water

Water for water supply in the United Kingdom originates from both surface water, as rivers and lakes, and from groundwater. Although most parts of the country draw water from both sources, the balance between the two varies markedly, depending on both the availability of adequate surface water supplies and the suitability of the underlying rocks to provide groundwater (Figure 2.4). In general, the wetter north and west parts of the UK, which tend to be underlain by rocks which are poor aquifers, rely predominantly on surface water, whereas the drier south and east, which also have more aquifers, are more reliant on groundwater.

Substantial amounts of water are also drawn from estuaries for industrial purposes, especially for cooling in electricity generation. It should be noted that many industrial uses of water, for example for hydroelectric power and for cooling, and fish farming, do not ultimately consume water resources, as the water is shortly returned to surface water within the hydrological cycle, albeit in some cases as of degraded quality. In contrast, some uses, such as spray irrigation and abstractions for evaporative cooling, do represent a loss to the water resource.

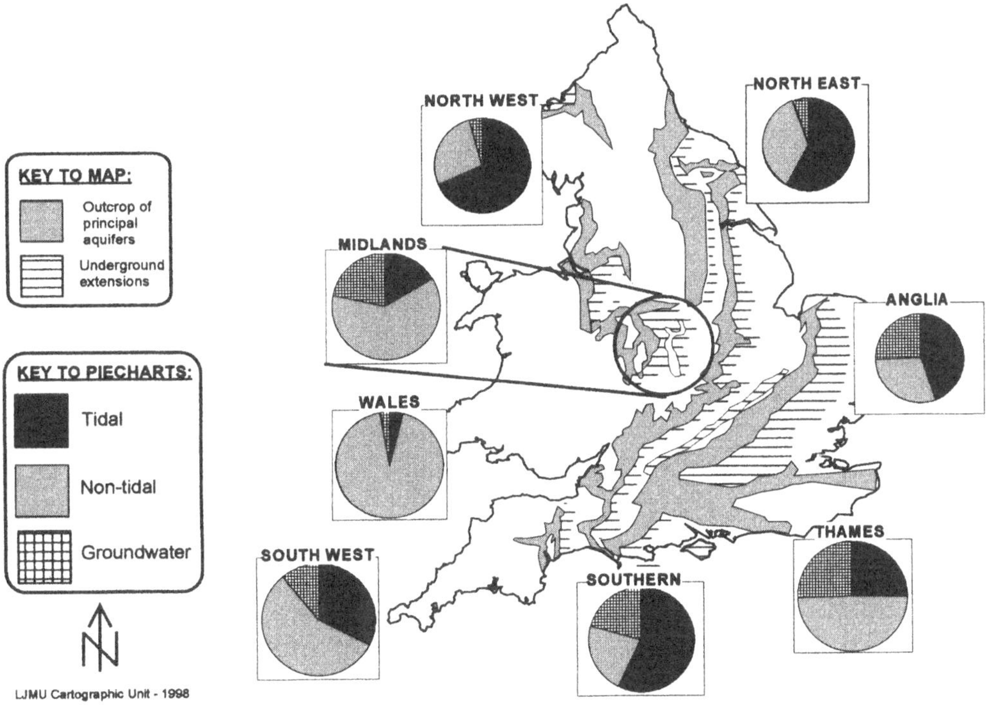

FIGURE 2.4 Estimated abstractions from various sources of water, 1995
Source: From data in the DETR World Wide Web pages.

The framework of public water supply

The present legislative, administrative and organisational framework for water resource management in the UK differs considerably from that which existed when the previous editions of this book were published. The public supply of water in the United Kingdom was first carried out by a multitude of private water companies and local authorities, with over 3,000 such bodies before the First World War. The 1945 Water Act encouraged amalgamations, and by 1963 the numbers had reduced to 100 water boards, fifty local authorities and twenty-nine privately owned statutory water companies – some of which had been operating since the seventeenth century. However, the need was recognised for larger units which could provide the integrated management of water resources within major river basins, and ten Regional Water Authorities (RWAs) were set up by the 1974 Water Act, as public undertakings. The twenty-nine statutory water companies then in existence acted as their agents for water supply, with the water supply and disposal functions of numerous local authorities being combined together into a river basin framework under the water authorities. The ten multipurpose water authorities, based on major river basins, had a wide range of responsibilities connected with the water cycle, including conservation, pollution, drainage, fisheries, water supply and sewerage. The majority of their members were from local authorities, and they calculated customers' bills on the basis of rateable values.

In the second half of the 1970s, however, there was insufficient maintenance of and capital investment in the water distribution system, whilst the RWAs were accused of individually building substantial empires, including capital investment in large supply schemes such as the Carsington (Derbyshire) and Kielder (Northumberland) reservoir schemes, which were arguably neither efficient nor necessary within the national context.

During the 1980s, the RWAs increasingly became subject to government targets and financial controls, similar to those of the nationalised industries. At a time of economic recession they were faced with a need for increased expenditure and investment in capital projects to meet increased water quality and environmental standards, mainly resulting from EC legislation. Their budgets were squeezed by the conflicting pressures of the need for investment and limits on public funding and borrowing. Prices inevitably rose, and there was increasing pressure to find alternative sources of finance for the required environmental and quality improvements.

A preliminary move away from public control occurred in 1983, when as a result of the 1983 Water Act the local authorities lost some rights of representation on the RWAs, and their meetings were closed to the press and public, although Consumer Consultative Committees were set up as some degree of compensation. A discussion paper on privatisation of the industry was published in 1986, which advocated privatisation of the RWAs as a means of freeing them from government control, and as a means of ensuring access to sources of funds for capital investments. Many organisations and individuals expressed concern about privatising the regulatory aspects of the water authorities, seeing possible conflicts in a role combining responsibility for water quality with that for sewage disposal. Despite concerted campaigns, especially from environmentalists and those opposed to the privatisation of the regulatory role of the RWAs, the 1989 Water Act allowed the RWAs to be sold off as part of the Conservative government's programme of privatisation. Independent regulatory bodies were set up: the National Rivers Authority to deal with water quality in natural water bodies; the Director-General of Water Services (OFWAT) to regulate prices in the industry; and the Drinking Water Inspectorate, to deal with drinking water quality. The then twenty-nine private companies were brought under the same regulatory controls

While privatisation was being discussed, the implications of EC legislation on water quality standards became more apparent. In recognition of this, the government wrote off £5 billion of the industry's debts before privatisation, and endowed them with a further £1.6 billion cash injection (the so-called 'green dowry'). Shares in the ten water companies which replaced the RWAs were offered for sale in November 1989, and the offer was over-subscribed. Most of the private water-only companies in existence before 1989 have since re-registered under the Companies Act 1985, so that their earlier restrictions on borrowing and paying of dividend no longer apply. Since 1989, twelve of these companies which were in common ownership have been brought together under five single licences.

Under the present system the privatised water companies, comprising the ten large water service (water supply and sewerage) companies, resulting from the sale of the former RWAs, and the much smaller water supply companies which had never been absorbed into the RWAs, have a statutory duty to maintain supplies. Under the 1989 Act regulatory roles were given to three new agencies, the National Rivers Authority, OFWAT and the DWI (see pp. 16–17). A regulator was felt necessary, as water formed a natural monopoly. Regulation relies on comparative competition and on competition for capital in the financial markets. OFWAT has allowed prices charged for water to be increased above the rate of inflation, but only in order to allow for investments necessary in order to meet raised standards, although there are those who believe that this is a failure of the regulatory system. It has been argued that the mergers which have taken place have reduced the scope for comparative competition.

Since privatisation, the industry has been affected by further legislation, which has consolidated existing legislation, and strengthened the powers of regulators. In addition, in 1994, the government relinquished its special ('Golden Share') holdings in the water and sewerage companies, which has exposed them to the potential of merger and take-over as for any other quoted company. However the Monopolies and Mergers Commission has to be consulted if any proposal would result in a new water enterprise with gross assets exceeding £30 million, and there is also European legislation governing certain large mergers. Notwithstanding this, since 1989 several of the water-only companies have been taken over by major shareholders, and their number has been reduced by amalgamations, to eighteen at the time of writing. Mergers have occurred between various utility companies of different kinds (e.g. North West Water and NORWEB, the electricity company, 1995); between water-only and water and sewerage utilities (e.g. Northumbrian Water and North East Water in 1995); and between water-only companies (e.g. East Surrey Water and Sutton Water in 1995). Some water and sewerage companies are now part of multi-utility groups (for example North West Water, Dwr Cymru and Southern Water Services are all joined to a regulated electricity business), and these multi-utility groups are increasingly moving into other competitive utility businesses such as telecommunications and gas. A number of the smaller water-only companies have also entered into the competitive gas business. Each company is required to operate at arm's length from its associates, and without cross-subsidy, and further investigations are continuing into trading relationships and the transparency of financial performance. Some of these groupings have an international dimension, for example as in the linking of Southern Water and Scottish Power, and the merger of Northumbrian Water and Folkestone and Dover Water Services Ltd with French companies. At the time of writing, Wessex Water is the subject of a take-over bid from the American utility company, Enron. Indeed, water privatisation has been seen as part of an inevitable global privatisation process by transnational companies and corporations

Undoubtedly, there has been a change in ethos in the water industry within the last decade, from that of a public service to that of a provider of a consumer need. This has been harshly underlined by the debate on disconnections of domestic supplies of those most disadvantaged in society who have been unable to pay for water, and by the enforced use of pre-payment meters. There have been many contentious and sometimes very controversial issues raised, and these are discussed below. However a crucial element in understanding these issues is the existence of regulatory bodies in the new structure of the water industry. These will therefore be described before addressing the controversial issues.

Regulation in the water industry

OFWAT (the Office of Water Services) is a non-ministerial central government agency employing (in 1997/8) 206 staff, and with running costs of £9.9 million per year, plus another £2.2 million for its consumer committees. Its head, the the Director-General of Water Services, is the economic regulator of the water and sewerage industry in England and Wales. His (the current Director-General is a man) primary duties are to ensure that the companies carry out their functions in accord with legislation, and that they are able to finance their functions. His secondary duties are to ensure that no undue preferences or discriminations are shown in charging – for example between urban and rural areas, large and small consumers, metered or unmetered consumers – and that customers' interests are protected. In some senses, he acts to even out geographical differences. However, the Director-General has called for his responsibility to protect the interests of customers to be made his single primary duty, putting customers' interests at the heart of water regulation. The Director-General has a duty to promote efficiency of water use, to facilitate competition in the industry, and a general environmental responsibility. He does not set environmental standards, but works closely with the Environment Agency (which is responsible for protecting the quality of rivers, estuaries and coastal waters) and the Drinking Water Inspectorate (which oversees the quality of tap water) (see pp. 18–21). He carries out his regulatory duties largely by means of setting price caps which allow the companies to finance their functions; he does not control profits or dividends directly. The current price limits were set in 1994, and will be reset in 1999. He also compares the performance of the companies, which helps the poor performers to rise to the standards of the best, and by setting targets, for example for leakage reduction. There are also ten Regional Customer Service Committees, and a National Customer Council. Strong and effective arrangements for the independent representation of the interests of customers are seen as vitally important in the regulation of a monopoly utility, where customers cannot take their custom elsewhere.

The ten regional Customer Service Committees of OFWAT have a statutory duty to investigate complaints made by customers. These can concern water quality, pressure, supply, sewerage, billing, charges and administration. Complaints about the industry have declined annually since 1992/3, although showing an upturn in 1997/8. There is some indication that the temporal, and perhaps spatial, distribution of complaints accords with media coverage of the industry, and changes in public confidence in and awareness of the complaints procedure. The proportion of complaints received in each category varies markedly with region. Of the ten combined water and sewerage companies, South West Water receives by far the highest rate of complaints per connection, many of which are about billing, charges or administration. Mid Kent receives a greater proportion of complaints than any other water-only company.

The Environment Agency is a non-departmental public body established by the Environment Act 1995. It is sponsored by the Department of the Environment, Transport and the Regions, with policy links to the Welsh Office and the Ministry of Agriculture, Fisheries and Food. It took over the functions of its predecessors, the National Rivers Authority, Her Majesty's Inspectorate of Pollution, Waste Regulation Authorities, and parts of the Department of the Environment. The Environment Agency has the central role in the planning of water resources at the national and regional levels, and in co-ordinating action between water undertakings; the industry's three regulators have the duty of ensuring that it is carried out properly.

The Drinking Water Inspectorate is responsible for checking that water companies in England and Wales supply water that is safe to drink and meets the standards set in the UK's Water Quality Regulations. Most of these come directly from an obligatory European Community Directive, but some UK standards are more stringent. Most standards are based on World Health Organisation guidelines, and generally include wide safety margins. There are standards for bacteria, chemicals such as nitrates and pesticides, metals, and the appearance and taste of water.

A tension necessarily exists between the need for regulation to ensure the safeguarding of the public and the environment, and the need for the water companies to be able to act freely within the market in terms of pricing and the adoption of technological innovation. It has been argued (see for example the debate in *Economic Affairs*, 1998) that there is a need for further deregulation of the industry, that price controls and rate-of-return regulation lead suppliers to lower the quality of services, that regulation can never be a substitute for competition, and that the price mechanism is the most effective means of matching supply to demand in the long term.

As well as there being debate about the need for regulation, public debate, controversy and undoubtedly sometimes resentment have focused on many other issues, including the need for integrated long-term planning and resource management; the impact of privatisation on quality of water in rivers, lakes and streams; the quality of water supplied; the impact of EC legislation; financial aspects, including the 'fat cat' salaries paid to the chairmen of the privatised water companies, the soaring stock market values of shares in utilities, and the high cost of water services to consumers; the absence of any real competition in the industry; and waste through pipe leakages and other inefficiencies. These are considered below.

Long-term planning

Prior to privatisation, strategic and integrated planning of water resources at the national level was the concern of the Water Resources Board. Its final report, in 1993, envisaged higher growth in the demand for water, to be met by large-scale responses including, for example, estuarial storage of water, increased use of river regulation, and inter-basin transfers of water. It paid relatively little attention to likely environmental impacts. By comparison, the National River Authority's report in 1994 (*Water: Nature's Precious Resource*) questioned whether major developments were in fact necessary for the next thirty years, and placed much more emphasis upon sustainability, the precautionary principle, and improved management of demand. By 1994 OFWAT was predicting no significant growth in demand for the next two decades, and suggesting that localised shortfalls in water could be met by increased sharing of resources and significant reduction in leakages. This report favoured universal water metering after the abolition of water rates at the end of the century, in order to contain demand, although the water companies opposed this strategy because they would have to bear the cost of installing meters. Strategic planning of water resources is now the responsibility of the Environment Agency, and their assessment of the long-term sustainability of the UK's water resource policies is summarised below.

In Scotland and Northern Ireland strategic water resource issues have been less critical, if only because these countries are rather better endowed with the basic natural resource – rainfall. In Scotland from 1974, water services were the responsibility of the Regional Councils. From 1996 three water authorities were formed in conjunction with local government reform, and these became responsible for the provision of both water and sewerage services. Monitoring of all aspects of water services in Scotland is carried out by the Scottish Water and Sewerage Customers' Council. The largest water authority is West of Scotland Water, which serves the areas formerly served by Strathclyde and Dumfries and Galloway Regional Councils – over 2.25 million people in an area of over 20,000 square kilometres, which includes not only the industrial heartland of Scotland and major cities, but also extensive rural areas. Supply is largely from surface water, and includes thirteen natural lochs, 134 impounding reservoirs, and ninety-five stream abstraction points, springs and boreholes. East of Scotland Water supplies water services to 1.58 million people in Edinburgh, Lothians, Borders, the Forth Valley, Fife and Kinross, and parts of Dunbartonshire and North Lanarkshire. This area has a mean annual precipitation of 111.5 cm and supply comes from 107 surface sources, including reservoirs, lochs, and rivers, and thirty-two groundwater sources. Water services for the rest of Scotland, including the Western Isles, Orkney, Highlands, Grampian, Tayside and Shetlands, are provided by North of Scotland Water.

In Northern Ireland, the Water Service both supplies water and collects and treats sewage. It supplies on average 150 million gallons per day to approximately 560,000 homes and 70,000 businesses, including farms, using reservoirs, pumping stations and treatment plants at 2,600 locations. A considerable amount of money is being spent in Northern Ireland on improvement projects to upgrade the standards of sewage and water treatment.

Quality of water

The quality of natural waters is covered in Chapter 18 of this book. Since privatisation the quality of drinking water in England and Wales has been monitored and reported upon by the Drinking Water Inspectorate (DWI), who carry out an annual assessment of the quality of drinking water supplied by each water company, and inspections of individual companies. Millions of tests are carried out each year by the water companies, and the DWI check these against the legal standards for drinking water embodied in the Water Quality Regulations. Most of these stem from an obligatory EC Directive, although some UK standards are more stringent, and most are based on World Health Organisation guidelines. There are standards for bacteria, chemicals, metals and the appearance and taste of water.

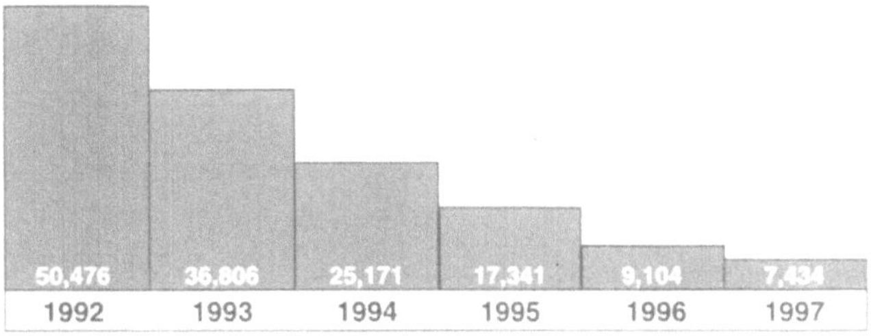

FIGURE 2.5 Tests on drinking water which failed to meet standards, 1992–7. In 1997 the water companies carried out nearly 3 million tests, of which 99.75 per cent passed – a considerable improvement over 1992
Source: Drinking Water Inspectorate Annual Report, on DWI World Wide Web pages.

Overall, drinking water in England and Wales is of very high quality, and has been improving since 1992 (Figure 2.5). In 1997, all but 0.25 per cent of the nearly 3 million tests carried out at treatment works, service reservoirs and in supply zones demonstrated compliance with the standards set. About 80 per cent of these tests were carried out on samples taken from consumers' taps. This improvement, compared with previous years, reflects the ongoing impact of the enforcement process which has resulted in improvement programmes. There is no evidence that any of the contraventions found during these routine compliance tests endangered the health of consumers, as they were of limited magnitude or duration, or concerned assessment parameters of only aesthetic significance.

Much concern has been expressed concerning pesticides in drinking water, originating from their use by farmers, gardeners and highway authorities. The standard for individual pesticides is very strict, but in recent years there has been a highly significant improvement in compliance. In 1997, 94.6 per cent of water supply zones met the standard, compared with 87 per cent in 1996, and 79.2 per cent in 1995. However it must be noted that if just one sample from a zone fails the test during the year, the zone is treated as having failed for the whole year. In 1997, 99.96 per cent of the 789,000 tests carried out for pesticides met the standards, as compared with 99.81 per cent in 1996 and 99.22 per cent in 1995. This improvement is the result of improvement programmes costing around £1 billion since 1989.

Overall, for eleven of the seventeen key parameters indicative of water quality, there were improvements in zone compliance in 1997 compared with the previous year. Compliance for faecal coliforms was unchanged, and there were small increases in zone non-compliance for taste and nitrite. However, there were significant increases for coliforms, colour and trihalomethanes.

The DWI also report upon the extent to which companies conform to monitoring requirements set by the EC's Drinking Water Directive. Generally, the companies conform, although action has been considered against seven companies in 1997 for minor shortfalls in sample numbers or inappropriate sampling methodology. Each company also met the Regulations in terms of making the data readily available to the public, although the DWI are concerned that water consumers make very little use of the public record of drinking water quality.

There has been some concern expressed about consistency in the notification of water quality incidents. Between 1990 and 1995 the DWI analysed 317 incidents. Over this period the number of incidents at service reservoirs decreased, but those arising in the distribution system increased. In 1997, ninety-five incidents were notified – a slight increase over previous years. However, this is thought to be due to increased diligence in reporting as well as to difficulties in the distribution system. Of the ninety-five incidents, fifty-two arose from the distribution system. Some were microbiological failures, but thirty-one stemmed from bursts or as a result of operational activities resulting in discoloured water in supply. Decisions on prosecution for alleged offences under the Water Industry Act 1991 of supplying water unfit for human consumption have been delegated to the DWI since 1996. In 1997 there were four successful, three of Dwr Cymru, and one of Yorkshire Water, with fines totalling £37,500. A case against South West Water was dismissed.

In 1997, nine incidents related to either an increase in the illness cryptosporidiosis or the detection of *cryptosporidium* oocysts in water. Five were considered to be associated with water supply. The most serious occurred in North London in February 1997, with 345 cases of the illness, showing a strong association with water from the Clay Lane works of the Three Valleys Water Company. Continuous sampling for cryptosporidium is now being proposed, and a treatment standard is likely to be introduced in 1999.

TABLE 2.3 Compliance for iron, 1991–7, for water companies in England and Wales

	Trend for period 1991–7		
Compliance in 1997	*Deteriorating trend over the period*	*No significant trend over the period*	*Some improvement over the period*
Above average performance		Mid Southern South Staffordshire Three Valleys Thames Wessex	Anglian Severn Trent Southern
Average performance	Essex & Suffolk	Bristol Bournemouth & West Hants Cambridge Chester Cholderton Folkestone & Dover Hartlepool North Surrey North West Portsmouth South East Wrexham York	Mid Kent Northumbrian Sutton & East Surrey Tendring Hundred
Below average performance	South West	Yorkshire	Dwr Cymru

Source: Drinking Water Inspectorate, 1998, Chief Inspector's Statement (on World Wide Web).

Is there a geography of drinking water quality in England and Wales? This is difficult to answer, as the Regulations embrace fifty-five different parameters. One approach is to examine the percentage of total determinations which fail to reach the compliance standard. No company has more than 1 per cent of determinations failing to comply, and only Sutton and East Surrey and Cambridge Water Companies and North West Water have more than half of 1 per cent non-compliance. At the other extreme, Chester, Cholderton, Folkestone and Dover, and North Surrey Companies have less than a tenth of 1 per cent non-compliance. However one incident may produce a number of different non-compliance determinations, and as water quality is dependent upon a range of different characteristics it is not felt meaningful to rely on such a simple single index. An alternative approach is to recognise that consumers' perceptions of quality depend very much on the aesthetic factors and chemical qualities which determine the appearance of water, especially iron content. Iron comes largely from rusty mains, but although it discolours water it is unlikely to be harmful to health. Table 2.3 summarises 1997 compliance data for iron, in terms of both average performance and trend over the period 1991–7, for all water companies for which data are adequate. Thus Anglian, Severn Trent and Southern are achieving above average performance, and are improving, whereas South West is below average, and deteriorating. This relative company performance is a complex situation, but reflects both the enforcement process and skill in operating the distribution systems, as well as many other factors, including the historical development of the system, and difficulties experienced during drier years. The standard for iron was met in 76.9 per cent of water supply zones in 1997, and overall the trend is an improving one, as many companies have major programmes to replace or reline corroded mains, which will take a further twelve years to complete.

The impact of EC legislation

The Drinking Water Directive was adopted by the EC in 1980, and sets out standards of water intended for human consumption. In 1995 the EC published proposals for a revised Directive, and negotiations are proceeding. The Directive sets out standards for a range of microbiological, physical and chemical properties to be met in drinking water. The requirements were incorporated into UK legislation in 1989, and in some respects this legislation is more stringent than the EC Directive. The financial impact on the UK water industry and individual companies is immense. For example, it is estimated that East Scotland Water requires £900 million in investment in the next few years in order to meet EC targets on water quality. Between privatisation and March 1996 the companies invested £9 billion on water services, including £3.4 billion specifically for improving drinking water standards. Investment continues, with £4 billion planned for 1995 to 2005.

The Urban Waste Water Treatment Directive of 1991 passed into UK legislation in 1994. Its purpose is to protect the environment from the adverse effects of sewage discharges, including an end to the dumping of sewage sludges at sea. Works must comply with the requirements by various dates, between 1998 and 2005. There are currently about 6,500 sewage treatment works in England and Wales, and the estimated cost of work necessary to comply with the Directive is £6 billion of capital investment, and £1 billion per year in operating costs. The impact of this legislation could fall differentially on the water companies, according to the amount of coastline and number of bathing beaches. In effect, residents in popular holiday areas are having to pay through their higher water charges to ensure that beaches remain clean for visitors. OFWAT have estimated that the Environment Agency's proposed wide-ranging environmental action plan could cost water companies up to £11 billion and could add an extra £46 to the average household bill between 2000 and 2005. For this reason, OFWAT have in 1998 proposed a 'bucket and spade' tax on tourism to help pay this huge cost. The tax would be paid to local authorities by local businesses that benefited from tourism, but would undoubtedly be passed on to the holidaymaker, for example as higher prices for services and goods so diverse as hotel rooms, fish and chips, donkey rides and flip-flops. Businesses reacted with alarm to this proposal, which at the time of writing is out for consultation. This debate illustrates some of the difficulties associated with multiple regulators of a privatised industry, whose objectives may conflict with one another.

Financial aspects

One major source of complaint following privatisation was the size of water bills. It is, however, important to note that average water bills are relatively stable or falling in real terms, although sewerage bills have continued to rise for some customers to pay for improved sewage treatment. The main factor which has forced water prices up has been a huge programme of spending, which will amount to about £24 billion in the ten years from 1994/5. At least half of this was to improve the quality of drinking water and sewage treatment. Most of this expenditure is to meet legally enforceable UK and EC requirements for better drinking water and a cleaner environment, as described above. Companies are also meeting a substantial element of the costs from their own resources, from efficiency savings, and from borrowing – in 1996/7 their spending exceeded income by over £700 million. The limits to prices companies are allowed to charge for water and services were set initially in 1989 on privatisation, and are reviewed periodically by OFWAT. OFWAT has been rigorous in restraining price increases wherever possible, consonant with allowing companies to realise their planned investment programmes, and in the realisation that companies have to make profits in order to attract investment to finance spending. OFWAT's report on the financial performance and capital investment of the water companies in England and Wales shows that capital expenditure is now approximately double the average level of that in the 1980s (inflation-adjusted), at over £3 billion per year, so to that extent it can be argued that one of the primary aims of the whole privatisation process has been achieved

The public perception of water company performance has been heavily influenced by the criticisms raised by both the media and politicians of the so-called 'fat cat' salaries and bonuses received by the chairmen and other senior executives of the privatised companies, despite a perceived lack of adequate performance. For example, in 1997 the chief executive of Yorkshire Water received an extra £55,000 on top of his basic salary, which with benefits brought his total remuneration to £298,000. This was set in the media against the background of the company having failed to maintain supplies to customers in 1995, of it telling customers that they faced being cut off if they had a bath, and that prices were being raised by 8.1 per cent for unmetered and 6.1 per cent for metered supplies. Water companies counter that their executives' salaries are set as the market average for equivalent jobs in the sector. OFWAT has no direct control over profits, share prices, dividends or chairmen's salaries, all of which have been bones of contention with the public. However, OFWAT and the government have urged the utilities to adopt best practice in the supervision of board-room pay by shareholders, and OFWAT are proposing to introduce objective comparative information on company performance, so that performance-related pay policies can reflect actual performance in the competitive market. The present government is also reported to be looking into ways in which utility executive salaries are set, and how they could be linked to standards of performance achieved by the companies.

It is important to note that the average cost of water in the UK is still low. For 1998/9 the average water and sewerage bill works out at around 66p per day – a bath costs about 11p, a cycle of a washing machine the same, and watering the garden costs 75p. By contrast heating the water for a bath or washing machine costs almost 20p, and a litre of bottled water costs typically more than 50p. Average bills for water and sewerage range from £201 (Thames Water) to £354 (South West Water). For water, averages range from £158 (South East Water) to £73 (Portsmouth), and for sewerage from £102 (Thames) to £229 (South West Water). In the four years since OFWAT set price limits, average household bills have increased by 4.5 per cent in real terms (17.8 per cent taking inflation into account). In the same period, average household sewerage bills have increased by 10.5 per cent (24.6 per cent with inflation), this reflecting the greater investment necessary to meet legal standards for treatment. For metered properties the average total bill is £223, compared with £245 for unmeasured ones.

Traditionally, most water supplies to domestic properties in the UK have been unmetered, and most charges for household water and sewerage services are still based on the rateable value of the property (despite the fact that the domestic rating system ended in 1990, with the consecutive advents of the Poll Tax, Community Charge and Council Tax). However, in the last two decades metering has become more common, with 14 per cent of households now having meters. Metering trials have been held, in which the majority of customers found metering led to lower bills, although about one in five paid more than previously. Metering has potentially an important role to play in conserving water resources, and OFWAT believes that it is sensible to meter where it is cheap or economic to do so – for example in new properties, for high non-essential users and where resources are limited. However, consumer bodies have expressed concern that some 95 per cent of domestic water consumption is essential, and therefore insensitive to changes in prices, and heavy reliance on charges to encourage customers to reduce consumption by metering is unlikely to be effective. OFWAT does not advocate either universal metering or a crash programme of metering. Thirteen companies (out of twenty-eight) now offer free meters and installation to every customer. There are some additional costs associated with metering, and OFWAT has issued guidance as to how much of these it is reasonable to pass on to customers. Initially the government planned to end rateable-value-based charging by 2000, but this deadline has been extended.

Companies in the south and east of the country, whose water resources are most short, are the most likely to adopt compulsory metering of existing properties; Anglia Water has announced plans to meter 95 per cent of households by 2014–15. However, installing meters in all homes would be uneconomic, and most companies are introducing meters selectively, to large users (e.g. those using sprinklers and hosepipes, hotels and guest houses). Alternatives to metering include a flat rate, or a banding system similar to that used for Council Tax, based on the type of property.

Since 1945 water suppliers have had the legal power to disconnect customers' water supply (but not sewerage services) for non-payment of charges. This continued after privatisation, and there was much public concern about the risk to public health caused by disconnection, and unfair debt recovery procedures. OFWAT issued guidelines in 1992 to ensure companies acted consistently and fairly, and since then disconnection figures have fallen consistently. For six years running disconnections from the water supply for non-payment of bills have fallen. In 1998 1,907 household disconnections were made (a 39 per cent reduction over the previous year); 1,774 non-household disconnections were also made. Water companies have increasingly differentiated between customers who are unable to pay, and those unwilling to pay – in 1991/2, 21,282 household disconnections were made. OFWAT believes that disconnection should be used only as a last resort.

A major review of water-charging policy and tariffs was initiated by the issue of a consultation paper in March 1998, and this is likely to result in more metering and more sophisticated tariff structures than at present, with, for example, more cost-reflective tariffs for large users, and seasonally variable tariffs.

Competition in the water industry

The Director-General of Water Services has duties to promote economy and efficiency, and to help create effective competition. As the water industry is made up of local companies set up in 1989 to provide water and sewerage services, as vertically integrated services with a regional monopoly, normal market competition cannot exist. Opportunities for direct competition are limited, as the cost of transportation of water across the clear geographical boundaries of the companies is high. Any attempt to duplicate the distribution system would lead to prohibitive increases in costs, and there is no independent production of water possible. However comparative competition can help to achieve results similar to those achieved in a competitive market. OFWAT does this by comparing the performance of individual companies and then setting price limits that give the companies incentive to increase their efficiency. Comparisons between companies can also lead to the adoption of best practice so that all customers can benefit. The consultation paper issued in March 1998 is also addressing issues related to competition, in increasing consumer choice and developing more cost-reflective tariffs.

In addition, a direct mechanism for at least limited direct competition and choice has been developed, as the so-called 'inset-appointments'. These can be granted to a company to provide water and/or sewerage services either on a greenfield site without water supply, within the existing appointee's area, or to a site supplied with 250 megalitres or more of water per year. An inset appointment can be granted to an existing company, or a new entrant to the industry; a large customer can effectively become its own supplier by setting up an affiliated company to be the appointee. Water for an inset appointee can come from a new source, but in practice is more likely to come from a connection to the existing appointee. There is some limited evidence that the procedure, or the threat of it, is beginning to be effective in driving down prices to large users. In the charging year 1998/9 some twenty-two (of twenty-eight) companies will have large user tariffs and three (of ten) will have large user sewerage tariffs, resulting in a reduction of 15–30 per cent in prices. The first two successful inset appointments were granted in 1997, to a large chicken processing firm and to a Royal Air Force station, both in Anglian Water's area. A third application, which will involve a new entrant to the industry, Albion Water Ltd, has been approved, to supply a paper manufacturer in the Dwr Cymru area. Other changes to legislation are currently under discussion in order to facilitate competition, including changes to abstraction licensing.

Leakage and efficiency

Considerable media and public concern has been expressed about leakage. Alarmingly, almost a quarter of the water which is expensively abstracted from sources or stored in reservoirs, and then treated to potable quality by the water companies, never reaches the consumers as it is lost from leaking mains en route. Losses for individual companies vary from 15 per cent to 38 per cent. In 1995/6, almost 5,000 ml/d were lost. OFWAT has monitored this closely, and asked the industry to set its own targets for leakage reduction. In the first year the water industry in England and Wales achieved a 9.5 per cent reduction, and from 1996/7 to 1997/8 it managed to reduce leakage overall by more than 12 per cent, with South West Water, Folkestone and Dover and Hartlepool companies reporting reductions of more than 20 per cent. Savings of over 1,000 ml/d have been made nationally in the last two years. However, three companies (Anglian, Portsmouth and Mid Kent) failed to meet their self-set leakage targets for 1997/8, and the mild winter probably helped many others, who only barely met their targets. Thames Water remains a particular concern, with leakage considerably above the industry average. Henceforth, more stringent OFWAT-set targets will be used to judge company performance in reducing leakage, and it is expected that the biggest fall in leakage will come in 1998/9.

Leakage is only one aspect of efficient use of water. At a Water Summit held in May 1997 companies were asked to review their water efficiency plans in line with the government's ten-point plan. Since 1996 companies have had a duty to promote the efficient use of water, although co-ordinated strategies have been slow to emerge in some cases. However, particular initiatives introduced by many include providing a freephone leakline number, free supply pipe repairs, free water audits, and a free cistern device (e.g. Thames' 'hippo') to reduce toilet flushing volumes (of which 4 million have been distributed). Companies now have to monitor and report to OFWAT on their progress against strategies to promote efficient water use.

There are sufficient sewers in England and Wales to go more than seven times around the world. Over 1,400 kilometres have been renewed since privatisation, and, despite some media campaigns, there is no evidence nationally of long-term deterioration in serviceability of sewer networks. Sewers can last a very long time. Indeed, the performance of the system is improving – since 1992/3 the number of properties flooded by sewers has fallen from 10,858 to 4,627, and there is no evidence of an increase in the number of sewer collapses over the last fifteen years. The companies spend about £80 million each year on maintenance, and about £150 million on renewals.

Sustainability

Are the water resources of the United Kingdom being developed in a sustainable way? The brief review which follows is largely based on the Environment Agency's analysis of this issue. Overall, the total available water resources of the United Kingdom significantly exceed demand for water. About 263,400 million cubic metres of precipitation fall over the UK. About 60 per cent, equivalent to over 2 million litres per person per year, forms runoff, and could be exploited as a water resource. The total demand for water is of the order of only 250,000 litres per person per year.

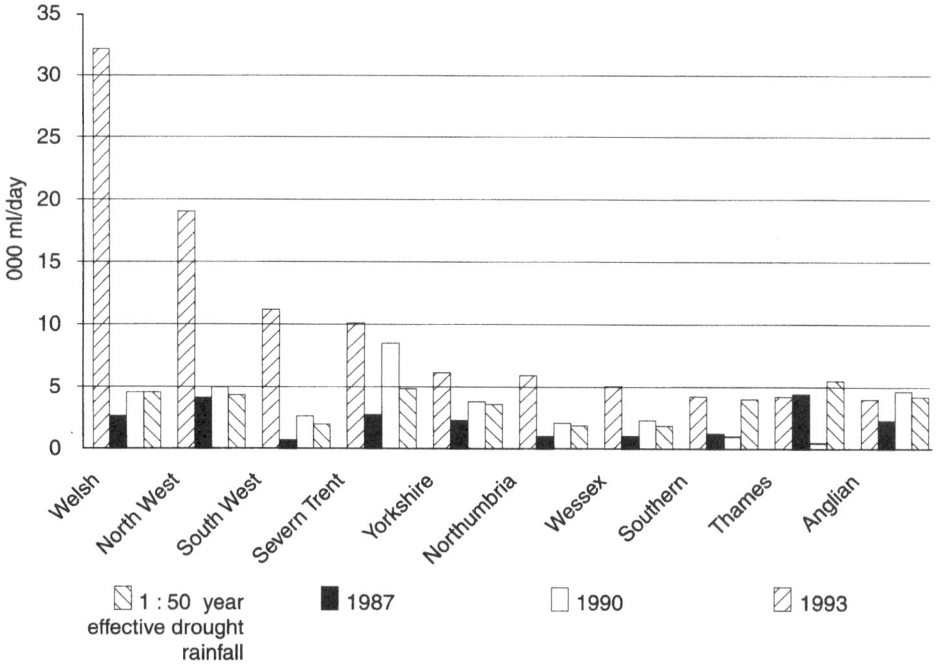

FIGURE 2.6 Geographical variation in resource potential: licensed abstractions compared with 1:50 year effective drought rainfall, England Wales. Note the high utilisation of resources in the Anglian and Thames regions

Source: Derived from figure in the DETR World Wide Web pages.

However, the balance between supply (from rainfall) and demand (from the multiplicity of human activities, including residential, agricultural and industrial use) varies widely from region to region. In broad terms, the south and east of the country is the driest, yet it is in the south and east of the country that demand is generally higher because of the centres of population, with associated industry. The geographical variation in resource potential is shown by Figure 2.6. This compares licensed abstractions, as an indicator of demand, with effective rainfall in a drought which can be expected to occur once in every fifty years, for the ten National Rivers Authority regions. In eight of the ten regions the ratio of abstractions to effective rainfall is less than one-third, whereas in the Thames and Anglian regions licensed abstractions can exceed the effective rainfall, with the deficit being made up in droughts by re-use, and storage in reservoirs and groundwater. Clearly, stress on the resource is greatest in these regions, and river basin management has had to ensure that water which falls as rain can be re-used several times before its eventual return to the sea. This is particularly the case for the Thames basin, where recycling of abstracted water takes place along the length of the river. Increasing demand and abstraction can cause unacceptably low flows in rivers, thus damaging habitats and the amenity value of rivers. Water transfer and river regulation schemes have been implemented in order to alleviate the worst affected reaches of river. Since 1990 £6 million has been spent, and this has been effective in reducing affected reaches from 335 km to a target of 79 km.

However, in years of exceptionally low winter rainfall and hot dry summers the margin between the available resource and the public demand for water is sufficiently small to require restrictions on water use. For example, in the prolonged fifteen-month drought of 1995 Yorkshire Water almost became the object of public ridicule when it was forced to transport water by tanker from Northumberland in order to maintain supplies.

The margin between average demand for public water supply and the resource available in dry years is reasonable in most areas. In 1993/4, a dry year, five of the ten NRA regions (Wessex, Thames, North West, Yorkshire and Severn Trent) used over 80 per cent of the available resource, and the last named used over 90 per cent. Only one, Northumbria, used less than 70 per cent. That there is no scope for complacency is underlined by the uncertainty over the possible effects of climatic change on water resources, which could increase demand by up to 4 per cent in the south and east of the country by 2021.

Under extreme conditions of stress on available resources Drought Orders can be made in order to introduce emergency measures, such as the imposition of bans on the use of hosepipes and the abstraction of water from rivers when flows are below normally permissible levels. Table 2.4 illustrates how these identify major drought years of 1976, 1984 and 1995, and the succession of dry years 1989–92. Those Environment Agency Regions mostly likely to have to resort to Drought Orders are the Southern and South West regions.

Any such estimates of sustainability could of course be undermined by changes in the climate, which provides the basic resource. Chapter 19 in this book examines in more detail the probable impact on the water industry of likely climatic changes.

Conclusion

The basic geography of water resources in the United Kingdom remains essentially unchanged from that described in earlier editions of this book, with the wetter northern and western parts of the country enjoying a relative wealth of the basic resource, as compared with the relatively drier south and east. This physical framework is not totally fixed, indications being that climatic change (see Chapter 19) is working to accentuate the basic dichotomy between the two halves of the country. However, the human participation in the water demand and supply relationship has changed markedly in the last decade. Changes in demand for water have resulted from many of the economic changes described elsewhere in this book, as outlined above, with (in general) a decline in demand occasioned by changing patterns of industry being a least partly counterbalanced by rising consumption associated with rising living standards, and uses such as irrigation and fish farming. The major change

TABLE 2.4 Number of drought orders in England and Wales by Environment Agency Region, 1976–95

	1976	*1977*	*1978*	*1979*	*1980*	*1981*	*1982*	*1983*	*1984*	*1985*	*1986*	*1987*	*1988*	*1989*	*1990*	*1991*	*1992*	*1993*	*1994*	*1995*
Anglian	15	0	0	0	0	0	0	0	0	0	0	0	0	1	3	0	0	0	0	0
North East	18	0	0	0	0	0	0	0	0	0	0	0	0	9	17	9	1	0	0	21
North West	0	0	2	0	2	0	6	0	31	0	0	0	0	21	0	1	0	0	0	23
Midlands	13	0	0	0	0	0	0	0	6	0	0	0	0	5	0	0	0	0	0	0
Southern	4	0	0	0	0	0	0	0	0	0	0	0	0	19	25	18	11	0	0	1
South West	58	0	15	0	0	0	7	5	45	0	0	0	0	21	13	0	1	0	0	1
Thames	8	0	0	0	0	0	0	0	0	0	0	0	0	0	2	0	3	0	0	7
Welsh	20	0	2	0	2	0	2	1	22	0	0	0	0	13	1	0	0	0	0	0
Total	136	0	19	0	4	0	15	6	104	0	0	0	0	89	61	28	16	0	0	53

Source: From data in the DETR World Wide Web pages.

BOX 2.1 Bristol Water

The nature of water supply in the UK may be illustrated by a more detailed examination of one area, which in many ways is fairly typical. Bristol Water is one of the UK's largest and most successful water supply companies, and since 1846 it has existed with the sole purpose of supplying water. The Company now supplies 1,049,000 people in an area of about 2,391 square kilometres, in most of Avon, large parts of Somerset, and small parts of Gloucestershire and Wiltshire, centred on the Bristol conurbation, and embracing Tetbury, Clevedon, Wells and Frome. Sewage disposal in the area is the responsibility of Wessex Water. The area is a varied one, having highland areas in parts of the Mendips and Cotswolds, with both dramatic limestone scenery and rolling chalk downs, as well as the lowlands of the Somerset levels.

The Company's founders included William Budd, the sanitation pioneer, Francis Fry, the industrialist, and George Thomas, the Quaker merchant who founded Bristol General Hospital. The town's main water supplies prior to this were based on medieval conduits and wells, and were sources of typhoid, cholera and dysentery. The Company first supplied clean spring water from the nearby Mendip Hills, through an impressive 16-km pipeline. A reservoir was opened in 1901, with water being pumped by huge steam-driven beam engines. Now, 6,472 km of mains supply (on average) 327 million litres of water per day to 444,000 properties. The water supplied comes from sixty-eight different sources, including fourteen reservoirs, as well as springs, wells and boreholes, with the largest reservoir being Chew Lake, and the largest single source being abstraction from the River Severn, which supplies more than half of the total, although this requires much more complex purification than Mendip water. The company has twenty-six treatment works, 164 pumping stations, and 139 covered treated water storage reservoirs. As well as supplying water, the reservoirs of the Company provide extensive conservation and recreation facilities. Ninety-nine point nine per cent of the Company's compliance tests meet the stringent water quality standards, and there are no restrictions on the use of water. Over 99.9 per cent of properties receive water at a pressure above standard. During the year 1997/8 less than 0.1 per cent of customers were without water for more than six hours due to burst pipes or overruns on planned maintenance.. Water charges are below average for the industry, equivalent to only 30p per day for each household.

The group contains an engineering subsidiary as well as Bristol Water plc. Its share capital comprises 7.2 million ordinary shares and 1,414,000 non-voting shares. The turnover of the Group was £78 million in 1997/8, and regulated pre-tax profits of Bristol Water plc were £13.8 million. For the year, a dividend of 57p was paid on each share. In the year there was capital expenditure of £29 million, including replacing or reconditioning more than 80 km of mains, as part of major water quality improvement. Key issue for the future will be the new water quality directive – especially the standard for lead, as Bristol has one of the highest incidences of lead pipes in the country. Four thousand new properties were connected to the mains, but impact on consumption was contained within the overall economies achieved, and consumption was at its lowest level for five years, partly as a result of increasing customer awareness of the need for economies, and partly as a result of the weather.

in the geography of water resources, though, is in the administrative and legislative framework for water services. In some senses, the water industry in the UK has now gone full cycle, from a plethora of private suppliers in the last century, through an essentially nationalised industry in the 1970s and 1980s, to a regulated privatised industry. The system of regulation is complex. Despite the doubts of many, the regulators, and especially OFWAT, have made what to many neutral observers appears to be an impressive start. However, it remains to be seen whether the regulatory system can cope adequately with the complex interrelationships between the economic, political and physical changes which will have repercussions for the water industry in the twenty-first century, or whether future editions of this book will again describe a completely changed system.

References

Economic Affairs (1998) 18(2) (a collection of papers on water policy).

National Rivers Authority (1994) *Water: Nature's Precious Resource*, London: HMSO.

Park, C. (1991) 'Water resources', in R.J. Johnston and V. Gardiner (eds) *The Changing Geography of the United Kingdom*, London: Routledge.

Water Resources Board (1973) *Water Resources in England and Wales, vols 1 and 2*, London: HMSO.

Further reading

A good introductory overview of issues involved in water resources and their management is A.T. McDonald and D. Kay (1988), *Water Resources: Issues and Strategies*, Harlow: Longman, which takes a global perspective but contains some good insights into the water resources of the UK. The development of water resources and the water industry in the UK is well described by Park in previous editions of this book. Other sources for this topic are K. Smith (1972) *Water in Britain*, London: Macmillan, and D.J. Parker and E.C. Penning-Rowsell (1980) *Water Planning in Britain*, London: Allen and Unwin, whilst a very basic introduction is C. Kirby (1984) *Water in Great Britain*, Harmondsworth: Penguin. A dissenting view of the water industry during the 1970s is F. Pearce (1982) *Watershed*, London: Junction Books. Water demand, and its forecasting, are considered by papers in V. Gardiner and P. Herrington (eds) (1986) *Water Demand Forecasting*, Norwich: GeoBooks. A discussion of issues relating to sustainable development of water resources is J. Rees and S. Williams (1993) *Water for Life – Strategies for Sustainable Water Resource Management*, London: Council for the Preservation of Rural England. A critical debate on privatisation is contained in the collection of papers in an issue of the journal *Economic Affairs* (volume 18(2), 1998), whilst regulation is considered by S.L. Holborn (1993) *Guide to the Economic Regulation of the Water Industry*, Oxford: Oxford Economic Research Associates.

Data on the water industry are now much more readily and conveniently available than before, through the Internet. The sites listed below were checked in July 1998. In using web-derived data readers should always be aware that the validity and integrity of the data always depend on the nature of the site. Official government sources include the Department of the Environment, Transport and Regions (DETR) (http://www.environment.detr.gov.uk). The DETR publish a very useful little free booklet, *The Environment in your Pocket*, annually, and the data and illustrations from this, and much more, are available to download through the Internet. A gateway site to many water industry sources is the STS Guide to the UK Utility Industry (http://www.sts.co.uk/utltysuk), which provides links to gas, electricity and water utility company home pages, and to those of various associated organisations. The sites of the individual water companies often provide much of interest. An immensely valuable source of information is OFWAT (http://www.open.gov.uk/ofwat), which makes available not only their own technical papers related to regulation, but also some of more general interest – for example, on the history of the water industry. Much data on the quality of tap water is available from the Drinking Water Inspectorate (http://www.dwi.detr.gov.uk). For keeping up to date with developments in the industry the quality newspapers are a good source, and these are also available on searchable CD-ROM in many librararies, and via the Internet. An Internet source offering a water industry perspective is *Water Magazine* (http://www.water-services.co.uk/magazine).

Chapter 3

Energy

Keith Chapman

Introduction

The production and consumption of energy have had a massive impact upon the geography of the United Kingdom (UK). This impact extends from the influence of water power on the location of the early textile mills, through the economic and social transformations deriving from the exploitation of coal during the Industrial Revolution to the relatively recent consequences of North Sea oil and gas. The pace of change has accelerated and this Chapter focuses upon developments in the relatively recent past which have moulded contemporary patterns. It is divided into two sections. The first identifies the principal factors shaping the evolution of energy markets in the UK and the second reviews some of the spatial implications of these changes.

Changing energy markets

The fundamental importance of energy resources to the quality of human life ensures that the interpretation of change in energy markets requires a very broad perspective. Economic considerations obviously influence decision-making within these markets, but social, political and environmental agendas also form part of the wider context shaping the actions of energy producers and consumers. This wider context has a geographical dimension in the sense that the energy market in the UK is not insulated from international pressures. The following section describes general trends in the UK energy market, as reflected in the changing contributions of the various fuels to total consumption, before identifying some of the key influences upon these trends.

Trends in the energy market

Before the Second World War the UK was a significant net exporter of energy in the form of coal, but the combined effect of inadequate investment during the war and personnel shortages resulted in a failure to meet even domestic demand when hostilities ceased and the coal industry's virtual monopoly as fuel supplier to the British economy was gradually weakened as the volume of imported oil increased. The gap between indigenous energy production and consumption widened steadily for almost twenty-five years (Figure 3.1) and it was not until 1975 that any perceptible narrowing of this gap occurred. By 1980, the rapid rise in North Sea oil production had transformed the energy gap into a surplus once more and it seems unlikely that the UK will return, in the foreseeable future, to the situation of the early 1970s when more than 50 per cent of its primary energy requirements were imported.

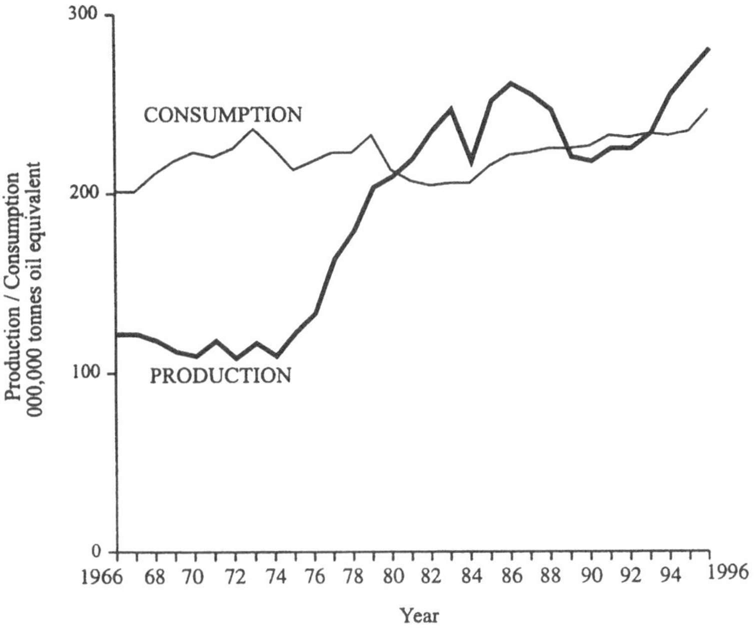

FIGURE 3.1 Production and consumption of primary energy, 1966–96

Associated with these changes in dependence upon internal and external sources of supply have been shifts in the relative contributions of the various fuels to total energy consumption (Figure 3.2). Coal consumption has been reduced not only by the substitution of imported and, more recently, domestic oil, but also by the introduction of natural gas and, to a much lesser extent, nuclear power. Thus whereas coal accounted for 95 per cent of primary energy consumption in 1950, its share had fallen to 73.9 per cent by 1960 and 20.1 per cent by 1996. This decline was initially related to the rapid growth of imported oil in the 1950s and the trend was reinforced by the rise of natural gas and nuclear power in the 1960s (Figure 3.3). Although oil accounts for a higher percentage of primary energy consumption in 1996 (33.9) than in 1960 (25.5), the sources of supply are very different and the UK has been a net exporter since 1981 (Figure 3.4).

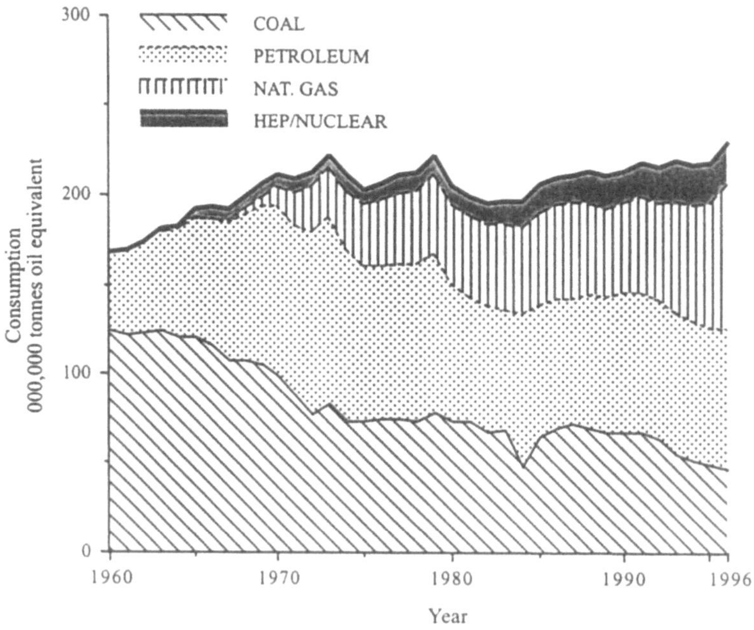

FIGURE 3.2 Consumption of primary fuels, 1960–96

Global influences

The international price of oil has been the most important single influence upon the evolution of the UK energy market over the last fifty years. Producers and consumers have responded to this external influence, which has also been a major factor shaping the economic and energy policies of successive governments. It would be difficult to overestimate the impact of the sharp and dramatic increases in the international oil price resulting from the actions of the Organisation of Oil Exporting Countries (OPEC) which saw the average f.o.b. (free on board) price of, for example, an Abu Dhabi crude rise from US$2.54 a barrel in 1972 to US$36.56 in 1981. In these circumstances, oil prices became one of the key global economic variables. The impact of the 1973 and 1979 oil price 'shocks' was all the more striking because of the contrast with the history of a gradually declining 'real' price of oil from the early 1920s.

It was this declining price which encouraged UK consumers to switch from domestic coal to imported oil during the 1950s and 1960s. However, the oil price increase in 1973 reversed these trends. The Plan for Coal, produced by the National Coal Board (NCB) in 1974, aimed to halt the decline in output and to reduce overall costs by replacing old capacity with an approximately equivalent volume of new production by expanding suitable existing mines and creating some new ones. Although the planning horizon of the document extended only to 1985, various factors contributed to a reappearance of the industry's secular decline by the early 1980s. The economic recession triggered by the second oil price shock in 1979 reduced overall energy demand and increased competition between the various fuels.

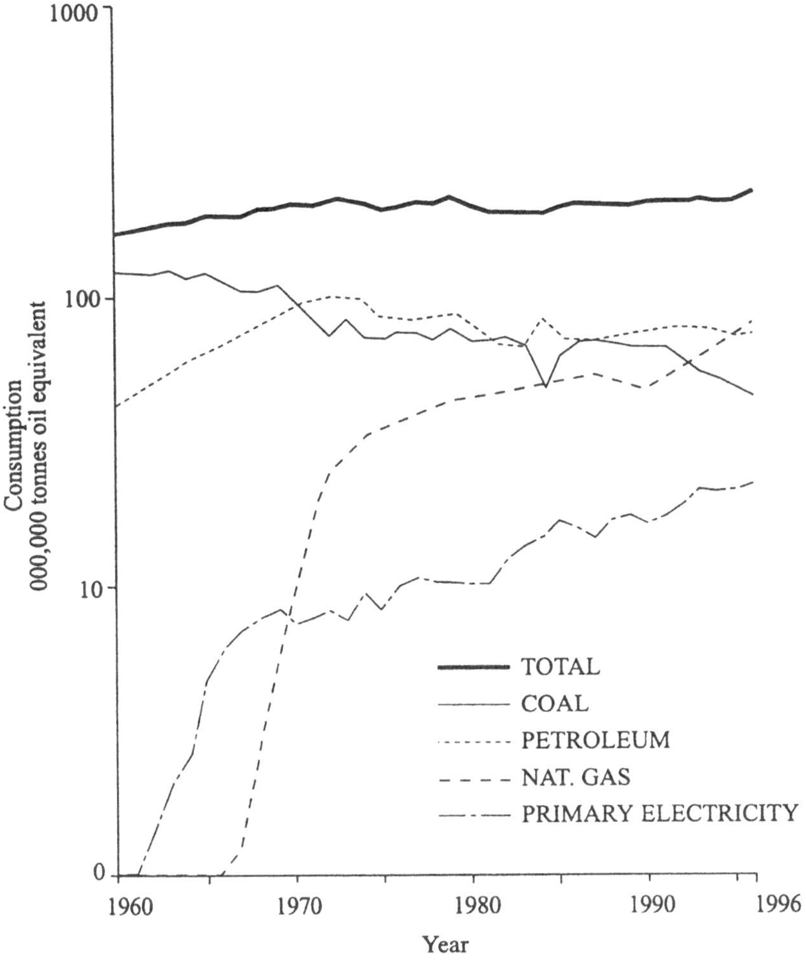

FIGURE 3.3 Rates of change in consumption of primary fuels, 1960–96

A relatively new element in this competition was introduced with the growth of coal imports, which had been negligible before 1970 (Figure 3.5). The international coal trade was boosted significantly by the increase in oil prices which made it possible for cheap, opencast coal to be delivered to UK ports at prices competitive with fuel oil and below those quoted by the NCB. The expansion of the deep-sea coal trade has, therefore, become another important global influence upon the UK energy market. It is secondary to the international oil price as an external 'regulator' of the UK energy system, but has certainly contributed to the decline of the coal industry.

Global influences upon the UK energy market have not been restricted to the coal industry. Although commercial interest in the hydrocarbon potential of the North Sea may be traced to the early 1960s, the oil price shocks provided a massive stimulus to this activity. This stimulus incorporated both political and economic elements. There had been an

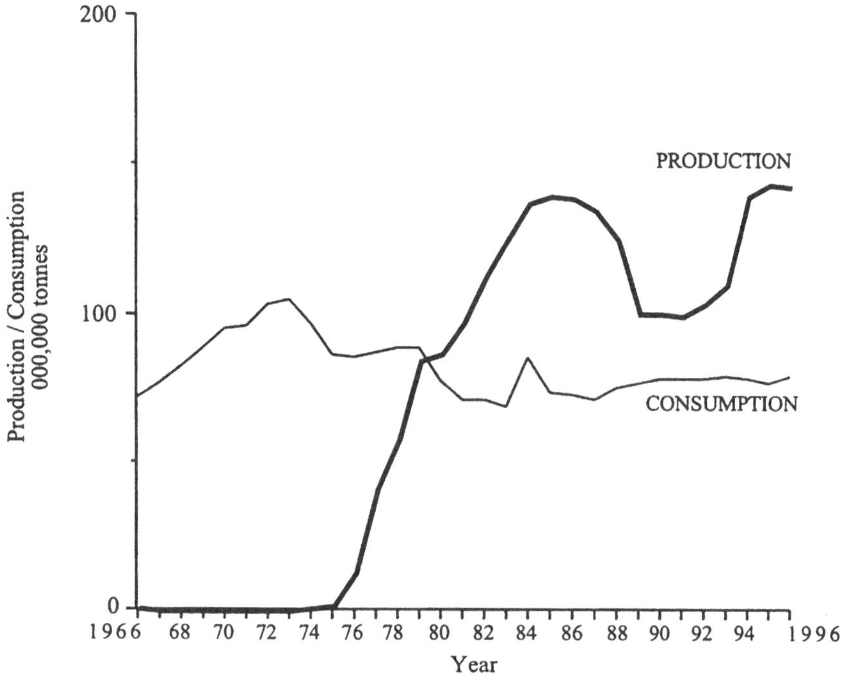

FIGURE 3.4 Oil production and consumption, 1966–96

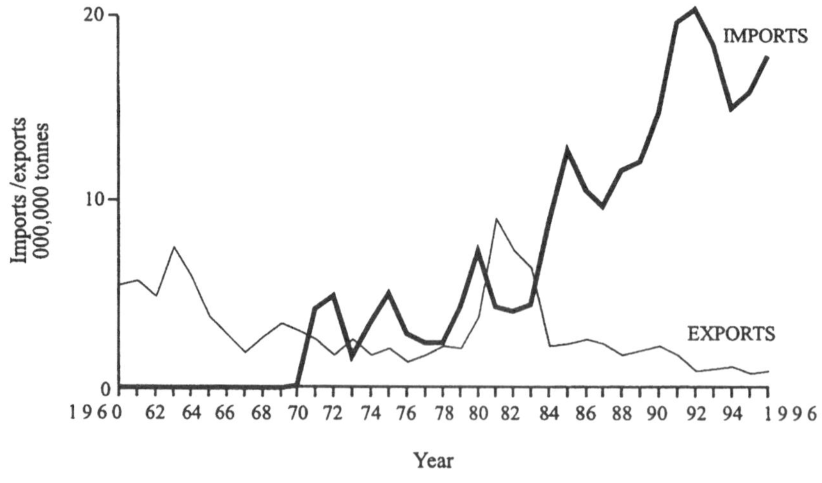

FIGURE 3.5 Coal exports and imports, 1960–96

awareness from the 1950s of the political risks deriving from increasing dependence upon the imported oil which accounted for almost 50 per cent of UK primary energy consumption by 1973. Events such as the Suez crisis in 1956 and the Arab–Israeli war of 1967 emphasised these risks, which were confirmed by the success of the OPEC countries in harnessing their latent power as suppliers of energy to Western Europe, Japan and the United States. Although supplies were generally maintained, the first and second price shocks stimulated interest in indigenous energy sources in all the major oil-consuming countries. In the case of the UK, an explicit and overwhelming policy commitment to encourage the rapid exploitation of the oil and gas resources of the North Sea was the most obvious expression of this interest. The international oil price has a continuing significance for events in the North Sea because it is the yardstick against which commercial decisions are made. The desire of successive governments to maintain the momentum of development has required periodic adjustments of the fiscal terms and conditions under which the oil companies operate. Many factors enter into these negotiations, but one of the most important is the international oil price.

Resource availabilities

Although energy developments in the UK have been strongly influenced by political and economic circumstances on the world stage, the country has been very fortunate in its endowment of fossil fuels. It is ironic that the UK's return to a position of energy surplus in 1980 coincided with the decline of coal which, in geological terms, is much more abundant than either oil or natural gas. Volumetric estimates in 1976 suggested the existence of around 190 billion tonnes ($190 \cdot 10^9$) of coal-in-place in the UK. Defining the recoverable portion of this gross figure is very difficult given the many variables which determine the economics of coal production. In the publicity surrounding the announcement of Plan for Coal in 1974, the NCB estimated recoverable reserves at 45 billion tonnes, which represented approximately 300 years supply at prevailing rates of consumption. By the end of 1996, the corresponding figure had fallen to only fifty years. This dramatic decline is, at first sight, hard to explain bearing in mind that coal consumption has fallen steadily in the UK since 1974 (Figures 3.2 and 3.3). The UK has obviously not removed the equivalent of 250 years' supply from its coal resource base in just over twenty years. The contradiction lies in the factors influencing the apparent elimination of reserves. Various economic, technical and geological assumptions are incorporated in the definition of proved reserves.[1] These assumptions interact with prevailing policies in the sense that colliery closures can effectively sterilise the resource by making it either technically impossible or prohibitively expensive to return to abandoned workings. This accounts for the continuous decline in the proved reserves of coal in the UK from 10,750 million tonnes in 1986 to 2,500 million tonnes in 1996.[2] This counter-intuitive relationship between falling production and shrinking reserves demonstrates, in a sense, how the demise of the coal industry has become a self-fulfilling prophecy.

As the proved reserves of coal have declined, so the equivalent projections for North Sea oil and gas have moved in the opposite direction. Almost 2,000 exploration wells had been drilled in British waters[3] by the end of 1996, plus a very much larger number of appraisal and development wells (Figure 3.6). This activity has established substantial

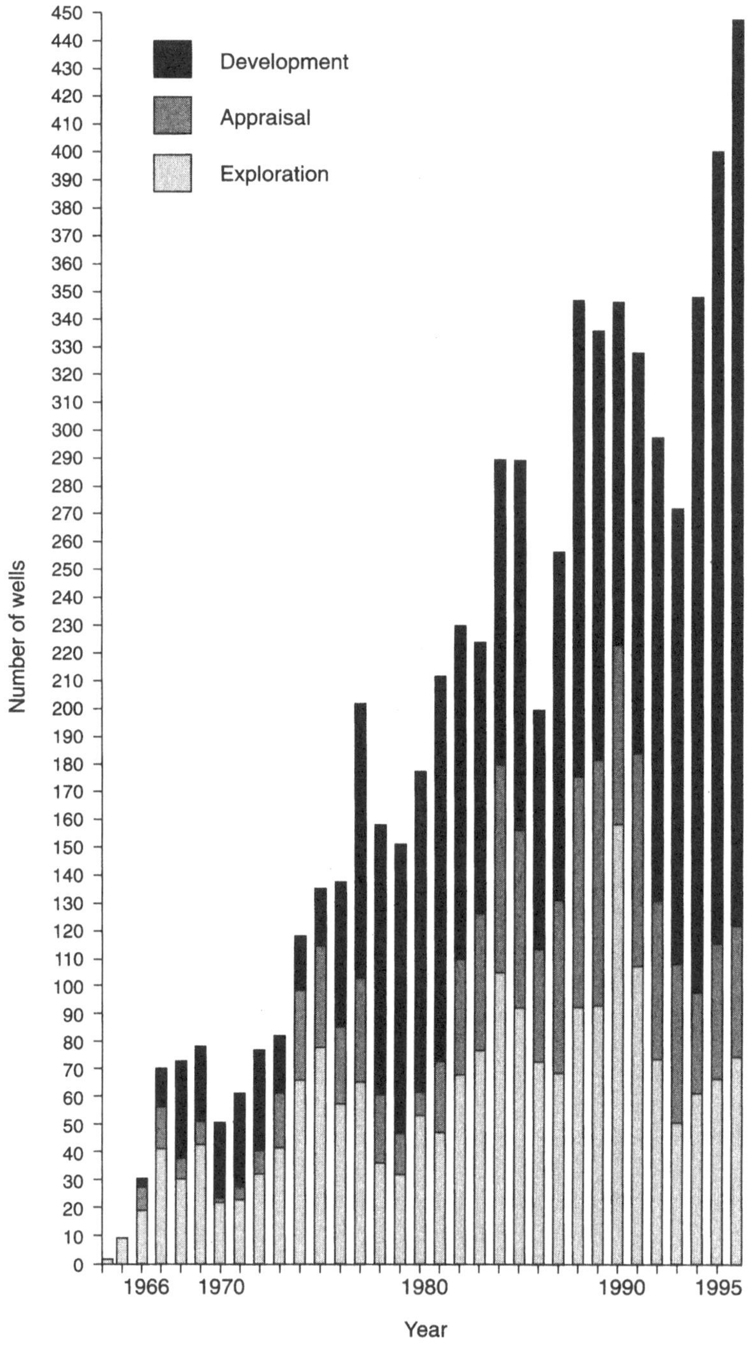

FIGURE 3.6 Offshore drilling activity, 1964–96

reserves of oil and gas. Commercial interest in the hydrocarbon potential of the North Sea was initially focused upon natural gas as a result of the discovery of the massive Groningen gas field in the Netherlands in 1959. This discovery drew attention to the possible existence of similar geological structures offshore. The validity of this inference has been confirmed by subsequent events, and estimates of natural gas reserves on the UK Continental Shelf (UKCS) have increased steadily since 1980. Total cumulative production of natural gas over the last thirty years represents approximately one-third of the current estimate of maximum recoverable reserves (Figure 3.7)[4] The inference that UKCS natural gas will last for another sixty years is not valid because it cannot be assumed that there will be no further discoveries or that the volume of production in each of the next two thirty-year periods will be the same as that between 1965 and 1995. Nevertheless, it can be concluded that there is no prospect of imminent exhaustion of the resource base.

Despite the importance of natural gas, oil has been the principal objective in the exploration effort since 1970. This is apparent in the northward shift of interest in successive licensing rounds from the primarily gas-bearing areas off the English coast to the predominantly oil-producing territories off Scotland and the Shetlands. The UK has been a net exporter of oil since 1980 (Figure 3.4) and, following a period of relative stagnation in the late 1980s/early 1990s, production reached its highest ever level in 1996. It is projected to remain at around this level to 2001. The immediate control upon output beyond this date is the timing of decisions to develop known fields. These decisions will be influenced by the interactions between the prevailing fiscal regime, which is controlled by government, the international oil price (see p. 34) and technical changes affecting the economics of exploiting small and/or inaccessible fields (see pp. 42–3). Taking a longer-term perspective, there are several indicators suggesting that the North Sea oil and gas province is at a mature phase in its life cycle (Figure 3.8). The declining average size of field is consistent with the expectation that the best prospects are targeted first, and the declining ratio of exploration to total wells drilled is also characteristic of a maturing hydrocarbon province. Similarly, abandonments in the early 1990s are precursors of a shifting balance between fields commencing and ceasing production. It is difficult to predict beyond 2010 because of the numerous uncertainties surrounding exploration and development decisions. Nevertheless, there are good reasons for caution in deriving pessimistic conclusions from the trends indicated in Figure 3.8. The UKCS is not restricted to the North Sea and several discoveries have been made to the west of the Shetlands. If the North Sea is at a mature stage, the Atlantic frontier is only just beginning its cycle of development. Figure 3.7 emphasises that the progressive enhancement of reserve estimates has more or less kept pace with the growth in cumulative oil production since 1980. Department of Trade and Industry (DTI) estimates in 1996 placed the total remaining reserves within the range 1,080–5,075 million tonnes. These figures acquire greater meaning when matched to the United Kingdom's annual consumption of approximately 78 million tonnes in the same year. A combination of simple arithmetic and heroic assumptions thus makes it possible to project fourteen to sixty-five years of self-sufficiency in oil. However, this calculation does not take account of the rapid build-up and subsequent long decline which is characteristic of the production history of any oil province, nor does it acknowledge the possibility of changes in demand. The definition of 'recoverable' adopted by the DTI is also no guarantee that oil-in-place will actually be extracted under future technical and economic conditions. On the other hand, historical precedent suggests that government agencies err on the side of caution when coping with the

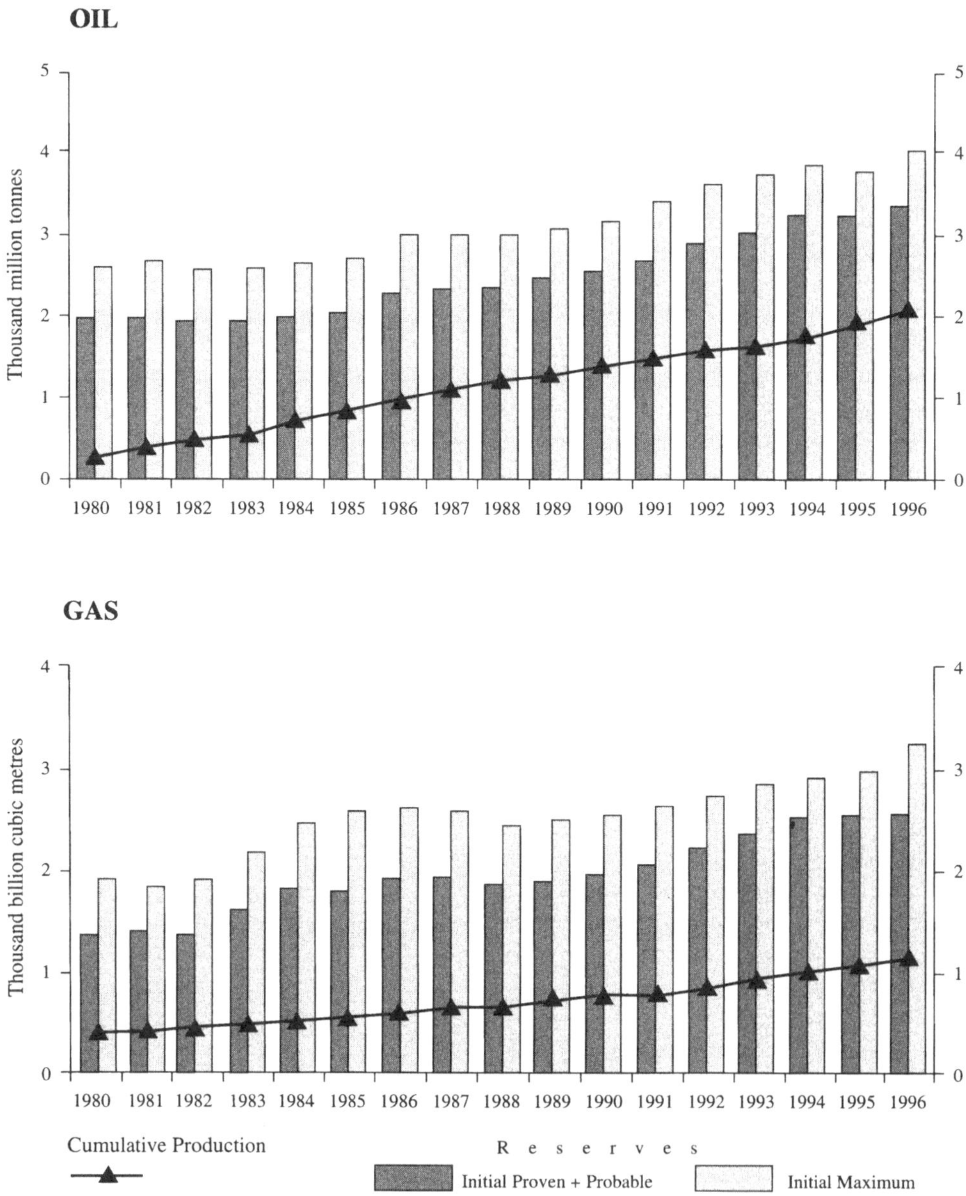

FIGURE 3.7 Discovered recoverable oil and gas reserves and cumulative production, 1980–96

substantial margins of error involved in making reserve estimates, and some observers believe that the DTI's upper figure should be at least doubled to obtain a truer indication of the oil potential of the North Sea. The picture will only become clear when the final well runs dry, but it is probable that this resource will remain a major national asset, even if much of it is not necessarily used directly within the UK, well beyond 2030.

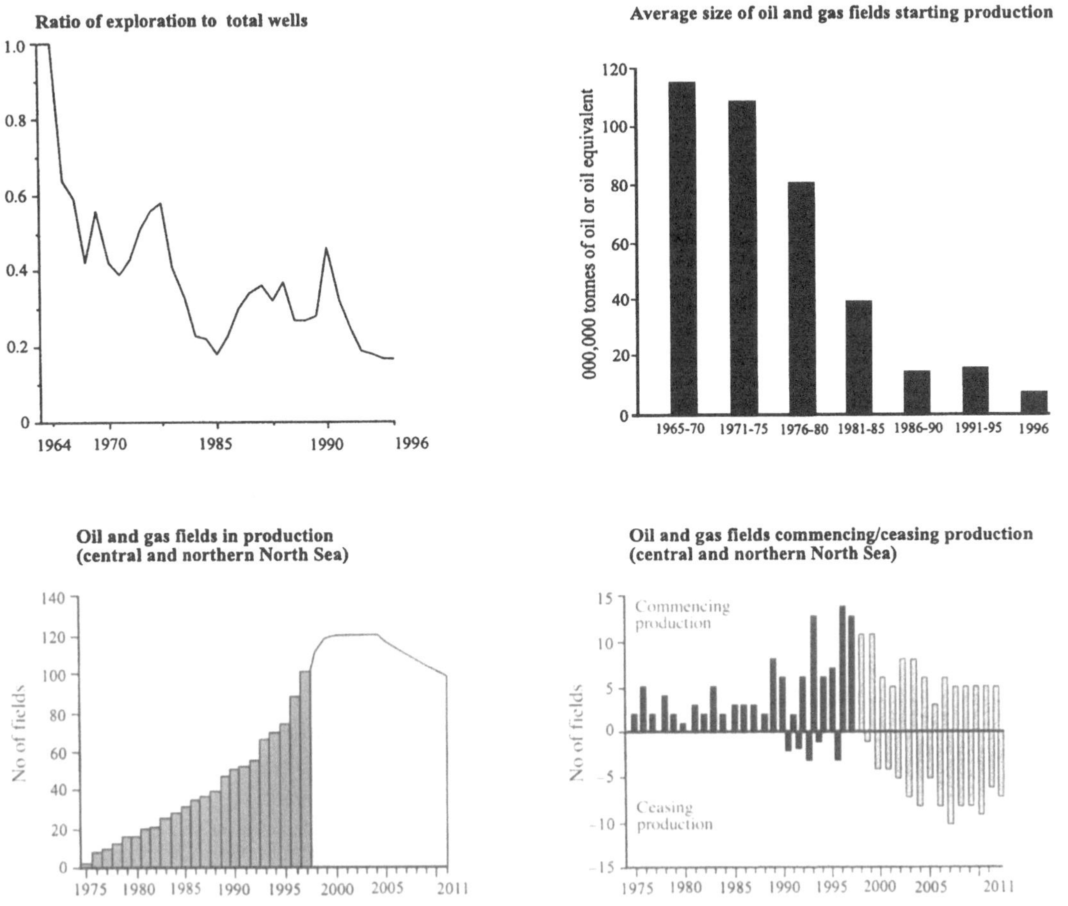

FIGURE 3.8 Life-cycle indicators for UK Continental Shelf oil and gas

Technical change

Technical change influences energy consumption, production and transportation. The significance of changes in the technology of moving or transporting energy, which have directly affected the geography of the UK energy industries, will be addressed later in the chapter, and immediate attention is focused on some of the principal developments affecting the consumption and production of specific fuels and the competition between fuels.

One of the most obvious impacts of technical change has been upon the overall level of energy consumption. Figure 3.9 plots trends in the ratio of energy consumption to gross domestic product (GDP) since 1970. It indicates a declining energy ratio, suggesting that less energy is being consumed per unit (by value) of economic output. This trend is usually interpreted as an indicator of more efficient energy use. The full explanation is more complex, but technical change is a common theme. In addition to improvements in the efficiency of end-users such as individual motor vehicles, the ratio also includes improvements in the efficiency of energy producers, most notably in electricity generation. Another important influence upon the energy ratio has been the changing structure of the economy and, in particular, the greater importance of services relative to energy-intensive manufacturing such as iron and steel. Attempts to disentangle the relative contributions of these various influences upon the energy ratio have suggested that improvements in energy efficiency, especially in manufacturing, were especially important between 1973 and 1989, but that the structural effect has been more significant in recent years. Although technical change, broadly conceived, has been responsible for a long-term downward trend in the energy ratio, the gradient has been steeper than anticipated twenty or thirty years ago and current UK energy consumption is well below the levels assumed in prudent strategic planning in the 1960s.

The aggregate demand-side effect of technical change is composed of impacts upon specific sectors such as coal. The decline of coal in the 1950s and 1960s was, as noted above, mainly due to a widening price differential in favour of imported oil. The problems of the industry were, however, compounded by certain technical changes which reduced the level of demand. Not only were the major coal-using industries heavily represented in the slow-growing and declining sectors of the economy, but also these large consumers were successfully developing more efficient methods of fuel use. Furthermore, several major markets were lost altogether. In 1950, the railways accounted for 7 per cent of domestic coal consumption: by 1965, the corresponding figure had fallen to 1.5 per cent as a result of the changeover to diesel and electric traction. A similar situation arose in the early 1960s as petroleum fractions replaced coal as the basic feedstock in town gas manufacture. The trend was repeated in the electricity industry as oil, natural gas and nuclear power challenged the monopoly position of coal as fuel for thermal power stations. In 1950, coal represented over 90 per cent of the fuel used for electricity generation in terms of energy content; by 1996 it accounted for only 20.1 per cent (i.e. approximately 46 million tonnes of oil equivalent).

Whereas technical change has primarily influenced the demand side of the equation in the case of coal, it has been more significant in affecting the supply of certain other fuels. For example, the exploitation of the oil and gas resources of the North Sea depends upon the ability to operate in such a difficult marine environment, and the necessary exploration – and, more especially, production technology – has been developed in response to this challenge. Future developments to the west of the Shetlands will depend upon continuing

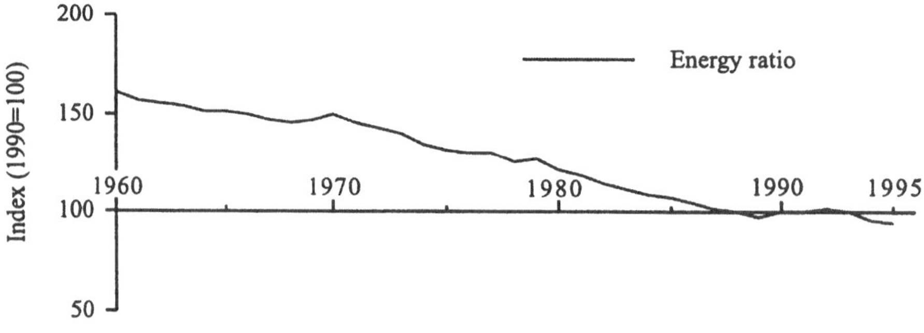

FIGURE 3.9 Energy ratio, 1960–95

the record of innovation which has been a feature of the North Sea experience. This record is reflected in the Cost Reduction Initiative for the New Era (CRINE) launched in 1993. This is a scheme, supported by government and the oil industry, aimed at reducing capital and operating costs, by a combination of technical and managerial innovations, with the overall objective of extending the economic life of the North Sea oil and gas province.

The history of nuclear power in the UK emphasises that technical change is often both complex and controversial. There was considerable optimism during the early 1950s regarding the future contribution of nuclear power to energy supply in the UK. This was encouraged when, in 1956, Calder Hall became the first nuclear plant in the world to supply power on a commercial rather than on an experimental basis. Similarly, the ten-year nuclear power programme announced in 1955 was the first national endorsement of this technology as a major source of energy supply. Projections made by the Department of Energy in 1978 and 1979 suggested that nuclear power would account for 17 to 22 per cent of primary energy demand by the year 2000. The actual figure in 1996 was 9.5 per cent and it seems likely that the late 1990s will witness the zenith of nuclear power in the UK as Sizewell B, commissioned in 1995, could be the only remaining operational nuclear power station in the UK by 2020. Some of the first generation Magnox stations have already been decommissioned and others are already beyond their original design lives. Thus the prognosis is poor and contrasts sharply with the enthusiasm surrounding the UK's perceived technical leadership forty years earlier at 'the dawn of the atomic age'.

Many factors have contributed to the fading of this vision, but there is no doubt that the technology has proved more difficult than anticipated. This is reflected in construction costs, which have been much higher than projected in both the Magnox and, especially, the Advanced Gas-Cooled Reactor (AGR) programmes. Site work at Dungeness B, for example, began in 1966 and the station only became fully operational in 1989. Operating experience was also disappointing for many years as individual stations failed to achieve their design output. Performance has improved in the 1990s, but there have consistently been serious doubts about the economics of nuclear power, which was originally justified in terms of its low cost relative to other fuels. These doubts have been reinforced by a lack of agreement on the average cost per unit of nuclear electricity. The complexity of assumptions underpinning these calculations has provided ample opportunity for creative accounting and the manipulation of figures to suit political agendas (see pp. 44–5). The Chernobyl incident in 1986 contributed to the collective loss of confidence in nuclear power and, at the same time, drew attention to the fundamental, unsolved technical problem of dealing with the radioactive waste products, especially at the decommissioning stage. The magnitude of this problem was acknowledged in 1990 with the introduction of the Fossil Fuel Levy which accounts for approximately 10 per cent of electricity prices paid by consumers in England and Wales. This tax is designed to create a fund 'to meet the higher than anticipated back-end costs of nuclear power' (Department of Trade and Industry 1995: 15).

Public policy

Government has been a major influence upon the evolution of the UK energy market since the Second World War. Until comparatively recently only oil, of the major energy supply industries, was not in public ownership. For more than thirty years there was a political consensus that these industries should remain in the public sector because of their central role in the national economy. Fiscal measures, such as taxation and subsidies, have also been frequently employed to manipulate the energy market. These measures have not, however, been incorporated within a coherent and consistent strategy despite a long history of energy policy reviews culminating in yet another appraisal initiated by the new government in 1997 and due for completion in 1998.

The Department of Trade and Industry (1995: 3) states that 'The aim of the Government's energy policy is to ensure secure, diverse and sustainable supplies of energy in the forms that people and businesses want, and at competitive prices.' This statement provides clues to some of the problems which have faced successive governments in regulating the energy market. In the absence of a clearly defined time-scale for their achievement, several of the aims are contradictory. 'Competitive' presumably means 'low' prices. However, cheap energy is not consistent with the promotion of efficiency and sustainability. Thus fossil fuels such as coal and oil may currently be available at more 'competitive prices' than renewables such as wind or wave power, but the latter will certainly need to be developed in the longer term. Such development requires investment which may be inhibited by a preoccupation with the short-term price relativities of the various energy sources. Similarly price-driven consumer choices may conflict with wider, strategic concerns such as the diversification of sources of energy supply. The dramatic decline of the coal industry has, for example, narrowed future options by making large quantities of coal in flooded pits inaccessible.

The changing priorities of energy policy in general and attitudes towards the coal industry in particular illustrate the essentially *ad hoc* approach. Successive White Papers on fuel policy in 1965 and 1967 made it clear that the principal objective was 'to make possible the supply of energy at the lowest total cost to the community'. The priorities changed after the first oil price jump in 1972/3, which resulted in a greater awareness of the political and economic implications of dependence upon external sources of supply. Accordingly policy statements published by the Department of Energy in 1977 and 1978 demonstrated a greater commitment to indigenous resources in the support for the expansion plans of the NCB, the accelerated development of nuclear power, and the promotion of research into renewable sources of energy. By the 1990s, environmental objectives had become more prominent, reflecting a prevailing public interest in the concept of sustainable development and the requirements of international treaty obligations to reduce atmospheric emissions from combustion processes.

The long-term decline of the coal industry has ultimately been determined by the preference of consumers for other fuels, but government policy has also been important. Apart from a relatively brief revival in the 1970s, stimulated by events in international oil markets (see pp. 34–5), public policy towards the coal industry since 1950 has largely been concerned with the management of decline. In the 1960s, several steps were taken to slow down the rate of pit closures to ameliorate the adverse economic and social consequences for coalfield communities. The most important measures were designed to sustain the demand for coal in its principal market – the power stations. A completely different agenda emerged in the 1980s when, influenced by the memory of the miners' strike in 1973/4 which effectively paralysed the country and brought down an earlier Conservative government, the Conservative Prime Minister, Mrs Thatcher, pursued policies intended to break the power of the National Union of Mineworkers (NUM) and to undermine the position of coal as the dominant fuel in electricity generation. These policies were motivated primarily by political considerations. These considerations found expression in the imposition of severe public expenditure controls upon the NCB, which accelerated pit closures and forced a confrontation with the NUM. They were also evident in a pro-nuclear policy which resulted in the approval in 1979 of Sizewell B, which is the most recent nuclear power station in the UK. Mrs Thatcher's governments imposed less demanding financial targets in the appraisal of nuclear investment proposals than applied to the operations of the NCB. This disparity of treatment did not convince commercial interests to bid for nuclear power at the privatisation of the electricity supply industry in 1989, but it did reverse the policies of previous governments by accelerating rather than retarding the decline of coal. The privatisation policies, which became a central feature of the Thatcher period, were not explicitly directed towards the energy industries. Nevertheless, the successive transfers of gas, a substantial proportion of the electricity supply industry and coal from public to private ownership between 1986 and 1994 collectively represented a significant element of the privatisation project. This project, which is linked to the promotion of competition (i.e. liberalisation), has radically altered the context of energy production and consumption in the UK.

The structure of the energy sectors following privatisation varies. In the case of coal, the bulk of British Coal's[5] assets, amounting to approximately 70 per cent of UK output in 1994, were sold to RJB Mining (UK) Ltd with smaller companies assuming responsibility for operations in Scotland and Wales. British Gas plc initially inherited the monopoly position of its state-owned predecessor at privatisation. However, various steps designed to create 'managed competition' were taken. These allowed new suppliers to deliver gas from the North Sea and obliged the transportation and storage arm of British Gas plc (i.e. Transco) to operate as a 'common carrier' and make its on-shore pipeline distribution system available to other companies. British Gas was itself 'de-merged' in 1997 to create two independent companies – BG plc, responsible for exploration, production and UK pipeline distribution, and Centrica plc, which is primarily involved in retailing and gas trading activities. The privatisation of the electricity supply industry has created the most complex structure, partly because of differences between England and Wales relative to Scotland, and partly because the nuclear sector remains in public ownership. In England and Wales, the system has been fragmented into generation, national transmission and regional distribution (Figure 3.10). The system remains vertically integrated in Scotland, where two area-based companies are responsible for generation, transmission, distribution and supply to final consumers. The Scottish generators also contribute electricity to 'the Pool' in England and Wales, which is a wholesale electricity market operated by the National Grid Company.

There is no doubt that privatisation has contributed to declining real energy prices in the UK since 1986, but there remain doubts about the ability of the new utility regulators, OFGAS and OFFER, to fulfil their responsibilities of protecting the interests of consumers whilst restraining and guiding the activities of the gas and electricity supply industries respectively. On a strategic level, there is a danger that the preoccupation with costs and prices will obscure the wider issues implicit in the concept of energy policy which rests upon the premise that the market mechanism does not necessarily guarantee an outcome consistent with the national interest, broadly conceived. Indeed, it could be argued that privatisation has rendered energy policy obsolete by sharply reducing the ability of government to influence the direction of change.

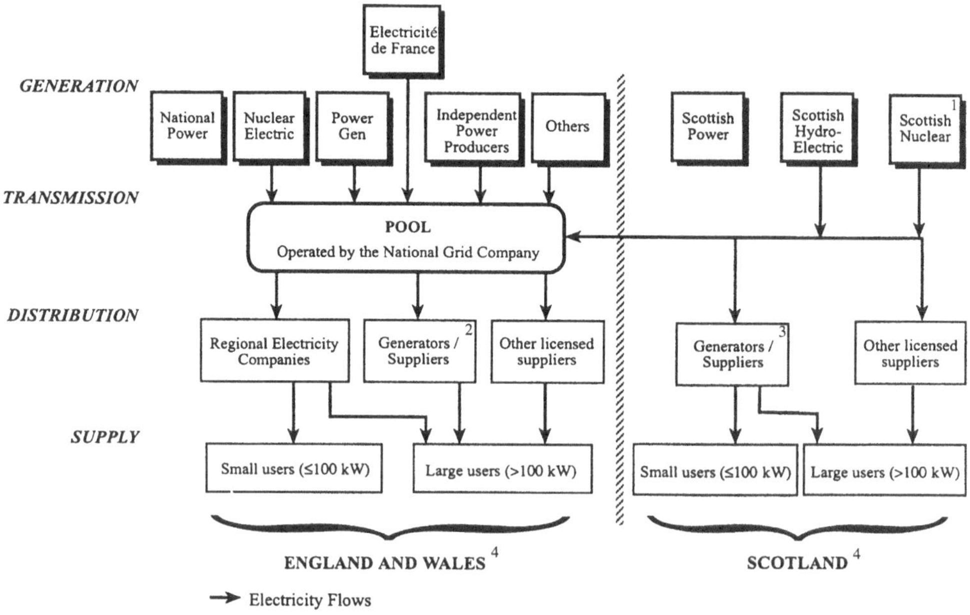

FIGURE 3.10 Structure of privatised electricity supply industry

1 Other small generators supply Scottish Power and Scottish Hydro-Electric.
2 i.e. National Power, Nuclear Electric, Power Gen, Scottish Power, Scottish Hydro-Electric.
3 i.e. Scottish Power, Scottish Hydro-Electric.
4 Northern Ireland system not interconnected with the rest of the UK.

'National interest' in energy policy has generally been defined in economic and political terms. There is, however, a growing recognition of the importance of environmental concerns. This is reflected in, for example, an admittedly belated incorporation of the costs of decommissioning and waste disposal in the debate on the economics of nuclear power, itself deriving from a wider acknowledgement of the impact of 'externalities' associated with producing and delivering energy. The accounting mechanisms for evaluating these externalities are poorly developed, but the principle is widely accepted. Perhaps the clearest evidence of this is provided by the UK's commitment to achieve specific targets in reducing its discharge of greenhouse gases to the atmosphere. This commitment, which is discussed in Chapter 18, has introduced a new 'external' constraint upon the operations of the energy supply industries in the UK. This constraint also affects energy consuming activities. Although power stations are by far the most important source of sulphur dioxide (which is largely responsible for the problem of acid rain), carbon dioxide (which is the principal 'greenhouse gas') is mainly a product of energy consumption in the domestic, commercial and transport sectors. Some of the environmental implications of energy use are considered in Chapters 17–19.

Changing geographies

Changing energy markets have created new geographies of energy production in the UK. The principal features of these geographies are described below. The approach is both systematic and temporal in the sense that it proceeds from a review of the geographical implications of the decline of coal, through a consideration of the legacy of the era of dependence on imported oil, to the impact of North Sea oil and gas and, finally, the prospects for renewable energy sources.

Coal: the geography of decline

The decline in the contribution of indigenous coal to national energy supply has already been noted. This decline is reflected in a fall of employment from approximately 600,000 in 1960 to 50,000 at privatisation in 1994 and in a reduction in the number of collieries from almost 700 to just nineteen (including five in the Selby complex) over the same period. The industry which shaped the cultures of communities from South Wales to North-East England and Central Scotland has become progressively concentrated in the Yorkshire/Derbyshire/Nottinghamshire coalfields. This contraction and geographical concentration has been heavily influenced by spatial variations in the costs of production. Geological circumstances have been the principal influences upon these costs, and the concentration of remaining mines in Yorkshire and the East Midlands reflects generally more favourable underground operating conditions, such as thicker seams, gentler dips, and less faulting. The pressure on costs which has driven the once-mighty coal industry to the edge of extinction is also evident in the relative growth of opencast production (Figure 3.11). Where coal can be exploited in this way, unit costs of production are much lower than for underground mining.

Regional differences in the profitability of coalfields have been influenced by market considerations as well as cost factors. For example, Tower, the last remaining coal-mine in South Wales, produces anthracite which commands a higher price than poorer quality coals. The high costs of transporting coal by comparison with other primary energy sources have ensured that certain coalfields have been better placed than others to cope with a contracting market. Distance helped, for a time, to protect Welsh and Scottish pits from the competition of the more efficient mines of Yorkshire and the East Midlands, but proximity to major markets reinforced the cost advantages of the latter. In particular, the construction, in the 1950s and 1960s, of a series of major coal-burning power stations along the valleys of the Aire, Calder and Trent was based on the combined attractions of abundant cooling water and the cheapest coal available in the UK. The dependence of the coal industry upon the power-station market, together with its diminishing share of this market, has already been noted. The massive power station at Drax in Yorkshire was the zenith of coal-fired capacity and no significant additions have been made since its commissioning in 1974, and several stations have been either closed or 'mothballed' since privatisation in 1990. The prognosis is even more discouraging to the end of the century following a three-year deal struck between RJB Mining and National Power at the end of 1997. This contract involves a sharp fall in power station coal consumption which, in the absence of government intervention, will inevitably result in further pit closures and redundancies.

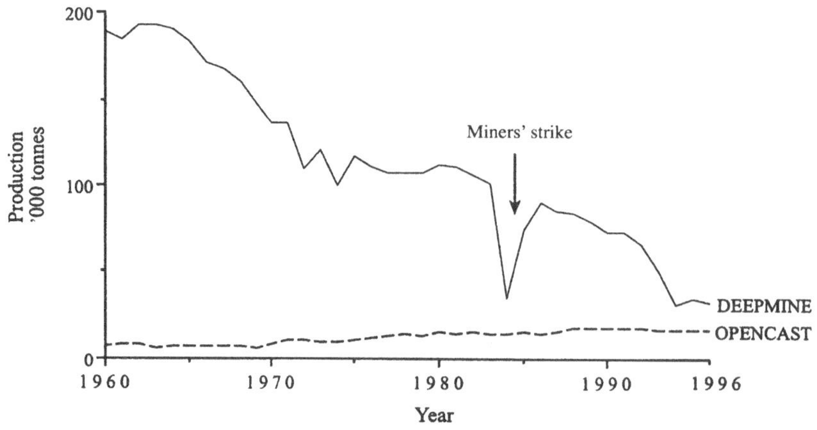

FIGURE 3.11 Trends in coal production, 1960–96

Imported oil: processing and distribution patterns

Imported oil provided the first serious challenge to the dominance of coal in the UK energy economy. A major programme of pit closures and associated job losses during the 1960s was, perhaps, the most visible impact of this challenge. A related consequence was a corresponding expansion in the UK oil refining industry which transformed the country's position with regard to its supplies of petroleum products. Domestic refinery output in 1938 accounted for only 27 per cent of the total consumption of oil products. However, the growth in the refining industry since the Second World War has enabled it not only to keep pace with the spectacular rise in demand but also to maintain an approximate balance between indigenous production and consumption of petroleum products. Self-sufficiency in petroleum products was first achieved in 1951 and has been maintained ever since.[6]

Although UK refining capacity expanded tenfold between 1939 and 1959, this increase was largely achieved by the expansion of existing facilities. The distribution of the industry in 1960 reflected location decisions taken before the Second World War, and almost 60 per cent of capacity was situated on the lower Thames and at the seaward and inland ends of the Manchester Ship Canal linking Liverpool and Manchester. A combination of commercial and technical factors contributed to the emergence of major new refining centres on Milford Haven, Humberside and Teesside during the rapid expansion of the 1960s. The attraction of Milford Haven lay in its advantages as a deep-water harbour with the advent of the 100,000-ton oil tanker towards the end of the 1950s. These vessels offered considerable economies of scale in the movement of crude oil, and Milford Haven is only one of a number of European refining complexes which owe their 1960s origins to the berthing requirements of the supertanker. The search for new sites, stimulated by advances in shipping technology, was also encouraged by changes in the organisational structure of the industry. Several new companies entered the UK market in the late 1950s and early 1960s, establishing refining capacity which reduced the proportion controlled by Shell, Esso and BP from 94 per cent in 1962 to 44 per cent by 1996. Each of these major companies only established one new site

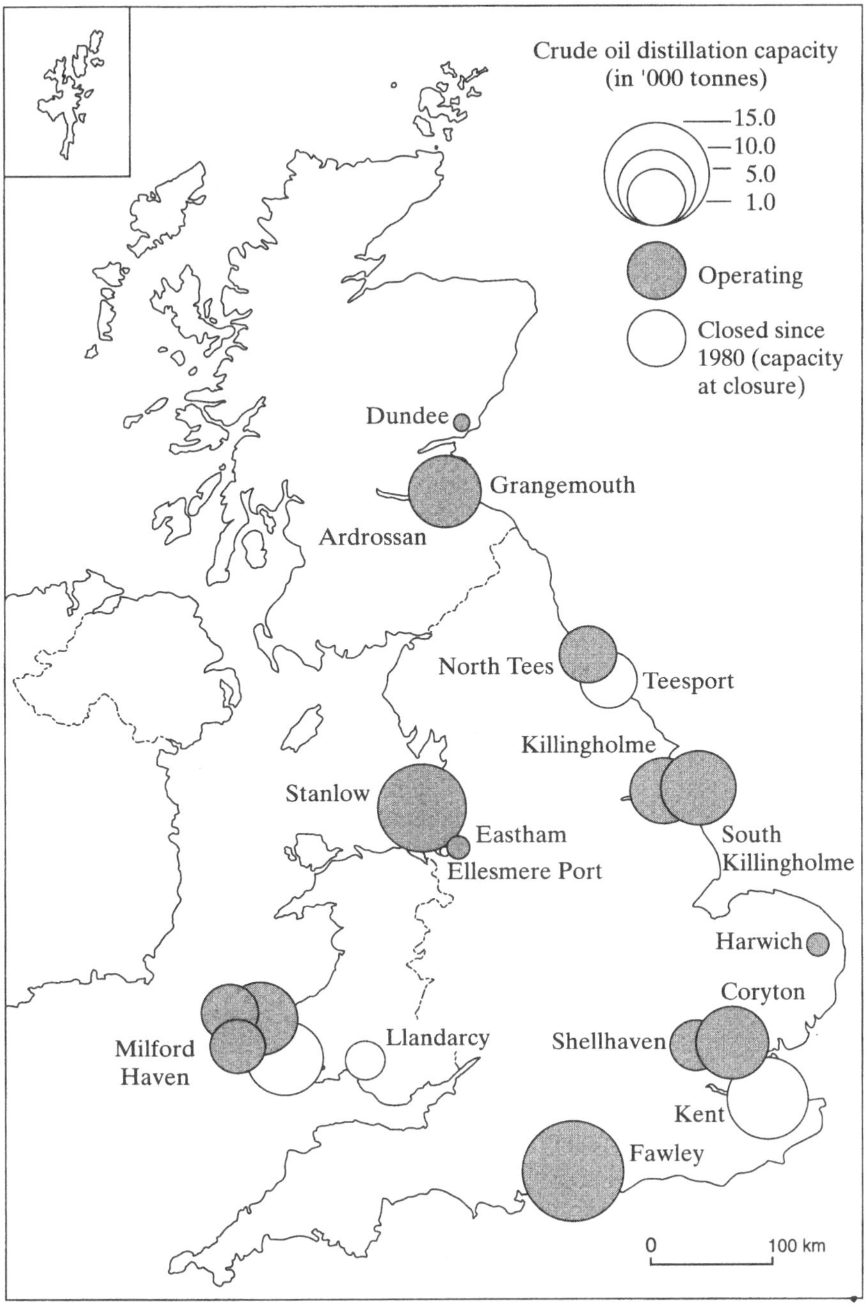

FIGURE 3.12 Oil refineries, 1996

after 1960 and tended to add any incremental capacity at existing refineries. On the other hand, the new refining companies were not restricted by the location of established facilities and therefore played an important role in the emergence of Milford Haven and Humberside as major refining centres.

The spatial pattern of oil refineries in 1996 (Figure 3.12) differs little from that established by 1970. The Amoco plant at Milford Haven is the only new installation, and the recession prompted by the big oil price increases of the 1970s invalidated the assumptions of growth upon which the investments of the 1960s were based. Individual units and entire refineries have been shut down, especially since 1980, as the oil companies have attempted to adjust to the new circumstances; total throughput capacity in 1996 was only 63 per cent of the 1974 peak. These events not only represent a radical departure from the previously continuous post-war growth in capacity but also emphasise that the availability of North Sea oil is no guarantee of renewed investment in refining facilities.

North Sea oil: exploitation and economic development

Many have argued, with some justification, that North Sea oil has been a missed opportunity, but its exploitation has undoubtedly had a major impact on the UK economy. The direct contribution of the offshore oil and gas sector to UK gross domestic product peaked at almost 7 per cent in 1984 and has stabilised at around 2–2.5 per cent since the mid-1980s. It has also accounted for substantial proportions of annual UK industrial investment, generated considerable tax revenues and was estimated in 1998 to be responsible for more than 350,000 jobs, including employment in the offshore sector and in related industries such as metal goods and engineering. Despite such indicators of importance, North Sea oil exerts little obvious impact upon existing processing and distribution patterns. It has, for example, stimulated relatively little new investment in the oil refining industry, and the principal geographical consequences of North Sea oil have been associated with its production rather than its consumption.

These consequences are most evident in Scotland where the economic impacts generated by the exploitation of North Sea oil have contributed to the growing political confidence of Scottish nationalists which has itself encouraged the devolution movement (see Chapters 15 and 16). Scotland has, because of its position with respect to most of the offshore fields (Figure 3.13), attracted a substantial proportion of the employment generated by the exploitation of North Sea oil. This has contributed to a significant improvement in Scotland's unemployment position relative to the rest of the country since the mid-1970s as compared with the preceding twenty years. Oil-related jobs are, however, unevenly distributed. In 1995, Grampian Region, focused on Aberdeen, accounted for approximately 86 per cent of a total of almost 64,000 jobs in Scotland estimated to be directly related to the North Sea oil industry.[7] The corresponding share in 1974 was only 36 per cent and the dominance of north-east Scotland in general and Aberdeen in particular has been accentuated as the service and administration functions of the production phase have replaced the manufacturing jobs associated with the construction of capital equipment during the development stage. Thus most of the new jobs have been created in areas outside the Central Belt, which has traditionally been the focus of Scotland's unemployment problem.

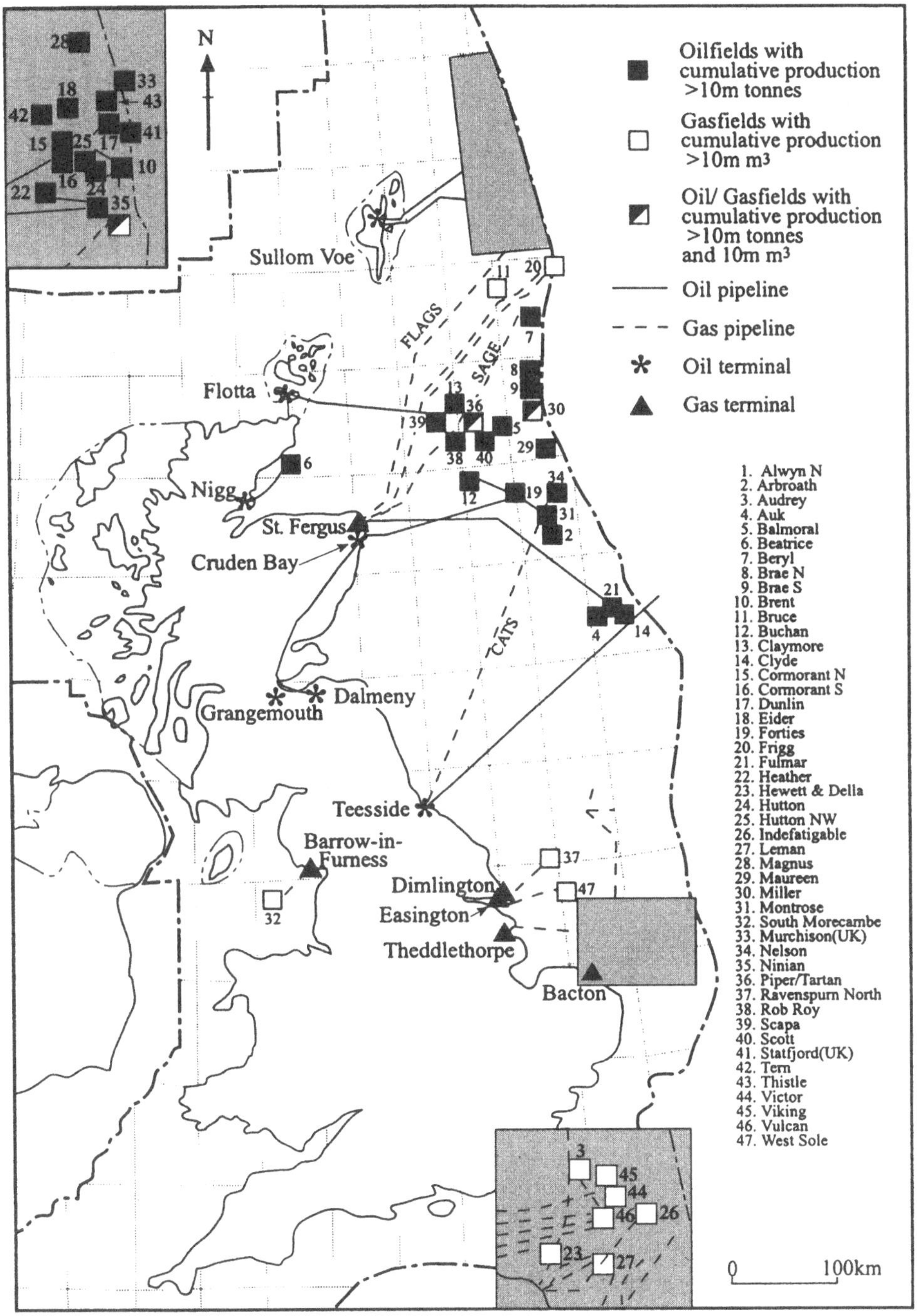

FIGURE 3.13 North Sea oil and gas, 1996 (principal fields only)

The progressive concentration of oil-related employment upon Aberdeen has not been the only consequence of the maturing cycle of North Sea oil exploitation. The key areas of local and regional policy concern have changed. During the 1970s, large-scale projects such as platform fabrication yards and reception terminals generated sometimes bitter environmental and land-use conflicts in which local preferences were frequently overwhelmed by arguments based on the premise that rapid exploitation was in 'the national interest'. By the mid-1990s, attention was turning to the implications of a gradual run-down of operations. It has already been noted that oil and gas seem likely to be produced from the North Sea for at least another forty years, but the number of fields ceasing production will gradually increase (Figure 3.8), raising important questions about the decommissioning phase and the removal of offshore structures. Projections suggest that direct oil-related employment in north-east Scotland, which peaked at approximately 54,000 in 1991, will gradually decline beyond the turn of the century. The extent and timing of this decline has obvious implications for the future of Aberdeen as one of the most prosperous cities in the UK. These implications have focused the minds of local authorities and development agencies on the problems of converting the North Sea oil and gas experience to long-term economic advantage. Scottish Enterprise, a development agency, for example, is promoting the concept of a 'cluster' of oil-related activities and technologies which will elevate the status of operations in Scotland and, especially, in Aberdeen beyond the support of the North Sea to serving global markets. This objective derives from essentially geographical ideas concerning the development of localised, place-based competitive advantages.

Natural gas: geographies of consumption

Activities associated with the exploitation of natural gas are similar to, and often indistinguishable from, the same activities related to oil and it has been in its distribution and consumption rather than its production that the geographical impacts of natural gas have been most apparent. An early consequence of the advent of natural gas was the disappearance of what had been one of the most distinctive features of the urban landscape for approximately a hundred years – the gas works. Town gas had traditionally been made from coke, but natural gas transformed the gas industry from a manufacturer and supplier of secondary energy to a distributor of primary fuel. It also led to the emergence of a gas industry that was truly national in its scale of organisation and pattern of distribution.

This transformation was initiated by imports of liquified natural gas from Algeria in 1964 which was distributed via a specially constructed transmission system to most of the major urban markets in England. It was progressed with the introduction of supplies from the North Sea in 1967 and more or less completed with the establishment of a national pipeline network by the late 1970s. The network developed with reference to the key inputs to the system at the coastal terminals and the principal centres of consumption oriented along the London–Birmingham–Manchester corridor. As originally conceived, the transmission system was in the public sector. Following privatisation in 1986, ownership and control passed to British Gas plc. Revised arrangements led to the creation of BG Transco in 1996, which is required to operate as a 'common carrier' at 'arm's length' from its parent. Some of the most significant geographical implications of privatisation for the UK energy market derive from this 'common carriage' obligation.

Some commentators in the 1960s predicted that the east coast landfalls would attract energy-intensive, gas-based industries such as fertilisers and petrochemicals. These predictions were based on the assumption that gas prices would be lowest at the landfalls and would increase with distance from these points. In fact, spatial variations in the prices charged by the state-owned British Gas Corporation were very limited and certainly did not reflect the substantial differences in the real costs of delivering gas within the UK. However, spatial variations in gas prices to industrial consumers (i.e. the 'contract' market) began to emerge following privatisation. These were market- rather than cost-based in the sense that higher prices were charged to users who, for technical reasons, were unable to switch to alternative fuels such as oil or coal. Legal challenges, supported by regulatory changes, forced British Gas plc to end geographical price discrimination in the contract market and to publish a national price schedule reflecting such factors as the volume of gas supplied and length of contract, but *excluding* the influence of location. Ironically, these changes have resulted in the introduction of geographical price differentials based on delivery costs because new suppliers, which include a multiplicity of commercial interests ranging from oil companies and electricity companies to multinational utility companies, were not obliged to publish national price lists. This, together with their right of access to the transmission system under the terms of BG Transco's 'common carrier' obligation, placed these suppliers at a competitive advantage relative to BG plc. They have exploited this advantage by adopting price structures which more accurately reflect the costs of supply and, therefore, favour customers located near the pipeline landfalls. Thus the cross-subsidies implicit in the national price schedule of BG plc and the objections to geographical price differentials which prompted these schedules have, encouraged by the intervention of the regulatory authorities, had the apparently perverse effect of promoting such differentials. The justification has changed from market power to costs of supply, but the overall effect seems likely to be the exaggeration of these differentials. The forces which have driven these changes have been associated with the evolution of the contract market, but the potential for competition between gas suppliers in the 'tariff' (i.e. domestic) market was in place across the UK by mid-1998. The extent and magnitude of any resulting price variations between households in different parts of the country is unclear, although it seems likely that social and political considerations will serve as moderating influences.

The clearest illustration of the possible geographical implications of spatial variations in gas prices has been the so-called 'dash for gas' in electricity generation. Peterhead was the only gas-fired power station in 1992, but twenty-two were operational by the winter of 1996/7. Further committed and approved projects suggest that Department of Trade and Industry projections in 1995, that just over one-third of the country's electricity could be generated from gas by the end of the century, may be a serious underestimate and others have suggested that the proportion could reach 70 per cent by 2010. These forecasts probably represent the upper and lower extremes of the range of possibilities, but there is no dispute that the extensive use of natural gas represents a very significant departure from previous practice in the UK. Several factors have contributed to this development. The technology of combined-cycle gas turbines (CCGT) offers many advantages; it has a higher thermal efficiency than traditional stations fired by coal and oil; capital costs per unit of output are low; construction times are short; and incremental additions to capacity are relatively easy. One of the most significant benefits of CCGT technology derives from the relatively low emissions of atmospheric pollutants associated with the combustion of natural gas. In particular, virtually no sulphur dioxide is emitted from gas-fired CCGT stations – a characteristic which is attractive both to the electricity supply industry, because it reduces the liabilities to retrofit expensive flue gas desulphurisation equipment to existing coal-fired stations, and to the government, because it will help the UK meet its international treaty obligations to reduce sulphur dioxide emissions.

The 'dash for gas' is not only related to the technical advantages of CCGT power stations. Organisational changes associated with the privatisation of both the gas and electricity supply industries also contributed as a result of the coincident commercial interests of the suppliers and purchasers of gas. In an attempt to counter the threat to its market share posed by the new suppliers in the contract market, British Gas plc pitched its average beachhead price at relatively low levels in the early 1990s. This, together with offers made by its new competitors, stimulated considerable interest in long-term contracts by the regional electricity companies. Their interest was reinforced by the desire of these companies to reduce their dependence upon the dominant generators (i.e. National Power, Powergen and Nuclear Electric). The regional electricity companies have, therefore, taken equity interests in various independent power producers whilst National Power and Powergen have responded by building their own CCGT stations. The influence of these strategic games deriving from the fragmented commercial structure of the electricity supply industry in England and Wales (Figure 3.10) is emphasised by the absence of CCGT investment in the vertically integrated Scottish industry.[8]

The effects of the 'dash for gas' upon the geography of electricity generation in England and Wales are evident in Figure 3.14. Several of the largest plants are located on the east coast to benefit from lower gas prices; others are adjacent to the main trunk pipelines of the transmission system; yet others are situated within the market of the south-east. These changes raise issues with a significance extending beyond the geography of the electricity supply industry. The fragmented structure of the privatised industry makes system-wide planning of the type which informed investment decisions under public ownership more difficult. In these circumstances, the evolving spatial pattern of the industry probably contains significant inefficiencies in the disposition of capacity relative to markets and the structure of the transmission system. Such inefficiencies may be replicated elsewhere in the energy sector as the opportunism necessarily present in the drive to lower short-term supply cost inhibits the adoption of a wider, longer-term perspective. This concern may be illustrated with reference to the present state and future prospects of renewable energy in the UK.

Renewable geographies?

Renewables incorporate a diversity of energy sources and technologies sharing the common fundamental property that, unlike fossil fuels such as coal, oil and natural gas, they do not involve the depletion of a finite resource. Despite their diversity, including tidal, wind, wave and solar power, 97 per cent of the energy output (as heat and electricity) from renewables in the UK in 1996 was derived from only two generic sources – biofuels and hydropower. The former has overtaken the latter as the dominant source of renewable energy, accounting for 80.3 per cent of energy output in 1996 as compared with 16.7 per cent from hydropower. A variety of approaches are employed in the generation of heat and, indirectly, electricity from

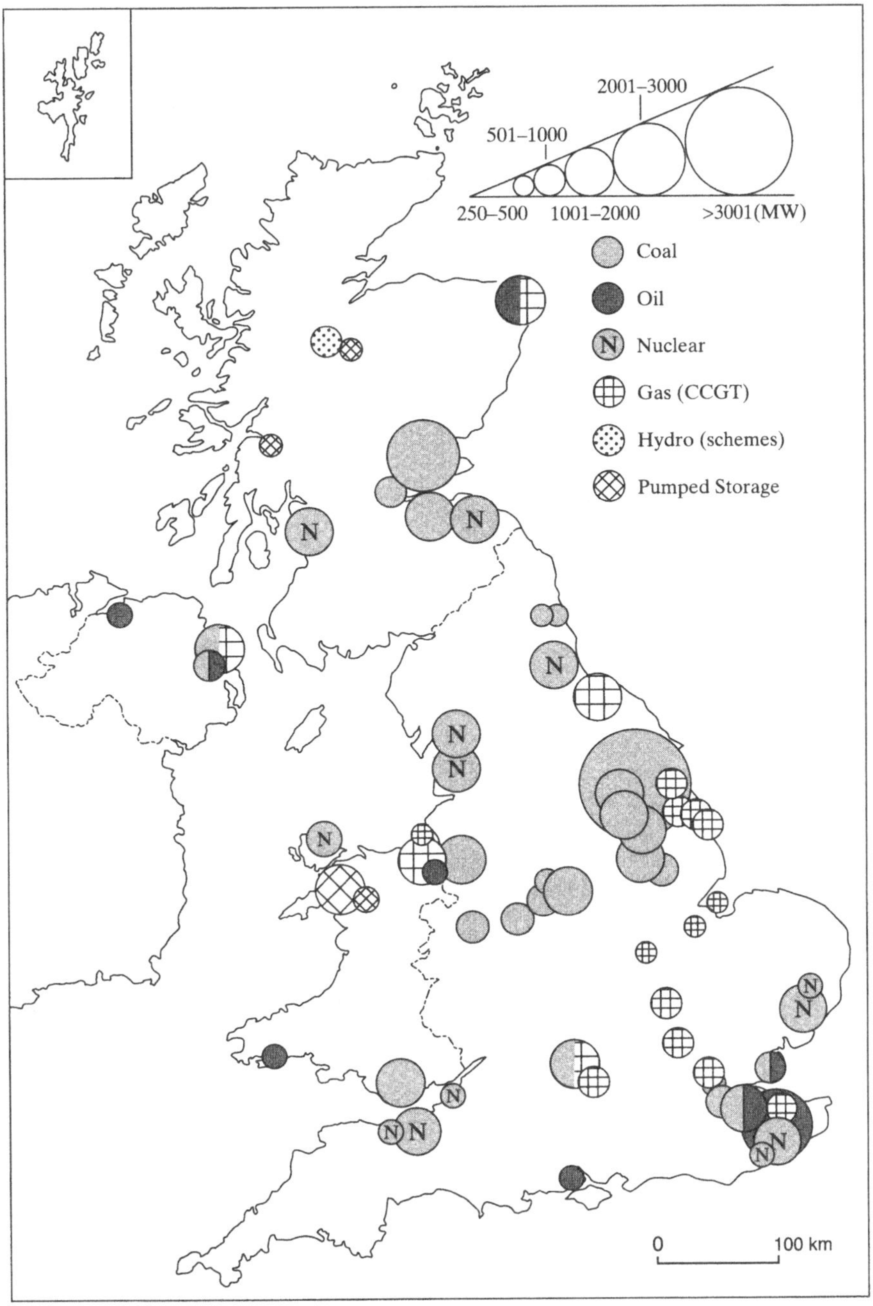

FIGURE 3.14 Electricity generating stations (>250 MW installed capacity only)

biofuels, including waste incineration, the recovery of methane-rich biogas from landfills and sewage gas from sludge digestion.

Many governments have paid lip-service to the promotion of renewable energy in the UK. The oil price-shocks of the 1970s provided a boost when renewables were seen as potential contributors to an energy strategy aimed at reducing dependence on expensive imported oil. The then Labour government initiated a programme of research and development co-ordinated by the newly created Energy Technology Support Unit (ETSU). Assessments of the potential for wind and wave power in the UK were published in 1977 and 1979 respectively. These were followed by various small demonstration projects, but there was little government encouragement during the 1980s until the issue of global warming appeared on the political agenda. This provided a different reason to reconsider renewable energy sources which offer a potential route to the achievement of the UK's target reductions in greenhouse gas emissions (see Chapter 18). The legislation privatising the electricity supply industry provided an opportunity to address the problem of greenhouse gases and, at the same time, to promote renewable energy. The 1989 Electricity Act introduced the Non-Fossil Fuel Obligation (NFFO) in England and Wales. This requires the regional electricity companies to purchase a specified proportion of their electricity from non-fossil fuel sources. Nuclear power was regarded as the most important of these sources, and the NFFO was originally conceived to protect the market share of the state-owned nuclear power stations. The objectives of the NFFO were, however, widened by allowing the government to require a certain 'tranche' of electricity to be generated from renewable sources. This 'tranche' is effectively subsidised by the 'Fossil Fuel Levy', which is a tax on the electricity bills of all consumers. Thus the NFFO and its equivalent in Scotland, the Scottish Renewables Obligation (SRO), now form the central mechanism for supporting renewable energy in the UK. Recognition of the principle that renewable energy needs subsidising to encourage its development in the longer term is a very important departure from previous government practice and there are indications that the Energy Review scheduled for completion in 1998 will endorse this principle by committing the government to annual subsidies of up to £300 million.

Despite these trends, the contribution of renewables to total energy consumption was only 0.3 per cent in 1995. This is lower than any other EU country and contrasts with the overall EU average of 6 per cent and a corresponding figure of 25.4 per cent for Sweden. These statistics reflect the combined legacy of an abundant fossil fuel resource base and an early commitment to nuclear power. This legacy suppressed the development of renewables, and an assessment of the geographical impact of any increased future contribution of these technologies to energy supply in the UK must, necessarily, be speculative. The largest renewable energy projects in the UK are the hydro-electric schemes established in Scotland as early as 1895. There is little prospect of further developments of this type, and the frequently discussed tidal barrage schemes for the estuaries of the Severn and Dee are unlikely to be reactivated. The future of renewable energy almost certainly lies in small-scale, localised schemes. Some clues as to the probable patterns may be gained by reference to the results of the introduction of the NFFO and the SRO. These are evident in the outcomes of a series of Renewables Orders or rounds starting in 1990. Companies proposing to establish renewable energy generating projects bid to supply electricity under a Renewables Order and the various proposals are approved or rejected by the Office of Electricity Regulation (OFFER), which specifies the price to be paid for electricity generated from different categories or 'bands' of technology. The identified 'bands' relate to electricity generated from wind, hydro, waste incineration, sewage gas, landfill gas, and 'biofuels' such as agricultural and forestry wastes. The applicants for Renewables Orders include a wide range of interests extending from the major electricity generators (i.e. National Power and Powergen), which have been especially active in developing wind farms, to small, project-specific groupings. Figure 3.15 indicates the locations of 388 separate projects contracted under the five NFFO rounds to 1995. 'Contracted' projects are not necessarily implemented, but the maps may be regarded as indicative of the spatial characteristics of renewable energy sources. The different sources have distinctive geographies. Wind and hydro projects tend to be associated with the west of the country, reflecting environmental influences such as wind-speed, rainfall and relief. By contrast, various schemes utilising waste products such as landfill and sewage gas are, for obvious reasons, urban-oriented. Despite these differences between the various sources of renewable energy, they share the common characteristics of being small-scale and widely scattered. These characteristics are obviously related in the sense that multiple supply points are a logical consequence of limited output. They represent further fundamental differences between renewable and fossil fuel energy sources other than the familiar distinction based on long-term sustainability. The dispersed pattern implied by Figure 3.15 represents a reversal of the dominant trend shaping the geography of energy supply in the UK over the last thirty or forty years towards the concentration of capacity into fewer, larger centres of production. It is, however, important to remember the minimal contribution currently made by renewables to UK energy consumption. There is no doubt that this contribution will grow, but great uncertainty about the rate of increase and, therefore, about the size of this contribution over the next twenty to thirty years.

Conclusions

In 1979 UK oil production exceeded the output of coal in terms of energy equivalent for the first time, and the country became a net exporter of oil in the following year. These events underline the major changes which have taken place in the energy market since the Second World War as the country has passed from almost total dependence on indigenous coal, through a period in which imported oil became the principal fuel, to the beginning of a new era of self-sufficiency based upon a more diversified range of energy sources. These changes in the structure of the energy market have had important spatial consequences as the country's key primary energy nodes have shifted successively from the coalfields to the oil refineries to the offshore fields and associated terminals. These shifts have in turn resulted in new distribution systems linking energy sources with their markets. Generally speaking, the scale at which these systems operate has increased through time. The principal energy nodes usually served local and regional markets in the immediate post-war years. This was certainly true of the coal and gas industries, and even the electricity system provided little opportunity for inter-regional transfers at the beginning of the 1950s. Despite improvements in the transportation of coal by rail, this fuel remains costly to move by comparison with other forms of energy and, because of its dependence upon the power-station market, much of its output is consumed on or adjacent to the coalfields themselves. Distance is a less significant constraint upon the other fuels. National grids exist for both natural gas and

FIGURE 3.15 Projects contracted under the NFFO, 1995

electricity, while pipelines enable petroleum products to be transported in bulk from coastal refineries to regional distribution terminals serving major inland markets.

The significance of the market as the most important single influence upon the contemporary geography of the energy industries is a sharp contrast to the traditional attractions of the coalfields to certain types of manufacturing. Despite the greater mobility of petroleum products and electricity, for example, the location of both oil refineries and power stations has been strongly influenced by market considerations. The refining complexes on the Thames and Mersey estuaries are well placed to serve the intervening axial belt of population. Similarly, the distribution of power stations broadly matches the regional pattern of demand, although local concentrations of capacity are related to the proximity of fuel sources such as coal and natural gas.

The mobility which characterises contemporary forms of energy is, to some extent, related to the costs of the fuels themselves. With the very significant exception of the oil price-shocks in the 1970s, the real costs of energy have tended to fall over the long term. More recently, privatisation has, for example, contributed to a substantial fall in the prices of gas and electricity to consumers in the UK. Although superficially attractive, such trends discourage conservation and sustainable patterns of consumption. Thus it might be argued that substantial price increases are needed to promote a more conservation-oriented society in which energy movements are minimised by adopting the local supply/demand systems associated with renewable sources. It would, however, be unwise to overestimate the political appeal of such an argument and a radical shift away from the long-established dependence of the UK upon fossil fuels seems unlikely. Nevertheless, the great changes which have taken place in the geography of UK energy during the last thirty years emphasise the risks involved in making even the most cautious predictions.

Notes

1 Proved reserves are those quantities which geological and engineering information indicates, with reasonable certainty, can be recovered in the future from known deposits under existing economic and operating conditions.

2 Applying a different set of assumptions, estimates of recoverable coal reserves in a report submitted to the government in 1992 were as low as 700 million tonnes.

3 British waters incorporate the UK Continental Shelf over which the UK exercises sovereign rights of exploration for and exploitation of natural resources. This area includes offshore areas to the west of the UK as well as the North Sea.

4 Maximum recoverable reserves are the sum of proven, probable and possible reserves where proven reserves are defined as having a better than 90 per cent chance of being produced, probable reserves have a better than 50 per cent chance and possible reserves have a 'significant' but less than 50 per cent chance of being technically and economically producible.

5 The National Coal Board adopted the title 'British Coal Corporation' in 1987.

6 The only exception was 1984 when substantial imports of fuel oil were made as a result of the miners' strike.

7 This estimate is made on a more restrictive definition of oil-related jobs than the 350,000 referred to earlier, which includes indirect employment effects.

8 The long-established surplus of generating capacity of electricity demand in Scotland has also discouraged CCGT investment.

Reference

Department of Trade and Industry (1995) *The Energy Report* (vol. 1): *Competition, Competitiveness and Sustainability*, London: HMSO.

Further reading

Cumbers, A. (1995) 'North Sea oil and regional economic development: the case of the North East of England', *Area* 27: 208–17.

Department of Trade and Industry (annual) *Digest of United Kingdom Energy Statistics*, London: HMSO.

Department of Trade and Industry (annual) *The Energy Report* (2 vols), London: HMSO.

Department of Trade and Industry (1993) *The Prospects for Coal*, Cm. 2235, London: HMSO.

Department of Trade and Industry (1994) *New and Renewable Energy: Future Prospects for the UK*, Energy Paper 60, London: HMSO.

Department of Trade and Industry (1995) *Energy Projections for the UK: Energy Use and Energy-related Emissions of Carbon Dioxide in the UK, 1995–2020*, Energy Paper 65, London: HMSO.

Department of Trade and Industry (1995) *The Prospects for Nuclear Power in the UK*, Cm. 2680, London: HMSO.

Fernie, J. (1980) *A Geography of Energy in the United Kingdom*, London: Longman.

Harris, A., Lloyd, G. and Newlands D. (1988) *The Impact of Oil on the Aberdeen Economy*, Aldershot: Avebury.

Hoare, A. (1979) 'Alternative energy; alternative geographies?', *Progress in Human Geography*, 3: 506–37.

Hudson, R. and Sadler, D. (1990) State policies and the changing geography of the coal industry in the United Kingdom in the 1980s and 1990s, *Institute of British Geographers Transactions New Series* 15: 435–54.

Manners, G. (1994) 'The 1992/93 coal crisis in retrospect', *Area* 26: 105–11.

Manners, G. (1997) 'Gas market liberalization in Britain: some geographical observations', *Regional Studies* 31: 295–309.

Mitchell, C. (1995) 'The renewables NFFO: a review', *Energy Policy* 23: 1077–91.

Mounfield, P. (1995) 'The future of nuclear power in the United Kingdom', *Geography* 80: 263–71.

Openshaw, S. (1986) *Nuclear Power: Siting and Safety*, London: Routledge and Kegan Paul.

Parker, M. and Surrey, J. (1995) 'Contrasting British policies for coal and nuclear power 1979–1992', *Energy Policy* 23: 821–50.

Spooner, D. (1995) 'The "dash for gas" in electricity generation in the UK', *Geography* 80: 393–406.

Stevens, P. (1996) 'Oil prices: the start of an era?', *Energy Policy* 24: 391–402.

Trade and Industry Committee (1996) *Energy Regulation*, House of Commons Paper 50, Session 1996–97, London: HMSO.

Trade and Industry Committee (1998) *Energy Policy*, House of Commons Paper 471 (3 vols), Session 1997–98, London: HMSO.

Walker, G (1997) 'Renewable energy in the UK: the Cinderella sector transformed', *Geography* 82: 59–74.

Chapter 4

Earth resources

John Blunden

Introduction

The UK contains rocks which are representative of almost all the identified geological systems, from the oldest which occur in the north and west and include many igneous and metamorphic types, to the youngest which predominate in the south and east and include the softer sedimentary rocks. Besides giving rise to a remarkably varied and often extremely attractive range of landscapes these rock types provide an extensive source of mineral products, some of which are widely mined or quarried and have important economic value (Figures 4.1a–c). These extend from the ubiquitous aggregate materials, sand and gravel, igneous rocks, limestone and sandstone; to other non-metalliferous minerals such as chalk, brick clays and slate; to more localised non-metalliferous minerals such as china clay, ball clay, gypsum and anhydrite, salt, fluorspar, barytes and potash; and to the metalliferous minerals, iron, lead, zinc, tin, copper, gold and silver (Dunham 1979).

Minerals exploitation: a changing milieu

Little more than a quarter of a century ago the United Kingdom was thought to be about to enjoy a period of unparalleled expansion in the exploitation of its minerals' potential. Not since the mid-nineteenth century when the UK had been a major world producer of copper, tin, iron and lead and zinc had it been possible to envisage for the UK a situation in which minerals might play such a key role in its economy (Blunden 1975). Experts, in reassessing the potential of Cornwall, one of the country's richest metalliferous regions in the last century, were able to suggest that its then somewhat moribund economy would be transformed by a 'considerable renaissance' in the production of tin and that as a result 'the prospects for the national economy as well as that of the South West of England could be exciting indeed' (Eglin 1969).

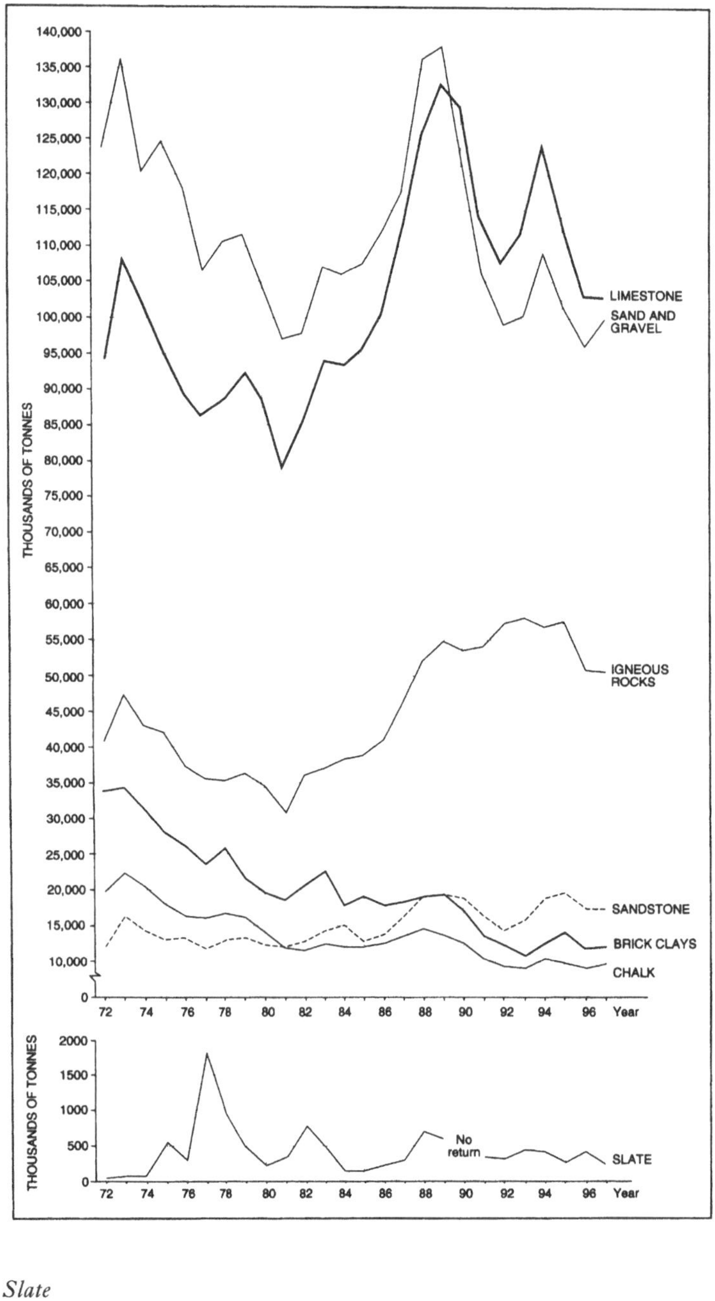

FIGURE 4.1a Ubiquitous non-metalliferous UK minerals: production levels, chief locations and utilisation, 1972–97
Source: British Geological Survey, *United Kingdom Minerals Yearbook*

Sand and gravel, igneous rocks, sandstone, limestone
The use of aggregates by the construction industry is notoriously susceptible to the state of the economy. Recent increases in the consumption of igneous rocks, sandstone and limestone are indicative of the growing problems associated with sand and gravel production, though the extraction of hard rock materials from traditional sources is becoming increasingly problematic. The substitution of igneous material from remote quarries is posited as one solution.

Some limestones, those of greatest purity, are also extracted for their chemical properties and are used in a number of industrial processes, including the manufacture of soda ash, caustic soda, organic acids and solvents, dyestuffs and bleaching powders, as well in the production of cement.

Chalk and brick clays
Both undergo secondary processing, the former being a key input to the manufacture of cement and the latter the chief element in brick-making. Although clays are common, brick-making has become concentrated on the Oxford clays of the Peterborough and Bedford areas. The production of both reflect the state of the UK economy.

Slate
Found widely in the Grampian Mountains of Scotland, the Lake District, North Wales and Devon and Cornwall, the number and the scale of workings has fallen dramatically as slate is no longer the universal roofing material. Although still used in this role to a limited extent, demand is now largely confined to architectural cladding. Penryhn (North Wales) and Delabole (north Cornwall) are the most significant producers. However, large quantities of waste (some twenty times that of usable product) remain at a number of sites and have a limited use as a filler in plastics and fertilisers, and in the manufacture of roofing felt.

Whilst at that time the traditionally locally traded low-cost building and construction materials were experiencing unprecedented growth in output on the back of a demand for new houses and offices and a rapidly expanding motorway programme, a different set of factors were influencing those more costly minerals marketed internationally. Abroad, many long-standing sources of supply appeared to be becoming problematic as reserves dwindled or the political climate became more uncertain. At the same time new exploration techniques were giving promise of the possibility of greatly increasing the UK resource base. This was not merely in terms of reviving the fortunes of previously worked minerals, but also of developing other non-renewable industrial minerals, some of which had made little or no economic impact previously. Thus by the beginning of 1971, twenty-eight companies had agents or staff geologists exploring eighty sites (Figure 4.2), ranging from one re-examining the potential of the old Levant tin mine on the Land's End Peninsula to another seeking to evaluate copper deposits in Ross and Cromarty (Blunden 1975). All were also assisted by government, first by investment grants and tax concessions and then in 1972 through its Minerals Exploration and Investment Grant Act. In moving to replace the previous arrangements the new legislation offered financial support of up to 35 per cent of total costs for new exploration initiatives, leaving regional investment grants to play a major part in helping to underwrite the costly early years of mining development and minerals production (Rogers *et al.* 1985).

However, in the intervening period between the mid-1970s and the present day, other exogenous factors have come to influence the exploitation of that resource base in terms of those minerals that can be traded internationally. Further developments in methods of locating and evaluating potential resources have achieved ever higher levels of sophistication and have been applied in the 1980s in many developing countries (Andrews 1992) whose former ambivalent political attitudes, particularly towards foreign investment, have undergone a remarkable transformation. This has often been as a result of pressure from the International Monetary Fund which sees mining development as an important tool in promoting economic growth. Many countries, particularly in South America, are now offering mining companies security of mining tenure; government joint ventures guaranteed through third parties; and a prompt government response to, and assistance with, development plan proposals, including those related to environmental controls. The result has been that newly discovered ore fields have rapidly been developed and then brought into production, usually utilising surface mining methods of the most capital-intensive and cost-effective kind (Gooding 1992).

But these intiatives have not been the result of the actions of the smaller, often national, companies that characterised the early 1970s. They have been largely replaced by multinational corporations operating at a global level and able to influence market conditions and raise venture capital on the international finance markets. As a result they look to achieve maximum economies of scale at large, long-life opencast operations, and, where possible, in a local milieu relatively free from onerous environmental constraints (Radetski 1992). This was a situation hardly likely to favour investment in the UK, with its rigorous system of planning controls and where mining development, if approved, might at best be operable only at the edge of profitability. Indeed, in the case of tin mining, when compared with the extraction of the disseminated placer or alluvial ores by capital intensive open pit methods in, for example, Indonesia and more recently Peru, its underground extraction in Cornwall involving the working of vein deposits was bound to be a more costly operation. Under the

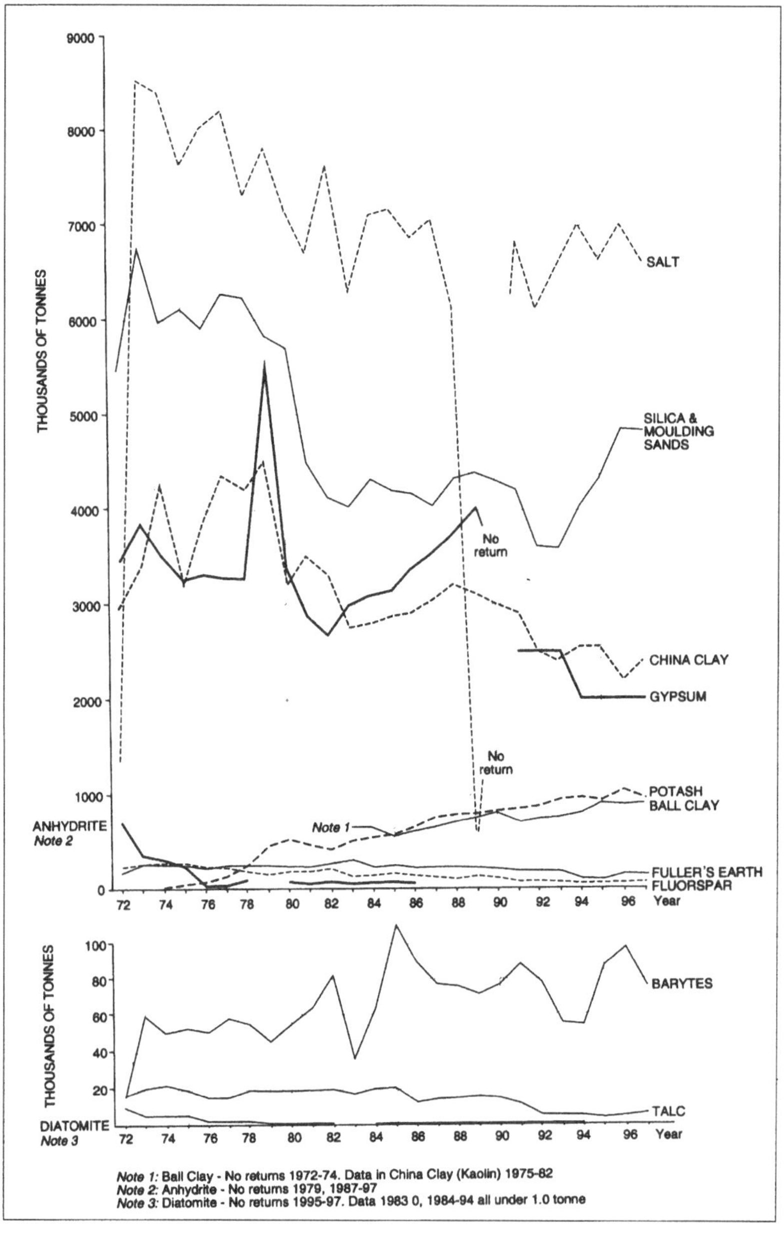

FIGURE 4.1b Localised non-metalliferous UK minerals: production levels, chief locations and utilisation, 1972–97
Source: British Geological Survey, *United Kingdom Minerals Yearbook*

China clay
The UK, together with the USA, produces 60 per cent of world output. Seventy-five per cent is used as a filler for fine papers. The rest is used to make ceramics and as a filler for rubber,

plastics, and paints. China clay extraction is now coming under increased overseas competition from substitute materials.

Ball clay
The UK may be only the world's third largest producer but the quality of its ball clay is outstanding. The main extraction area is located in south Devon with some additional workings in east Dorset. Open pit working predominates, although there are some underground workings. Ball clay is used in the manufacture of earthenware, tableware, tiles and sanitaryware.

Fuller's earth
Used as a binder in foundry sands, in the bleaching process, for domestic water treatment and for the drilling of oil wells; the primary source of supply is Woburn in Bedfordshire.

Salt
Heavily concentrated in Cheshire where five major corporations operate, substantial quantities are, however, also available from the Cleveland Potash mine at Boulby. The UK is a major world producer. More than 70 per cent of output is used in the chemicals industry in the production of chlorine, caustic soda and soda ash.

Gypsum and anhydrite
Mainly worked in the East Midlands but also in Cumbria and Co. Monaghan, gypsum is used to make plasterboard and plaster products. Production is around 6 per cent of the world total. Anhydrite, found in association with gypsum, is used to make sulphuric acid.

Fluorspar
The UK, formerly a significant producer in global terms, still has a world presence even if it is rather diminished. Demand for the product is closely associated with the construction industry and major capital goods. Although it is extracted in Weardale in the northern Pennines, the Peak National Park is a key production area.

Barytes
Like fluorspar, this mineral is often associated with lead and zinc and is therefore mainly worked in the southern Pennines. However, Scotland now has a single producer at Aberfeldy. Although it has a number of uses related to the production of ceramics and glass, and in one form finds its way into the production of paints as well as a compound used for the medical examination of the digestive tract, around 90 per cent is employed in the oil and gas drilling industries the activities of which largely control its market.

Potash
Although in the world context, the UK is a small producer of a mineral vital to agriculture, output has risen throughout the period. Extraction takes place solely in the North York Moors National Park at Boulby, but the workings are underground and cause minimal disturbance to visual amenity.

Silica and moulding sands
Used in glassmaking and foundry mouldings respectively, the chief production areas for the former are at Blubberhouses, west of Harrogate, Reigate in Surrey, Loch Aline near Oban and Alloa, and for the latter in Cheshire.

Diatomite and talc
Both are mined for their unique physical and chemical qualities. Diatomite is used mainly as an insulating material and in filtration processes, although if its quality is poor it can be employed as a filler in paints and rubber. Talc is largely employed in cosmetics and as a dry form of lubricant. Both have been produced in relatively small quantities from deposits in the far north and west of Britain.

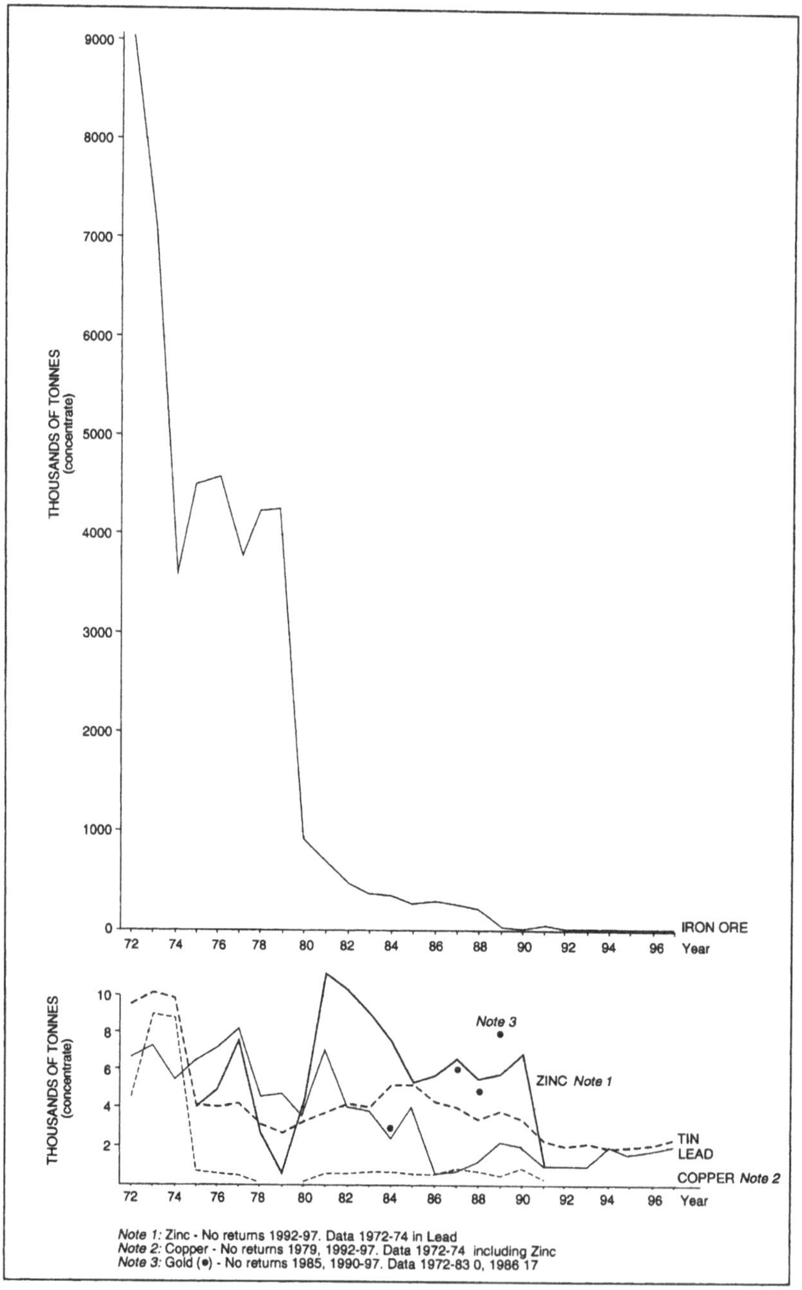

FIGURE 4.1c Metalliferous UK minerals: production levels, chief locations and utilisation, 1972–97
Source: British Geological Survey, *United Kingdom Minerals Yearbook*

Iron ore
Used in steel manufacture, the primary outlets are the construction and transport industries. In 1966 nearly 14 million tonnes (metal content) were produced, but in that year the British Iron and Steel Federation decided to withdraw from domestic supplies because of their low metal and high sulphur content. The downturn in 1980 marks the closure of the main orefields of Northamptonshire. The small quantities currently produced are from north Oxfordshire and Cumbria.

Lead
Used mainly in the manufacture of batteries, it is mined as a by-product of fluorspar. The reduction of output over the period therefore parallels that of fluorspar. Small quantities of lead and zinc were produced at the Parys Mountain mine in Anglesey in 1991, but a reappraisal of the site has given rise to the expectation that production will soon recommence at a substantially greater level.

Zinc
Primarily used in galvanising, die-casting and as an alloy, it is worked only in association with lead.

Tin
It is mainly used as a coating for thin steel cans and in solder alloys, but also as an alloy with copper to form bronze, and most recently in organo-tin compounds. Production reached its peak of 11,100 tonnes (metal content) in 1871, a figure which was approached again only in 1973 (10,100 tonnes), at the height of the revival of interest in tin mining in Cornwall. In 1993 output had dropped to 1,900 tonnes, a figure from which it had barely recovered when production ceased in 1997.

Copper
Used mainly in electrical goods, its UK production is only very small and as a result of its presence with other valuable ores. Production figures for most of the 1990s are not available. However, recent exploration work at Parys Mountain in Anglesey suggests that this may become, within UK terms, a significant source of copper, as well as lead and zinc. Gold and silver are also present in the deposit.

Gold
Mined only in North Wales in small amounts, output is confined to the jewellery market. Recent production figures are not available but there is much exploratory activity involving both gold and silver at Cononish near Tyndrum in the Scottish Highlands which is now well advanced, at Loch Tay and Aberfeldy, in South Devon, in Co. Tyrone, and Parys Mountain, Anglesey.

circumstance of considerable overseas competitive advantage the very survival of tin mining, let alone its revitalisation, would have to have depended ultimately on the maintenance of buoyant world prices. But this was a situation which was to disappear in 1986 when the Association of Tin Producers, which had largely been a controlling agency with regard to price, was found to have built up over many years an untenably high strategic stock, news which brought about a collapse in the world tin market (Shrimpton 1987). This blow to high-cost operators such as those in Cornwall was one from which they could not recover except with considerable state aid. Thus the 1990s have witnessed the closure of Geevor, the cessation of underground working at Wheal Jane with the withdrawal of its last government loan, and finally the one remaining mine that had been in continuous production since the nineteenth century, South Crofty, in February 1998.

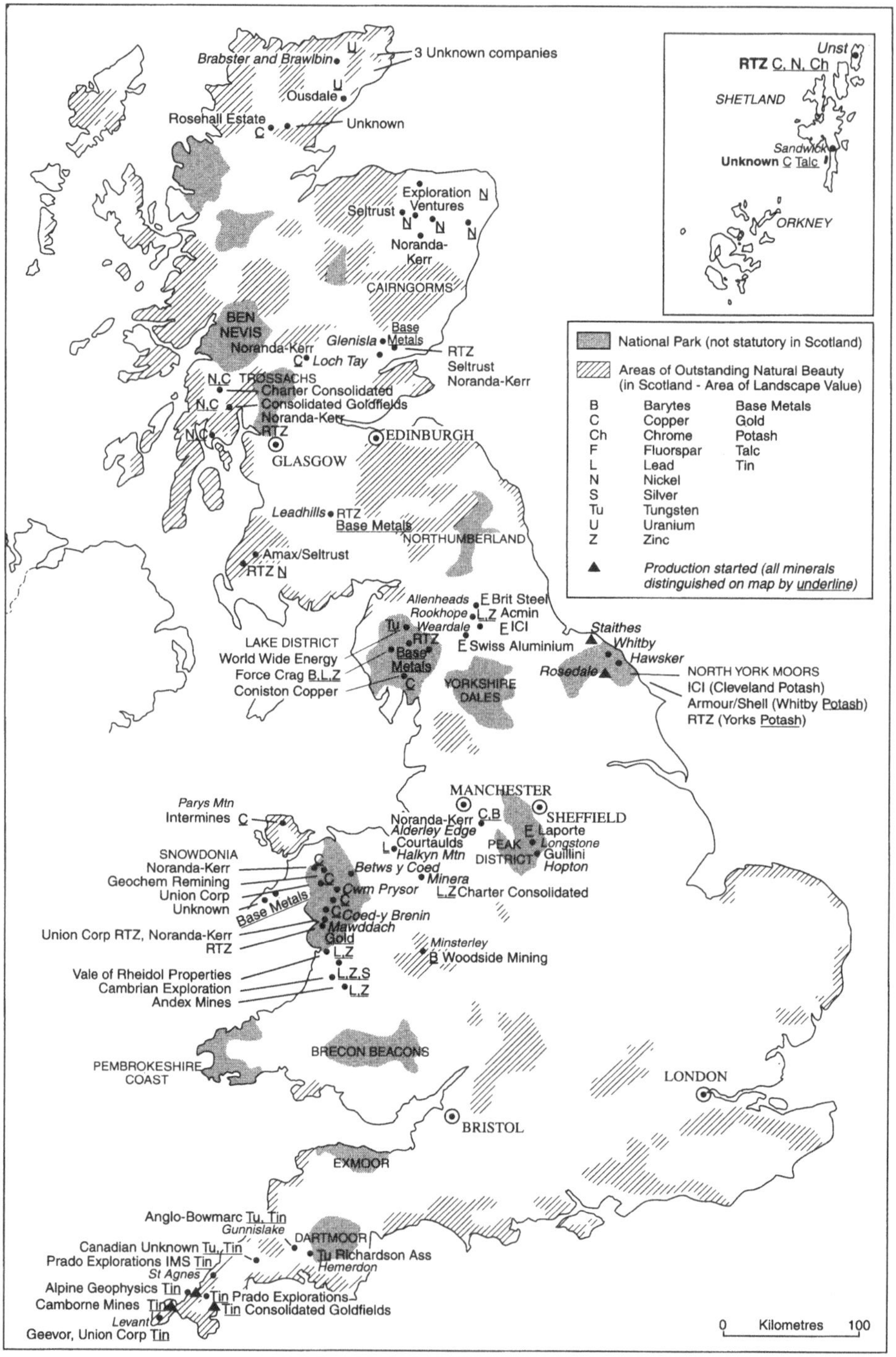

FIGURE 4.2 Minerals prospecting in Britain, 1972
Source: Blunden (1975).

Minerals and the environment: the legislative framework

However, it was in the wider field of minerals development that another factor, that of the environment, was to prove crucial in damping down much of the earlier optimism. Given the densely populated nature of much of the UK and the smallness of its size, it was recognised by government, even in the early 1970s, that expansion in minerals output would have to meet the need for the highest possible environmental standards and, as if to provide a counterbalance to its own new more permissive enabling legislation, it set up three groups of experts to advise it in the 1970s: the Verney Committee, whose remit was to investigate the supply of aggregates at minimum economic and social cost but with the recognition that every effort would need to be made to reduce environmental problems; the Stevens Committee on planning control over mineral working including after-care and site restoration; and the Department of the Environment's (DoE) Mining Environmental Research Unit at Imperial College, directed by the author, whose task it was to consider all the environmental aspects of expanding indigenous minerals output, except those concerned with energy. All three in the four years leading to the completion of their work took evidence from a wide range of sources representing both industry and the conservation movement in the guise of non-government organisations (NGOs), with the research group looking particularly at other countries that had already experienced major mining operations, especially those relating to industrial minerals. Both Stevens and Verney had the findings of their respective committees published in 1976, whilst the work of the research group was completed with its series of private reports to the DoE in the same year.

Their success in effecting change was partial but none the less meaningful and was incorporated in the Town and Country Planning (Minerals) Act, 1981. This achieved a significant extension of the Mineral Workings Act, 1951, with respect to post-operational site requirements, since it included what is termed 'after-care' whereby minerals companies sustain an ongoing management regime, normally to be for five years after minerals extraction had ceased, rather than a once and for all rehabilitation programme which could easily be cosmetic. Indeed, the regime imposed might also require of the mining company that the land be made suitable for agriculture, forestry, or amenity. The Act also determined that county planning authorities should be the minerals planning authority with a duty to make regular reviews of areas which have been, are being or are to be used for mineral working. Following a review, the authority could revoke or modify a planning permission or prevent further working of the land (Roberts and Shaw 1982).

Mining and sensitive environments

However, it may be seen in retrospect that the Act has not worked as well as it might have done with regard to curtailing damage to especially sensitive environments which can result from the process of mining or quarrying itself, and with respect to the disposal of wastes where these can become a significant factor of land-use and amenity adjacent to workings. This has been particularly so in the case of National Parks where *old* consents to work minerals often exist which pre-date the 1951 Minerals Working Act. The Environmental Protection Act, which became law in 1990, however, allowed planning authorities to review such workings with a view to either modifying the consent or permitting its continuation as before. Few planning authorities used this power until the Peak Park Joint Planning Authority was faced in the mid-1990s with a situation posed by Ready Mix Concrete (RMC) which might have inflicted severe damage on the landscape at Longstone Edge. At their recently acquired quarry at this location, the company attempted to extract limestone in a situation where the consent had specified that fluorspar and barytes might be worked along with other minerals, but made no specific reference to limestone. RMC's plan to remove 15 ha to a depth of 60 m would not only have produced 19 million tonnes of limestone, but have severely impaired the appearance of the skyline to the extent that the planning authority decided that it would use its powers to revoke the original consent. The company at first stated that it would appeal against the decision, but it has now agreed not to contest it, and showing great pragmatism perhaps in the light of its wider interests as a leading aggregate producer, expressed its keenness to work with the Peak Park. Apart from the decision being welcomed by environmental NGOs, the action taken by the planning authority is being seen as a test case by other National Parks in England and Wales which currently face more than a hundred similar ambiguous quarrying consents (Jury 1998; Bent 1998).

This particular problem apart, the extraction of limestone for chemical or aggregate purposes from its ten active quarries has been a major source of conflict in an area which is supposed to be devoted to conservation and recreation. However, the largest of these, Tunstead, near Buxton, which was formerly outside the boundaries of the Park, has been allowed a major extension into it. The approval of this development was given on appeal because the national need for high purity limestone for chemical use was considered as more important than its environmental impact and the designation of the area as a national park. The quarry came into production in the mid-1980s and over its predicted life of sixty years it is likely to have an average output of 10 million tonnes per year of chemical limestone, together with some aggregate material.

As for fluorspar, of which the UK is an exporter, 70 per cent is won within the boundaries of the Peak Park. When the market was at its most buoyant in the late 1970s and early 1980s, this mineral, along with barytes, was extracted at a number of sites some of which were open pit operations rather than underground workings. This in itself caused widespread damage to the visual qualities of the landscape, as did the disposal of waste from its primary processing (about 3 tonnes to every one of valuable material). This was largely accomplished through the provision of tailings ponds where the wastes, suspended in water as a result of processing, settle out. Even though these may be ultimately drained and grassed over, they remain an alien feature in the landscape. However, the environmental situation has recently improved. There are now fewer open pit operations, with a greater concentration on underground working and with minerals processing limited to one plant. Moreover, the backfilling of waste, combined with cement, into the worked-out areas has reduced the need for tailings ponds but not eliminated them. The only way in which this might be achieved would be to use the remaining wastes for the manufacture of aggregate materials. Unfortunately, these would still be produced in quantities well in excess of the needs of local markets (Peak Park Joint Planning Board 1998).

Waste materials: the case of china clay

Of the remaining highly localised non-metalliferous minerals extracted in the UK, all have some interface with the environment even if, like salt and potash, they are worked underground and have a minimal waste problem. However, in terms of the scale of waste production and its visual impact, by far and away the most impressive of UK minerals is china clay (kaolin). This, above other industrial minerals, is the nation's chief export in terms of both tonnes extracted and value, with 80 per cent of output going overseas. For all these reasons it demands special attention here. China clay is removed from roughly circular pits the largest of which can be around 100 ha, or about twenty times the size of Wembley football stadium. These are frequently around 60 m deep, and roughly two-thirds the width of the grassed area of the same arena. Of the material extracted seven out of every eight tonnes are quartz waste, which is piled up in immense white tips. In addition there are also extensive ponds in which the very fine micaceous waste materials from processing are placed to de-water now that these are no longer discharged into local rivers. Not surprisingly, the land requirements for the industry in the area are of gigantic proportions, as Table 4.1 demonstrates (Blunden 1996a).

Given high levels of production over many years and the concentration of the main Cornish workings in the St Austell area they represent a major visual intrusion in a rural landscape otherwise largely exploited by agriculture and tourism. Travellers approaching the area from the north-east along the A30 road can catch their first glimpse of the waste tips,

TABLE 4.1 Land requirement (in hectares) for the china clay industry in the St Austell area

	Land occupied by industry 1969	*Land occupied by industry 1984*	*Land occupied by industry 1991*	*Anticipated land occupied by industry 2011*	*Anticipated increase in land occupied by industry 1991/2011*
For pits	645	992	1,108	1,232	124
For tips (quartz sand, overburden)	830	1,318	1,287	1,649	362
For micaceous residue disposal	77	301	295	383	88
For plant	310	378	345	338	–1
Total	1,862	2,989	3,035	3,602	567

Source: ECC International.
Note: The problems of waste disposal for the china clay industry are particularly problematic. The fourfold increase in land used for the dumping of micaceous residues between 1969 and 1984 reflects the phasing out of marine disposal in favour of tailings ponds. In the years since 1977 when a reclamation scheme was instituted, 348 ha of the land taken for tipping and 80 ha of that given to tailings have been reclaimed, mainly for amenity purposes. Open pits remained unreclaimed until recently but now 162 ha have been back-filled with waste materials from processing. In the period 1995 to 2011, 34 per cent of tip requirements will come from back-filling. The relatively static requirements for plant over a period of increasing product output are an indicator of improved processing efficiency.

sometimes referred to as the 'Cornish Alps', at least 12 km away. The same is true of the only other important clay workings which abut the Dartmoor National Park in Devon. The waste tips at this location can be seen from Plymouth 16 km away, as well as from a number of tors (rocky hills) inside the south-western boundary of the Park.

Since the St Austell china clay workings represent the only major industrial development in Cornwall, with all that follows from that in terms of employment, the planning authority has granted it a special status in order to secure its long-term interests in land-use terms. But whilst the operating company's consents for working are long standing and largely pre-date planning legislation, the planning authority has negotiated the closing down of many small clay pits in the county in order to concentrate environmental intrusion and damage into the one area. Moreover, it has accepted English China Clays' desire not to sterilise its pits by back-filling with waste since none of them has yet been bottomed, but has otherwise negotiated with the company a long-term programme of waste revegetation and general landscaping work. However, its other site at Lee Moor on the edge of Dartmoor has long been a contested area. Although the exhaustion of the working can be foreseen in the twenty-first century and for which rehabilitation schemes have been agreed with the planning authority, the Dartmoor Preservation Society has vigorously resisted all applications for additional land for waste tipping for the reasons of visual impact on the boundary of the National Park described above.

Aggregates: some problematic issues

In the case of non-internationally traded aggregate minerals, it is the environmental factor that has also proved particularly problematic. In the first place the appetite for sand and gravel has been particularly voracious (see Figure 4.3). Moreover, sources of sand and gravel are usually synonymous with areas that are sensitive to such extractive activities; for example they are likely to have to be developed close to urban areas, or on high-quality grade 1 agricultural land. In London's Green Belt alone in excess of 3,300 ha have been actively worked, with more than another 3,000 ha approved by planning authorities for the same purpose. Nuisance caused by the unsightly appearance of such workings, and by dust and noise at them, together with increased traffic moving to and from these sites, is often the source of considerable dissatisfaction amongst the general public. Sand and gravel extraction is therefore increasingly opposed whenever a planning application is made.

But since such workings often take place at or below the water table, as they are most frequently located in valley floors, their eventual rehabilitation can usually be readily achieved. Processing plant can easily be removed and any waste sand or gravel poses no threat to the environment since it can readily be landscaped by using earth-moving equipment and is often graded so that it slopes gently to a central lake. In the UK the best-known examples are along the River Thames, both in its lower reaches west of London and near its headwaters on the Wiltshire/Gloucestershire border; near Olney in Buckinghamshire; and at Holme Pierrepoint in Nottinghamshire. Most of these areas have been reclaimed as recreational facilities in the form of water parks catering primarily for water sports such as sailing, water skiing and fishing. Nevertheless, the demand for such facilities is far from inexhaustible and other forms of reclamation have been tried. Some planning authorities have thought them to be appropriate as landfill sites for domestic,

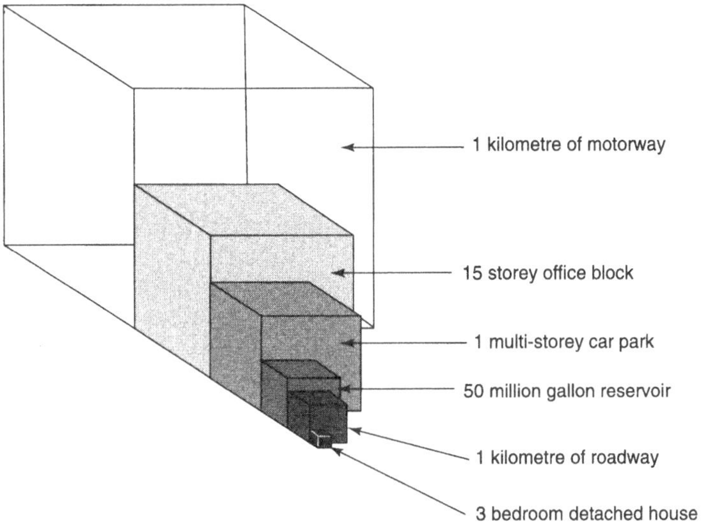

FIGURE 4.3 Aggregates and construction

Aggregates are the cheapest of all minerals to exploit and vast quantities are used in construction work. On average 50 tonnes are needed for a three-bedroom house and 62,500 tonnes for every kilometre of a six-lane motorway.

Source: Blunden (1996b: 160).

industrial and other wastes. Several hundred hectares of wet sand and gravel pits have been filled in the Greater London area where the demand for land for development has been strong. However, where these wet sites are concerned, the main type of material used is rubble or inert waste, for in spite of encouraging research into disposing of domestic refuse into wet pits, no satisfactory economic method of eliminating the risk of polluting groundwater supplies has been found (Blunden 1996a).

Such a generally less than encouraging situation, together with a falling supply of sites that are likely to prove unproblematic, has tended to direct the pressure for aggregate elswhere, especially sources of hard rock generally to be found in upland areas. This has led to greatly increased quarrying activities in places such as the Mendip region of Somerset, the Craven district of North Yorkshire, the Brecon Beacons of South Wales and the Charnwood Forest district of Leicestershire, all of which supply centres of consumption in the south-east of England. But these are valued for their landscape qualities, with many of them designated as National Parks or Areas of Outstanding Natural Beauty (see Chapter 20). Open pit operations of the sort required here (or for that matter for the extraction of metallic ores and clays) give rise to many of the same problems as lowland sand and gravel workings. They may also require the removal of considerable volumes of rock overburden in the course of their development, and, since they are generally distant from major population centres, refilling quarried-out areas with suitable urban waste when working ceases may not be economically feasible. Thus in spite of efforts to restore voids to their original surface levels, this remains the exception rather than the rule. In the UK, of the annual take of land for limestone, only 5 per cent is completely restored. In such situations attempts at rehabilitation are confined to trying to mellow as rapidly as possible the disused vertical working bluffs. Here the natural or artificial regeneration of flora can be assisted by creating screes at their bases and by layering their faces with ledges on which plant life can take hold (Clouston 1993). Apart from the presumption that the primary purpose of National Parks is conservation and the encouragement of recreation, quarries may still continue to damage the landscape since, as already noted, some of these pre-date the designation of those parks. Indeed, in exceptional cases consents are still given in such areas, in spite of fierce resistance from the many conservation NGOs, such as the Council for the Protection of Rural England (CPRE), Council for the Protection of Rural Wales (CPRW), and the Council for National Parks.

It was the Mining Environmental Research Unit at Imperial College that in the mid-1970s recognised these difficulties and suggested that one solution might be found in the development of remote coastally located aggregate superquarries in Scotland (Figure 4.4). Through the use of inexpensive sea transport to the major market of south-eastern England, these might economically subsume the role of many smaller traditional local operations and assume a lower environmental profile. Only very recently has government come to support this notion actively, acknowledging that, given its demand-led policies for aggregates, if it allows rising output to be met in such a way it might be less exposed in terms of public concern. However, it is an approach that has come under attack because it cannot be consistent with the government's acceptance of the principles of sustainable development laid down at UNCED in Rio in 1992, which demands alternative approaches that embrace aggregate conservation and recycling (CPRE 1996).

Minerals and sustainable development

Definitions of sustainability are, of course, arguable. Some suggest that it must preclude any form of growth. Others have put forward a rather more pragmatic approach. English Nature, for example, has sought to express its own view of sustainability in looking 'to establish limits of human impacts based on carrying capacity . . . promote demand management so as to keep development within carrying capacity', and 'oppose development and land use which adversely and irrevocably affects critical natural capital' (1993). Such a view, if not precisely the same in every particular, has been generally shared by other statutory agencies such as the Countryside Commission (1993), NGOs such as the Royal Society for the Protection of Birds (1993), as well as the House of Lords Select Committee on Sustainable Development. Differences of opinion, where they do exist, are to be found amongst the environmental organisations, where some feel that the concept of development within environmental capacities might be interpreted as a presumption in favour of development *up to* capacity in all areas. This might threaten some of the wildest and most unspoilt places. Others have reservations about the utility of the concept of 'critical environmental capacity', fearing that even if intentions are good it could undermine existing designations.

However, the question arises as to how far the Department of the Environment (DoE, now DETR) has managed to accept at least a mediated definition of sustainable development? Certainly, since the Verney and Stevens Reports of 1976 it has accepted that the

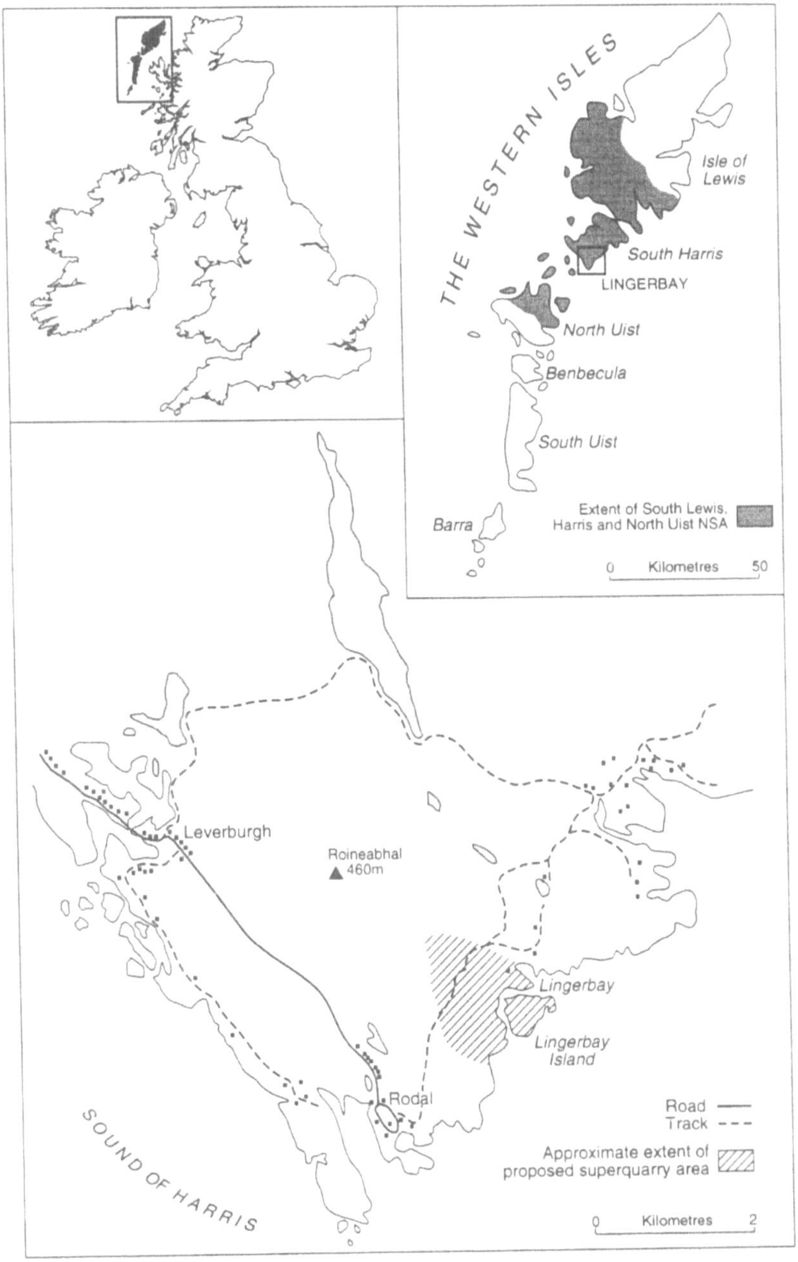

FIGURE 4.4 The Lingerbay superquarry proposal, South Harris. One of a number of proposals from minerals companies for a superquarry. Such an operation must have aggregate reserves in excess of 150 million tonnes and an annual output greater than 5 million tonnes, which is mostly exported by sea. This proposal came from Redland Aggregates Limited
Source: CPRE (1996).

adverse impacts of minerals extraction should be minimised and in its Minerals Planning Guidance note of 1989, MPG6, whilst striving to make provision for adequate supplies of aggregate, has advised minerals planning authorities that in the interests of the environment they should influence the location, timing and methods of extraction, as well as imposing under the relevant Town and Country (Minerals) Planning Act (1981) conditions of restoration and after-care. Yet attempts to impose restrictions on the use of National Parks as sources of aggregate have long been unsuccessful (McEwen and McEwen 1982; Elcoate 1996), whilst as recently as 1996 MPG6 has suggested a virtual doubling of mid-1990s aggregate consumption by 2011.

That the Department of the Environment, Transport and the Regions seems to accept willingly this forecast presumably lies in the expectation that the environmental impact of such an increased level of demand can be met via the superquarry concept. A number of these have been suggested, some of which seem to show scant regard for the fact that the sites are also in areas which enjoy protected status. For example, one of these in the Outer Hebrides lies within the South Lewis, Harris and North Uist National Scenic Area, designated for its landscape quality. Fierce opposition to this development has already been expressed by the Island Council and in a referendum on Harris. However, even if the reason for having minerals extracted from an NSA were arguable on the grounds of overriding 'national benefit', its *raison d'être* remains that of demand for aggregate materials without properly addressing the extent to which demand-led forecasting is compatible with sustainable development (CPRE 1996). This is not to argue that since aggregates are non-renewable resources such policies, for this reason, should curtail their use, but rather that it is the impact of aggregates extraction, their transport and their end use that are matters of concern in relation to sustainable development.

In fairness to the DoE, efforts to mitigate these impacts are evident in a revised MPG1 of 1995 which exhorts mineral planning authorities, as never before, to pursue the objective of sensitive working, restoration and after-care, and the protection of nationally designated areas of landscape or conservation value, except *in extremis*. In addition it also now calls for the minimisation of impacts caused by the transport of materials, as well as resource conservation, the better use of materials, and the recycling of aggregates and allied waste materials.

Minerals waste: its positive utilisation?

Although government agencies such as the Building Research Establishment have long advocated the positive use of waste materials for construction purposes (Gutt and Nixon 1980) remarkably little has been achieved. In the early 1970s when the coastal site of Maplin at the head of the Thames Estuary was a short-listed location for a third London Airport, one study suggested that the reclamation of the land for this purpose might be achieved by moving 400 million tonnes of waste from two sites much affected by dumping of unwanted material. One is in Snowdonia where it estimated that broken slate and slate chippings, largely a relic of the importance of the industry in the nineteenth century, amount to between 400 and 500 million tonnes. The other is in west Devon, that of Lee Moor, where about 28 million tonnes of waste quartz from china clay processing lies in tips on the edge of Dartmoor National Park (see p. 72). But in spite of the virtues of the scheme, not least in returning land on which the waste is dumped to some useful pupose, greatly enhancing local amenity in two scenically sensitive areas, and reducing the take of 'new' minerals, it failed to be adopted because of transport costs, and the idea of an airport on reclaimed land was abandoned.

Also in the 1970s a major study was undertaken by Devon County Council into the recycling of 28 million tonnes of waste quartz from Lee Moor, since only about 8 per cent of each year's production of waste was being consumed locally. The idea was to use it in the manufacture of artificial aggregate and sell it in south-east England where demand was high. Not only did the costs of this run out at four times those for natural aggregate available in the south-east, but it was also unable to compete on transport costs compared with its nearest rival, newly won limestone from a Mendip location some 140 km closer to the market (Corner and Stafford 1971).

When considered in conventional economic terms, the long-distance transport of large quantities of waste is seldom likely to be attractive. For this reason it is unlikely that mineral operators will ever voluntarily undertake such movements. Conventional economics does not take account of the unquantifiable benefits mentioned above, to which might be added the lessening of nuisance and the removal of a danger to the public. But it is when these factors appear important that there is scope for government intervention on the basis of a policy that could decide whether the benefits of waste removal are worth the apparent financial 'loss' in doing so.

There is also much to be said for government attempting to identify and quantify all existing stocks of wastes, as well as current and likely future production, along with their properties and their suitability for different purposes, including the refilling of identified voids caused by minerals extraction. A complete computerised data bank listing voids and wastes, along with a program which is available to associate one with another according to suitability, exists in Canada and in at least one region in the UK, the West Yorkshire conurbation (Blunden 1996a). Until such information is readily available on a UK-wide basis, proper planning for waste utilisation will not be possible. Both of the above approaches to waste management and utilisation are fundamental to any real identification with the notion of sustainability, as is the DETR's final break with strategies for the supply of aggregate materials which are demand-led.

Minerals and regional development

Apart from the environmental aspects of minerals exploitation, plainly there are strong economic reasons for their production which can be identified at regional and national levels. As a more formal way of attempting to quantify the likely stimulus to regional economies of the opening up of new mining operations, the concept of the multiplier model has been developed. Although the practical application of multiplier models involves many complex problems, the basic concept is simple and deterministic. It may be supposed, for instance, that there is an initial injection of incomes into an area through the development of a mining project that also requires several hundred persons, mainly local manual workers and perhaps 20 to 30 other people for clerical and management tasks. At least some of the money generated by this enterprise would be spent locally and would find its way into the pockets of local business people who, in turn, would spend some of it locally with other enterprises, and so on. The total amount of local income generated will be greater when more is spent locally and less allowed to leak away through the purchase of goods and services outside the area. Thus it is possible to estimate in any situation the hypothetical size of the local multiplier effect, given assumptions about local spending and after making allowance for rates of taxation and savings.

In the UK, studies using this methodology have largely concentrated on examining the development impact of offshore oil on north-east Scotland, and, more particularly the Shetland Islands (Mackay and Mackay 1975; McNicholl and Walker 1978; McNicholl 1980). However, there have been some examples of the use of multipliers in connection with the extraction of other mineral resources. For example, a study of the likely impact of a large copper mine in north-west Wales was prepared by the Department of Economics at the University College of Wales, Bangor, at a time when, in the early 1970s, it seemed likely that such development might take place. Based on a model of the economy of Anglesey, it was hypothesised that an assumed £6 million of investment at the construction stage of the mine might raise the total output of the island economy by over £8 million, while at the operational stage the £5 million output of the mine might improve the overall performance of the economy by over £6 million (Zuckerman 1972). Another exercise, conducted in relation to the economic significance of an existing working, that of china clay extraction at Lee Moor in south-west Devon (see p. 72), looked only at the propensity of the enterprise in terms of job creation in the Plymouth area. It concluded that the 950 persons employed on site resulted in creating half as many jobs again in the local economy.

A much more detailed exercise using multiplier models, again related to the Plymouth area, was carried out in the 1980s with regard to the development of a tungsten working in order to serve the needs of a planning inquiry. To this end four types of relationships were investigated once the study area had been delineated. These were (1) purchases of inputs by the mine (transport, plant hire, construction and engineering services, etc.); (2) purchases by the labour force employed (for everyday needs, recreation, etc.); (3) induced investment in the infrastructure which may occur as a result of employee demands for extra services (schools, hospitals, social services, etc.); and (4) downstream linkages (local processing activities and smelting; manufacturing activity based on the mine output, etc.), which can involve a further build-up of employment.

From evidence of local consumer spending collected in the study area, and a comparison with other studies concerning the impact of other similar developments both in the South West region and other areas of the UK, a multiplier of 1.7 was suggested as appropriate for the mine. Thus for the 340 workers expected to be involved in the operation (including forty-six clerical and supervisory staff), with an annual payroll estimated at £2.8 million for the first year of operation, it was anticipated that the multiplier would transform this sum into a local impact of £4.8 million per annum. However, because a third of the workforce was recruited from outside the area and would 'export' some of their earnings, and assumptions regarding rates of taxation etc. needed to be taken into consideration, a modest downward adjustment of this figure was required. As to the non-labour expenditure by the operation, since not all goods and services could be supplied from within the area (for example, major items such as electricity and fuel), a considerable leakage of expenditure (more than 80 per cent) was allowed for before arriving at a figure to which the multiplier could be applied. This gave a total of just under £1 million for the first year of operation.

No allowance was made for infrastructural investment because investigations showed provision in the area had substantially moved ahead of other similar areas in the South West over the previous decade. 'Downstream' linkages also failed to materialise because of the limited size of annual mine production and the specialised nature of its output. Nevertheless, the study concluded that, at the time the multiplier model was applied (1982), the mine was likely to generate just under an additional £6 million in the study area in its first year of production. The mine was predicted to have an active life of twenty years, assuming an average annual output of 4,500 tonnes of tungsten concentrate. Inputs to the model were subsequently adjusted to permit forecasts of the impact of the mine in later years (Blunden 1985).

Operations such as this are unlikely to result in a massive change in the fortunes of a local economy, but in areas undergoing considerable structural change in their economic base and thus suffering above average unemployment, their impact cannot be disregarded. Indeed, at the planning inquiry which had to consider the impact of this operation in an area adjacent to the borders of Dartmoor National Park, the multiplier exercise proved an important element in the inspector's recommendation that development be approved. In other areas of even greater economic marginality such as the Highlands and Islands of Scotland, the agency responsible to government for its socio-economic well-being, Highland and Islands Enterprise (formerly the Highland and Islands Development Board), has given specific backing to the realisation of a number of extractive operations, albeit mostly in areas of outstanding visual amenity. This support rests on the capacity of such developments to offer an element of diversification in economies that are all too narrowly based and to provide additional assistance to what are extremely isolated communities with high levels of unemployment. It is an approach which first found expression in the early 1970s in the backing the then Board gave to feasibility studies into mining a number of diatomite deposits, and continues into the late 1990s in the assistance it has given to the development of a 1.2 million tonne talc quarry at Cunningsburgh in Shetland (Highland and Islands Enterprise 1996).

Minerals: their national significance

Nothwithstanding the local/regional significance of some extractive operations, they all make a significant contribution to the national economy as a whole. Figures are difficult to establish with certainty, but mining activity probably directly employs about 45,000 persons. The materials that have been the subject of this chapter are worth considerably less in terms of monetary value than energy minerals (oil, natural gas and coal), which in 1995 amounted to £17,043 million. This figure is nearly eight times that of the combined value of industrial, construction and metalliferous minerals. But in spite of such a disparity this does not reflect their ultimate importance in the national economy. Whilst most of the UK's metals are now imported, a situation exacerbated by the closure of the last of the Cornish tin mines in February 1998, it does export a number of industrial minerals, notably china clay, ball clay, potash and salt. Net exports of minerals-based materials are dominated, apart from oil, by chemicals and steel. However, of the export of minerals in their strict sense, china clay is dominant with a value of £200 million in 1995 – a significant part of the 19.9 per cent of total exports that minerals-based materials make up.

In terms of their impact internal to the UK, industrial minerals, in serving as a basic raw material for the manufacturing sector, prove to be much more valuable than their worth per unit as raw materials. When the value of those sectors of manufacturing that are based on domestic and imported mineral materials is added to the value of mining and quarrying, the result is a total of £51.5 billion or 8.5 per cent of total GDP. To this the construction industry itself, which is almost entirely minerals based, was estimated to have contributed in 1995 a further £31.8 billion to total GDP for the UK.

Conclusion

Minerals produced in the UK are an essential input to the construction industry as well as contributing to a number of other industrial activities. Even though the total land area involved in their extraction is relatively small, perhaps not in excess of 100,000 ha, workings

FIGURE 4.5 A new approach to hard rock quarry rehabilitation

Bare vertical faces of stone are often left as an alien aftermath of quarrying in areas of outstanding landscape value. However, experimental work in the Peak District attempts to create varied rock slopes consisting of rock scree, butresses, head walls and vegetational cover similar to that of the surrounding countryside, as the above picture indicates. This approach to rehabilitation, whilst effective, involves additional costs in terms of the need for extra land for the production of the new landforms and the blasting and rock-moving operations involved in their creation; for hydroseeding and tree planting; and for after-care.

Source: J. Gunn (1998) 'Limestone landform simulation: results of trials undertaken in the Peak District', *Mineral Planning* 76 (September): 16–17.

often occur in circumstances where they are likely to cause disturbance to the public either because of their physical or their aesthetic impact. Whilst government policies towards private transport may now militate against the need for aggregate materials for motorway construction, the rising, and so far unmet housing needs requirement over the next twenty years, whether these are built on greenfield sites or are part of schemes of urban renewal, would appear to suggest that the need for supplies will remain firm. The view that the DETR now espouses, whereby the national need for minerals is balanced by attempts to mitigate the damage to landscape that their extraction causes, requires to be more firmly mediated by the contribution that can be made by recycled materials, well above the 12 per cent suggested for the period 1992–2006 in MPG6. This, as part of a demand-managed rather than a demand-led policy, will surely be essential if sustainability is to be a meaningful part of policy which government appears to have signed up to internationally.

What also seems to be required under the banner of sustainability is a much more proactive approach towards remedial action with regard to the results of those extractive activities that have yet to be brought to a conclusion. Whilst this may imply something more than water-parks for disused sand and gravel pits, it also means a much more comprehensive, if not bolder approach to other worked-out sites, particularly those resulting from the removal of hard-rock materials. The cosmetic solution is inadequate and might well be replaced by one which returns to the countryside a quarried landscape more in keeping with its surroundings. Interestingly, it is the commercial companies that have collaborated with the Department of the Environment, Transport and the Regions (until June 1997 the DoE) in demonstrating what might be achieved by such an approach in and adjacent to the Peak Park. As Figure 4.5 shows, a typical quarry face can be replaced with a landform in keeping with the surrounding landscape. As one commentator has stated 'more land will be taken, but the result will be an enduring and most interesting landscape' (Mabey 1996: 6).

References

Andrews, C.B. (1992) 'Mineral sector technologies: policy implications for the developing countries', *Natural Resources Forum* 16(3): 212–20.

Bent, D. (1998) Personal communication, Peak Park Joint Planning Board, 3 April.

Blunden, J. (1975) *The Mineral Resources of Britain – a Study in Exploitation and Planning*, London: Hutchinson.

Blunden, J. (1985) *Mineral Resources and their Management*, London: Longman.

Blunden, J. (1996a) 'The environmental impact of mining and mineral processing', in J. Blunden and A. Reddish (eds), *Energy, Resources and Environment*, Hodder and Stoughton.

Blunden, J. (1996b) 'Mineral resources', in J. Blunden and A. Reddish (eds) *Energy, Resources and Environment*, London: Hodder and Stoughton.

Clouston, B. (1993) 'Landscape restoration of quarries', *Quarry Management*, March: 23–59.

Corner, D.C. and Stafford, D.C. (1971) *China Clay Sand*, Exeter: Devon County Council.

Council for the Protection of Rural England (CPRE) (1996) *Rocks and Hard Places – Mineral Resource Planning and Sustainability*, Report by S. Owens and R. Cowell for CPRE, London: CPRE.

Countryside Commission (1993) *Position Statement: Sustainability and the English Countryside*, CCP432, Cheltenham: Countryside Commission.

Dunham, K.C. (ed.) (1979) *Atlas of Earth Resources*, London: Mitchell Beazley.

Eglin, R. (1969) 'Mr Shore, the miners' friend', *Observer* 29 June.

Elcoate, V. (1996) Comment in R. Spence and N. Woodcock (eds) *Aggregates and Sustainability: the Technology, Economics and Environmental Impact of Future Policy Options*, Knapwell, Cambridge: White Horse Press.

English Nature (1993) *Position Statement on Sustainable Development*, Peterborough: English Nature.

Gooding, K. (1992) 'Latin American Mining', *Financial Times Survey*, 17 September.

Gutt, W. and Nixon, P.J. (1980) 'Use of waste materials in the construction industry', *Resources Policy* 6(1): 71–3.

Highland and Islands Enterprise (1996) *Annual Report*, Inverness.

Jury, L. (1998) 'Firm pulls out of Peak Park quarry', *The Independent*, 23 February.

Mabey, R. (1996) 'Bold vision needed on minerals mining', *Countryside*, March/April, no. 78, Cheltenham: Countryside Commission.

Mackay, D.I. and Mackay, G.A. (1975) *The Political Economy of North Sea Oil*, London: Robertson.

MacEwen, M. and MacEwen, A. (1982) *National Parks: Conservation or Cosmetics?*, London: Allen and Unwin.

McNicholl, I.H. (1980) 'The impact of oil on the Shetland economy', *Managerial and Decision Economics*, 1(2): 91–8.

McNicholl, I.H. and Walker, G. (1978) *The Shetland Economy 1976–77, Structure and Performance*, Shetland Times.

Peak Park Joint Planning Board (1998) *Peak National Park Structure Plan*, Bakewell, Derbyshire.

Radetski, M. (1992) 'The decline and rise of the multi-national corporation in the metal mining industry', *Resources Policy* 18(1): 2–8.

Roberts, P.W. and Shaw, T. (1982) *Mineral Resources in Regional and Strategic Planning*, Aldershot: Gower.

Rogers, A., Blunden, J. and Curry, N. (1985) *The Countryside Handbook*, London: Croom Helm.

Royal Society for the Protection of Birds (1993) *UK Biodiversity and Sustainability Plans: Comments on their Preparation and Basis*, Sandy, Bedfordshire: RSPB.

Shrimpton, G.J. (1987) 'Western Europe – United Kingdom', *Mining Annual Review – 1987*, 490–5.

Stevens Report (1976) *Planning Control Over Mineral Working*, London: HMSO.

Verney Report (1976) *Aggregates: The Way Ahead*, Advisory Committee on Aggregates, London: HMSO.

Zuckerman, S. (1972) *Report of the Commission on Mining and the Environment (the Zuckerman Commission)*, London: The Commission.

Further reading

For the geographer the last decade and more has witnessed a dearth of texts relating to the minerals of the UK. One needs to look back more than twenty years to find a comprehensive review of earth resources in a context which largely relates to this chapter. This is to be found in the now out-of-print text by myself, *Mineral Resources of Britain – a Study in Exploration and Planning* (London, Hutchinson, 1975). Access to a library copy does, however, provide a detailed background to the key post-war issues – those of a strongly rising demand, especially for construction materials; a rapid expansion in quarry size and output made possible by the development of new extractive technologies substituting for labour; and problems over the environmental impact of such workings, both in their proximity to urban areas and their presence in or adjacent to protected landscapes.

These themes are also dealt with in a contemporary context in J. Blunden and A. Reddish, *Energy, Resources and Environment* (London, Hodder and Stoughton, 1996), but the chapters in question take a world-view and lay emphasis on the interface between minerals extraction and processing and the environment. A companion volume from the same publisher edited by P. Sarre and J. Blunden, *Environment, Population and Development* (1996) offers a brief summary of the minerals industry and its impact on rural Britain in the chapter 'Competing demands in the countryside'.

Another recent world-view, *Mining the Earth* by J.E. Young (Washington, DC, World Watch Institute, Washington, DC, 1992) also strikes a strong environmental note, and, although somewhat polemical in style, usefully acknowledges the importance of demand management in the quest to limit the ecological damage resulting from mining. An edited extract of key parts of Young's text also appears in *An Overcrowded World?* edited by P. Sarre and J. Blunden (Oxford, Oxford University Press, 1995), a volume which contains two chapters on the exploitation of natural resources (including minerals), and questions of sustainability – but again in the world context.

The *UK Minerals Yearbook* (British Geological Survey, London, HMSO) contains a comprehensive, if not definitive, annual review of production statistics together with a commentary on the background to the figures, including the part played by minerals in the UK economy. *Mining Annual Review* (London, Mining Journal Ltd.) performs a somewhat similar function but has a much more producer-oriented stance.

Notwithstanding the limitations of books about UK earth resources, two journals provide useful coverage of pertinent issues, although more often than not taking an international view. These are *Natural Resources Forum* and *Resources Policy*, both published by Heinemann, London, four times a year. However, the best of journals, not least because of its almost exclusive concern with the UK, is *Mineral Planning*. Published at 2 The Greenways, Little Fencote, Northallerton and appearing quarterly, it largely divides its attention between a consideration of aggregates and energy minerals, but it also offers good coverage of industrial minerals as well as problems relating to quarry rehabilitation and minerals planning. Its Review section of new publications provides a comprehensive synopsis of papers and reports on all aspects of UK minerals.

Chapter 5

Agriculture

Ian Bowler

Introduction

During the last fifteen years, agriculture in the United Kingdom (UK) appears to have entered a new phase in its development, termed variously 'the post-productionist transition' (Lowe *et al.* 1993), 'the third food regime' (Le Heron 1993) and 'rural restructuring' (Marsden 1997). Irrespective of the term used, recognition is given to the emergence of an agricultural sector increasingly influenced by economic and regulatory conditions that are qualitatively different from previous decades. For example, in common with other developed countries, agriculture in the UK is having to operate within a global rather than national market for food, under regulations oriented towards reducing state-financed price supports and output maximisation, and with farming practices and land uses reshaped to meet environmental concerns.

Three theoretical perspectives inform this account of the changing geography of agriculture in the UK. The first perspective is that of '*food regimes*', implying the existence of a number of distinctive, historical periods in international agricultural development, each regime being characterised by particular farm products, food trade structures linking production with consumption, and regulations governing capitalist accumulation. Three food regimes are commonly recognised: the first from 1870 to the First World War; the second from the 1920s to the mid-1980s; and the third from the mid-1980s to the present. The second food regime, for example, was characterised by the production and trade of grain, meat and durable foods, the restructuring of agriculture to supply mass markets, and the growth of forward and backward linkages within the food supply system. The term 'productionism'(or 'productivism') is used to describe this period of modernisation and industrialisation within farming. The third food regime is commonly termed 'the post-productionist transition', with the origins of many of its characteristics traceable to the early 1980s. This regime is characterised by the global financing and linking of production with consumption, the reregulation of agriculture, the production of fresh, organic and reconstituted foods and the 'greening' of farming practices. The following analysis is mainly concerned with the development of agriculture in the UK under this most recent food regime.

The second theoretical perspective is *'real' regulation*; that is, the 'administrative manner, style and logic by which the state regulates society in general and the economic landscape in particular' (Clark 1992: 616). Under the second food regime, UK agriculture experienced two phases of 'real' regulation, each with long-term consequences for the geography of the industry. Prior to 1973 agriculture was subject to regulation by UK agricultural policy; between 1973 and the early 1980s productionist regulations were applied under the Common Agricultural Policy (CAP) of the European Union (EU). Since the early 1980s a post-productionist reregulation of agriculture under the CAP has become increasingly influential.

The third theoretical perspective is provided by *political economy*. This perspective 'locates economic analysis within specific social formations and explains the development process in terms of the benefits and costs they carry for different social classes' (Redclift 1984: 5). In particular, political economy focuses attention on the role of non-farm capitals in the restructuring of agriculture and the consequences for different fractions of the farm population. Such a focus takes the analysis beyond the farm gate to consider the influence on both the farm sector and the food supply system of financial institutions, food processors and food retailers.

The following analysis concentrates on reregulation under the third food regime and the associated changes in the geography of UK agriculture. The broad argument to be deployed is that under reregulation, increasingly influential changes can be detected in farming *practice* and food *marketing*, but only marginal and spatially differentiated adjustments are observable in the *location and structure* of agriculture. While portents of significant changes in the agricultural geography of the UK can be identified, many productionist processes and structures persist into the third food regime.

The reregulation of UK agriculture

Membership of the European Community (now EU) in 1973 brought about the phased replacement of national agricultural policy by the CAP (Bowler 1985). The detail and contradictions of the CAP are too numerous to discuss here, but a number of features need to be examined so as to understand the reregulation of agriculture that occurred after the mid-1980s. Central to this understanding is the way in which the full impacts of the productivity gains from agricultural technology were being experienced in the UK and other member states by the late 1970s, such that the growth in the supply of farm products was beginning to outstrip domestic demand. Policy-makers in the EU, therefore, were faced with the problem of managing agricultural surpluses as well as farm incomes, whereas previous national policies, as in the UK, had been concerned mainly with raising food production while managing the transition of agriculture from a position of a high to a low employer of labour.

For a variety of reasons, the CAP continued with measures inherited from the original members of the EU. These reasons included: (a) defence of the CAP and its support of farm incomes by politically powerful farm unions in a number of member states, such as the UK,

France and Greece; (b) national self-interest in maintaining the benefits of substantial direct and indirect financial transfers under the CAP, for instance the Netherlands, Denmark and Ireland; and (c) the labyrinthine policy-making process of the EU, its numerous 'checks and balances' favouring incremental adjustments to the status quo rather than radical reform. Thus the founding principles of the CAP were continued into the mid-1980s, including maintaining common guaranteed prices above international market levels, intervention buying of the main farm products, protection by variable levies against imports from non-EU producers, and common funding. Running in parallel with price policies were measures designed to raise the efficiency of farming. EU Regulations 17/64, 355/77 and 1932/84, for example, offered grant aid to improve agricultural marketing and processing, while Directives 72/159, 72/160 and 72/161 sought to further modernise individual farm businesses, pay early retirement pensions and provide socio-economic guidance and agricultural training to farmers. Like their counterparts in the other member states, UK farmers responded by increasing their output of production per hectare of farmland, particularly milk, wheat and oilseed rape. By the early 1980s, however, the financial burden of agricultural support on the member states had become so great as to threaten the very existence of the EU. Consequently, during the 1980s, price support levels for all farm products were allowed to fall in real terms by between 2 and 5 per cent each year, while expenditure from the European Agricultural Guidance and Guarantee Fund (EAGGF) was capped by a series of measures, including co-responsibility levies, production quotas and maximum guaranteed quantities (stabilisers).

The financial crisis facing the EU formed part of the context for a reregulation of agriculture under the third food regime. In a broader context, however, reregulation should be interpreted as a renegotiation of the relationship between the state and agriculture, not just in the EU but internationally. Three arenas of renegotiation can be identified in a process more accurately termed the 'reregulation' rather than 'deregulation' of agriculture: state intervention in the market, agri-environmental relations and food quality/safety.

State intervention in the market

On the first arena, the member states of the EU are attempting to reduce their level of intervention in the market for agricultural produce so as to: (a) reduce the output of 'surplus' farm products; (b) reduce the financial cost to the EAGGF of 'productionist' support policies; and (c) open the market of the EU to global competition. The need to reduce, or at least limit the increase in production of farm products underpinned a 1992 package of CAP measures, commonly known as the 'MacSharry reforms' after the incumbent EU Commissioner for Agriculture. These reforms cut the support prices of cereals by 29 per cent and beef by 15 per cent over three years, placed individual farm quotas on subsidies in the beef and sheep sectors, reduced the price support on milk by 5 per cent but extended the time limit on the 1984 milk quota scheme, introduced area-based direct income payments for arable crops (the Arable Area Payments Scheme in the UK – AAPs), made set-aside of arable land compulsory for the receipt of the AAPs (cross-compliance) and introduced 'accompanying measures' for the afforestation of farmland and early farmer retirement. In sum, an attempt was made through direct income aids to decouple the link between farm incomes and the volume of food produced. In practice intervention stocks and their associated costs have fallen, while the burden of farm support has been shifted from consumers to taxpayers but with little impact on the overall cost of the CAP.

However, the MacSharry reforms must also be placed in the context of parallel international disputes and negotiations under the General Agreement on Tariffs and Trade (GATT: now the World Trade Organisation – WTO). The EU had come under considerable political pressure from the United States and the 'Cairns Group' of primary produce exporting countries to reduce the level of protection afforded to its agricultural sector. Negotiations leading up to the conclusion of the Uruguay Round in 1993, therefore, shaped the outcome to the MacSharry reforms and began the process of opening EU agriculture to global competition through a significant lowering, if not total elimination, of barriers to trade with food producers from outside the EU. Food wholesalers, processors and retailers in the UK, as well as the other member states, now have increasing access to lower-cost sources of raw materials and food products outside the EU market.

Agri-environmental regulations

Turning to the second arena of reregulation, the 1992 MacSharry reforms also recognised the need to introduce programmes of environmental conservation into farming. For over three decades evidence had been mounting of the environmentally damaging consequences of productionist agriculture as regards habitat loss, resource depletion and pollution (soil and water) and landscape degradation. The activities of environmental pressure groups, together with the development of national agri-environmental programmes as in the UK (Potter 1997), prompted action at the EU level. Under the heading of 'accompanying measures' (Regulation 2078/92), member states were required to develop national agri-environmental action policies (AEP) to encourage farmers to adopt environmentally sensitive farming practices. Fifty per cent of the expenditure was to be provided from EAGGF, but at a cost estimated at only 2 per cent of its budget.

However, the implementation of these agri-environmental policies in the UK should be placed in the context of another political event, namely the 1992 'Earth Summit' at Rio de Janeiro and its resulting 'Agenda 21'. Although the core concept of 'sustainable development' has proved to be both contested and chaotic, and despite the absence of a clearly defined programme of action, UK agriculture has been committed politically to 'Agenda 21' (Munton 1997). Nevertheless, 'Agenda 21' validates the reduction of farm inputs (such as fertilisers and agri-chemicals), the production of environmental goods (such as wetlands, moorlands and herb-rich pastures), and the protection of valued landscapes.

Food quality and safety

The third arena of reregulation in agriculture concerns food quality and, more recently, food safety; here the issues include animal welfare and *how* food is produced on the farm (i.e. farming practice). Increasing numbers of consumers in the UK are prepared to pay a premium price to assure the welfare conditions under which meat, eggs and milk are produced, as well as guarantee chemical-free fruit, vegetables, meat and dairy products. On the other hand, awareness of the health risks attached to food has increased. Risk is now interpreted not only in terms of how different foods and qualities of food impact indirectly on individual health (for example as regards heart disease and obesity), but also directly in terms of salmonella poisoning, pesticide residues, *E. coli* infection and the potential link between bovine spongiform encephalopathy (BSE) in beef cows and new variant Creutzfeldt-Jacob disease (nvCJD) in humans. Thus consumers are increasingly concerned with the origin of the food they eat, the farming practices by which food is produced, the 'freshness' of food and the processing food is subjected to after leaving the farm.

While the MacSharry reforms of 1992 recognised the issue of food quality indirectly by identifying organic farming as a sector to be developed, more direct recognition was given in 1992, under EU Regulations 2081/92 and 2082/92, which provided protection to farm products characterised by particular modes of production and regions of origin respectively. By 1997, for example, the following farm products in the UK had been registered under the Regulations: Orkney Beef, Scotch Lamb, White Stilton Cheese, Swaledale Cheese, Buxton Blue Cheese, Jersey Royal Potatoes, Worcestershire Cider and Rutland Bitter. The development of high-quality, niche market products is increasingly viewed as a way of combating a dependency on mass-produced food products in increasingly competitive global markets, with regulations protecting the premium prices available to the producers of such products. The development of 'green' consumerism underpins this arena of reregulation of agriculture, including the market demand for farm products and food carrying an assurance of quality and safety. Large food retailers have responded to the market opportunity offered by these consumer concerns by establishing their own networks of farmers who are prepared to conform to inspected standards and practices of crop and animal production. Such 'assurance' schemes are viewed as a marketing strategy at a time when new food health scares appear with regularity.

The restructuring of UK agriculture

Regulation is but one of the processes bringing about agricultural change (Bowler 1996). Other processes include agricultural technology, trade on national and international food markets, the behaviour of non-farm capitals and consumer preferences. Given the confines of space, this analysis now turns to the combined expression of these processes in empirical measures of agricultural structures in the UK. Two groups of measures are identified: first, features of productionist agriculture that are persisting from the second into the third food regime; second, features that appear to presage the emergence of post-productionism in agriculture.

The persistence of productionism

Tables 5.1 to 5.4 present a selection of indicators of national agricultural trends for the transition between the second and third food regimes. The tables illustrate the continuity of many productionist trends from the 1970s and 1980s into the 1990s. For example, Table 5.1 shows the broad structure of national farm output (by value) to be relatively stable between 1975 and 1995, with evidence of a gradual increase in the relative importance of milk, poultrymeat, sheepmeat and potatoes at the expense of wheat, pigmeat, eggs and barley.

When measured in terms of national self-sufficiency (Table 5.2), most products have continued with trends established during the second food regime. Increasing national self-sufficiency is evident for wheat, sheepmeat, beef and butter, for example, whereas falling levels have continued for eggs, poultrymeat and potatoes. Only barley and pigmeat show changes in trend: self-sufficiency in barley has fallen in line with its reduced area under cultivation (see p. 96), while self-sufficiency in pigmeat is rising after three decades of decline. Overall, UK self-sufficiency in temperate foods and animal feeds has increased to approximately 73 per cent and to 58 per cent for all food and feed supplies. These national trends must be placed in the context of the UK's trading relationships with other members of the EU. Table 5.3, for example, shows how the growing competition in the UK market for agricultural products has come from other member states, with national agricultural imports reoriented from non-EU to EU sources. On the other hand, two-thirds of UK agricultural exports are now marketed within the EU. As membership of the EU has been enlarged, so the UK's contribution to EU production of most products has fallen (Table 5.4); but the country still accounts for over a quarter of the EU's production of sheepmeat and significant proportions of the production of barley, potatoes and wheat.

TABLE 5.1 UK national farm output, 1975–95 (per cent value of final production)

Product	*1975*	*1985*	*1995*
Milk	21	20	24
Beef and veal	14	14	14
Wheat	8	13	10
Fruit and vegetables	9	10	10
Pigmeat	10	9	7
Poultrymeat	6	6	7
Sheepmeat	4	4	6
Potatoes	4	2	5
Eggs	8	5	4
Barley	8	8	3
Other	8	9	10

Source: Commission of the European Communities, *The Agricultural Situation in the Community – Report* (various years), Office for Official Publications of the European Communities.

TABLE 5.2 UK self-sufficiency in agricultural production, 1961–95 (production as per cent total new supply for use in United Kingdom)

Product	*1961–4*	*1968*	*1978*	*1988*	*1995*
Barley	96	90	106	145	126
Wheat	42	42	68	102	110
Sheepmeat	44	40	58	89	108
Skimmed milk	75	90	145	102	102
Eggs	94	99	102	97	95
Poultrymeat	99	100	101	98	92
Potatoes	91	91	93	90	89
Beef	75	77	85	75	89
Pigmeat	98	101	98	72	76
Cheese	45	45	67	72	72
Sugar beet	28	33	41	67	67
Butter	11	12	39	49	65

Source: Commission of the European Communities, *The Agricultural Situation in the Community – Report* (various years), Office for Official Publications of the European Communities.

TABLE 5.3 UK agricultural trade with members of the EU, 1971–96

	Total imports from EU (%)	*Total exports to EU (%)*
1971*	14.0	21.9
1976*	32.7	45.9
1981†	41.1	51.8
1986‡	51.6	59.8
1992	60.1	61.5
1996§	60.8	66.7

Source: Commission of the European Communities, *The Agricultural Situation in the Community – Report* (various years), Office for Official Publications of the European Communities.
Notes: *EC6; †EC10; ‡EC12; §EU15.

The broad continuity of trends within national agriculture reflects the persistence of many productionist processes into the post-productionist period. These processes can be summarised using the terms 'intensification', 'concentration' and 'specialisation' (Bowler 1996: 4), each dimension contributing to the uneven or spatially differentiated development of UK agriculture. Table 5.5 summarises the *intensification of farm inputs* (expenditure) between 1950 and 1996. The capitalisation of agriculture includes a continuing if declining reliance on purchased livestock feed (cheaper imports at the expense of domestic production, especially barley) and the gradual decline in expenditure on farm labour. Table 5.6 shows the changing structure of the declining labour input in more detail. Capital continues to be used to purchase labour-saving plant and machinery (10 per cent of national farm expenditure by 1996), allowing a reduction in the total workforce from approximately 803,000 workers in 1939 to 240,000 by 1996. Within this total figure, lower-cost 'intermittent' or 'flexible' labour forms, such as seasonal/casual workers, have increased to 34 per cent and part-time workers to 24 per cent of the workforce. The main reduction has been in the number of more expensive whole-time hired workers, especially males; by contrast, the number of family workers (i.e. farmers, spouses, partners and directors) has fallen at a much lower rate and still provides about 63 per cent of the total labour force. In addition agricultural contractors supply a wide range of specialised machinery and labour for farming

TABLE 5.4 The UK's changing share of EU output, 1969–95 (by volume)

Commodity	*Total (%) 1969/70**	*Total (%) 1978/9**	*Total (%) 1986/7†*	*Total (%) 1995/6‡*
Sheepmeat	50	47	30	26
Barley	28	28	30	19
Potatoes	14	17	21	19
Wheat	9	14	19	18
Poultrymeat	22	20	17	14
Eggs	25	22	17	11
Milk	14	16	12	11
Beef	16	16	10	11
Sugar beet	10	8	8	8
Pigmeat	13	9	8	6

Source: Commission of the European Communities, *The Agricultural Situation in the Community – Report* (various years), Office for Official Publications of the European Communities.
Notes: *EC9; †EC10; ‡EU15.

operations, ranging from ploughing, fertilising and crop spraying to harvesting and land drainage. Just as farm labour inputs concerned with animal feed have been replaced by non-farm workers in compound feed manufacturing firms, so other labour inputs continue to be removed from the production sector and substituted by off-farm specialised labour and capital.

Table 5.5 also indicates the fluctuating importance of interest payments over the last three decades, set against a long-term decline in real value. Advances to agriculture by the London Clearing Banks, for example, rose in just four years from £2,205 million to £4,234

TABLE 5.5 UK farm expenditure, 1950–96 (per cent total expenditure in each year)*

Item	*1950*	*1960*	*1969*	*1979*	*1988*	*1996†*
Feeding stuffs	19	28	29	35	29	27
Labour	33	24	18	21	22	15
Machinery	16	17	15	10	11	10
Fertilisers	7	9	8	9	7	7
Rent & interest	8	8	10	5	9	6
Seed				4	3	3
Maintenance				3	3	4
Miscellaneous				11	14	24‡
Livestock	17	14	20	3	2	2
Total value (£m)	737	1,263	1,876	5,707	9,758	11,654

Source: *Annual Review White Papers* (1950–88), HMSO; MAFF/SOAFF/DANI/WO, *Farm Incomes in the UK* (1996/7), HMSO.
Notes: *Rounding errors; †Forecast; ‡Of which pesticides 4%.

TABLE 5.6 Workers employed on agricultural holdings in the UK, 1939–96 (per cent total workforce in each year)*

Type of worker	*1939*	*1944†*	*1958*	*1968*	*1977*	*1987*	*1996*
Total whole-time	83	77	74	72	56	48	42
male	74	63	67	66	51	43	37
female	9	14	7	6	5	5	5
Total part-time‡	17	23	11	11	18	21	24
male	11	13	7	6	9	11	14
female	6	10	4	5	9	10	10
Total seasonal/casual			15	17	25	32	34
male			9	9	14	19	23
female			6	8	11	13	11
Total workers ('000)	803.5	902.1	730.3	450.1	370.5	297.3	239.9

Source: MAFF/SOAFD/DANI/WO, *Digest of Agricultural Census Statistics* (various years), HMSO.
Notes: *Excludes farmers, partners, directors and spouses but includes 'regular' family workers; †Excludes Women's Land Army and prisoners-of-war; ‡Part-time workers in Northern Ireland included with casual workers.

million between 1980 and 1984, an increase of 7.2 per cent in real terms. The Agricultural Mortgage Corporation, mainly concerned with loans for the purchase of farmland, also increased its lending to farmers, from £393 million in 1974 to £945 million in 1985. Indeed the value of farmland increased ahead of other assets with the flight of finance capital into agricultural land under the 1973 oil-price crisis and the capitalisation of increased CAP price supports into farmland values after the mid-1970s. Thus between 1970 and 1979, the farmland capital value index increased by 273 points, compared with an equities price index gain of 73 points, and between the mid-1970s and mid-1980s the average price of farmland with vacant possession in England rose from £1,643 to £3,789 per hectare. By 1988 UK farmers were paying £684 million in annual interest charges, compared with £216 million in 1978.

The variability in farm indebtedness and farm incomes has been magnified in the 1990s both from year to year and between production sectors, but again set against the background of a continuation of the long-term decline in the real value of farm incomes. There have been a number of contributory factors, including falling real price levels under the CAP; the opening of UK markets to greater competition; the devaluation and then revaluation of sterling, leading to increased and then decreased EU farm support prices within the UK; year-to-year variations in the level of payments under the Hill Livestock Compensatory Allowance (HLCA); restructuring of the dairy processing sector; and the banning of beef exports as a result of BSE in the national beef herd. Thus, aggregate net farming incomes fell by 113 per cent in real terms between 1977 and 1987, bringing bankruptcy to many individual farm businesses. But then average farm incomes rose by 65 per cent in real terms between 1990 and 1995, after which falling trends were again experienced under a period of revaluation after 1996. In the beef sector, for instance, production fell by 29 per cent between 1995 and 1996 under the BSE crisis, with beef farm incomes reduced by 38 per cent.

Turning to the *intensification of farm output*, crop and livestock yields have maintained their upward trend. Table 5.7 shows the increasing yields of a selection of crops, supported

TABLE 5.7 UK production and yield of selected crops, 1939–97

Crop		*1939*	*1944*	*1958*	*1968*	*1978*	*1988*	*1997**
Wheat	P	1,672	3,189	2,755	3,469	6,613	11,750	15,130
	Y	2.0	2.2	2.7	3.5	5.3	6.2	7.4
Barley	P	906	1,780	3,221	8,270	9,848	8,773	7,850
	Y	2.2	2.2	2.9	3.4	4.2	4.7	5.8
Oats	P	2,035	3,001	2,172	1,224	706	540	535
	Y	2.0	2.0	2.4	3.2	3.9	4.5	5.4
Sugar beet	P	3,586	3,320	5,835	7,119	6,382	8,152	9,555
	Y	25.8	19.3	33.1	38.1	31.7	41.3	48.0
Potatoes†	P	5,302	9,243	5,646	6,872	7,331	6,479	6,697
	Y	18.6	16.1	17.1	24.6	34.2	39.6	41.0
Hops	P	14.6	12.9	15.3	10.1	9.4	5.0	5.8
	Y	1.9	1.7	1.8	1.4	1.6	1.3	1.7

Source: MAFF/SOAFD/DANI/WO, *Digest of Agricultural Census Statistics* (various years), HMSO.
Notes: P = production in '000 tonnes; Y = yield in tonnes per hectare; *Forecast; †Maincrop.

TABLE 5.8 Farm-size structure of UK agriculture, 1976 and 1996 (per cent of holdings in each year)*

Country	*Under 20*		*20–100*		*Over 100 hectares*		*Holdings ('000)*	
	1976	*1996*	*1976*	*1996*	*1976*	*1996*	*1976*	*1996*
England	44	43	41	40	15	18	171.8	145.6
Wales	40	39	50	48	10	13	32.6	28.1
Scotland	38	36	46	36	16	28	30.9	33.0
N. Ireland	74	42	25	53	0.5	5	51.8	27.5
United Kingdom	50	41	40	42	10	17	287.1	234.2

Source: MAFF/SOAFD/DANI/WO, *Digest of Agricultural Census Statistics* (various years), HMSO.
Note: *Change in minimum size threshold of holdings enumerated in the census and differences in threshold size between countries.

by more controlled applications of fertilisers (7 per cent of national farm expenditure in 1996: Table 5.5). Yields of wheat, for example, increased by over threefold between 1939 and 1997, while the yields of barley and oats more than doubled. Similar tendencies are evident in yields from all classes of livestock. Taking an example from the dairy sector, the average annual milk yield per dairy cow (recorded herds) in England and Wales increased from 4,668 to 6,269 kilograms between 1975 and 1995; comparable figures for Northern Ireland were 5,029 and 6,163 kilograms, with 4,602 and 6,354 kilograms for Scotland.

Looking now at *concentration* in UK agriculture, the development of fewer but larger farm holdings has continued (Table 5.8), including the concentration of individual farm products into fewer production units (Tables 5.9 and 5.10). It is worth noting that measuring farm size structure is problematic, because the threshold size for holdings included in the annual agricultural census has been revised upwards several times during the last forty years (i.e. to exclude the smallest holdings). Taking the census statistics at face value, however, the restructuring of agricultural holdings has continued, the total recorded number falling from 303,600 to 145,600 in England between 1939 and 1996; comparable figures for Scotland were 74,300 and 33,000, for Wales 58,000 and 28,100, and for Northern Ireland 88,900 and 27,500. Within these global figures, there is evidence of a greater rate of attrition amongst smaller farms to the relative advantage of larger farms; that is to say, agricultural holdings over 120 hectares in the context of the UK. However, there is little evidence of a 'disappearing middle' of mid-sized family labour farms; indeed the main factor determining survival amongst mid-sized and larger farms during the 1980s and 1990s appears to have been the level of indebtedness of the farm business rather than its area size. Between 4 and 6 per cent of farms change occupiers each year, with approximately 60 per cent of these businesses falling vacant on the death or retirement from agriculture of the occupier. Between a third and a half of farms changing hands are purchased by existing farmers or landowners with a view to increasing the size of their farm business, thereby spreading production costs over more units of output. Scotland appears to have experienced the greatest rate of increase in the number of large farms, but the caveat regarding the changing basis of the agricultural census remains.

TABLE 5.9 Enterprise concentration in the dairy sector, 1975–95

Country	*1975*			*1985*			*1995*		
	A	*B*	*C*	*A*	*B*	*C*	*A*	*B*	*C*
England and Wales	46	26	8	67	39	17	72	47	22
Scotland	71	36	18	89	54	35	87	55	33
Northern Ireland	20	n/a	n/a	36	17	5	44	18	7

Source: National Dairy Council, *Dairy Facts and Figures 1996*, NDC.
Notes: A = average number of cows per herd; B = proportion of cows in herds above 100 head; C = proportion of herds over 100 head.

TABLE 5.10 Enterprise concentration in the UK, 1996

Enterprise	*England*		*Wales*		*Scotland*		*N. Ireland*	
	A	*B*	*A*	*B*	*A*	*B*	*A*	*B*
Laying hens	1	77	0.2	57	0.3	81	2.2	68
Total pigs	19	79	2	60	19	82	9	60
Table fowl	9	55	4	65	15	80	3	21
Breeding ewes	11	46	8	34	16	56	18	52
Beef cows	13	48	14	46	13	42	7	28
Dairy cows	25	50	14	34	33	58	8	23
Wheat	12	43	7	32	17	47	7	31
Barley	11	40	7	27	18	50	4	22
Potatoes	16	59	5	37	9	40	5	36
Oilseed rape	11	35	4	14	7	26	—	—
Horticulture	4	41	2	32	12	54	25	75

Source: MAFF/SOAFD/DANI/WO, *Digest of Agricultural Census Statistics* (various years), HMSO.
Notes: A = per cent largest producers (varying definitions); B = per cent area or livestock number (varying definitions).

Enterprise structure is an alternative way of measuring concentration in agriculture. During the second food regime, enterprise sizes were increased so as to gain economies of scale – for instance, the area planted to potatoes on individual farms – and a small number of larger farm enterprises increasingly accounted for a higher proportion of the output of each product. Table 5.9 shows the situation for the dairy sector. In England and Wales, in 1975, 8 per cent of farms with the largest herds (over 100 cows in milk) contained 26 per cent of all dairy cows; by 1985, 17 per cent of herds had over 100 cows, accounting for 39 per cent of all dairy cows; while by 1995 the respective figures were 22 per cent and 47 per cent. Indeed the process of enterprise concentration has continued into the third food regime and Table 5.10 sets out the situation for a range of enterprises in 1996. Intensive livestock (pigs and poultry) exhibit the greatest levels of concentration throughout the UK: in England, for instance, the

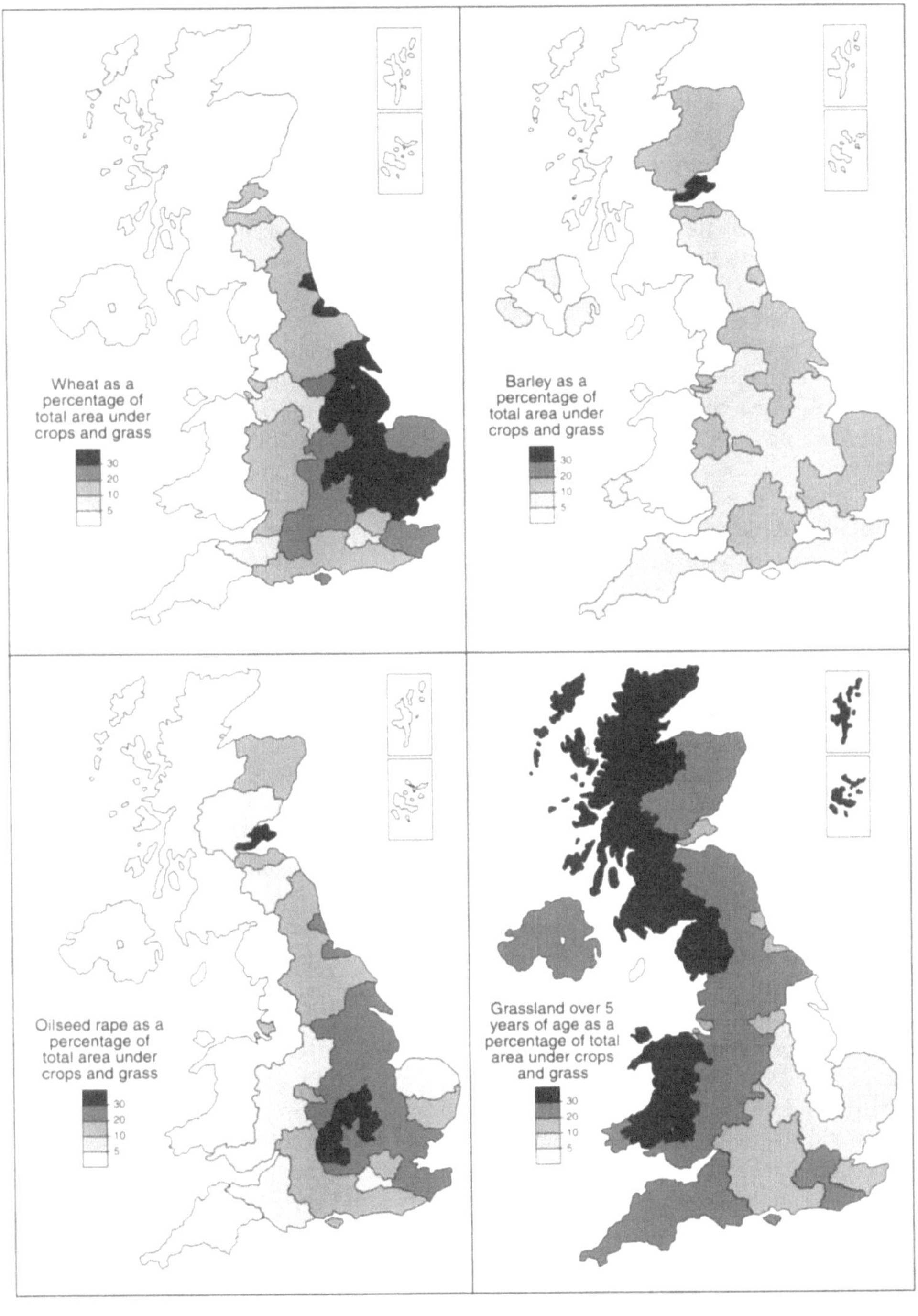

FIGURE 5.1 Regional distribution of selected crops, 1996
Sources: Agricultural Statistics, MAFF; SOAFD; DANI; WO.

1 per cent of largest enterprises account for 77 per cent of all laying hens. Other enterprises, and other parts of the UK, have experienced lower levels of concentration: in Wales, for example, the 14 per cent of largest producers farm only 46 per cent of the beef cows.

On *specialisation*, the broad regional contrasts of specialisation in crop and livestock production remain from the productionist era, with only changes in detail. Under the increasingly competitive market conditions of the 1990s, and in common with producers in other EU states, individual farm businesses in the UK have continued to obtain economies of scale by specialising in a limited number of farm products. Taking some examples from *crop farming*, and comparing the mid-1970s (Bowler 1982) with 1996, wheat production remains localised in the eastern counties of England, with the area of the crop contracting in north-east Scotland, Somerset and Hampshire (Figure 5.1). Barley, more widely distributed than wheat throughout England and Scotland, has lost its significance in most counties in competition with imports of livestock feed, but especially in north-east Scotland and north-east England. The area of oilseed rape, now concentrated in east-central England, has increased in Fife and Lothian but has decreased in the Ridings of Yorkshire and in the East Midlands.

Similarly, regional differences in specialisation in *livestock farming* have continued (Figure 5.2). Market trading in milk quotas, for example, has continued the relocation of dairying into western parts of the UK under the forces of comparative economic advantage, especially to the benefit of Northern Ireland, Cheshire and counties in south-west England. During 1995/6, for example, Northern Ireland became a net gainer of 85.7 million litres of purchased quota (mainly the counties of Down and Tyrone), whereas England suffered a net loss of 41.7 million litres (mainly West Yorkshire, Norfolk and the West Midlands). Beef cows, although widely distributed throughout the UK, remain most localised in Northern Ireland, the Scottish Borders and central Wales, but with numbers reduced by the late-1990s under the slaughter programme designed to eliminate BSE from the food chain. Sheep densities remain highest throughout Wales and northern England, Devon and west-central Scotland; but production has spread out into the lowlands, partly to compensate for the reduction in dairying but mainly to meet the demand for sheepmeat on the EU market. Specialised areas of pig production have been further consolidated in the Ridings of Yorkshire, and in Lancashire, East Anglia, Fife, Lothian and Armagh (Figure 5.2).

Table 5.11 and Figure 5.3 summarise the impacts of these recent adjustments in UK agriculture as regards farm-type structure. For example, the number and proportion of specialist dairy farms declined between 1975 and 1996, with compensating gains in the proportion of cattle and sheep farms. The latter now dominate the type of farm structure in western counties of the UK, from Cornwall and Devon in the south-west, through Wales and Northern Ireland, to northern England and Scotland. The proportion of cropping farms increased slightly, to the advantage of most counties in east and central England; whereas the relative importance of horticulture and pig and poultry farms declined, mainly through competition with low-cost imports from other countries both inside and outside the EU.

Portents of post-productionism

Despite the continuation of many features of productionist agriculture, a number of post-productionist trends can be identified in UK agriculture and summarised using the terms

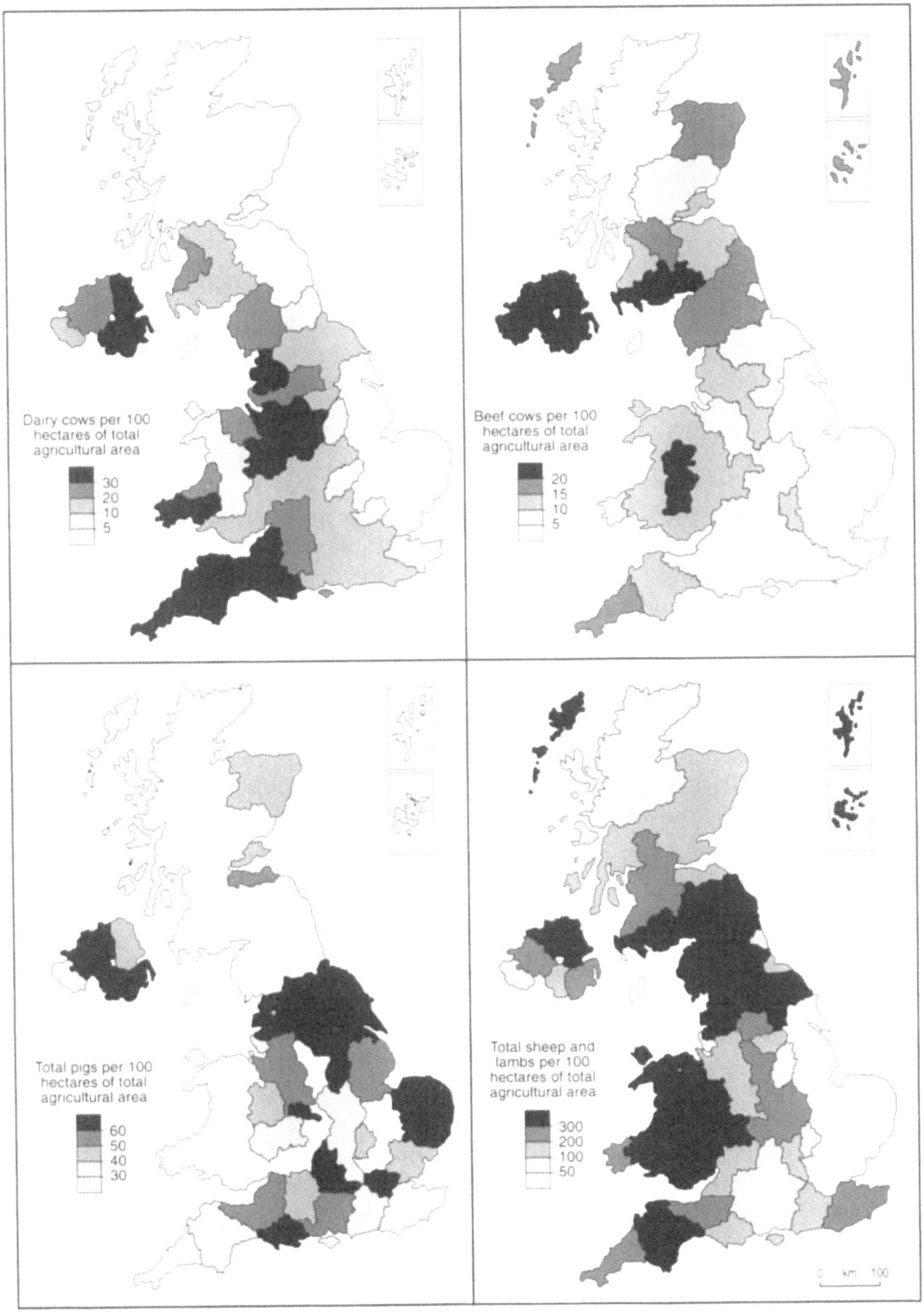

FIGURE 5.2 Regional distribution of selected livestock, 1996
Sources: Agricultural Statistics, MAFF; SOAFD; DANI; WO.

TABLE 5.11 Type of farm structure in the UK, 1975 and 1996 (per cent full-time farms)*

Type of farm	*England*		*Wales*		*Scotland*		*N. Ireland*		*UK†*	
	1975	*1996*	*1975*	*1996*	*1975*	*1996*	*1975*	*1996*	*1975*	*1996*
Dairying	35	13	43	15	20	6	36	19	34	13
Cattle & sheep	18	28	48	59	53	42	24	65	36	49
Cropping	20	23	1	2	15	18	5	4	10	12
Horticulture	12	3	2	1	—	2	—	1	4	2
Pigs & poultry	8	6	2	2	6	1	5	2	5	3
Mixed & others	7	27	4	21	6	31	20	9	9	22

Source: MAFF/SOAFD/DANI/WO, *Digest of Agricultural Census Statistics* (various years), HMSO.
Notes: * Definitions vary in detail over time and between counties; †Rounding errors.

'extensification', 'diversification' and the 'greening' of farming practices. Looking first at *extensification*, this tendency is consistent with the EU's agricultural objectives of reducing total farm production while increasing environmental benefits. Evidence is growing of lower fertiliser inputs to agriculture, partly as a result of grant-aided schemes such as Environmentally Sensitive Areas (ESA) and Nitrate Sensitive Areas (NSA), and partly as a result of the economic pressure on farmers to increase the cost-effectiveness and efficiency of their expenditure on fertilisers. Evidence of reductions in agri-chemicals is less clear (pesticides comprised 4 per cent of national agricultural expenditure in 1995), but the wider acceptance of integrated crop management (ICM) practices suggests that the upward growth in the application of herbicides and pesticides may have been checked.

These developments have been associated with a significant downturn in the area of cropland in the UK (Table 5.12) as a result of the compulsory set-aside programme introduced in 1993. The earlier voluntary set-aside programme of 1988 (20 per cent of arable land) had a very limited impact within the UK: by 1992 only 4 per cent of the 1988 arable area had been retired, and then mainly in the marginal cereal growing areas of central and southern England. The 1993 programme, with its rotational (initially 15 per cent of arable land) and non-rotational (initially 18 per cent of arable land for five years) options, by contrast, has had a greater impact. The volume and location of the set-aside land follows the prior distribution of crop farming in the eastern counties of England and Scotland. However, with 80 per cent of the land in rotational set-aside, and the percentage of arable land needing to be set-aside falling during the 1990s (e.g. 5 per cent in 1998), environmental benefits have been limited.

Another contribution to the extensification of agriculture has been made by capping the number of livestock eligible for financial support under the CAP. For example, individual farm quotas apply to the Sheep Annual Premium Scheme and the Suckler Cow Premium, while 'regional ceilings' apply to payments under the Beef Special Premium and the HLCA. Restraints have been placed on the expansion of subsidised production through these measures, including the further over-grazing of upland pastures.

A second feature of post-productionism is the development of *diversification* in agriculture; here three dimensions can be identified: farm diversification, other gainful activity (OGA) and farm woodland. Farm diversification involves the introduction of a new,

TABLE 5.12 UK agricultural trends in selected crops and livestock, 1938–96 (crop areas or livestock numbers as per cent of 1958)

Crop or livestock	*1938*	*1944*	*1958*	*1968*	*1978*	*1988*	*1996*
Beef cows*	(80)	(90)	100	136	186	158	250
Wheat	80	146	100	109	141	223	228†
Total sheep and lambs	103	77	100	107	114	148	160
Barley	37	72	100	215	211	164	156†
Heifers in calf*‡	(80)	(90)	100	100	104	94	147
Total fowls	73	52	100	127	133	134	131
Cattle under 1 year‡	63	66	100	118	133	121	125
Total crops and fallow	79	103	100	110	109	155	117
Total pigs	68	29	100	114	119	122	116
Cattle 1–2 years‡	70	66	100	96	123	103	113
Sugar beet	79	98	100	106	118	114	112
Grass over 5 years§	139	87	100	90	92	94	97
Dairy cows*	(80)	(90)	100	102	103	96	95
Rye	61	548	100	48	96	75	88
Vegetables	70	112	100	104	125	78	78
Soft fruit	112	80	100	88	82	72	62
Cattle over 2 years‡	94	102	100	70	75	56	57
Grass under 5 years§	65	76	100	94	82	67	54
Potatoes	86	172	100	84	64	53	53
Hops	90	95	100	86	71	47	40
Orchard fruit	104	108	100	71	46	36	24
Oats	109	165	100	43	20	11	11
Mixed corn	30	149	100	39	15	5	2

Source: MAFF/SOAFD/DANI/WO, *Digest of Agricultural Census Statistics* (various years), HMSO.
Notes: * Before 1960, beef cows, dairy cows and heifers in calf were collected together; †1997; ‡Dairy and beef livestock; §Changed census definitions in 1959 and 1975.

non-traditional enterprise into the farm business, resulting in a redeployment of land, labour and capital. A distinction is often drawn between agricultural and non-agricultural diversification, the former including new crops or livestock, such as linseed and deer respectively. Non-agricultural diversification includes enterprises that add value to products on the farm (e.g. farm-made yoghurts and cheeses), direct marketing (e.g. pick-your-own crops and farm shops), farm accommodation, recreational activities (e.g. sport fishing and horse livery), and catering services (e.g. farm restaurants). Farm diversification has been grant-aided in the UK under the Alternative Land Use and the Rural Economy (ALURE) programme, and more specifically through the Farm Diversification Grant Scheme (FDGS), between 1988 and 1993. In the UK, diversification has affected approximately 20 per cent of farms but more so in the urban fringes of large urban centres (for instance, London and Birmingham where market opportunities exist in serving urban consumers), and in areas favoured by tourism (for instance, agricultural areas in the south-west and north-east of England). Under the FDGS, farm tourism was the most important enterprise to be developed, with such diversification greatest in the north-east, south-east and south-west counties of England. But

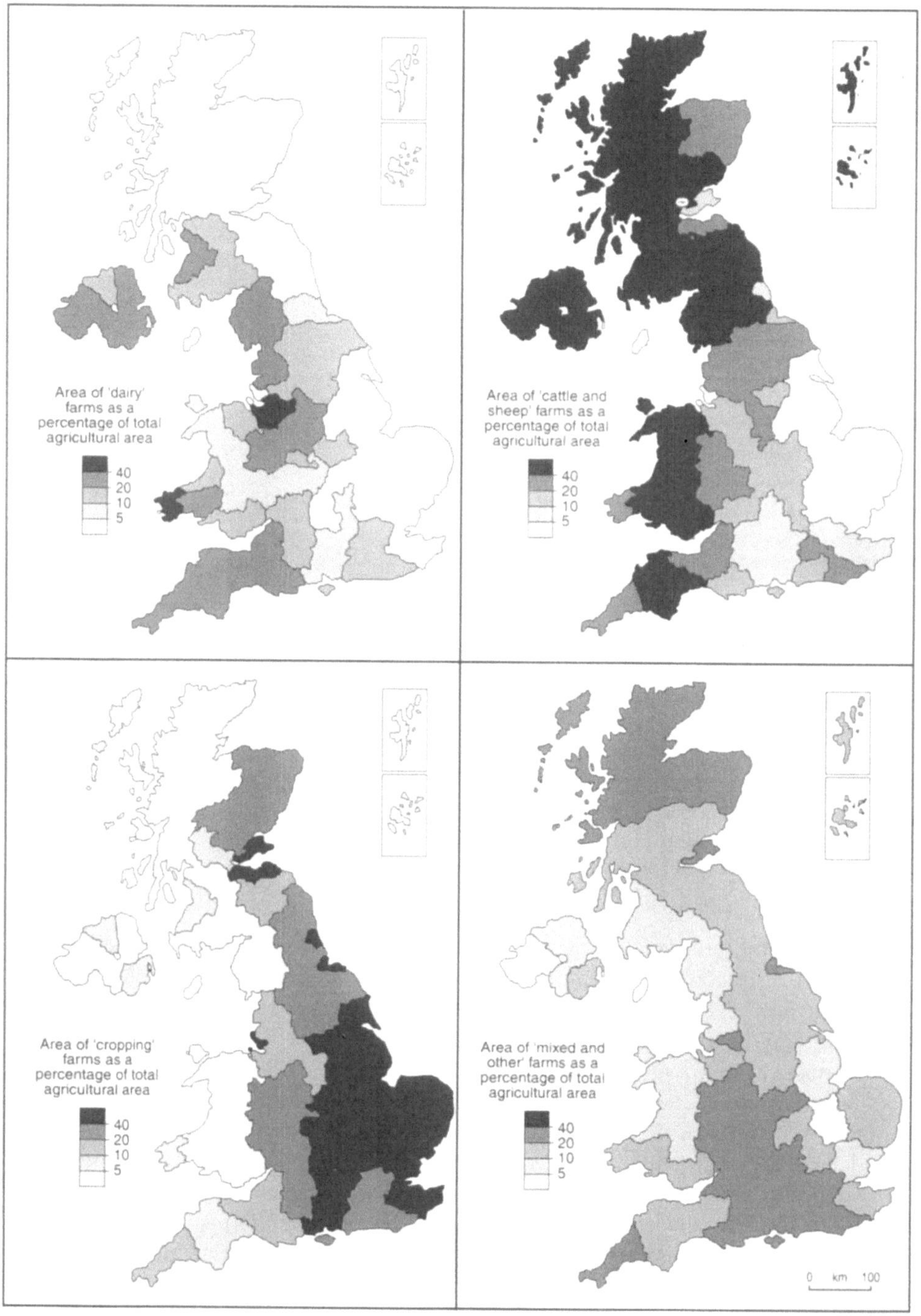

FIGURE 5.3 Regional distribution of selected types of farming, 1996
Sources: Agricultural Statistics, MAFF; SOAFD; DANI; WO.

the process of farm diversification has been socially selective to the advantage of younger, better educated and trained farmers, farm families in which farm women have become involved in managing the farm business, and larger farms where there is greater access to, and ability to finance, borrowed capital. The increased importance of 'mixed' farming in the type of farm structure of the UK, especially in south and central England (Table 5.11 and Figure 5.3), provides an indication of greater agricultural diversification, although the category also reflects the introduction of additional but traditional enterprises onto formerly specialist farms in the 1990s (e.g. beef onto cropping farms or sheep onto dairy farms).

Seeking employment away from the farm (OGA) is increasingly important, partly as a survival strategy amongst farm families on smaller farms but partly as a means of personal development or capital accumulation on larger farms. Approximately 24 and 29 per cent of adult males and females, respectively, in UK farm families appear to be involved in OGA, together providing a conservative estimate that one-third of farm households have a second source of earned income. This percentage increases in areas adjacent to urban-based employment opportunities and in peripheral regions where multiple job holding is a cultural tradition, for instance in the Highlands of Scotland. Depending on their educational qualifications and non-farming skills, both men and women gain employment in a wide range of occupations, varying from jobs in services, factories and offices, through professional occupations such as teaching and nursing, to management in companies sometimes owned by the farm family. On almost half of smaller farms, OGA compensates for low farm incomes and is significant in enabling them to survive as businesses. Indeed on a majority of farms with OGA, half or more of the family income is generated by off-farm work and forms an important component of the diversification of agriculture. Nevertheless, it is important to recognise that nearly half of farmers with OGA have entered agriculture from a non-farming background, so that taking an off-farm job is more than a survival strategy by 'traditional' farm families.

Farm woodland provides the third example of diversification in agriculture. Although often identified as the main alternative use for 'surplus' farmland, farm woodland shows only a modest and spatially uneven development in the UK over the last decade, despite financial incentives from the state. For example, grant aid has been available to farmers for planting, managing, compensating for loss of agricultural income and improving farm woodland under a succession of schemes, including the 1988 Farm Woodland Scheme (£10 million from MAFF under ALURE) and the Woodland Grant Scheme (Forestry Commission), the non-rotational element of the 1988 and 1993 set-aside programmes, and the 1992 Farm Woodland Premium Scheme (under EU Regulation 2080/92). Entry into these schemes has been voluntary for participating farmers and has favoured younger, better-educated farmers in the central counties of England, especially in localities having a tradition of farm woodland. The limited development of farm woodland is illustrated by the 14,000 hectares planted out of a target of 36,000 hectares under the 1988 Farm Woodland Scheme, although most of the trees were broadleaf varieties (oak, ash and beech). At issue is the setting of grant aid at too low a level to compensate farmers for 'lost' agricultural production, the long payback time of the investment, and inexperience with farming woodland.

The third dimension of post-productionism is the '*greening*' of farming practices, and two main features can be identified as regards *how* farming is practised: the development of organic farming and farmer response to state-aided agri-environmental schemes. Organic farming implies production without inorganic fertilisers, agri-chemicals and intensive livestock techniques and it has developed to a lesser extent in the UK compared with many other European countries. On the one hand, the growth in consumer demand for organic produce has not been as great as anticipated, despite a number of surveys showing consumer support for such food. In practice, consumers have been resistant to the price premium placed on organic farm produce, with multiple retailers in turn proving cautious in marketing organic produce because of its lower turnover on supermarket shelves. On the other hand, potential producers have been unsure of the economic returns to be obtained from organic as compared with conventional crop and livestock production. In addition there has been an 'income gap' during the two-year conversion period from conventional to organic production, although this has been partly addressed by the Organic Scheme introduced under the AEP. The marketing problem for producers has been compounded by retailers importing lower-cost organic produce from other European countries. The outcome for organic farming in the UK is a concentration of development on small farms in west Wales, the Vale of Evesham and Sussex, and on larger crop farms scattered throughout eastern England.

On agri-environmental schemes to stimulate the 'greening' of farming practices, the UK took a lead in the EU in developing the concept of ESA, introduced into the EU in 1985 under Regulation 797/85 (Article 19) and into the UK under the 1986 Agriculture Act. For England, five ESA were designated in 1986/7 and a further seventeen by 1994. In common with all agri-environmental schemes to date, participation by farmers has been voluntary, producing an uneven pattern of involvement. By 1995, ten-year management agreements had been reached with 7,700 farmers, covering an area of 400,000 hectares. Research in the ESAs has found that the annual payments to farmers succeed in reducing levels of fertiliser use, lead to pasture management practices that are more sensitive to wildlife, especially ground-nesting birds, have little effect on pre-existing livestock densities, but maintain landscape features such as field barns, walls and hedges. In 1989, the Farm and Conservation Grant Scheme provided grant aid for capital works on hedges, stone walls, shelter belts, repairs to traditional buildings and farm waste handling facilities. Also in 1989, a Countryside Premium (for set-aside land) scheme was introduced into seven counties in eastern England: this scheme was funded by £13 million over three years from the Department of the Environment (DoE) and aimed at grant-aiding crop farmers in wildlife and landscape management. A national Countryside Stewardship (Pilot) Scheme funded by the Countryside Commission followed in 1991 (upgraded to a full scheme by MAFF in 1996), with grant aid this time helping farmers (re)create environmental features such as riverside water meadows, lowland heath, chalk and limestone grassland, moorlands and coastal vegetation in 'target landscapes'. By 1996, over 5,000 management agreements were in place at a cost of £11.7 million each year. A parallel Tir-Cymen Scheme has operated in Wales under the aegis of the Countryside Council for Wales. The UK has also taken part in the EU's programme to reduce goundwater pollution from farm fertilisers (Directive 91/676) through the designation of ten (pilot) nitrate sensitive areas (NSA); twenty-two further NSA were designated in 1994 with financial compensation available to farmers who agreed to change their farming practices over a five-year period so as to reduce nitrate leaching. Most recently, grant aid has been provided under the UK's AEP under titles such as the Habitat Improvement, Moorland, Organic and Countryside Access Schemes, but with consequences that have yet to be fully researched. Taking the UK as a whole, these agri-environmental schemes have had only marginal impacts on farming practices and environmental outcomes.

As shown by Battershill and Gilg (1996) and Potter (1997), such impacts are confined to a limited range of farming areas and individual farms, where pro-conservation attitudes of farmers and a commitment to 'traditional' farming act selectively on farmers participating voluntarily in the schemes.

Developments in agricultural marketing

An important feature of the third food regime is the further transformation of how food is marketed to the consumer. Marketing at the farm gate, as well as down the marketing chain, is being affected by a further tightening of control by non-farm capitals in the food processing and retailing sectors, and also by an increasing reaction to such control by both farmers and consumers.

Looking first at the role of the EU, intervention agencies under the CAP still underpin the market for the main agricultural products. In the UK, the Intervention Board for Agricultural Produce acts as the agent of the EU by purchasing, storing and subsequently marketing surplus farm products, often through subsidised exports. Considerable sums of money are expended, despite a general reduction in price support levels. Whereas expenditure under the CAP on cereals, and to a lesser extent sugar beet, has fallen considerably under the revised 1992 pricing arrangements (see pp. 86–7), greatly increased expenditure is evident in the marketing of sheepmeat and beef/veal (Table 5.13).

The significance of food processors in UK agriculture, as established during the second food regime, has continued into the 1990s. Most food reaching the consumer is subject to some form of industrial processing, including washing, grading and packaging.

TABLE 5.13 UK agricultural marketing of selected products through co-operatives, contracts and CAP intervention, 1982–94

Product	*Co-operatives (%)*		*Contracts (%)*		*Intervention (£m)**	
	1982	*1994*	*1982*	*1994*	*1987/8*	*1995/6*
Pigmeat	13	28	50	70	1.9	2.1
Poultrymeat	2	25	95	95	—	—
Eggs	28	—	65	70	—	—
Potatoes	—	98	13	35	—	—
Milk	—	—	—	98	153.4	163.5
Sugar beet	—	—	100	100	151.1	115.6
Cereals	20	21	—	—	229.4	12.2
All vegetables	17	34	95†	60†	—	—
All fruit	33	25	—	—	—	—
Beef & veal	—	—	—	—	201.4	515.7‡
Sheepmeat	—	—	—	—	—	423.5

Source: Commission of the European Communities, *The Agricultural Situation in the Community – Report* (various years), Office for Official Publications of the European Communities.
Notes: *Expenditure by the Intervention Board for Agricultural Produce; †Peas; ‡Non-BSE.

For example, until recently the Milk Marketing Board for England and Wales (now Milk Marque Limited) controlled over 75 per cent of manufacturing capacity for butter and 50 per cent of the capacity for hard-pressed cheese. In common with other parts of the manufacturing sector, food processing companies have experienced downsizing and restructuring in recent years; sectors showing the highest levels of concentration (i.e. fewest competing firms) are brewing, biscuits, margarine, starch and bread/flour; the least concentrated are animal/poultry food and vegetable/animal oils and fats. Most companies hold back from direct ownership of farms, finding it more cost-effective to obtain their raw materials through forward production contracts placed with a small number of larger producers, including farmer co-operatives. By the mid-1990s, for example, all sugar beet, 98 per cent of milk, 95 per cent of poultrymeat and 70 per cent of eggs and pigmeat were produced under contract (Table 5.13). In these circumstances, nearness to the relevant processing factory has become an important factor in the location of production. At present most UK food processors obtain their raw materials from domestic producers or from within the EU; however, the prospective enlargement of the EU could well widen the area over which UK food processors place their contracts.

By far the most significant development in the marketing of farm products in the 1970s and 1980s, however, was the growth in the economic power of a relatively few but large multiple retailers through their chains of supermarkets (Wrigley 1987). By the early 1990s, large food retailers such as Sainsbury (approximately 18 per cent of market share), Tesco (15 per cent), Gateway (13 per cent), the Co-op (12 per cent) and ASDA (8 per cent), dominated food sales to consumers, becoming the price setters for the products of the food processing sector. Those prices then influenced price levels in contracts offered to farmers by individual food processors. In addition, large multiple retailers are able to offer contracts directly to those larger farm businesses and farmer co-operatives which can meet exacting standards of volume, quality, timeliness, price and packaging for certain products – for instance potatoes and eggs. By the mid-1990s, the top six retailers controlled 86 per cent of the grocery market in the UK, leading to the further demise of the small grocer and greengrocer in urban and rural areas alike.

So as to obtain a degree of 'countervailing power', farmers with larger businesses continue to group together to develop marketing co-operatives for their contract negotiations with food processors and retailers. By offering a product of the necessary assured quantity and quality, farmer co-operatives attempt to negotiate more favourable contract prices, although with variable results. Probably of greater importance for the economics of individual farm businesses is the passing of responsibility for marketing from the individual farmer to the specialised staff of the co-operative. Potatoes, vegetables, pigmeat, fruit and poultrymeat are the products most associated with the development of co-operative marketing in the 1990s (Table 5.13). The distribution of farmer co-operatives around the UK reflects the prior regional location of these different products (Bowler 1982: 96).

Many smaller farmers, however, are seeking other ways of by-passing the marketing chain and gaining direct access to consumers, thereby achieving a higher price for their products. For example, experience in the 1980s with the development of pick-your-own and farm shops (Bowler 1982: 96) is being extended to the development of farmers' markets, as found already in North America. In the organic farming sector, smaller producers are selling their fruit, vegetable and livestock products locally through retail outlets or increasingly popular 'vegebox' schemes. Under this latter system a network of local customers pay a fixed amount to an organic producer to receive a weekly box of seasonal fruit and vegetables. Other smaller producers are addressing a further aspect of 'green' consumerism, namely the increasing niche market for speciality or 'quality' foods, such as farm cheeses, yoghurts and meat products. These markets tend to be local in extent and supplied mainly through retailers. Producers are having to group together to be able to advertise and supply larger regional and national markets.

Multiple retailers have responded to the commercial potential of 'green' consumerism by introducing organic and 'conservation grade' foods into their stores and developing networks of farmers who are able to comply with their own schemes of assurance as regards food safety, including the traceability of food products. Competition under the third food regime between farmers and food retailers for the growing number of 'green' consumers is underway.

Conclusion

Two concluding observations can be made as regards the likely condition of UK agriculture in the early decades of the next millennium. First, just as a spatially differentiated agriculture developed under the productionist imperatives of the second food regime, so varying regional combinations of productionist and post-productionist agricultural trends will continue to evolve within the UK. How far and how fast such trends as the increased extensification, diversification and 'greening' of farming practices and structures take place will be conditioned in large part by the next round of reforms to the CAP ('Agenda 2000'), the further enlargement of the EU, and the WTO negotiations of 2002/3. The promised deepening and extension of the 1992 reforms, especially the further scaling down and decoupling of financial support from production, seem likely to strengthen the development of post-productionist over productionist developments in agriculture.

Second, just as agriculture is subject to its own 'internal' restructuring, so the sector is coming under increasing 'external' pressures from other users of rural land, including urban, recreation, tourism and conservation interests. A number of different regional contexts have been identified in this restructuring of rural space (Marsden 1997), including 'preserved', 'contested', 'paternalistic' and 'clientelist' countrysides. The economic and social roles of agriculture will vary within and between such new rural spaces: in particular farm production could become increasingly concentrated within the UK in the most productive areas as regards climate, soil and large-farm structure, with farming in the remaining 'marginalised' areas oriented more towards housing, recreation, tourism and conservation functions. In these circumstances, agriculture's place in the economy and society will be interpreted increasingly within the wider processes of rural development and in regional (i.e. endogenous) rather than national and international (i.e. exogenous) contexts.

References

Battershill, M. and Gilg, A. (1996) 'Environmentally friendly farming in south west England', in N. Curry and S. Owen (eds) *Changing Rural Policy in Britain*, 200–24, Cheltenham: Countryside & Community Press.

Bowler, I.R. (1979) *Government and Agriculture: a Spatial Perspective*, Harlow: Longman.

Bowler, I.R. (1982) 'The agricultural pattern', in R.J. Johnston and J.C. Doornkamp (eds) *The Changing Geography of the United Kingdom*, 75–104, London: Methuen.

Bowler, I.R. (1985) *Agriculture under the Common Agricultural Policy: a Geography*, Manchester: Manchester University Press.

Bowler, I.R. (1996) *Agricultural Change in Developed Countries*, Cambridge: Cambridge University Press.

Clark, G.L. (1992) '"Real" regulation: the administrative state', *Environment and Planning A* 24: 615–27.

Le Heron, R.B. (1993) *Globalized Agriculture: Political Choice*, Oxford: Pergamon Press.

Lowe, P., Murdoch, J., Marsden, T., Munton, R. and Flynn, A. (1993) 'Regulating the new rural spaces: the uneven development of land', *Journal of Rural Studies* 9: 205–22.

Marsden, T. (1997) 'Theoretical approaches to rural restructuring: economic perspectives', in B. Ilbery (ed.) *The Geography of Rural Change*, 13–30, Harlow: Longman.

Munton, R. (1997) 'Sustainable development: a critical review of rural land-use policy in the UK', in B. Ilbery, Q. Chiotti and T. Rickard (eds) *Agricultural Restructuring and Sustainability: a Geographical Perspective*, 11–24, Wallingford: CAB International.

Potter, C. (1997) 'Environmental change and farm restructuring in Britain', in B. Ilbery, Q. Chiotti and T. Rickard (eds) *Agricultural Restructuring and Sustainability: a Geographical Perspective*, 73–87, Wallingford: CAB International.

Redclift, M. (1984) *Development and the Environmental Crisis: Red or Green Alternatives?*, London: Methuen.

Wrigley, N. (1987) 'The concentration of capital in UK grocery retailing', *Environment and Planning A* 19, 1283–8.

Further reading

Early surveys of productionist agriculture are provided in J.T. Coppock (1976) *An Agricultural Atlas of England and Wales* (London: Faber and Faber) and (1976) *An Agricultural Atlas of Scotland* (Edinburgh: Donald); later interpretations can be found in G.M. Robinson (1988) *Agricultural Change – Geographical Studies of British Agriculture* (Edinburgh: North British Publishing); I.R. Bowler (1991) 'The agricultural pattern', in R.J. Johnston, and V. Gardiner (eds) *The Changing Geography of the United Kingdom*, (London: Routledge, 83–114); I.R. Bowler (ed.) (1992) *The Geography of Agriculture in Developed Market Economies* (London: Longman); and D. Britton (ed.) *Agriculture in Britain: Changing Pressures and Policies* (Wallingford: CAB International).

The restructuring of agriculture, including agri-environmental policy, is considered by different authors in various chapters of the following edited books: B. Ilbery (ed.) (1997) *The Geography of Rural Change* (London: Longman); B. Ilbery, Q. Chiotti and T. Rickard (eds) (1997) *Agricultural Restructuring and Sustainability: a Geographical Perspective* (Wallingford: CAB International); N. Curry and S. Owen (eds) (1996) *Changing Rural Policy in Britain* (Cheltenham: Countryside & Community Press); and I. Bowler, C. Bryant and M.D. Nellis (eds) (1992) *Contemporary Rural Systems in Transition: Agriculture and Environment* (Wallingford: CAB International).

More specific topics are covered by individual papers in journals, such as: A. Flynn and T. Marsden (1992) 'Food regulation in a period of agricultural retreat: the British experience', *Geoforum* 23, 85–93; T. Marsden, R. Munton, N. Ward and S. Whatmore (1996) 'Agricultural geography and the political economy approach: a review', *Economic Geography* 72, 361–75; I. Bowler and B. Ilbery (1989) 'The spatial restructuring of agriculture in the English counties', *Tijdschrift voor Economische en Sociale Geografie* 80: 302–11; A. Errington and R. Gasson (1996) 'The increasing flexibility of the farm and horticultural workforce in England and Wales', *Journal of Agricultural Economics* 12, 127–41; H. Edmond and R. Crabtree (1994) 'Regional variation in Scottish pluriactivity: the socio-economic context for different types of non-farming activity', *Scottish Geographical Magazine* 110: 6–84; and C. Potter (1986) 'Processes of countryside change in lowland England', *Journal of Rural Studies* 2: 187–95.

Chapter 6

Transport

Brian Turton

Introduction

During the 1990s the main developments in transport have been concerned with policy (in particular the privatisation of large sectors of the industry), with attempts to stem car ownership and ridership and to promote public transport, and with an acceptance of the need to take the environmental impact of transport activities more fully into account. Additions and improvements to the national road and rail networks were largely completed with the opening in 1991 of the last section of the M40 motorway between Oxford and Birmingham and the electrification of the final section of the London to Edinburgh trunk railway in 1994. The most significant individual development of the 1990s, however, was the Channel Tunnel link, opened in 1994 between Folkestone and Calais to provide through rail facilities between the United Kingdom and the Continent and a shuttle service between the two tunnel terminals for road traffic.

Between 1945 and the late 1970s much of the transport sector was in public ownership, but since 1980 components of this state-owned system have been progressively dismantled and transferred to the private sector. This process was largely completed in the mid-1990s with the break-up of British Rail, one of the last survivors of the nationalised enterprises established in the early post-war period. Privatisation and deregulation of major transport enterprises were two of the principal objectives in the former Conservative administration's free market policy initiated in 1980, and although the present Labour government has retained some of its predecessor's privatisation plans it has also introduced a new transport policy which is directed much more towards the integration and co-ordination of the various inland modes (Transport Policy 1998).

Both Conservative and Labour governments have attempted to stem the ever-growing tide of private car ownership, and the increasing use of road transport in general, by attempting to create a more attractive public transport system coupled with restraint upon car drivers, especially within inner towns and cities. However, despite massive investment in

TABLE 6.1 Passenger traffic in Great Britain by mode of transport (percentage distribution of total passenger km by mode)

Year	*Road*	*Passengers* *Rail*	*Air*
1965	89	11	<1
1975	92	8	<1
1985	92	7	1
1996	94	5	1

Sources: HMSO, *Transport Statistics Great Britain* (London, 1997); HMSO, *1993–95 National Travel Survey* (London, 1996).

TABLE 6.2 Freight traffic in Great Britain by mode of transport (percentage distribution of total freight tonne km by mode)

Year	*Road*	*Freight* *Rail*	*Other modes**
1965	57	21	22
1975	66	15	19
1985	56	9	35
1996	66	6	28

Sources: As Table 6.1.
Note: *Includes coastal shipping until 1975 and also pipelines.

TABLE 6.3 Passenger journeys by mode and purpose, 1993–5 (percentage of total journey distance per person per year)

Purpose	*Walk*	*Car (includes drivers and passengers)*	*Bus (includes urban and non-urban buses)*	*Train (includes main line and urban metro services)*	*Other modes (includes cycles, motorcycles and taxis)*
Journey to work	9.6	18.2	17.7	40	13.6
Business	1.2	11.2	>1	10.9	14.8*
Education	12	2.6	13.9	4.3	10
Shopping	20.5	11.9	21.1	5.6	5.0
Social	19.3	28.7	20.8	21.3	24.9
All other purposes	37.4	27.4	26.5	17.9	31.7

Sources: As Table 6.1.
Note: *Includes air transport.

inter- and intra-urban public transport the imbalance between road and rail traffic remains (Tables 6.1, 6.2 and 6.3). Serious congestion on many sections of the principal road network indicates that the demands of road transport for additional capacity have still to be adequately satisfied, but the policy of continual road construction has been recognised as one that will not achieve the desired objective of providing a transport system that is efficient and capable of meeting the needs of all sections of society.

During the last two decades there has also emerged a growing concern for the effects of the existing and proposed elements of the transport infrastructure upon the environment, coupled with the recognition that sustainable transport is also an essential issue to be considered in future planning exercises. The strength of public concern over new road construction schemes which have damaged or threatened many tracts of high-quality landscape has been demonstrated both by action groups dedicated to halting or at least delaying controversial new road projects and by wider campaigns which have succeeded in their aim of securing environmental impact assessments as an integral part of transport planning processes.

This chapter discusses these various policy and related issues in the context of rail, road, urban, air and maritime transport, identifying the extent to which specific transport improvement programmes have been successful and, where appropriate, drawing comparisons between policy and planning in the United Kingdom and other European Union countries.

The railway system

Since the mid-twentieth century the railway industry has adopted a variety of strategies to meet the challenge of road transport and to adapt its services to suit the changing demands of its freight and passenger markets. This section outlines recent trends in rail traffic and then reviews the various organisational, structural and technological changes made in an attempt to regain increased shares of traffic.

Although the proportions of national passenger and freight traffic carried by rail have continued to decline in the 1990s a distinction can be drawn between the passenger sector, where the amount of traffic has increased, and freight haulage, which is still falling in both absolute and relative terms.Passenger traffic has been the leading source of railway revenue since 1975. Receipts between 1986–96 , using the 1996–7 price base, rose by 14.8 per cent, with an increase overall of 4.5 per cent in passenger km. The length of route open to passenger train services was increased by 5 per cent to 15,034 km, the highest total since 1969, and 110 new or reopened stations have been added since 1986 (Table 6.4). These positive developments have largely been jointly financed by British Rail and local authorities in urban areas where rail commuter services have been improved. Passenger journeys on local and inter-urban routes also rose to 788 million in 1996–7, the highest total recorded since 1971 (Table 6.5). In 1998 31 per cent of the national network is electrified, although some of the earlier routes, notably the heavily used west coast main line where electrification dates from the mid-1960s, are in urgent need of upgrading. Despite these improvements in passenger services and the absolute increases in the numbers carried, in 1996 the railways

TABLE 6.4 Railways open for traffic in Great Britain, 1965–96

Year	*Total length open (km)*	*Length open to passenger traffic (km)*	*Length of route electrified (km)*
1965	24,012	17,516	2,886
1970	18,988	14,637	3,162
1975	18,118	14,431	3,655
1980	17,645	14,394	3,718
1985	16,752	14,310	3,809
1990	16,584	14,317	4,912
1996	16,666	15,034	5,176

Source: HMSO, *Transport Statistics Great Britain* (London, 1997).

TABLE 6.5 Changes in rail traffic in Great Britain

Year	*Passenger journeys (millions)*	*Passenger km (millions)*	*Freight tonnes (millions)*	*Freight tonne km (millions)*	*Coal as percentage of total freight tonnes*
1965	865	30,116	232	25,229	61
1970	824	30,408	209	24,500	54
1975	730	30,300	175	20,900	55
1980	760	31,700	154	17,600	37
1985	697	29,700	122	15,400	27*
1990	762	33,200	160	13,820	54
1996	788	32,200	130	10,100	50

Source: HMSO, *Transport Statistics Great Britain* (London, 1997).
Note: *Period of national coal industry dispute.

still accounted for only 5 per cent of all passenger travel, whereas a decade ago the proportion was 7 per cent (Table 6.1).

Changes since 1986 in the freight traffic situation have been less promising, with a decline of 31 per cent in the length of routes open to freight trains and a fall of 27 per cent in the amount of goods hauled. Coal traffic fell by one-third in the 1986–96 decade with the continued contraction of the mining industry, although rail is still the dominant means of transport for this commodity. The rail market for petroleum and its products also contracted by 22 per cent, and in terms of the transport of all basic energy sources pipelines now carry three and a half times as much as the railways. Only 5 per cent of all freight by tonnage is now carried by rail, compared to a proportion of 8 per cent in 1986.

During the period between 1948 and 1993, when the railway system was state-owned, various changes in organisation were made within the overall framework of British Rail to establish a trunk route network (the InterCity), a provincial or regional sector incorporating many of the subsidised routes, a south-east England commuter zone and a freight sector. In early 1993 the government decided to transfer the railways to the private sector, and the Railways Act passed in December 1993 enabled the process of dismantling British Rail and awarding franchises for specific areas of passenger and freight operations to come into effect. The current situation of the railways must be examined in the light of these far-reaching changes, which are the most significant in British railway history since the amalgamations of 1923.

The principal aims of the 1993 Act were to secure both an improvement in rail passenger services and in the financial performance of the train operators. In turn it was hoped that these changes would create a more attractive market for travellers. These objectives were to be achieved by franchising passenger and freight operations to a group of private sector companies and establishing other new companies responsible for the track, supporting infrastructure, rolling stock provision and maintenance. Subsidies would be provided to those companies electing to operate loss-making passenger services, and franchisees were required to guarantee specified levels of service or be subject to financial penalties (Eaton 1997). An initial aim was to transfer 51 per cent of the national network by

TABLE 6.6 Principal train operating companies (TOCs) in 1998

TOC and principal area of operation	*Passenger journeys (millions)*	*Passenger km (millions)*	*Passenger km as percentage of total passenger km*
Great North East Railway: east coast main line	11.9	3,352	10.4
South West Trains: south-west of England	110.7	3,276	10.17
West Coast Trains: west coast main line	13.2	2,917	9.05
Connex South Eastern: south-eastern England	113.4	2,768	8.59
Connex South Central: southern England	86	2,091	6.49
Great Western Trains: west of England	15	2,043	6.34
Cross Country Trains: intercity routes	12	1,928	5.98
Scotrail: Scotland	53.4	1,673	5.19

Source: Office of Passenger Rail Franchising, *Annual Report 1996–7* (1997).

revenue to the new private train operating companies by April 1996. The actual process took much longer and was not completed until 1997. The first set of franchises were for three small units (the Gatwick Airport Express, the Isle of Wight, and the London and Southend line), two large sectors (Scotrail and part of the Network Southeast system), and two trunk routes (Intercity East Coast and Intercity Great Western); others covering the rest of the passenger network were awarded later.

By 1998 services had been franchised to twenty-five principal train operating companies (TOCs), with the largest, in terms of their share of total passenger km, being the Great North East Railway, South West Trains, West Coast Trains and Connex South Eastern (Table 6.6). The first three of these provide services over sections of the former Intercity network whereas Connex has a mainly suburban traffic in the South East. The eight TOCs awarded franchises covering inter-urban routes now generate 53 per cent of all passenger traffic, and about one-quarter of all passenger km are accounted for by the seven companies operating in and around Greater London. The highest total of passenger journeys per annum in 1997–8 was recorded by Connex South Eastern, reflecting its intensive commuter traffic in Kent and parts of Surrey and Sussex.

Freight haulage has also been franchised to the private sector and plans have been announced to invest in new rolling stock and improved operating methods. One of the more important initiatives involves upgrading the trunk route between the Channel Tunnel terminal, London and Scotland by raising bridges and lowering some sections of track to enable road trailers to be carried on low-loader rail wagons in the 'piggyback' intermodal system.

Competition between the train-operating companies is governed and moderated by an independent Rail Regulator, whilst the Office for Passenger Rail Franchising (OPRAF) is responsible for contracts between each company and Railtrack, the new undertaking which owns and maintains the track, passenger stations and freight depots and signalling equipment. OPRAF also administers the subsidies, currently about £1.45 billion, to companies which have taken over loss-making regional services, and oversees the maintenance of the agreed performance levels. Railtrack plays a significant role in the newly privatised structure of rail services as it is responsible for investment in the infrastructure, using revenue from the franchisees to benefit both passenger and freight operators.

The acquisition by several major bus companies of some of the rail passenger franchises has raised the issue of restricted competition over intercity corridors where both bus and rail services are controlled by the same undertaking. Such a situation arose in the East Midlands, where National Express has the Midland Mainline franchise for trains between London, Nottingham and South Yorkshire, and also runs coaches in the same area (Harnden 1997). Potential monopolies of this nature are issues for the Rail Regulator to resolve and at this early stage of the privatisation exercise it is not possible to provide a full picture of the eventual benefits and disadvantages of this railway reorganisation.

Structural changes, principally the closure of many loss-making passenger lines, date from the recommendations of the Beeching Report in 1963, and between 1960 and 1968 the total length of railways in Britain was reduced by 32 per cent. After 1968, following the government's decsion to subsidise lines which were acknowledged to be uneconomic but which were identified as worthy of retention on social grounds (Department of Transport 1963, 1965), far fewer closures were authorised and since 1986 the network has been stabilised at about 16,600 km (Table 6.4).

Technological changes have been based upon the electrification of heavily used passenger trunk lines, the introduction of diesel traction through the network and the use of improved rolling stock. The electrification programme was begun in the 1960s and finally completed with the opening of the east coast route between London , the North East and Edinburgh (Figure 6.1). Diesel power is employed for freight haulage and trunk and local passenger services, with the replacement of the original stock by more efficient 'Sprinter' trains. There are also plans to upgrade the west coast main line services with the introduction of new tilting high-speed trains to be purchased by the franchisee.

The opening of the Channel Tunnel to through rail passenger and freight trains in 1994 has to date had only a limited impact upon the British system, as the projected high-speed rail link between London and Folkestone has still to be built. The Eurostar passenger expresses to Paris and Brussels currently terminate at Waterloo station in London and have to make use of the conventional railway to Folkestone, whereas in France and Belgium high-speed lines are available for much of the route between the Tunnel terminus, Paris and Belgium. With privatisation of the railways the new British link is to be built by London and Continental Railways, a consortium which includes National Express and Virgin, but a start date depends upon revenue from the Eurostar service and final agreement upon the amount of financial aid from the UK government.

The current status of the United Kingdom's railway system as a complex mosaic in which train services, rolling stock, track, and maintenance facilities are all owned by different

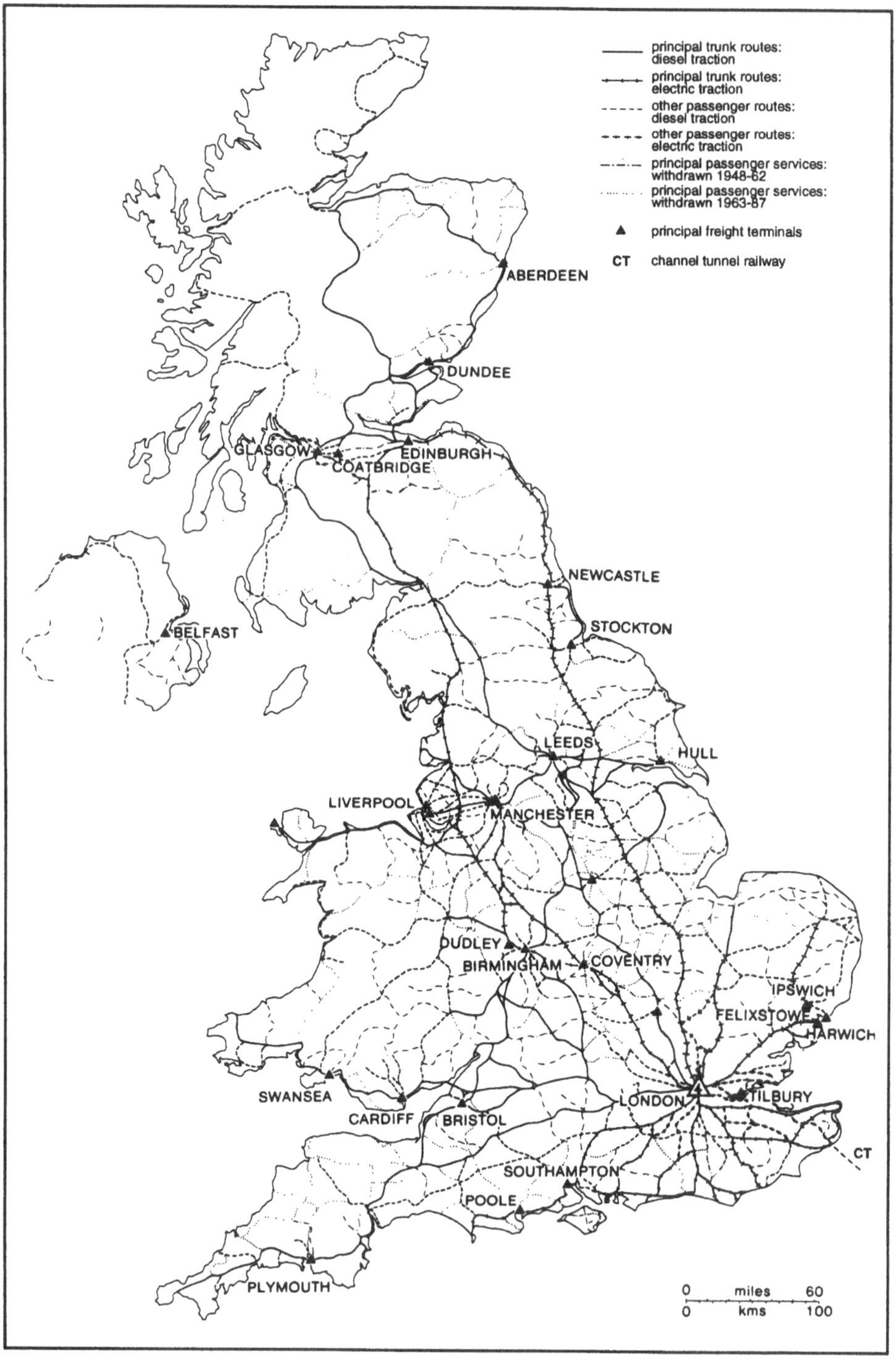

FIGURE 6.1 The railway network in 1998

undertakings is in marked contrast to systems in Europe, where state ownership with large subsidies from the government is still the most common situation.

Roads and road traffic

During the 1990s road traffic has continued to increase, with a consequent rise in levels of congestion on the road network. A demand-led approach to new road construction is no longer seen as the most appropriate or effective solution to congestion, however, and where road schemes have been agreed the government is actively encouraging private finance intitiatives to contribute to the considerable costs involved.

In terms of vehicle km all traffic increased by 36 per cent over the decade 1986–96, with cars and taxi traffic rising by 37 per cent. In 1996 there were 21.17 million private cars

TABLE 6.7 Changes in road passenger traffic in Great Britain, 1965–96

	Cars and taxis		*Buses and coaches*	
Year	*Vehicle km ('000 millions)*	*As a percentage of all road traffic*	*Vehicle km ('000 millions)*	*As a percentage of all road traffic*
1965	115.8	71	3.91	2.4
1970	161.3	78	3.62	1.7
1975	194.4	79	3.78	1.5
1980	197.2	80	3.06	1.2
1985	229.4	81	3.07	1.0
1990	335.9	82	4.6	1.0
1996	362.4	82	4.8	1.0

Source: HMSO, *Transport Statistics Great Britain* (London, 1997).

TABLE 6.8 Changes in numbers of principal types of vehicles, 1986 and 1996

	1986		*1996*		
Vehicle	*'000s*	*Percentage of all vehicles*	*'000s*	*Percentage of all vehicles*	*Percentage change in nos. of vehicles, 1986–96*
Private cars	16,981	78.2	21,172	80.5	+ 24.7
Buses	68	0.3	77	0.3	+ 13.2
Heavy goods	484	2.2	413	1.6	– 14.7
Light goods	1,880	8.8	2,267	8.6	+ 20.6
Other vehicles	2,286	10.5	2,373	9.0	+ 3.8
Total vehicles	21,699		26,302		+ 21.2

Source: HMSO, *Transport Statistics Great Britain* (London, 1997).

registered in Great Britain, and trips made by car drivers and their passengers accounted for 84.7 per cent of all journeys made by road and rail transport (Tables 6.7, 6.8). There has also been a 32 per cent rise since 1986 in the average distance travelled by car by each person per year, whereas the distances covered by bus, motorcycle and cycle have all fallen. Data from the 1993/5 National Travel Survey also indicates that car travel now accounts for 81 per cent of all road traffic and for 94 per cent of all national passenger traffic. Large increases have also taken place in road freight transport, with a 46 per cent rise in the amount of roadborne freight traffic since 1986 and a 30 per cent increase in the distance travelled by commercial vehicles.

These increased flows are carried on an infrastructure which has been progressively improved in terms of new and upgraded trunk and principal roads, but congestion remains a serious problem on both urban and national highways. The rate of motorway construction has gradually declined since the ambitious building programmes of the 1960s and 1970s, and in the 1986–96 decade only 306 km were added. Completion of the M40 between London and the M42 orbital road around Birmingham in the early 1990s marked the end of the major phase of inter-urban motorway building and recent additions have been confined to branches linking towns to the national system (Figure 6.2). Motorways now account for 20 per cent of the national trunk road network, but by 1996 just over a half of all trunk road traffic was making use of them (Table 6.9).

Motorway traffic is still frequently impeded by chronic congestion however, especially on the approaches to London, on the M25 and on the M6 around Birmingham, and there is now ample evidence that the major road investment programme published in 1989 will not meet the demands of traffic levels in the twenty-first century (Department of Transport 1989). In early 1994, however, it was announced that forty-nine of the major road schemes contained in this programme were to be abandoned and 242 further projects were to be suspended as part of a new policy initiative designed to encourage the use of alternative means of transport and reduce road-building expenditure. The inadequacy of the existing road network was admitted by the Department of Transport in evidence to the Royal Commission on Environmental Pollution in 1996, and since 1993 the government has been actively encouraging the involvement of the private sector in new road construction and, where appropriate, operation and maintenance (Department of Transport 1993). To date the only significant results of private investment have been the Queen Elizabeth II bridge carrying the M25 over the Thames to supplement the capacity of the earlier tunnel, and the second bridge over the Severn, opened in mid-1996 and linked with the M4–M5 motorway network in this area. In the early 1990s it was proposed that a 43-km privately funded toll road to relieve congestion on the Birmingham section of the M6 should be built to the north of the conurbation as a three-lane motorway. Traffic volumes of up to 50,000 vehicles per day were forecast and could yield a daily income of £100,000. This route is still under consideration, but if completed it would be the longest privately financed road in the United Kingdom. The summer 1998 White Paper on transport indicates that efforts to expand privately funded and operated toll roads will be continued, but it is unlikely that such routes will provide an effective solution to congestion. Proposals for the introduction of tolls on the national motorway system have also been considered, and if these were to be introduced it would bring about a system of road pricing similar to those which have operated for many years in France, Spain and Italy.

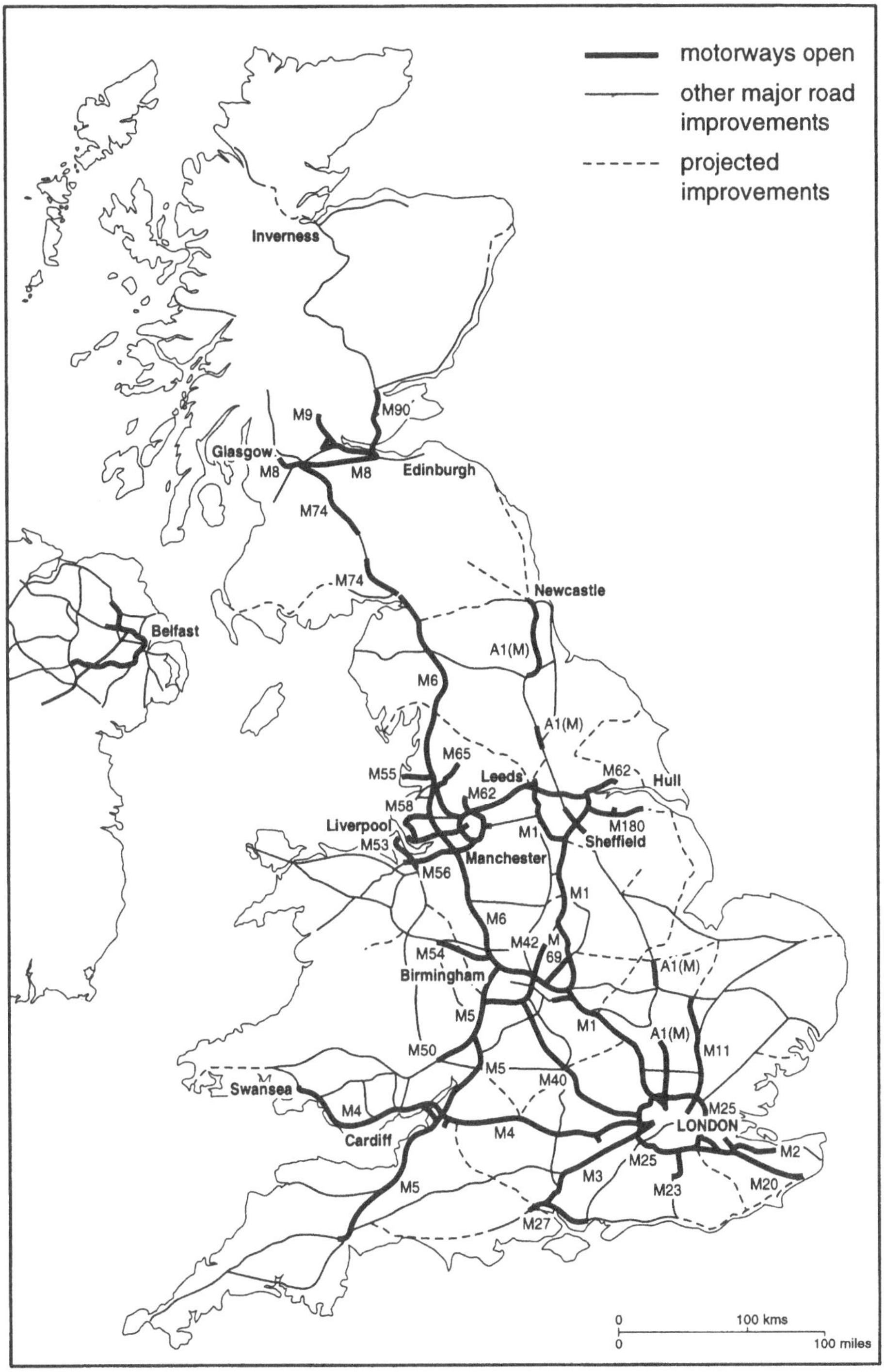

FIGURE 6.2 The motorway and principal road network in 1998

TABLE 6.9 Changes in traffic distribution by types of road, 1986 and 1996

Type of road	*Percentage of all traffic, 1986*	*Percentage of all traffic, 1996*	*Percentage change in amount of traffic, 1986–96*
Motorway	12.5	16.6	+ 81.6
Principal roads in urban areas	21.0	18.3	+ 18.4
Other principal roads	27.7	28.4	+ 39.6
Minor roads	38.8	36.7	+ 28.6

Source: HMSO, *Transport Statistics Great Britain* (London, 1997).

Urban transport

During the 1990s there has been a continuation of plans introduced in the 1980s designed to combat congestion on urban roads by improving public road and rail transport and discouraging the use of private cars in inner cities. This has involved the deregulation of bus services in order to promote competition and thus lower fares, investment in new public transport modes, such as light rail, and experimenting with physical and fiscal restraints upon cars in urban areas.

Journey-to-work traffic is one of the principal contributors to daily congestion on roads and on public transport, and in the large provincial conurbations most trips are still reliant upon the private car (Table 6.10). Patronage of public transport (buses, surface rail and light rail) is highest in Tyne and Wear, where the Metro was opened in 1984, and lowest in Greater Manchester, where the Metrolink was completed in 1992. Walking still accounts for between 10 and 15 per cent of all journey-to-work trips, a share that reinforces the current campaign for improved facilities in towns for pedestrians and cyclists. However, non-mechanised transport in towns is unlikely to become more popular unless the present levels of atmospheric pollution caused by motor vehicles can be reduced and safety levels improved.

The deregulation of bus services which came into effect in 1986 was intended to promote competition between urban bus companies, increase the quality of service and thus attract car users back to public transport. Trends since then indicate that the aims of the 1986 Transport Act have been only partially realised in terms of urban bus services. Competition between bus companies already operating in the 1980s and the new undertakings, attracted to the market as a result of deregulation, has produced higher service frequencies on many routes, reversing a pattern of declining frequencies that began in the 1950s. However, the number of passenger trips carried out by bus in the metropolitan areas outside London fell by 35 per cent in the period 1986–96, with an average fares rise of 17 per cent between 1987 and 1994, so that the only real change in service quality offered to the urban traveller has been an increase in frequency (White 1995). Bus operations in urban areas, excluding London, as measured by annual bus km, increased by 21 per cent between 1986 and 1996 so that the resultant pattern is one of more buses providing more services for a declining market. However, there is no firm evidence to date that the increased level of

TABLE 6.10 Commuter traffic in major conurbations, 1996

Area of workplace	*Number employed ('000s)*	*Main mode of transport to work (as percentage of all work journeys in area)*				
		Car, van, minibus	*Bus, coach*	*National rail*	*Urban rail, metro*	*Walk*
Central London	1,000	16	8	38	29	4
Outer London	1,399	70	9	3	3	11
Greater Manchester	1,029	73	10	2		11
Merseyside	474	69	12	4		10
South Yorkshire	476	69	13			12
Strathclyde	831	69	14	3		11
Tyne and Wear	460	65	17		3	11
West Midlands	1,129	73	13	1		10
West Yorkshire	879	70	13	2		13

Source: Labour Force Survey, Office for National Statistics (1997).

service has succeeded in encouraging more use of urban buses by members of non-car-owning households.

Investment in new urban rail systems as a means of revitalising public transport in cities was pioneered by the Tyne and Wear Passenger Transport Authority in 1984 with the opening of the Metro, followed in 1987 by the Docklands Light Railway serving the redeveloped dockland area of east London. Plans for similar light rail networks were actively considered by many other towns and cities in Britain, but the only systems to be opened to date are in Manchester and Sheffield (Figure 6.3). The Manchester Metrolink was opened in 1992 and makes use of the former British Rail suburban routes between Bury and Exchange Station and Altrincham and Deansgate station, with a new link along city-centre streets and a branch to Manchester Piccadilly station. This 31-km network will eventually be extended to serve Oldham, Rochdale and the Dumplington retail complex to the west of the city centre (Knowles 1996). The Sheffield Supertram network was opened in stages between 1994 and 1997 to link the city centre with the Meadowhall retail park and the south-eastern and north-western suburbs.

The traffic on these new systems in Manchester, London, Tyneside, Sheffield, and on the older but modernised Glasgow underground, shows a varying pattern of ridership. Passenger traffic on the Tyneside Metro has declined by 14 per cent since 1986 and risen only slightly on the Glasgow system. On the Docklands Light Railway traffic grew by 168 per cent between 1988 and 1996, but the network was extended east to Beckton during this period. Both the Manchester and Sheffield systems recorded substantial increases in traffic during their first few years of operation, but a longer time period is required before more informed judgements on their success in promoting public transport usage can be made. Many of the plans for light rail transit in other major cities such as Edinburgh, Nottingham and Bristol are in abeyance pending guarantees of funding from a combination of private and public sources, and the only new project to be completed (in mid-1999) will be the first route of the Birmingham network between Wolverhampton and the city centre.

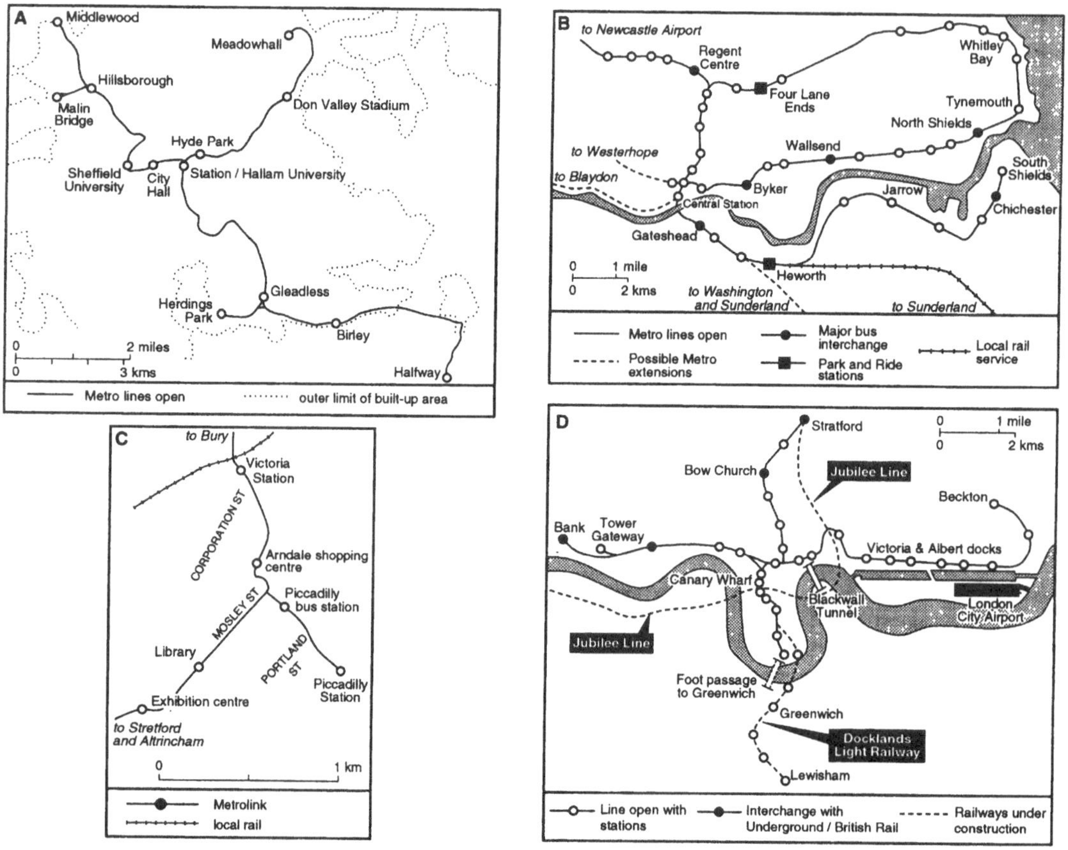

A Sheffield Supertram system
B Tyne and Wear Metro
C Manchester Metrolink system
D Railways in East London

FIGURE 6.3 Light rail transport systems in major cities

In London there has been a decline in the volume of daily commuters since 1986 and corresponding falls in the numbers using suburban rail, London underground and bus services during peak hours. However, there is still an urgent need to augment the present central area rail system, which is severely overloaded, and several schemes are nearing completion involving the extension or upgrading of existing railways. In east London an extension of the Jubilee underground line from the city centre at Green Park station to Stratford via the docklands development at Canada Water will probably be completed in 1999. This will be complemented by a southward extension of the Docklands Light Railway under the Thames to connect with the surface railway network at Greenwich. Many of these new lines will relieve chronic congestion on the existing Underground network and possibly attract car drivers and passengers back to the train.

Strategies to contain and eventually reduce the volumes of cars entering city centres have involved both physical and fiscal restraint measures. Progressive increases in car-parking charges as the centre is approached, coupled with priority for buses in these areas, have been combined with an extension of pedestrianised shopping zones in an effort to discourage access by car. Many towns and cities, such as Oxford, Shrewsbury and Chester, have introduced permanent park-and-ride facilities on their margins, enabling low-cost car parking to be combined with the efficiency of the bus for those wishing to use city-centre facilities. Partial restrictions on private cars are also imposed within traffic calming schemes in residential areas, whereby through traffic is discouraged by means of speed control humps and other devices (Tolley 1990). In early 1998 the government approved the setting of speed limits lower than the standard 30 m.p.h. within urban areas where buildings of historical and architectural interest are prone to damage by motor vehicles.

Fiscal constraints upon urban traffic include tolls on vehicles entering defined inner zones or a premium tax payable to enable cars to use such areas at any time. Recent experiments in Leicester indicate that a daily toll of £3 per vehicle could reduce car commuter traffic by 20 per cent, but that a £6 toll would only effect a further 5 per cent fall. Many other urban authorities are actively considering these schemes as part of a combined strategy of improving public transport with income from the tolls obtained from car restraint schemes, although electronic tolling may not be available until about 2003. If a policy of car restraint, combined with the upgrading of public bus and rail transport, is successful the current vehicle congestion in inner cities may be reduced to a more acceptable level. The experience of light rail transit in many European cities indicates that the expense of such systems can eventually be justified in terms of a greater use of public transport, but the United Kingdom has yet to adopt light transit on the scale displayed by many European countries.

Air traffic and airports

Congested runways and passenger terminals, and the lack of adequate high-speed connections with city centres, remain the principal problems facing British airports in the 1990s. The steady growth of air passenger traffic that began in the early 1950s continued into the 1990s, with rises between 1986 and 1996 of 84 per cent in international traffic and 74 per cent in internal flows. Charter flights, operated on behalf of the major holiday tour companies, now account for 41 per cent of all international traffic carried on United Kingdom airlines compared with 29 per cent in 1976. Heathrow and Gatwick airports continue to handle over two-thirds of all international traffic, and Stansted, opened as London's third airport in 1985, has seen its annual passenger totals rise from 500,000 in 1986 to 4.8 million in 1996. The largest increases at provincial airports were recorded by Birmingham, Edinburgh, Newcastle and Manchester (Table 6.11).

The improvement of terminal facilities for handling passengers is seen as an essential objective in both the London area and at principal regional airports, but no major projects have been completed since the opening of Manchester's second terminal and the Eurohub facility at Birmingham, which has increased capacity for passengers on short-haul flights to Europe. Heathrow, which currently is served by over ninetey airlines and is linked to 200 destinations worldwide, is one of several leading hub airports in Western Europe which act as transfer centres between short- and long-haul flights. Competition between these hubs for passengers is intensifying, and the British Airports Authority is determined to maintain

TABLE 6.11 Changes in air traffic, 1965–96
(a) Scheduled services operated by United Kingdom airlines

Year	*Passengers (millions)*		*Freight ('000s tonnes)*	
	International	*Domestic*	*International*	*Domestic*
1965	6	4	241	54
1970	8	5	204	64
1975	11	6	199	47
1980	15	7	264	31
1985	16	9	313	46
1990	25	12	374	50
1996	37	15	567	61

(b) Changes in passenger traffic at United Kingdom airports, 1986–96

Airport	*1986*		*1996*		*Percentage change in total traffic (all +)*
	International (millions)	*Domestic (millions)*	*International (millions)*	*Domestic (millions)*	
Heathrow	25.7	5.6	48.3	7.5	78
Gatwick	15.2	1.1	22.0	2.1	48
Manchester	6.0	1.5	12.0	2.5	93
Glasgow	1.4	1.7	2.4	3.0	77
Birmingham	1.6	0.5	4.4	1.0	157
Stansted	0.5	0.1	3.8	1.1	860
Edinburgh	0.2	1.4	0.8	3.0	137
Luton	1.9	—	2.0	0.5	20
Belfast	0.4	1.4	0.7	1.7	26
Aberdeen	0.6	0.9	0.8	1.6	60
Newcastle	0.8	0.4	1.6	0.8	100
East Midlands	0.9	0.3	1.4	0.4	64

Source: HMSO, *Transport Statistics Great Britain* (London, 1997).

Heathrow's status as a leading airport. The addition of a fifth terminal would increase the airport's capacity by 30 million passengers per annum and would make the requirement for a third runway less urgent. The Authority would like to see this new terminal open by 2015, when London's four major airports will need to cope with an estimated annual total of 160 million passengers, but the proposal has been strongly opposed on environmental grounds and no decision has yet been reached. However, the first high-speed rail link between Heathrow and central London was completed in summer 1998 to relieve pressure on the Underground and road links. At Manchester the demand for additional capacity created by the 68 per cent rise in aircraft operations between 1986 and 1996 has been met by the construction of a second runway, due to be completed in 1999, and links with Manchester city centre have been improved with the opening in 1993 of a branch from the main rail network.

The extension of high-speed rail services on the electrified east coast main line between London, Newcastle and Edinburgh has produced some competition for domestic airlines operating between these cities, but air traffic is still increasing in this corridor. In the South East the air passenger market between London, Paris and Brussels has been substantially challenged with the introduction of the high-speed Eurostar rail expresses through the Channel Tunnel, which have reduced the rail timings between these cities to a level which is offering strong competition with overall flight times. Elsewhere in Britain, however, domestic air services can still offer the shortest journey timings between major cities, and seasonal holiday traffic between the mainland and offshore islands is still important.

Seaports and international trade

Maritime traffic patterns and the distribution of trade by ports during the 1990s confirm and strengthen the trend for continuing growth of the east coast ports trading with the European Union. In addition the initial effects of the opening of the Channel Tunnel upon the Channel ferry ports were felt in the period after 1994.

During the period since 1979 British ports have undergone several changes in terms of ownership and organisation. These have been made in order to improve the performance of a port system, which in the late 1970s was seen as possessing excess capacity and perpetuating several inefficient cargo handling practices (Bassett 1993). The 1981 Transport Act led to the privatisation of the British Transport Docks Board and the creation of the Associated British Ports company which now controls over twenty-six ports. In 1991 legislation was passed requiring the leading trust ports, which each had an annual turnover of at least £5 million and were owned and operated by various types of local boards, to prepare for privatisation. The majority of larger British ports are therefore now within the private sector, but the pattern is still subject to take-overs and mergers as trading fortunes vary from port to port.

Other states in the European Union now receive 70.5 per cent of all United Kingdom exports by tonnage (63 per cent by value), and account for 37 per cent of all imports by tonnage and 64.5 per cent by value. Trade in bulk fuel, principally petroleum, has stabilised following the rapid rise in oil exports in the period 1965–85, but the tonnage of all incoming and outgoing trade has risen by 15.8 per cent in the decade 1986–96 (Figure 6.4). The most significant trend has been the steady rise since 1986 in the volume of unitised traffic to and from continental ports, including containers and roll-on/roll-off commercial vehicles. Of the total increase in port traffic of 78.25 million tonnes over the decade, 24.3 million tonnes were containerised and 20.7 million were carried on roll-on/roll-off trucks.

During the 1990s the fastest-growing ports continue to be those associated with European trade, and particularly those specialising in container and roll-on/roll-off traffic. Tonnage handled at Felixstowe and Harwich rose by 69 per cent between 1986 and 1996, and Portsmouth and Southampton recorded increases of 29 per cent. The group of ports

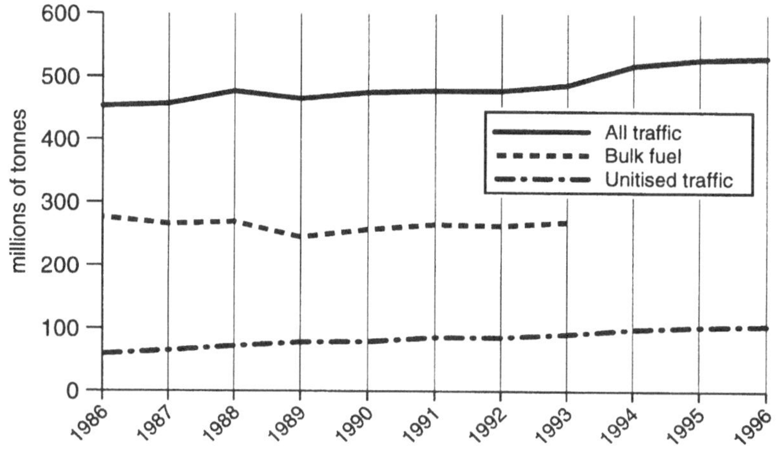

FIGURE 6.4 Changes in maritime trade, 1986–96

between the Medway and Southampton Water account for one-third of all unitised traffic, a share unchanged since 1986. Stability is in fact a feature of most ports in terms of their shares of total overseas trade during the last decade. Of the deep-sea ports concerned more with long-distance traffic, the largest increase has been at Liverpool, whose total tonnage has risen by 73 per cent since 1986.

The period since agreement to build the Channel Tunnel was reached has been marked by a vigorous programme of investment by port authorities and ferry operators in the English Channel corridor in order to compete with the new fixed link. Large ferries in excess of 20,000 tonnes have come into service between Dover and Calais and vehicle handling facilities have been upgraded at port terminals in order to reduce overall transit times. In 1996, two years after the Tunnel opened for traffic, 25.5 million passengers used the Channel ferries between Britain and France, with 18.8 million of these passing through Dover. Ferry ports along the Thames estuary and west to Weymouth now account for 80 per cent of all seaborne passenger traffic between Britain and European destinations, and within this group the flows are dominated by Dover (68 per cent), Portsmouth (11 per cent) and Ramsgate (10 per cent), the latter offering links with Dutch and Belgian ports.

The Channel Tunnel was opened in stages to passenger and freight traffic in 1994 and the annual traffic totals for 1995 and 1996 indicate that this addition to cross-Channel capacity has already had a significant impact upon the established traffic patterns. In 1996, 12.79 million passengers were carried on the shuttle service between the Folkestone and Calais terminals and on the through Eurostar express trains, compared with 19.69 million using the ferries on the short sea crossings. A total of 2.13 million passenger vehicles used the shuttle, whereas 3.15 million were conveyed on the ferries on the Straits of Dover routes and 1.29 million on other English Channel crossings to France.

The former British Rail undertaking and the French SNCF have negotiated a concession with Eurotunnel to operate up to thirty-five trains per day through the Tunnel, although this capacity is not always fully taken up. In 1996 freight trains carrying 2.33 million tonnes made use of the Tunnel, connecting freight terminals in Britain with those in

Europe. The freight shuttle trains carried 51,900 vehicles in 1996 but the Tunnel was closed in the later part of the year following a fire. Examples of traffic on the through trains includes cars from the Rover plant in Birmingham to northern Italy, Fiat vehicles from Turin to Avonmouth, and various Ford vehicle parts and sub-assemblies from Dagenham to destinations in Spain.

Insufficient data are available to date on the market served by the Tunnel to discover the extent to which traffic has been captured from surface modes and from air services between London and Paris, but the new fixed link has established itself as a strong competitor with the ferries, even if its financial viability gives cause for concern. There is at present excess capacity on the corridor between Dover and northern France, but if traffic using this route continues to rise at the rate displayed in recent years both the Tunnel and the ferries will continue to generate business.

Transport and the environment

It is inevitable that almost all new transport infrastructure projects will involve disruption in terms of visual intrusion on the landscape, an increase in noise levels and, in many cases, severance of communities and natural habitats through which a new road or railway has been designed to pass. In the United Kingdom and other European countries the late twentieth century has seen an increased recognition of these problems by governments and a widespread and strengthening opposition to the more controversial schemes by environmental protection groups.

The implications of the damaging impact that transport has upon the environment have been investigated by many different government and independent organisations. Royal Commissions have reported upon the major problems associated with consumption of energy by the major transport modes, and the atmospheric pollution caused by motor vehicles has also been the subject of several enquiries (Royal Commission 1994; Banister 1996). In the case of landscape impacts the public responses to proposed new road or airport runway building proposals have ranged from reasoned discussion in the media to the more vigorous demonstrations mounted by groups opposed to particular schemes such as the M3 Winchester bypass, the Newbury bypass, or the second runway at Manchester Airport.

The amount of disruption caused by new projects has to be measured against the perceived social and economic benefits of the proposed scheme; taking the example of an urban bypass, these benefits would include reductions in travelling time and accident rates and an improvement in the quality of life of the town from which through traffic is to be diverted. This process of measurement is the basis of the cost–benefit analysis used by the Department of Transport to compare the relative merits of different variants of a proposed road scheme. With a recognition and acceptance of the need to include environmental factors in this analysis, the technique of environment assessment has been adopted in Britain, following its initial application in the United States (Farington and Ryder 1993). This form of appraisal attempts to quantify a wide range of environmental and social effects of a road scheme, including land-take, visual intrusion, impacts upon urban and agricultural land uses, and heritage and ecological impacts, and then incorporates the results within the cost–benefit analysis which is made during the consultation stages of the scheme.

The spirited attempts to delay construction of the A34 Newbury bypass in Berkshire by environmental groups illustrate the diversity of opinions on the benefits that the new road is designed to yield when opened. Over 50,000 vehicles per day passed through Newbury, which is on the trunk Euroroute EO5 for traffic between the Midlands and the Channel ports. The £101 million bypass now diverts this traffic to the west of the town, producing estimated savings of 15 minutes for long-distance vehicles passing through the town at peak periods but savings of only 2 minutes at other times. Opposition to the bypass was based upon its damaging effects upon the Kennet valley and neighbouring sites of ecological and historical value, which were seen as outweighing the benefits that the road would confer on traffic, whereas much of the support came from residents and businesses within Newbury itself.

Securing an improvement in the quality of life by reducing the volume of motorised traffic in our society is now acknowledged to be a key environmental objective, and this aim is best achieved by reducing the level of vehicle emissions, pursuing existing plans to reduce traffic flows in town centres and providing, wherever possible, the necessary protection for areas of landscape value threatened by new transport projects.

Although the 1992 European Union transport policy document (EC Commission 1992) sees environmental issues as crucial to transport and mobility, the current situation is one of contrast in terms of efforts made by individual member states. The Netherlands and Denmark, together with certain cities in Germany and Switzerland, have devised realistic and effective means of reconciling transport with the environment, but other states have yet to implement such policies and the United Kingdom's record to date in this area cannot be regarded as wholly satisfactory.

Conclusions

The closing years of the twentieth century have seen several initiatives in the transport sector designed to increase the overall efficiency of the system, to achieve a greater use of public transport in towns and cities and to reduce the growing conflict between transport and the environment. Many of the policies of the 1990s were inherited from the previous decade, notably the deregulation and privatisation of bus undertakings, the progressive reduction in state subsidies for rail and bus operations, and the investment in light rail transit schemes for large cities. The most significant advance in the process of privatisation in the 1990s was the completion of the franchising programme for rail passenger and freight operators and the transfer of the railway and its supporting services to the private sector. In terms of infrastructure the most significant events were the opening of the Channel Tunnel and the completion of the major framework of the national motorway system, but a large part of the major road investment programme announced in the late 1980s was halted in early 1994 pending a review of the overall effectiveness of road building.

A major new direction in policy has been the encouragement of private investment in new road projects, although to date progress has been limited to the Birmingham Relief Road, now under construction. The White Paper of summer 1998 advocates a new transport policy based upon principles of integration and sustainable development. The last comparable document dates back to June 1977, when the declared objectives were to develop transport facilities to promote economic growth, to meet social needs and to take account of environmental protection (Department of Transport 1977). These aims have been met with varying degrees of success in the past two decades, but the new White Paper sets out a more positive approach to the transport industry. A prime objective is to consider all new transport investments in the context of sustainable development and to ensure the effective integration of the different modes. The existing situation of competition between and within modes is unlikely to receive much support in the future and a recurrent theme in recent government statements is the need to give public transport a much higher priority and to strengthen existing strategies designed to discourage the use of cars for trips that could be carried out by bus or rail. One of the most challenging areas for action is the management of demand for road space in terms of both future road-building policy and plans to restrain private motoring through increased taxes on fuel or higher vehicle licensing costs.

The wider issue of promoting regional development, especially in areas of economic decline, through the medium of improved transport facilites will also need to be addressed. The contributions that new motorways or electrified railways can make to industrial revival are not as significant as was once thought, and the effects of the Channel Tunnel, in terms of favouring the South East's economy at the expense of already declining regions in the periphery of Britain, are still being evaluated. Making use of the Tunnel to promote closer integration with European transport systems will depend upon to the extent to which the road and railway systems within Britain can be improved to meet the demands of domestic and international traffic.

References

Banister, D. (1996) 'Energy, quality of life and the environment: the role of transport', *Transport Reviews* 16(1): 23–35.

Bassett, K. (1993) 'British port privatization and its effect upon the port of Bristol', *Journal of Transport Geography* 4(1): 255–67.

Department of the Environment, Transport and the Regions (1998) *A New Deal for Transport: Better for Everyone*, London: HMSO.

Department of Transport (1963, 1965) *The Re-shaping of British Railways, Parts I and II* (Beeching Report), London: HMSO.

Department of Transport (1977) *Transport Policy*, London: HMSO.

Department of Transport (1989) *Roads for Prosperity*, London: HMSO.

Department of Transport (1993) *Paying for Better Roads*, London: HMSO.

Eaton, B.D. (1997) 'Passenger services under the Railways Act', *Proceedings Chartered Institute of Transport* 6(1): 14–28.

EC Commission (1992) *The Future Development of the Common Transport Policy*, Com 92 454, Brussels.

Farington, J. and Ryder, A. (1993) 'Environmental assessment of transport infrastructure', *Journal of Transport Geography* 2(1): 102–18.

Knowles, R.D. (1996) 'Transport impacts of Greater Manchester's Metrolink system', *Journal of Transport Geography* 4(1): 1–14.

Harnden, M. (1997) 'The rise and rise of the bus industry', *Global Transport* 9: 76–9.

Royal Commission on Environmental Pollution (1994) *Transport and the Environment*, London: HMSO.

Tolley, R.S. (1990) *Calming Traffic in Residential Areas*, Brefi: Coachex.

White, P. (1995) 'Deregulation of local bus services in Britain', *Transport Reviews* 15(2): 185–209.

Further reading

D. Banister and K. Button (eds) (1993) *Transport, the Environment and Sustainable Development* (London: Spon) contains comprehensive discussions of transport policy and environmental issues in Britain. R. Gibb (ed.) (1994) *The Channel Tunnel: a Geographical Perspective* (Chichester: Wiley) provides a detailed account of the the origins of the Channel Tunnel and its implications within Britain for tourism, freight and passenger traffic patterns, industrial location, the environment and regional economic development. B.S. Hoyle and R. D. Knowles (eds) (1998) *Modern Transport Geography*, 2nd edn (London: Belhaven) presents a worldwide perspective but contains numerous illustrations from the United Kingdom in the context of transport deregulation and privatisation, transport and the environment, and inter-urban transport. Problems of rural transport are also discussed. R. Tolley and B.J. Turton (1995) *Transport Systems, Policy and Planning: a Geographical Perspective.* (London: Longman) examines the transport system of the United Kingdom along with those of other industrial nations. Urban transport problems and remedies are discussed, as are the environmental and social implications of transport in the United Kingdom. *Transport Statistics, Great Britain* (annual publication) (London: HMSO) is accepted as the standard source for data relating to personal travel and road, rail, urban, air and maritime transport. P. White (1995) *Public Transport*, 3rd edn (London: UCL Press) offers a detailed account of all aspects of public transport in Britain, including inter-urban, urban and rural systems. Passenger transport policy issues are also examined, together with the organisation, control and technology of the principal modes. Various authors (1991) 'British passenger transport into the 1990s', *Geography* 77(334), Part 1, is a useful review of rail, bus, air and local urban transport developments.

Chapter 7

Manufacturing industry

David Sadler

Introduction

In the second edition of this collection, Bull (1991) highlighted the significance of the late 1970s and early 1980s as a period of major transformation for manufacturing industry in the United Kingdom. This chapter picks up on that story, focusing in particular on developments within the manufacturing sector in the UK since the mid-1980s and their regional implications. It looks first at the record of manufacturing activity during those years, then goes on to consider two significant aspects of recent debate concerning that performance. These are the extent and role of foreign direct investment (FDI), and changing patterns of work organisation and labour relations. It then explores the nature of national state policies towards the sector, concentrating on some of the consequences of those policies that have encouraged massive inward flows of foreign direct investment. The evolution of manufacturing within the UK is increasingly dependent upon such global corporate strategies. Finally, the conclusions look to the future and seek to consider the implications of events since the mid-1980s for manufacturing within the UK in the first decade of the next millennium.

The performance of manufacturing in the UK since the mid-1980s and its regional implications

It is impossible to understand the recent performance of manufacturing in the UK without brief reference to a longer-term picture. Employment in manufacturing in the post-1945 period peaked in 1966 at 8.5 million (although in relative terms, as a proportion of the UK workforce, it peaked in the mid-1950s at around 35 per cent). Since 1966 employment has declined almost relentlessly, particularly from 1979 to 1981 when over a million jobs were shed. Output, however, continued to grow until 1973 (so that from 1966 to 1973 there was a period of job-less growth) when it too began to decline, particularly from 1979 to 1981. In those two years output fell by 17 per cent and capital investment by 30 per cent; the million jobs lost represented a fall of 16 per cent. After 1981 output and investment both began to recover, whilst employment continued on a downward path, reaching 5 million by 1985. Thus the period after 1981 witnessed a second phase of job-less growth.

The immediate causes of the collapse in manufacturing in the UK from the late 1970s to the early 1980s included a government policy of high interest rates (in pursuit of the control of inflation and a reduced Public Sector Borrowing Requirement), and a relaxation of financial controls (which encouraged capital flight from the UK), in a context where a weak manufacturing base was already struggling to cope with a second deep international recession. The consequences were dramatic. In 1983 the UK recorded its first-ever balance-of-trade deficit in finished manufactured goods, and that indicator has remained in deficit to the present day. Output in real terms did not recover to pre-1979 levels until the late 1980s. Severe contraction in certain sectors of manufacturing, particularly those under state ownership such as coal and steel, was reflected in intense regional and sub-regional economic downturns, especially in the old industrial regions of central Scotland, north-east England and South Wales. With a concentration of growth in the early to mid-1980s in a broad arc stretching from the 'M4 corridor' between Bristol and London and round to East Anglia, concern was expressed over the resurgence of a North–South divide, evident in contrasting economic conditions. The national economy appeared to be in the midst of a major economic transformation, in which individual places and regions had sharply diverging profiles and prospects.

Around the mid- to late 1980s, there was considerable debate about the underlying reasons for the poor performance of manufacturing industry in the UK relative to other countries. Several competing explanations were put forward involving the links between manufacturing and other sectors of the economy, ranging from (what was seen as) an undue dominance of the financial services sector, to the nature of the relationship between finance capital and manufacturing. Some other contrasting explanations stressed labour productivity and poor management as key explanatory factors, while still others focused on the role of innovation and R&D, and the under-representation of small firms within the national economy. Evaluating the causes of the UK's manufacturing decline became a growth industry in its own right.

At the same time too, manufacturing in the UK – and the UK economy more generally – was in the midst of a recovery, albeit one which was hesitant at first. By the close of the 1980s an economic boom, partly engineered by national government policies (in a context of renewed international growth), had seen manufacturing recover its levels of output of ten years previously. The cycle peaked at the end of that decade, however, and a slump set in again in the early 1990s, with gradual recovery only apparent in the mid-1990s. This picture – of boom, bust and recovery, in a setting of long-term decline – underpins the performance of manufacturing in the UK since the mid-1980s.

In that period employment in manufacturing initially continued a steady decline, falling from 5 million in 1985 to 3.9 million in 1993, although it began to reverse that long-standing trend from 1994 onwards, growing to 4.1 million by 1997 (Table 7.1 and Figure 7.1). Thus absolute employment in manufacturing contracted during the boom years of the late 1980s – continuing the phase of job-less growth – and only began a recovery with the conditions of the mid-1990s. In relative terms however, manufacturing employment

TABLE 7.1 Employment, output and investment in manufacturing in the UK, 1985–97

	Employment ('000s)	*Percentage of total*	*Output (£bn)*	*Percentage of total*	*Net capital investment (£m)*
1985	5,002	23.4	77.8	25.3	8,742
1986	4,881	22.8	81.3	24.8	8,705
1987	4,815	22.3	88.6	24.6	9,754
1988	4,858	21.8	98.8	24.6	12,168
1989	4,851	21.4	107.2	24.3	14,499
1990	4,733	20.7	111.3	23.2	14,309
1991	4,319	19.4	106.9	21.5	13,100
1992	4,096	18.7	109.8	21.2	12,094
1993	3,913	18.1	115.7	21.1	12,583
1994	3,928	18.1	123.9	21.4	13,724
1995	4,026	18.3	131.7	21.7	
1996	4,067	18.2	137.0	21.3	
1997	4,111	18.0			

Sources: Calculated from *Labour Market Trends; UK National Accounts 1997* (The Blue Book).
Note: 1992 Standard Industrial Classification (SIC)

contracted continually, from 23 per cent of the UK's workforce in 1985 to 18 per cent by 1997.

Manufacturing output grew strongly during the late 1980s boom, reaching £111 billion at current costs in 1990 before falling back in 1991 and 1992. When standardised to 1990 costs though, output peaked in 1989 and it was not until 1994 that it regained such a level (Figure 7.1). On this basis the extent of the growth in output in the mid- to late 1980s is clearly apparent, although manufacturing continued to account for a declining share of national output (falling steadily from 25 per cent in 1985 to 21 per cent in 1996). An even more marked picture was apparent for investment, which grew exceptionally strongly from 1986 to 1989, only to fall back just as dramatically from 1990 to 1992.

These changes had an impact on the UK's balance of trade in finished manufactured goods. One distinctive aspect of the late 1980s boom in the UK economy was the extent to which it encouraged increased volumes of imports of manufactures, reaching over £60 billion in 1989 and again in 1990 (Figure 7.2). This surge of imports had a devastating effect on the UK's trade deficit, which grew markedly to reach £11 billion in 1988 and £13 billion in 1989 (Figure 7.3). The trade deficit in manufactured goods fell back equally sharply during the early 1990s recession (to £2 billion in 1991) but thereafter expanded once again, to a range of £6–9 billion during the mid-1990s.

The national changes in employment, output and investment also took on distinctive regional profiles.[1] *Male manufacturing employment* as a proportion of the regional total (Table 7.2) remained at its highest in 1995 in the West Midlands, even though it had fallen by then to 38 per cent from a level of 50 per cent in 1981. Other regions with a high proportion in 1995 included the East Midlands (35 per cent) and the North East, Yorkshire and the Humber, the North West and Merseyside, and Wales, which were all in the range of 30 to 31 per cent – well above the UK average of 26 per cent. Of these, all except Wales had seen

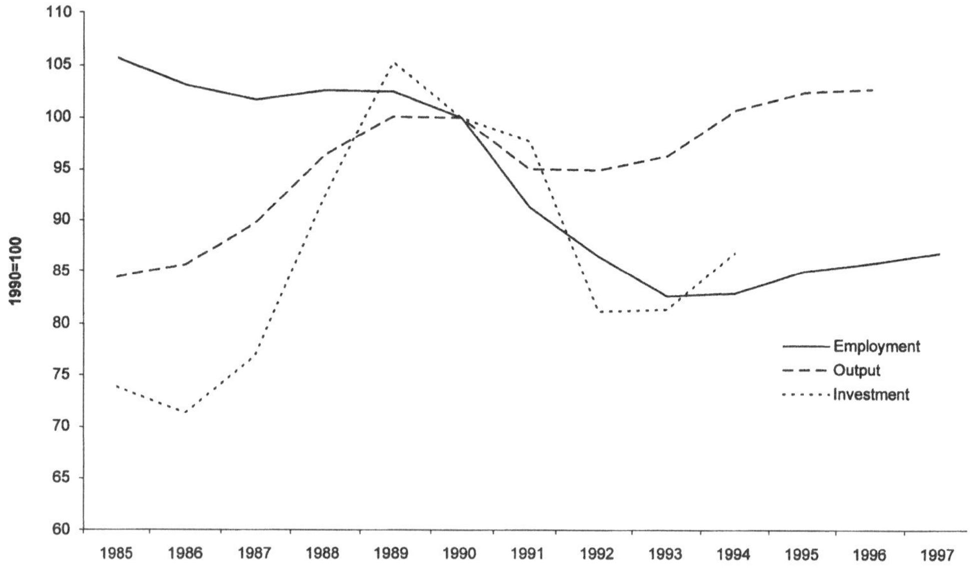

FIGURE 7.1 Employment, output and investment in manufacturing in the UK, 1985–97 (at constant 1990 costs)

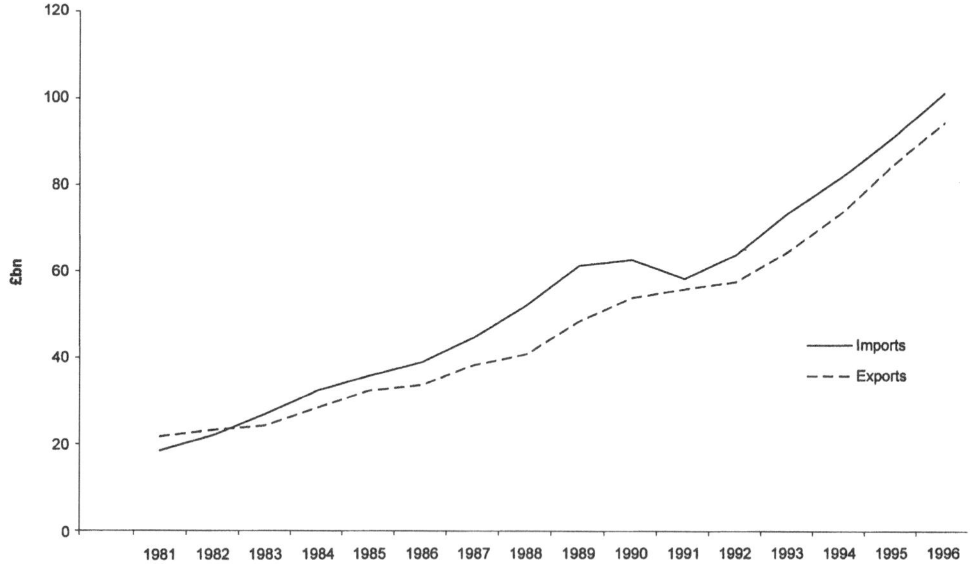

FIGURE 7.2 UK imports and exports of finished manufactured goods, 1981–96

FIGURE 7.3 Balance of trade in finished manufactured goods, 1981–96

a fall of between seven and eleven percentage points since 1981; by stark contrast, the relative proportion in Wales remained virtually unchanged during this period. Such was the extent of the long-term collapse of the manufacturing base in Scotland and Northern Ireland that throughout this period they had proportions below the UK average. The distinctive economic structure of the broader south-east of England was reflected in very low proportions of male manufacturing employment: a bare 11 per cent in London and 21 per cent in the South East region – well below the UK average.

Female manufacturing employment as a proportion of the regional total was also highest in 1995 in the West Midlands and the East Midlands, although in this case the East Midlands had the very highest level of 17 per cent, well above the UK average of 11 per cent (Table 7.2). Regions clearly above the mean included the North East, Yorkshire and the Humber, the North West and Merseyside, and Wales, which were all in the range of 12 to 13 per cent. The most severe falls (of between ten and twelve percentage points) since 1981 had taken place in Yorkshire and the Humber, the East Midlands, the West Midlands, the North West and Merseyside, and Scotland. Wales had seen only a limited fall of four percentage points, and that took place in the period after 1990. There were four regions below the UK average: London, the South East, the South West, and Scotland.

Average weekly male earnings in manufacturing also varied significantly from region to region (Table 7.3). Only three regions were above the UK average in 1996 – London, the South East and Eastern; all others were within eight percentage points of the mean with the exception of Northern Ireland, where earnings were just 79 per cent of the UK average. Several of these regions had converged slightly towards the mean since 1991, with gains of between one and two percentage points in Yorkshire and the Humber, Wales, and the North East; Scotland had seen a gain of four percentage points. The West Midlands by contrast had

TABLE 7.2 Percentage of employees in manufacturing, by region, 1981, 1990 and 1995

	1981	*1990*		*1995*
Males				
North	38.5	34.3	North East	30.8
Yorkshire and Humberside	37.9	35.1	Yorkshire and the Humber	30.6
East Midlands	41.0	39.1	East Midlands	35.1
East Anglia	32.7	29.4	Eastern	25.7
			London	10.7
South East	29.0	21.3	South East	20.9
South West	33.6	30.1	South West	25.3
West Midlands	49.8	42.0	West Midlands	38.1
North West	41.4	36.7	North West and Merseyside	30.5
Wales	32.1	32.8	Wales	31.6
Scotland	31.6	27.5	Scotland	22.3
Northern Ireland	28.8	24.2	Northern Ireland	24.2
UK	35.2	29.8	UK	25.7
Females				
North	18.5	15.5	North East	12.8
Yorkshire and Humberside	22.3	16.8	Yorkshire and the Humber	12.1
East Midlands	29.6	22.3	East Midlands	17.2
East Anglia	19.5	14.2	Eastern	10.9
			London	6.1
South East	15.6	10.6	South East	8.4
South West	15.3	11.9	South West	9.3
West Midlands	25.8	18.5	West Midlands	14.7
North West	21.8	17.4	North West and Merseyside	12.3
Wales	16.3	16.0	Wales	12.4
Scotland	17.7	14.4	Scotland	9.7
Northern Ireland	18.4	15.1	Northern Ireland	11.9
UK	19.2	14.5	UK	10.9

Sources: *Regional Trends* 26, Table 10.7; *Regional Trends* 32, Table 5.8.
Note: 1981 and 1990 data, 1980 SIC; 1995 data, 1992 SIC.

slipped slightly further away from the mean. A pattern of even wider inequality was evident in terms of *average weekly female earnings* in manufacturing (Table 7.3). The highest rate, for London at 148 per cent of the UK average, was practically double that for Northern Ireland at 80 per cent, and two other regions (the East Midlands, and Yorkshire and the Humber) were below 90 per cent. Most regions, however, had improved their position relative to the mean since 1991, particularly Scotland which had seen a gain of four percentage points. The notable exception to this trend was the East Midlands, which had seen a fall of over two percentage points.[2]

Manufacturing output as a proportion of regional Gross Domestic Product in 1995 ranged from 20 per cent in Northern Ireland to 31 per cent in the West Midlands (Table

TABLE 7.3 Average weekly earnings in manufacturing by region, 1991 and 1996

	1991		*1996*
Males			
North	95.0	North East	97.2
Yorkshire and Humberside	91.6	Yorkshire and the Humber	92.9
East Midlands	93.1	East Midlands	93.7
East Anglia	98.7	Eastern	105.8
		London	133.6
South East	115.1	South East	112.4
South West	96.7	South West	96.8
West Midlands	93.0	West Midlands	92.1
North West	98.1	North West and Merseyside	98.4
Wales	93.2	Wales	95.1
Scotland	95.2	Scotland	99.4
Northern Ireland	79.5	Northern Ireland	79.4
UK (£/week)	308.1	UK (£/week)	378.1
Females			
North	90.7	North East	93.1
Yorkshire and Humberside	88.0	Yorkshire and the Humber	89.1
East Midlands	88.8	East Midlands	86.6
East Anglia	99.2	Eastern	107.9
		London	147.6
South East	121.5	South East	112.8
South West	97.0	South West	99.5
West Midlands	90.5	West Midlands	91.8
North West	93.1	North West and Merseyside	96.2
Wales	91.6	Wales	93.0
Scotland	89.9	Scotland	93.8
Northern Ireland	78.4	Northern Ireland	79.6
UK (£/week)	192.9	UK (£/week)	245.1

Sources: Calculated from *Regional Trends* 27, Table 8.5; *Regional Trends* 32, Table 5.16.
Note: Expressed as a percentage of the UK average (1996) and GB average (1991); 1980 SIC for 1991, 1992 SIC for 1996.

7.4). It was well above the UK mean of 22 per cent in the North, Yorkshire and Humberside, the East Midlands, the North West, and Wales (which were all in the range from 26 to 30 per cent), and markedly below in the South East at 16 per cent. Both Scotland and Northern Ireland – traditionally regarded as industrial regions – had below-average proportions of output derived from the manufacturing sector. Put another way, manufacturing was proportionately most significant to the output of only part of the classic industrial heartland of the UK – the North of England, Wales and the West Midlands (but excluding Scotland and Northern Ireland) along with an area less frequently seen in this light, the East Midlands. The relative position of each region had remained quite constant since 1987,

TABLE 7.4 Manufacturing as a percentage of regional Gross Domestic Product, 1987, 1991 and 1995

	1987	*1991*	*1995*
North	28.7	28.0	30.4
Yorkshire and Humberside	27.4	26.0	25.7
East Midlands	31.2	27.7	29.7
East Anglia	23.3	21.2	22.7
South East	18.5	15.7	15.8
South West	22.0	19.4	19.3
West Midlands	33.1	29.3	31.3
North West	30.3	28.3	27.4
Wales	27.3	27.9	28.2
Scotland	23.1	20.2	20.1
Northern Ireland	17.9	20.1	19.5
UK	24.2	21.8	22.2

Sources: Calculated from *Regional Trends* 27, Table 12.2; *Regional Trends* 32, Table 12.4.
Note: 1980 SIC for 1987; 1992 SIC for 1991 and 1995.

although notable changes included an increased share of output derived from manufacturing in the North and in Northern Ireland (both up by two percentage points, against the national trend) and an above-average decline (around three percentage points) in the South East, South West, North West and Merseyside, and Scotland.

Net capital expenditure in manufacturing – expressed as a percentage of the UK total – is intrinsically more variable from year to year than output and employment, so that it is presented here in the form of an average over the period from 1985 to 1993 (Table 7.5). This shows that despite the South East's relatively low dependence upon manufacturing industry, so great is the weight of this region within the UK economy that it accounted for almost a quarter of all investment in manufacturing, well above its nearest rivals the North West and Merseyside (13 per cent), the West Midlands (11 per cent) and Yorkshire and the Humber (10 per cent).

On this basis too – expressed as a proportion of the UK total rather than of individual regional totals – the distributions of manufacturing *employment* and manufacturing *output* also largely reflect the differing weights of each regional economy (Table 7.6). The South East accounted for a quarter of manufacturing output in 1995, well above the West Midlands and the North West with 12 per cent each, whilst the small size of the manufacturing base in Northern Ireland and East Anglia was particularly evident, with these regions accounting for just 2 and 4 per cent of the UK total respectively. The North had a slightly higher share of output than of employment (even after allowing for some discrepancy introduced by differing regional bases for data collection), whilst Yorkshire and Humberside and the West Midlands had a higher share of employment than of output.

These, then, were the broad regional dimensions of UK manufacturing employment, output and investment in the period since the mid-1980s; years which saw continued relative decline in the sector, particularly in the early 1990s. That recession sharpened earlier

TABLE 7.5 Net capital expenditure in manufacturing by region as a percentage of UK total, 1985–94

	1985	*1986*	*1987*	*1988*	*1989*	*1990*	*1991*	*1992*	*1993*	*1985–93*		*1991*	*1994*
North	7.2	6.7	6.2	6.8	6.3	6.5	7.5	6.9	7.7	**6.9**	North East	6.3	4.5
Yorkshire and Humberside	7.8	9.8	11.1	10.2	9.2	9.2	9.3	10.3	10.0	**9.7**	Yorkshire and the Humber	9.3	9.3
East Midlands	7.8	7.8	7.6	7.2	6.9	7.3	9.2	10.6	9.0	**8.2**	East Midlands	9.2	7.7
East Anglia	3.5	3.4	3.4	3.3	3.4	3.2	2.7	3.5	3.6	**3.3**	Eastern	7.3	8.7
											London	10.0	6.9
South East	25.5	27.0	25.5	24.8	24.4	22.3	24.4	21.1	23.2	**24.2**	South East	9.9	12.6
South West	6.9	6.8	6.4	6.9	6.0	5.8	5.4	6.2	6.7	**6.3**	South West	5.5	6.8
West Midlands	10.2	9.9	11.0	10.8	11.3	10.6	10.0	10.2	10.5	**10.5**	West Midlands	10.0	12.1
North West	12.5	11.4	12.3	12.1	14.3	15.7	13.3	13.7	12.5	**13.1**	North West and Merseyside	14.5	14.3
Wales	6.4	6.0	6.5	7.0	7.7	8.0	6.4	6.0	5.6	**6.6**	Wales	6.4	5.8
Scotland	9.9	8.9	7.5	8.4	8.5	9.0	9.2	8.9	8.7	**8.8**	Scotland	9.2	9.0
Northern Ireland	2.3	2.1	2.4	2.3	2.1	2.3	2.6	2.6	2.5	**2.4**	Northern Ireland	2.6	2.3
UK (£m)	8,742	8,705	9,754	12,168	14,499	14,309	13,100	12,094	12,583		UK (£m)	13,100	13,723

Source: *Regional Trends*, various dates.
Note: 1980 SIC (SSRs, 1985–92); 1992 SIC (SSRs, 1993 and GORs, 1991 and 1994).

TABLE 7.6 Employment, output and investment in manufacturing by region as a percentage of UK total, 1995

	Employment	*Investment*		*Output*
North East	4.6	4.5	North	6.4
Yorkshire and the Humber	11.9	9.3	Yorkshire and Humberside	9.0
East Midlands	9.8	7.7	East Midlands	9.2
Eastern	8.5	8.7	East Anglia	3.8
London	6.5	6.9		
South East	10.1	12.6	South East	25.4
South West	7.3	6.8	South West	6.8
West Midlands	13.1	12.1	West Midlands	12.2
North West and Merseyside	12.9	14.3	North West	12.2
Wales	5.0	5.8	Wales	5.3
Scotland	7.7	9.0	Scotland	7.7
Northern Ireland	2.5	2.3	Northern Ireland	2.1

Source: *Regional Trends* 32, Tables 5.8, 12.4 and 13.4.
Note: 1992 SIC. Investment data for 1994.

concerns about the state of manufacturing in the UK and led to a full-scale investigation by a House of Commons Select Committee into the competitiveness of UK manufacturing (Trade and Industry Committee 1994). This report argued strongly that the declining relative significance of manufacturing to the UK economy – which had been more severe than in practically all other developed economies – should be a cause for concern. This was so not least because the sector accounted for over 60 per cent of UK exports (making the country the world's fifth largest exporter of manufactured goods), and a significant proportion of the service industries depended to some extent on manufacturing. The prospect of renewed extensive growth in *employment* in manufacturing was, however, discounted:

> Given the continuing rise in productivity, the jobs lost in manufacturing in developed countries in recent years are unlikely ever to be fully restored, even if output rises. Consequently we believe that growth in the UK's manufacturing sector should be viewed primarily as a way of increasing the creation of national wealth and thus of jobs elsewhere in the economy *rather than of creating jobs in manufacturing itself.*
>
> (Trade and Industry Committee 1994: 15; emphasis added)

The report went on to stress that there were, none the less, prospects for renewed growth in *output*, partly resulting from a short-term advantage of low labour costs but in the longer-term dependent upon continued good industrial relations, new management practices, and the UK's attractiveness as a location for foreign direct investment in manufacturing (themes which are expanded upon in the following sections). In addition, it highlighted two areas in which the UK possessed specific weaknesses that needed to be addressed: the size distribution of manufacturing firms, and the nature of the links between manufacturing and finance capital.

Whilst the UK was seen as being particularly strong in large international manufacturing companies – with forty-three of the world's 500 largest such firms in terms of turnover based there, against thirty-three in Germany and thirty-two in France – it was described as 'exceptionally weak' (p. 28) in medium-sized firms (those employing from 100 to 500 people), comparable to Germany's *Mittelstand*. Such firms accounted for just 16 per cent of manufacturing employment in the UK in 1990, against 28 per cent elsewhere in Western Europe. The UK also had an under-representation of smaller-sized enterprises, with firms of from 20–100 employees accounting for just 15 per cent of manufacturing employment against 23 per cent in the rest of Western Europe.

The UK's pattern of ownership of larger firms was also addressed. Pension funds alone held 35 per cent of the value of the shares in UK quoted companies in 1992; together with insurance companies, that proportion increased to 60 per cent. This was in contrast to the USA, where private shareholdings were more significant, and to Germany, where banks were more influential. In both Germany and Japan, large firms tended to have a small number of dominant shareholders, whilst in the USA and the UK, individual shareholdings were much smaller. Thus in the UK, large manufacturing firm ownership structures were distinctively characterised (amongst large industrial economies) by a large number of small holdings in the hands of pension funds and insurance companies, which collectively accounted for a significant part of the total value of these manufacturing companies. This, it was argued, had led to a degree of 'short-termism' (the favouring of short-term profits which could be distributed in the form of dividends to shareholders) at the expense of longer-run strategies. This had come about not necessarily because of the role of the financial institutions, but due to the nature of the relationship between them and the manufacturing sector.

Such concerns – about the degree of under-investment in UK manufacturing, and the competitive position of UK-based small- and medium-sized manufacturing firms – were increasingly regarded as significant policy issues in the course of the 1990s. This was in contrast to the previous decade when government policy had seemed to prioritise the service industries, particularly financial services. At the same time it was very clear that the prospects of manufacturing in the UK now depended heavily upon the UK's continued attractiveness as a location for inward investment. The following section goes on to explore the debate over the broader impact of such inward investment.

The extent and role of foreign direct investment in manufacturing

This section has two objectives: to describe some of the main regional impacts of flows of FDI in manufacturing within the UK, and to consider the debate surrounding such inward investment flows. The UK has been remarkably successful in attracting FDI. Over the period since 1951 it has received two-fifths of all Japanese and one-third of all US inward investment to the European Union. The absolute significance of FDI to the UK's manufacturing sector is widely attested. By the mid-1990s, foreign-owned manufacturing firms accounted for 25 per cent of the output of manufacturing industry in the UK and 20 per cent of manufacturing employment, whilst 30 per cent of the country's manufacturing exports consisted of trade *within* international firms. Investment from overseas has led to dramatic changes in the structure of many sectors, notably the automotive industry (Box 7.1).

BOX 7.1 Case-study: the automotive industry

The fortunes of the automotive industry in the UK epitomise many of the changes that have taken place within manufacturing since the depths of the early 1980s. Production of passenger cars collapsed from 1.9 million in 1972 to a low of 0.9 million in 1982, and hundreds of thousands of jobs were shed, notably in the West Midlands. Thereafter, output began a recovery, reaching 1.7 million in 1997, with growth forecast to continue to around 2 million by the early 2000s (although – in common with the rest of the manufacturing sector – employment continued a steady decline). In part this expansion came about because the UK was selected by three of the leading Japanese automotive firms for their major inward investment sites in Europe. Nissan began producing cars at Sunderland from 1986 onwards, whilst Toyota and Honda commenced production at Burnaston in Derbyshire and Swindon respectively in 1992. By the mid-1990s these firms had developed substantial manufacturing operations (Nissan had invested over £1.5 billion, for example) and they were manufacturing 0.5 million cars a year between them. The vast majority of these were exported, contributing substantially to the UK's trade balance in manufactured goods. Nissan and Honda had both announced plans to add a third model range, whilst Toyota had chosen instead to develop an additional plant for its third model to be produced in Europe (at Lens in northern France). Additionally, Rover – bought by BMW in 1994 from its previous owner British Aerospace – was in the midst of a £3 billion investment programme to the year 2000, whilst both Ford and General Motors (Vauxhall) had committed fresh investment for the production of new models in the UK.

The significance of the investment from Japan in this sector went well beyond the jobs created directly (amounting to no more than 12,000). All three firms sought rapidly to reach a minimum of 80 per cent local content (meaning sourced from within Europe) in the cars produced in their new factories, so that they could be sold as European-produced rather than Japanese vehicles (thereby circumventing quotas on imports from Japan). This brought Nissan, Toyota and Honda into close contact with existing UK component suppliers. Considerable improvements in the performance of these firms was both demanded and actively encouraged. Whilst they remained behind the standards of Japanese suppliers in Japan, firms in the UK in particular (and in Europe more generally) did succeed in closing the gap and securing a continued stream of business. This not only ensured that jobs were created elsewhere in the economy, it also further enhanced the UK's attractiveness as a location for investment in the automotive sector.

The proportion of value-added accounted for through FDI was at its highest in Scotland and Northern Ireland (both at 35 per cent), Wales, the North East, and the South East (all at 30 per cent), and at its lowest in the South West and East Midlands (16 per cent) and Yorkshire and the Humber (17 per cent). The jobs created through inward investment have also been unevenly distributed. From 1982 to 1992 over 50 per cent of the jobs arising from FDI were located in Scotland, Wales and the North, compared to only 16 per cent in southern England, whereas the respective shares of manufacturing employment in 1992 were 18 per cent and 35 per cent (Trade and Industry Committee 1995: para 43). This is partly in response to the range of incentives available in these areas from central government (an issue which is considered in a subsequent section). Thus inward investment in manufacturing has not only underpinned the very existence of the sector in the UK, but also contributed in some measure to a dampening of the regional inequalities described in the previous section.[3]

Some of the wider impacts of FDI – in particular the extent to which different waves of investment have succeeded one another over time – can be illustrated with respect to the case of the North of England (see Stone 1995). In 1978 there were 49,000 jobs in foreign-owned manufacturing establishments in the North, representing 12 per cent of all employment in manufacturing in the region at that date. By 1993 there were 56,000 such jobs: at first sight, only a small absolute increase, even if collapse of UK-owned firms meant that foreign-owned firms now accounted for 23 per cent of all manufacturing employment there (a position comparable to Scotland, but below the 33 per cent recorded in Wales). This slight net gain however represented the sum total of two very different processes: *disinvestment* by an earlier wave of American-owned firms (leading to a net loss of 16,000 jobs) and *investment* by firms from Japan (leading to 9,000 new jobs, mainly at new greenfield plants) and from the rest of Europe (a net gain of 13,000 jobs, mainly due to the acquisition of existing businesses). Thus it was very evident that jobs created through FDI were by no means a permanent feature on the economic landscape.

More generally, debate on the broader significance of FDI for regional and national growth prospects intensified in the UK during the 1990s. The House of Commons Select Committee enquiry into the competitiveness of manufacturing industry was quietly cautious in its appraisal:

> Inward investment offers considerable opportunities to the UK, but is unlikely to involve the most up-to-date technology (other than production technology) and is insufficient in scale to offset weaknesses in UK-owned manufacturing. It should therefore be seen as a contribution towards a competitive UK manufacturing sector but not as an alternative to improving the competitiveness of the sector as a whole.
>
> (Trade and Industry Committee 1994: 32)

Such caution reflected several aspects of emergent concern, including the extent to which at least some inward investors had transferred only the simplest of production processes to the UK. Some critics argued that the country's key attraction lay purely in low labour costs and that the production facilities being established there represented little more than a final assembly point (sometimes described as a 'screwdriver plant' in recognition of the limited extent of value-added locally) or a warehouse for components which were still largely manufactured abroad. It was also argued that branch plants established in the peripheral regions such as South Wales and north-east England were merely reproducing earlier rounds of dependency, particularly in the extent to which these plants possessed only very limited R&D capabilities.

Partly in response to these concerns, attention increasingly focused on a search for 'quality' or 'performance' plants. These were held to be sites at which inward investors had developed high levels of local procurement, sophisticated recruitment and training practices, and a good level of integration or 'embeddedness' with local public and private development organisations. Such ideas were expressed in a subsequent report from the Trade and Industry Committee (1995: para 45). This recommended that getting the best from inward investment involved developing a UK supply chain, ensuring that UK-owned firms learned from the new working practices being adopted, targeting those firms likely to be of greatest benefit to the UK (although in practice this was somewhat idealistic), and ensuring that reinvestment by existing foreign-owned firms came to the UK.

New management styles and working practices

Many reasons have been advanced to account for the extent to which inward investors, particularly those from Japan, favoured the UK as a base for inward investment in Europe during the 1980s and 1990s. Of these, one of the most significant factors was the extent to which the country welcomed the introduction of new working practices and management styles such as Just-in-Time production, quality circles and the use of team-working. In part, this openness to change rested on and helped to reinforce a government policy to reduce the power of trade unions, one that was never more clearly expressed than in the course of the 1984/5 strike in the coal-mining industry. One measure of the extent of the changes that took place with respect to the broad climate of industrial relations was that by the mid-1990s governments could point to *good* management–employee relations as underpinning the prospects of the manufacturing sector, in stark contrast to the problems which had been diagnosed just a decade previously. This is not to suggest that all the changes which took place in terms of production organisation and industrial relations during this period were due to the wave of FDI from Japan; but there is no doubt that this inward investment (and its success) had a very important demonstration effect.

The new industrial relations framework was described by the Trade and Industry Committee (1994) as one of 'empowerment', built on dialogue and trust. The report went on to argue that:

> 'Empowerment' does not remove the role of trade unions . . . Nonetheless the role of trade unions is changing significantly as direct communication between management and employees increases . . . We endorse the view that successful manufacturing companies are likely to be those which have constructive relationships both with their workforces and with the trade unions representing them, and that trade unions can contribute to the successful implementation of change within companies.
>
> (Trade and Industry Committee 1994: 89)

This rosy picture of a harmonious alliance between capital and labour had its detractors, some of whom argued that in essence not much had changed: that the 'new' management style rested on little more than high unemployment and the enforced disciplining of wage labour which that entailed.

These processes of change *within* the workplace were deeply significant. One frequently cited example was the agreement between Nissan Motor Manufacturing (UK) Ltd and the Amalgamated Engineering Union (AEU) made in 1985. This incorporated several features which were new (as a package) to the automotive industry in the UK, including single-union recognition, a company council elected from all employees, pendulum arbitration, complete task flexibility, and common conditions. Not all FDI projects recognised trade unions – in the electronics industry in particular, non-union plants were commonplace – but those that did ensured that the union's role was tightly delimited and that other channels of management–employee communication were put in place. In this respect – as in other fields, such as component purchasing and JIT buyer–supplier arrangements – the new plants arising from inward investment looked less like transplants and more like hybrids, adopting variants of best practice from the source country along with experience from other European countries (especially Germany), all tailored to the UK setting.

The broader context for this process of transformation was an increasingly globalised economy, and there was a lively debate in this period over the appropriate characterisation of the changes that were taking place. Some argued there had been a transition from an era of Fordist mass production to one of post-Fordism; others stressed the continued vitality and adaptability of mass production. Some emphasised a shift towards a world of flexible specialisation in which big firms no longer dominated and economies of scope had replaced economies of scale; others re-affirmed the significance of mass markets. It is, however, well beyond the scope of this present chapter to review these debates in detail. Instead the following section considers the evolution of state policies towards manufacturing in the UK.

The evolution of state policies towards the manufacturing sector

This section aims to explore two significant aspects in the evolution of UK government policies towards the manufacturing sector. These are the pattern of expenditure of national and European regional policies, and the intensification of regional rivalries in the pursuit of FDI. Whilst all forms of public expenditure (and other government policies including privatisation) have potentially uneven geographical implications, regional policies are of particular significance in that they are expressly targeted at geographical outcomes and have tended to focus upon manufacturing rather than service sector firms. There have been major changes in the longer term to the system of regional policy in the UK. These include the abolition of location controls through Industrial Development Certificates in 1982, after they had fallen into disuse in the 1970s, and the introduction of Regional Selective Assistance (RSA) in 1972 to supplement automatic grants on investment in the form of Regional Development Grant payments (RDG). Automatic grants were abolished in 1988 leaving RSA as the main instrument of regional industrial policy. There have also been many changes over time in the map of areas where firms are potentially eligible to receive regional policy assistance, notably those in 1979, 1984 and 1988 which drastically curtailed the extent of the Assisted Areas, although a subsequent revision in 1993 added several districts in southern England which had been particularly badly affected by the early 1990s recession.

Regional policy expenditure was cut back in the 1980s, along with many other components of public expenditure, and preferential assistance was increasingly targeted instead on urban measures. In 1993 this culminated in the creation of the Single Regeneration Budget (SRB), which was led by the Department of the Environment and brought together twenty separate programmes formerly administered by five different government Departments, including the Urban Programme and the Urban Development Corporations (see Chapter 10). Its total funding of £1.6 billion in 1993/4 dwarfed expenditure on regional preferential assistance in that year of £0.3 billion, and the transfer to the SRB of responsibility for Regional Enterprise Grants and English Estates (subsequently absorbed within English Partnerships) represented a significant reduction of the Department for Trade and Industry's regional policy role. The subsequent creation of Government Offices for the Regions in 1994 brought together what were then the Departments of Employment, Environment, Trade and Industry, and Transport. The Offices for the Regions were given a remit of carrying out the regional activities of these Departments, which included the administration of RSA and the newly created SRB. These changes took place at a time when there was mounting concern over the effectiveness of regional policy expenditure, leading to a review by the Trade and Industry Committee (1995). This report concluded that there was still a strong case for the continuation of regional policy,[4] particularly in order to improve the competitiveness of individual regional economies, although there were several ways in which its delivery could be improved. Not least among these was a need for more regional co-ordination, an issue considered in more detail later in this section.

Despite continued contraction since the mid-1980s, government expenditure on regional preferential assistance in Britain from 1988 to 1996 still amounted to £3.6 billion, with an additional £1 billion spent under a different system in Northern Ireland (Table 7.7). This expenditure was very heavily concentrated in a small number of regions. Wales and Scotland both received just over £1 billion, or almost 30 per cent of the total each, followed by £600 million in the North East (17 per cent) and £400 million in the North West (11 per cent). Expenditure from the EU's Structural Funds within the UK had by contrast grown markedly since 1989, with £3.4 billion at 1995 prices allocated in the programming period from 1989 to 1993 (Trade and Industry Committee 1995: para 18). A further £9 billion was committed from 1994 to 1999, including (under Objective I) £830 million for Northern

TABLE 7.7 Government expenditure on preferential assistance to industry by DTI region, 1988–96

	£m	*Percentage of GB total*
North East	585.7	16.5
North West	397.4	11.2
Yorkshire and Humberside	222.2	6.3
East Midlands	42.0	1.2
West Midlands	126.9	3.6
East	2.8	
South East	7.4	0.2
South West	77.5	2.2
Wales	1,034.1	29.1
Scotland	1,054.4	29.7
GB total	3,550.4	100
Northern Ireland	1,022.8	

Source: Calculated from *Regional Trends* 32, Table 13.7.
Note: Different system of assistance in Northern Ireland.

Ireland, £550 million for Merseyside and £210 million for the Highlands and Islands of Scotland. Main recipients under Objective II included the West Midlands (£650 million), Manchester, Lancashire and Cheshire (£580 million), Yorkshire and Humberside (£540 million), the North East (£530 million), West Scotland (£470 million) and South Wales (£320 million).

EU Structural Funds differed from national regional policy in that the latter was targeted at individual firms, whilst the former was delivered through a series of area-based or thematic regeneration programmes. Much of the expenditure on national regional policy thus represented a subsidy to inward investors, one which was offered as part of the bidding process for FDI between competing potential locations. In the mid-1990s, a succession of major success stories saw the UK seemingly re-confirm its position as Europe's prime inward investment location (Box 7.2). At the same time, however, the increasingly intense competition between locations within the UK led to concern that a degree of 'bidding-up' might be taking place, in which one UK location sought to outbid another for the investment and jobs on offer, resulting in excessive public funds being disbursed. The whole process by which inward investment enquiries were processed and co-ordinated was explored in a report from the Trade and Industry Committee (1997). This concluded somewhat equivocally that in a relatively small number of cases – mainly larger projects – the extent of public assistance on offer could play a crucial, even determining role in a company's choice of location, but that there was 'little substantial evidence' that bidding-up was taking place between locations in the UK. Equally, it found claims that Welsh and Scottish agencies were attempting to 'poach' existing firms by encouraging them to relocate there from sites in England to be 'not very persuasive'.

This issue of co-ordination was of particular significance in the context of the newly elected Labour government's commitments to devolution in Scotland and Wales, and to the creation of a network of Regional Development Agencies in nine of the English regions by 1999 (the exception was Merseyside, which was to be subsumed within the North West). The Treasury wished inward investment to be controlled by one central body, the Department of Trade and Industry's Invest in Britain Bureau, but the Welsh and Scottish Offices had already successfully resisted such a move in the last days of the Conservative government. The role of the English RDAs was also the subject of political debate within the Labour government, with the new Department of the Environment, Transport and the Regions keen to take over co-ordination through its control of the RDAs. In the event, however, the Department of Trade and Industry secured continued control in the co-ordination of assistance to FDI within England, although assistance to inward investors would be delivered through the network of RDAs.

Conclusion

This chapter first explored the performance of manufacturing industry since the mid-1980s, concentrating on its regional implications. It went on to examine the extent and role of foreign direct investment, identifying its uneven geographical impacts. The introduction of new management styles and working practices within manufacturing was also chronicled, in the context of changing trade union roles and management–employee relationships. This was followed by a consideration of two aspects of state policy towards manufacturing:

BOX 7.2 The competition for large inward investment projects

It is important not to overstate the significance of large inward investment projects. The bulk of public assistance to inward investors is allocated to smaller projects and to reinvestment by existing firms. On the other hand the competition for large projects has become increasingly intense and public, and in such cases it is possible that the amount of financial assistance made available by one location as opposed to another could be decisive. In the mid-1990s the UK secured a number of very substantial projects within the electronics industry involving competition between potential sites within Europe. Some of the consequences are explored below.

In 1995 the German electronics firm Siemens selected North Tyneside in north-east England from a shortlist of six locations for a new £1.2 billion semiconductor plant, which was estimated eventually to create 2,000 jobs. The following year a semi-public contest between locations within the UK – particularly Wales and Scotland – culminated in a decision by South Korean firm LG to establish its £1.7 billion combined semiconductor and consumer electronics complex at Newport in South Wales. This was hailed as the biggest inward investment project in Europe, one that would lead eventually to the creation of 6,000 jobs. Whilst details of the public subsidy were not officially released, it was believed to be in the order of £200–250 million (the largest ever offered to an inward investor in the UK), equivalent to about £30–40,000 per job created. Just months later, another South Korean firm – Hyundai – announced that it too had favoured the UK, choosing to build its new £1 billion semiconductor plant at Dunfermline in Scotland. This was expected to create 1,000 jobs, with a possible second phase involving a further investment of £1.4 billion and an additional 1,000 jobs.[5]

Amidst the delight at these inward investment successes, there were several notes of concern. The LG case in particular prompted expressions of dismay from promotional agencies within England, which argued that they faced unfair competition from the Scottish and Welsh Development Agencies, with their larger budgets and greater control of a range of government activities (see p. 145). The cost-per-job created was high even by comparison with the subsidies awarded to other large inward investors, let alone those given to smaller firms. The low local content of existing semiconductor manufacturers in the UK suggested that the spin-off impacts would be limited. The extent to which the advanced capital equipment required would be purchased outside the UK was identified as an indication of weakness within the country's manufacturing base.

The loss of local control created through inward investment was highlighted the following year in the wake of the Asian financial crisis. First Hyundai announced that it was reviewing the time-scale for its whole project, and had not yet begun raising the finance to purchase capital equipment, as opposed to the buildings (which were already under construction). This would delay the start-up of the project by twelve months; the second phase was put on indefinite hold. Then in 1998 LG confirmed speculation that it had postponed the opening of its semiconductor plant by at least six months. Later that year both LG and Hyundai warned of still further delays from the rescheduled start-up dates (mid- and late 1999 respectively), as they sought international partners to enable them to bring their projects to completion. Hyundai subsequently went so far as to halt all construction work on the site. Just weeks later, Siemens announced the closure of its two-year-old factory in north-east England (which had received around £40 million in government aid) with the loss of 1,100 jobs, as it sought to find a buyer for the plant. This blow to the regional economy was further compounded by Fujitsu's announcement that it too was closing its £0.4 billion semiconductor plant in north-east England, at Newton Aycliffe, with the loss of 700 jobs, just seven years after it had opened. These events graphically illustrated the extent to which an economy built on inward investment depended on processes and decisions well beyond national borders.

regional policy, and the recent intensification of inter-regional competition in the pursuit of inward investment.

The stress on FDI running through this chapter is deliberate, but the absolute significance of inward investment should not be overplayed; after all, foreign-owned plants only account for around a quarter of output in the manufacturing sector. On the other hand, inward investment has been decisive in transforming the shape of a large part of manufacturing industry in the UK since the mid-1980s, as demonstrated in the case-study of the automotive industry. Future prospects for manufacturing in the UK rest upon *both* foreign-owned and indigenous firms, which face a number of issues in common, including the continued process of European integration. At the same time, reform in Eastern Europe is posing new challenges to inward investment agencies in the UK, as a whole new range of competing locations has opened up. The future structure of the sector in the UK will thus depend in part on the way in which the issue of policy co-ordination in the pursuit of inward investment is tackled.

Notes

1 Regional statistics were formerly presented on the basis of Standard Statistical Regions (SSRs). In 1994, new Government Offices for the Regions (GoRs) were established in England which, with effect from 1997, became the primary basis for the presentation of regional statistics. This introduced some discontinuities to data series. The main changes were that Cumbria ceased to be classified as part of the North (which became the North East GoR) and instead was part of a new North West GoR (which is here combined with the new Merseyside GoR); and the former South East and East Anglia SSRs were divided into three GoRs. The former SSR of East Anglia became the Eastern GoR with the addition of Bedfordshire, Essex and Hertfordshire, whilst a separate London GoR was established.

2 The ratio between male and female average weekly earnings in manufacturing narrowed slightly but the differential remained substantial during this period at 1.60 in 1991 and 1.54 in 1996.

3 This pattern is even more evident on a longer-term basis. From 1963 to 1987 the South East's share of UK employment in foreign-owned manufacturing establishments declined from 51 to 32 per cent, whilst the North's share increased from 2 to 6 per cent, that of Wales from 4 to 6 per cent, of Scotland from 8 to 10 per cent, and that of the South West from 1 to 6 per cent. Over that period the total volume of employment in foreign-owned manufacturing in the UK grew from 540,000 to 620,000 (see Hill and Munday 1992).

4 It was estimated that regional policy had led to unemployment rates in the Assisted Areas being 1 per cent lower than they would otherwise have been in the period from 1980 to 1988. The cost of RSA was calculated at £500 to £700 per job from 1985 to 1988, with 0.7–1.0 million job-years created (75–100,000 jobs over thirteen years). Previous evaluations of regional policy had extrapolated employment trends from the 1950s (when regional policy was much smaller in scale); the report concluded that there needed to be much better mechanisms for the evaluation of the effectiveness of policy.

5 The other major South Korean electronics project in the UK at this time – and the first to be established – followed Samsung's decision in 1994 to build a £450 million semiconductor and consumer electronics plant on Teesside in north-east England, scheduled eventually to create 3,000 jobs there.

References

Bull, P. (1991) The changing geography of manufacturing activity', in R. Johnston and V. Gardiner (eds) *The Changing Geography of the United Kingdom* (2nd edn), London: Routledge.

Hill, S. and Munday, M. (1992) 'The UK regional distribution of foreign direct investment: analysis and determinants', *Regional Studies* 26: 535–44.

Stone, I. (1995) *Inward Investment in the North: Patterns, Performance and Policy*, NERU Research Paper 14, Newcastle upon Tyne: NERU.

Trade and Industry Committee (1994) *Competitiveness of UK Manufacturing Industry*, London: House of Commons Paper 41, session 1993/4.

Trade and Industry Committee (1995) *Regional Policy*, London: House of Commons Paper 356, session 1994/5.

Trade and Industry Committee (1997) *Co-ordination of Inward Investment*, London: House of Commons Paper 355, session 1997/8.

Further reading

On the performance of manufacturing in the UK over a longer-term period until the mid-1980s, see D. Massey (1988) 'What's happening to UK manufacturing?', J. Allen and D. Massey (eds) *The Economy in Question* (London: Sage). On the North–South divide debate in the late 1980s see R. Martin (1988) 'The political economy of Britain's North–South divide', in *Transactions of the Institute of British Geographers* 13: 389–413. For a review of the different explanations for the collapse of manufacturing as seen from the mid-1980s, see B. Rowthorn (1986) 'De-industrialisation in Britain', in R. Martin and B. Rowthorn (eds) *The Geography of De-industrialisation* (London: Macmillan). For an account which examines the roles of labour productivity and management, see T. Nichols (1986) *The British Worker Question: A New Look at Workers and Productivity in Manufacturing* (London: Routledge and Kegan Paul).

On the extent of FDI in the UK, see I. Stone and F. Peck (1996) 'The foreign-owned manufacturing sector in UK peripheral regions, 1978–93: restructuring and comparative performance', *Regional Studies* 30: 55–68. For the debate surrounding the impacts of FDI in the UK, see M. Munday, J. Morris and B. Wilkinson (1995) 'Factories or warehouses? A Welsh perspective on Japanese transplant manufacturing', *Regional Studies* 29: 1–17. On 'performance' plants see A. Amin and J. Tomaney (1995) 'The regional development potential of inward investment in the less favoured regions of the European Community', in A. Amin and J. Tomaney (eds) *Behind the Myth of European Union: Prospects for Cohesion* (London: Routledge). The broader issue of the interplay between inward investors and receiving regions is considered in P. Dicken, M. Forsgren and A. Malmberg (1995) 'The local embeddedness of transnational corporations', in A. Amin and N. Thrift (eds) *Globalisation, Institutions and Regional Development in Europe* (Oxford: Oxford University Press).

For a discussion of the role of new forms of labour relations as epitomised by the Nissan plant at Sunderland see chapters 4 and 5 of D. Sadler (1992) *The Global Region: Production, State Policies and Uneven Development* (Oxford: Pergamon). For a critical view of the significance of new management practices for trade unions, see R. Hudson (1997) 'The end of mass production and of the mass collective worker? Experimenting with production and employment', in R. Lee and J. Wills (eds) *Geographies of Economies* (London: Arnold). On the adaptability of inward investors with respect to component suppliers, see D. Sadler (1994) 'The geographies of "Just-in-Time"; Japanese investment and the automotive components industry in Western Europe', *Economic Geography* 70: 41–59.

On the broader debates concerning changing forms of production organisation, see M. Storper (1995) 'The resurgence of regional economies, ten years later: the region as a nexus of untraded interdependencies', *European Urban and Regional Studies* 2: 191–221. For a selection of papers which address different aspects of regional policy in the 1990s, see R. Harrison and M. Hart (eds) (1993) *Spatial Policy in a Divided Nation* (London: Jessica Kingsley); P. Townroe and R. Martin (eds) (1992) *Regional Development in the 1990s: the British Isles in Transition* (London: Jessica Kingsley).

The likely structure of EU regional policies after 1999 is considered in a debate between R. Hall (1998) 'Agenda 2000 and European cohesion policies', *European Urban and Regional Studies* 5: 176–83 and S. Fothergill (1998) 'The premature death of EU regional policy?', *European Urban and Regional Studies* 5: 183–8.

One response to the intensified competition between different locations for inward investment was a policy of 'targeting' certain forms of FDI: see S. Young, N. Hood and A. Wilson (1994) 'Targeting as a competitive strategy for European inward investment agencies', *European Urban and Regional Studies* 1: 143–59. Another response was to focus on developing the capacities of existing foreign-owned firms through 'after-care' strategies: see K. Morgan (1997) 'The learning region: institutions, innovation and regional renewal', *Regional Studies* 31: 491–503. For a critical assessment of the impact of the LG project in South Wales, see N. Phelps, J. Lovering and K. Morgan (1998) 'Tying the firm to the region or tying the region to the firm? Early observations on the case of LG in Wales', *European Urban and Regional Studies* 5: 119–37.

Chapter 8

Labour markets

Jamie Peck and Adam Tickell

Introduction

It is difficult to exaggerate the extent to which the UK labour market has been restructured over the period since the 1970s. Unemployment has risen to unprecedented levels and has stayed high; income inequalities have widened; female participation rates have increased significantly, such that women now constitute more than half the waged workforce; 'contingent employment' in temporary, part-time and contract jobs has continued to expand; union membership densities and strike rates have both fallen, especially in the private sector; manufacturing employment has collapsed and now eight out of every ten workers are employed in 'service' industries, such as retailing, tourism and finance. Perhaps the key change, however, has been the emergence of 'flexible' labour markets which are now celebrated and promoted by politicians of all stripes. The Conservatives came to power in 1979 determined to attack what they saw as labour market 'rigidities' – powerful trade unions, institutionalised employment practices, 'passive' benefits regimes, 'bureaucratic' training systems, and low levels of labour mobility. Under the banner of 'deregulation', Conservative governments set about restructuring the institutions and remaking the rules of the UK labour market, as a barrage of legislative and policy changes in every area from employment rights to benefit entitlements actively facilitated the shift towards the more flexible labour markets of the 1990s. Although the election of a Labour government in 1997 saw changes in emphasis, as social partnership and welfare-to-work were established as new priorities, advocacy of flexible labour markets remains a central element of the New Labour credo. In most areas of labour market policy, Labour would build upon the Conservatives' reforms; they would not reverse them.

These interrelated processes of flexibilisation and deregulation had widely diverging effects in regional and local labour markets across the UK. Labour market restructuring has been associated with a range of complex shifts in the geographies of employment change, in local labour market governance and in regional economic fortunes. Take, for example, the two major recessions of the early 1980s and early 1990s. These events both accelerated the pace of economic restructuring and consolidated shifts in the prevailing form of labour market regulation. The recession of the early 1980s was in many ways a 'traditional' recession: a massive shake-out of manufacturing jobs took its toll on the northern and western regional economies of the UK which had traditionally relied disproportionately on factory employment, adding to the structural problems of joblessness and poverty in these regions. The recession of the early 1990s, in contrast, was very much a 'flexible' recession: with its origins in the overheating *regional* economy of the South East in the late 1980s, this downturn began with the financial and business services sectors in the South East itself, only later spreading to the traditional unemployment 'problem areas' of the North and West. In the process, long-established patterns of regional inequality in the UK have been disrupted and distorted, though they have not been fundamentally remade. Regional differentials in unemployment were wider in the booming economy of the mid-1980s than at any time since the 1930s depression. For a brief time, the early 1990s recession brought about a perverse form of regional convergence, as the once-buoyant regions of the South and East experienced sharp rises in unemployment. But the flexible recession was also as quick to end as it had been to arrive: the South East and adjacent regions led the way in the subsequent recovery period, quickly re-establishing their positions at the core of the UK economy.

This chapter explores the geography of labour market restructuring in the UK in two ways. First, the dimensions of change are established by looking at some of the more significant spatial patterns and trends in labour market restructuring during the 1980s and 1990s. Second, we explore some of the ways in which the shifting policy and institutional environment has variously responded to, and contributed to, these developments, in the wake of the breakdown of the Keynesian-welfarist orthodoxy of full employment during the 1970s. The chapter concludes with a brief discussion of emergent patterns of spatial restructuring in the UK labour market under New Labour.

Geographies of labour market 'flexibility'

The underlying weakness of the UK labour market since the 1970s is revealed in the fact that while employment growth has tended to be fitful and short-lived, the problems of unemployment and poverty have remained stubbornly entrenched. The official unemployment count remained above 2 million in 1998, even as the national economy reached the top of the business cycle and the Bank of England raised interest rates to slow down the economy. Meanwhile, the job growth that occurred over the previous two decades was both fragile and highly uneven in character. Between 1981 and 1997, total employment rose by 1.18 million. Yet this national picture conceals wide regional disparities. Some 1.07 million new jobs were located in the three 'core' regions of the South East, East Anglia, and the South West (although London actually lost 237,000 jobs). While London and the South East accounted for a stable share of national employment – containing just over one-third of the country's jobs in both 1981 and 1997 – the three adjacent regions have clearly been beneficiaries of job growth 'spilling over' from the South East: East Anglia, the South West and the East Midlands accounted for 17 per cent of national employment in 1981, yet almost two-thirds of jobs created in the 1980s and 1990s were located in these regions. In contrast, the North

West and the North experienced an overall shrinkage in their employment base during this period.

In important respects, these divergent regional economic fortunes can be explained in terms of the *combined* effects of (uneven) manufacturing decline and (uneven) service sector growth (Figure 8.1). Some 2 million manufacturing jobs were shed between 1981 and 1997 (one-third of the nation's factory employment base), while the service sector added 4 million new jobs (a growth rate of 30 per cent). Yet the process of labour market 'conversion' from manufacturing to services has been highly geographically uneven. Manufacturing decline (or 'deindustrialisation') has primarily disadvantaged the manufacturing heartlands of the North and West, while service sector growth has disproportionately benefited the South and East (see Martin and Rowthorn 1986; Marshall *et al.* 1988). While the heaviest manufacturing jobs losses occurred in the South East and the North West, the South East gained 1.2 million new service jobs, more than compensating for its manufacturing losses, while the North West suffered from a form of 'non-compensatory' restructuring – losing more jobs in factories than it gained in offices and shops. This said, services now dominate the employment profiles of every region in the UK – ranging from a low of 69 per cent of total employment in the North to 82 per cent in the South East – but the nature of service employment remains spatially differentiated. The South East tends to be home to high-level service jobs, while adjoining regions have experienced very strong growth in relatively low-productivity, low-wage services (Dunford 1997). The regions of the North and West, meanwhile, are disproportionately dependent on low-level services, particularly public sector employment.

Clearly, the 'new service economy' is a highly differentiated phenomenon, both sectorally and geographically. A great many service jobs are low-paid, insecure and/or part-time, but there is also a significant layer of high-level service occupations paying professional

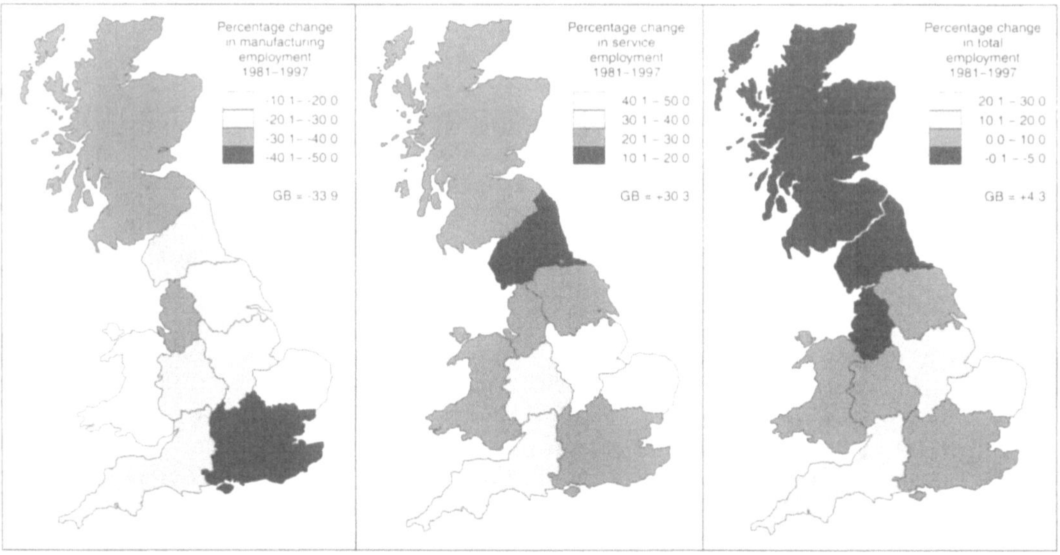

FIGURE 8.1 Change in employment by sector and region, 1981–97
Sources: Office of National Statistics (ONS); National Online Manpower Information Systems (NOMIS).

salaries. While the former are quite widely distributed across the UK (in part reflecting the dispersed demand for personal services), the latter are disproportionately concentrated in the South East. This is one of the factors behind the persistent patterns in regional income differentials. For both men and women, in both manual and non-manual occupations, earnings are higher in the three core southern regions than anywhere else in the United Kingdom. In 1995/6 men in non-manual jobs in London, for example, had an average weekly income of £586 (or 127 per cent of the average male non-manual income in the UK as a whole), whilst their counterparts in the North East had an average income of £406 (or 88 per cent of the UK average). Furthermore, the gap has widened during the past two decades for all groups of workers in both absolute and relative terms, but particularly for women with manual jobs: in 1978/9 no regional average income deviated from the UK average by more than 5 per cent, by 1995/6 women manual workers in London earned 125 per cent of the UK average, while their counterparts in Northern Ireland received 90 per cent of the UK average.

Such income disparities reflect both regional inequality and gendered pay rates. Over two decades after legislation which abolished differential pay rates for work of 'equal value', women's wages remain significantly lower than those of men. Although in one sense women have been the beneficiaries of recent labour market trends (in as far as they tend to be concentrated in the fastest-growing areas of service employment), in another sense they are the main 'losers' in that these jobs tend to be characterised by low-pay, insecure employment and limited promotion opportunities. In all regions, women working in both manual and non-manual jobs have an income of approximately two-thirds that of men in the same categories – although this represents progress of a sort. In 1978/9 the equivalent figure was half the male average. Women now represent just over half the national workforce (from 44 per cent in 1981), but tend to be crowded into those segments of the labour market with the worst employment conditions, such that their experience of 'flexibility' is quite different to that of most men (see McDowell 1991). There is geographical unevenness here too. Dunford's (1997) analysis of changing regional *employment* rates reveals that, during the period 1981–91, female employment rose by 3.7 per cent while male employment fell by 4.8 per cent. Increases in male *non*-employment occurred in every region except ROSE (the rest of the South East), but were most marked in the North, London, Wales, Yorkshire and Humberside and the North West. Moreover, female job growth has been weakest in the North, the North West, Scotland and Yorkshire and Humberside, while it has been strongest in the South West and East Anglia (Figure 8.2).

Low pay, job insecurity and other exploitative labour conditions are so much more prevalent in the northern and western regions in large part because these areas have for many years suffered the effects of structural unemployment and weak economic growth. Large-scale unemployment tends to act as a drag on pay and conditions because it swells the ranks of the low-wage labour supply, increases the substitutability of labour and tips the balance of power in the labour market in favour of employers (Peck 1996). Consequently, adverse labour market conditions have a tendency to be self-perpetuating, as regions with a legacy of unemployment and weak labour demand tend to attract mostly low-paying, contingent jobs which in no sense compensate for the jobs lost. This was the situation which confronted many northern and western regions during the 1980s and 1990s. At the start of the early 1990s recession, these regions were still carrying structural unemployment accumulated during the 1970s. The unemployment rates in the Northern region and Northern Ireland,

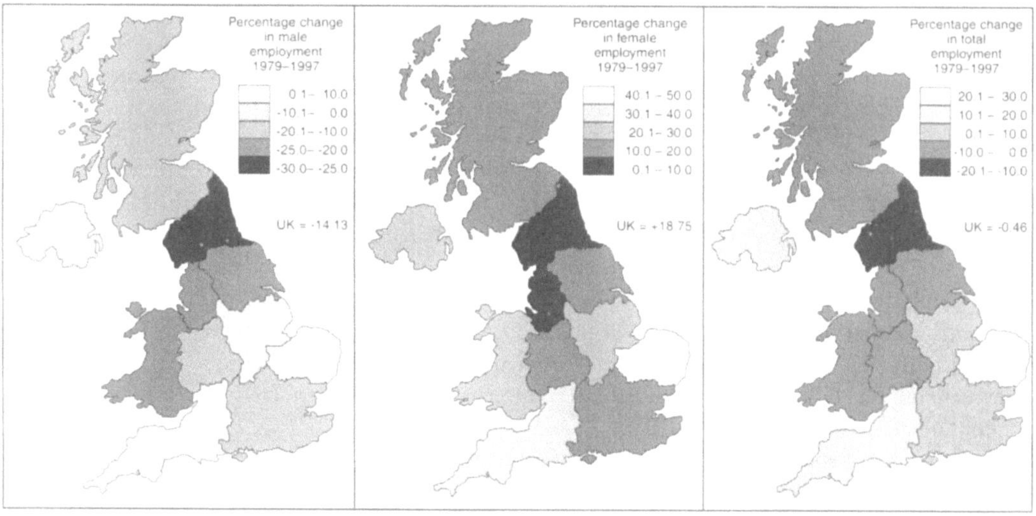

FIGURE 8.2 Change in employment by gender and region, 1981–97
Sources: ONS; NOMIS.

for example, were both more than 50 per cent above the national average in March 1979. As Figure 8.3 shows, this regional bias only deepened during the 1980s. When national unemployment peaked at over 3 million in 1986, the jobless rate in ROSE was only two-thirds the UK average, while in the Northern region it was 40 per cent higher and in Northern Ireland it was 50 per cent higher. So it was the economically weakest regions which bore the brunt of the job losses during the 1980s. The early 1980s recession put 2 million additional workers on the dole, 59 per cent of which were in the northern and western regions. This is not to say that the economic core in and around the South East was unaffected: unemployment also rose sharply in London and the East Midlands.

As Figure 8.4 shows, the South East-based recession of the early 1990s disrupted, but did not overturn, the broad regional pattern in unemployment differentials. In 1998, unemployment remained highest in Northern Ireland, the North, Scotland, Wales, the North West and Yorkshire and Humberside. In fact, the pattern of regional unemployment differentials was almost the same in 1998 as it had been in 1979. While the overall rate of unemployment was slightly higher in 1998, a trend reflected in most regions, the one significant change was in London, where the unemployment rate has more than doubled, from 2.8 per cent in 1979 to 5.7 per cent in 1998. Whereas London ended the 1970s with an unemployment rate 1.4 percentage points below the national average, following the ravages of the early 1990s recession its 1998 rate was 0.8 percentage points above the national average. Given that London also has the highest wage levels of any region, it is worth recalling Massey's (1987) comment, made in the context of mounting political-economic exuberance in the 1980s, that the country's most economically dynamic region is also its most socially divided and unequal.

The recovery period following the early 1990s recession was substantially South East-centred (Figure 8.5). Nationally, some 1.126 million new jobs were created between 1993 and 1997, but three-quarters of these job gains have been in the core regions of the 'greater

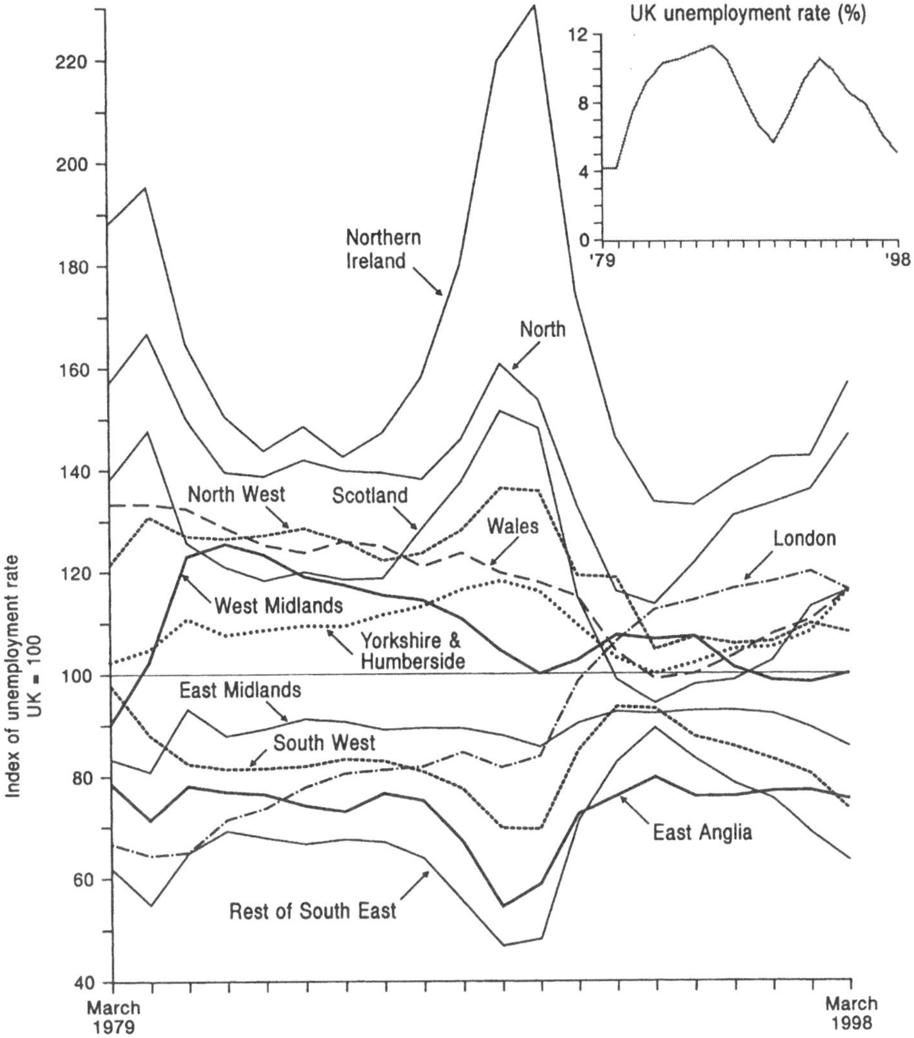

FIGURE 8.3 Regional unemployment differentials, 1979–98
Sources: ONS consistent unemployment series; NOMIS.

South East': the South East and its adjacent regions accounted for 54 per cent of national employment in 1993 but 74 per cent of subsequent job growth occurred in these regions. Yet just as the South East experienced a fragile form of growth during the 1980s, predicated as it was on consumer debt, housing market inflation and the short-term gains of privatisation, deregulation and middle-class tax cuts (Peck and Tickell 1995; Allen *et al.* 1998), so also the recovery appears less than robust in many respects. Indeed, there is a sense in which the fragile pattern of growth exhibited by the South East has become generalised into a fragile national recovery. First of all, fully 63 per cent of the new jobs created in the UK between 1993 and 1997 were part-time. In Scotland, Wales, the North and Yorkshire and

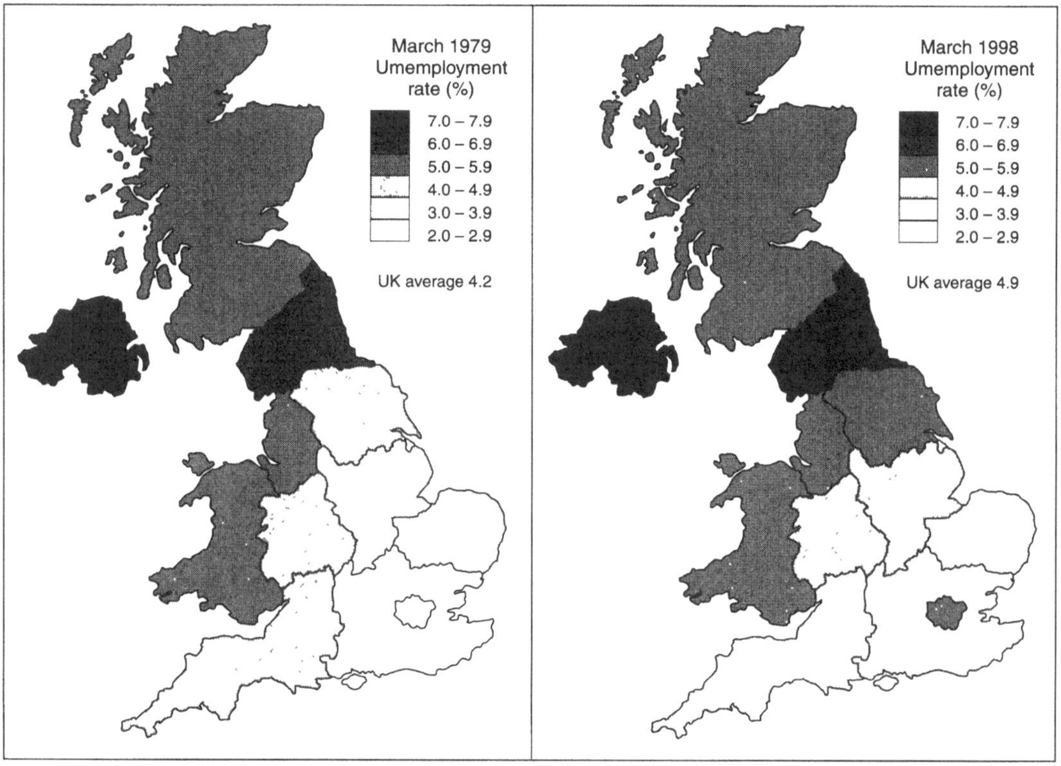

FIGURE 8.4 Unemployment rates by region, 1979 and 1998
Sources: ONS; NOMIS.

Humberside, full-time employment continued to contract during a period widely portrayed as a 'boom', so *all* net new job growth in these regions was part-time. Even in London, 82 per cent of post-recession employment growth has been in part-time jobs.

Second, recorded unemployment has been falling at a faster rate than job growth. There is evidence in some regions of a significant 'job gap' between the off-flow from unemployment and the creation of new jobs. While falling unemployment captured the headlines, job growth in many parts of the country remained sluggish. In Scotland, for example, less than 4,000 net new jobs were created in the 'recovery' period 1993–7 – all of which, as we have seen, were part-time – yet registered unemployment fell by 85,000. This means that, for every 100 people leaving the unemployment register in Scotland since the early 1990s recession, only 4.5 net new jobs were being created. This phenomenon is not restricted to Scotland. In fact, Yorkshire and Humberside, Wales, the North and the North West also exhibit a significant jobs gap, while in contrast, the 'core' regions have witnessed employment growth at a *faster* rate than the dole queues have been shrinking.

This suggests that regional labour market disparities were, if anything, widening during the mid-1990s recovery, which has done little to redress the structural weaknesses of the UK's lagging regional economies. On the contrary, it would seem that the increasingly

FIGURE 8.5 The geography of economic 'recovery', 1993–7
Sources: ONS; NOMIS.

strict benefit regime, reinforced by the introduction of the Jobseekers' Allowance (JSA) in 1996, has been acting to 'de-register' unemployed people in depressed regions in the absence of real employment opportunities (see Peck and Tickell 1997). This underlines the need for broader measures of the 'real' level of unemployment than is suggested by claimant statistics. These might take account of those on government training and welfare-to-work schemes, those 'involuntarily' working in temporary or part-time jobs, the unregistered unemployed and others discouraged from seeking work by the non-availability of jobs. More nuanced analyses of the wider levels of non- and underemployment in regional and local labour markets suggest that the 'real' unemployment rate is significantly higher across the board than the claimant count, but is *substantially* higher in precisely those areas which have suffered deep and persistent problems of economic decline and high registered unemployment – the northern and western regions in general, and northern cities and coalfield areas in particular (see Green 1996; Green and Hasluck 1998). This means that official unemployment statistics – including the revised, broader count adopted in 1998 – tend to understate levels of local and regional labour market inequality in the UK. As Green and Hasluck (1998: 556) comment, 'there appear to be reinforcing feedbacks between non-employment, underemployment, insecure employment, and a lack of investment in human

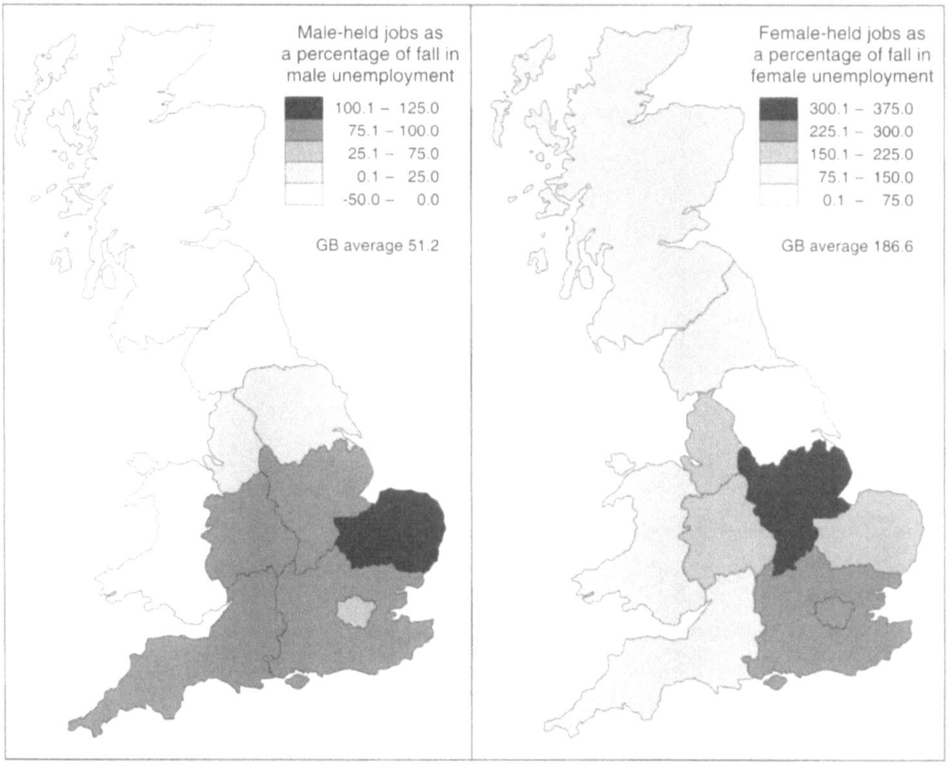

FIGURE 8.6 The job gap by gender, 1993–7
Sources: ONS; NOMIS.

capital' which lead to the entrenchment of regional labour market disadvantage and to the perpetuation of 'cycles of high unemployment'.

While an expanding labour market may be serving to 'pull' people from the unemployment register in the more buoyant areas of the country, there is no such demand-side pull in the depressed regions. Here, claimants are experiencing a concerted 'push' from unemployment in the context of weak or even negative employment growth, producing increased flows into economic inactivity, onto long-term sickness benefits, and doubtless also into the cash-and-crime economy. As Figure 8.6 shows, regional job gaps tend to be much wider for men than for women, reflecting the continued contraction of traditionally male-dominated employment in manufacturing and the relative buoyancy of service labour markets. Across the country as a whole, male unemployment fell by 981,000 between 1993 and 1997, but total male employment only rose by 502,000 during the same time period. Meanwhile, female unemployment fell by 334,000, while women's employment grew by 624,000. This represents a continuation of trends established in the 1980s, when 1.3 million male-held, full-time jobs were lost and when new employment opportunities were mostly part-time (71 per cent) and filled by women (83 per cent) (Dunford 1997). These developments underline the fact that processes of labour market restructuring in the UK are both geographically differentiated and strongly gendered.

Geographies of labour market 'deregulation'

Punctuated by two major recessions, the last two decades in the UK labour market have been characterised by low growth, widening social and regional inequalities, and a fraught transition from male-dominated manufacturing employment to a strongly services-oriented economy heavily reliant on part-time and female employment. Even at the peak of the business cycle, problems of low pay, unemployment, social exclusion and inadequate investment remain endemic, leading Dunford (1997) and others to suggest that the UK labour market has shifted to a new, low-employment/low-productivity 'equilibrium'. Dunford argues that this offers little long-term job potential or scope for sustainable economic development, due fundamentally to key weaknesses in the wider regulatory framework in which the labour market is embedded. In contrast to the Keynesian-welfarist order prevalent in the UK from the Second World War through to the mid-1970s, under which there was full male employment, rising incomes and productivity, expansive welfare provision and regional economic convergence, the low-employment/low-productivity 'equilibrium' of the 1980s and 1990s has been predicated on the deregulation of labour markets, polarising incomes, persistently high 'real' levels of unemployment and regional economic divergence.

While debate continues over whether the current alignment of 'flexible' labour markets and 'deregulationist' policies is a sustainable one (see Peck and Tickell 1994), it is clear that the old order of Keynesian-welfarism has fallen under concerted political attack, and may indeed be in permanent retreat. According to Jessop (1994), the Keynesian welfare state served to regulate and underwrite the post-war growth pattern in four ways: first, relative macro-economic stability was secured though labour market policies, the management of aggregate demand and the regulation of industrial relations; second, mass production and mass consumption – which were crucial to maintaining the virtuous cycle of mutually reinforcing growth in investment, productivity, incomes, demand and profit – were underpinned by a range of measures, from competition policy to infrastructure development and housing policies; third, consensus support was constructed around a programme of full employment and universal welfare; and fourth, social and welfare service needs were met through an expansive local state. In the British case, Kavanagh (1987) argues that the post-war political-economic settlement exhibited six key institutional features:

- full employment: both Labour and the Conservatives accepted the doctrine of full (male) employment as a fundamental policy goal; male unemployment rarely exceeded 3 per cent from the late 1940s to the early 1970s;
- the mixed economy: extensive state intervention in the economy garnered bipartisan support, as nationalised industries (such as steel, coal, gas, electricity and rail) involved the state directly in the production process and played a key role in underpinning weaker regional economies;
- interventionist government: British governments played an active role in the regulation of the domestic economy through the Keynesian techniques of demand management and the pursuit of industrial and labour market policies;
- welfare: both major parties accepted the need for extensive welfare provisions – ranging from unemployment and sickness benefits to state pensions, from the NHS to public housing – which were however predicated on healthy economic growth and high levels of male employment;

- corporatism: the trade unions were incorporated into the post-war political settlement, when along with business representatives they were afforded access to a range of decision-making and advisory bodies in areas like industrial training and labour market planning;
- technocracy: this was an era in which the government knew best – the dominant ethos underpinning the state's activist and interventionist approach to the regulation of economic and social life combined scientific management techniques with bureaucratic rationality and paternalism.

Geographically, the Keynesian-welfarist pattern of regulation was also distinctive in a number of ways (Martin and Sunley 1997). First, the domestic or national economy was privileged as the most important scale of regulation, domestic policies requiring a degree of 'insulation' from international economic forces. Second, the nation-state acted as the dominant agent of regulation, utilising centralised powers of regulation, control and intervention to manage the domestic economy; in particular to underpin high levels of aggregate demand and employment. Third, extensive state intervention and growing public expenditure led to spatial integration, as a standardised 'floor' of social welfare entitlements and infrastructure provisions was gradually established across all parts of the country. Fourth, these developments had the effect of countering spatial unevenness in wealth and welfare, as a range of measures including progressive tax systems, regional policies and needs-based benefits brought about income transfers from rich to poor areas.

The Keynesian-welfarist pattern of regulation went into crisis in the 1970s, following a slowdown in economic growth and an unprecedented rise in both unemployment and inflation ('stagflation'). In the UK, where these problems were especially acute, the rise of Thatcherism in the 1980s heralded a far-reaching shift in the prevailing political-economic orthodoxy which favoured market forces over state intervention, individualism over collectivism, privatisation over nationalisation, decentralised over centralised policies, and entrepreneurial zeal over corporatist bargaining. This new philosophy represented both a critique of, and a putative alternative to, the discredited welfare-statist orthodoxy.

> [T]he inability of the Keynesian state to resolve supply-side economic rigidities as evidenced by inflation, unemployment and 'lame-duck' industries, its endemic 'fiscal crisis' of rising costs of social welfare and public resistance to higher taxation, and its inability to control the demands of organized labour had culminated by the end of the 1970s in a crisis both of economic management and social legitimation. In addition, the very bases of national Keynesian interventionism have, it is widely argued, been completely undermined by changes in the international economic context, in particular the collapse of the international regulatory regime (Bretton Woods) that underpinned financial stability and, more recently, the 'globalization' of economic activities and economic spaces . . . [A] political response . . . epitomized by Thatcherism in the UK and Reaganism in the USA . . . has been to roll back national systems of regulation, intervention and welfare support in an attempt to give national economies the 'flexibility' needed to compete in today's global markets.
>
> (Martin and Sunley 1997: 281)

In this context, it should come as no surprise that the labour market has been one of the central terrains of struggle and reform (see Robertson 1986). Mrs Thatcher's first Conservative government was determined to reduce the power and influence of the trade unions, having been elected in the aftermath of the 'Winter of Discontent', a wave of strikes triggered by the breakdown of Labour's incomes policy. Priorities for the Conservatives were a package of industrial relations reforms, designed to restore the 'right to manage', and the installation of a stringent regime of macro-economic management – 'monetarism' – intended to restore the international competitiveness of the British economy. 'Deregulation' and 'flexibility' would be the new rallying calls. Yet while the Conservatives spoke of liberating market forces, decentralising power and removing the 'dead hand' of government interference in the marketplace, the irony was that their policy programme would require *new forms* of state intervention, typically backed up by the firm hand of central authority, an approach succinctly summarised by Andrew Gamble's (1988) phrase, 'the free economy and the strong state'. So while the Conservatives' programme of labour market reforms was far-reaching and in many ways radically transformative, it was also in some senses fragile and contradictory (see Peck and Jones 1995).

In order to illustrate this point, we comment briefly here on four key policy fields amongst the barrage of labour market reforms introduced by the Conservatives, highlighting in each case their associated spatial consequences: macro-economic policy; unemployment and welfare; training and workplace preparation; and, finally, industrial relations and trade unions. When the Conservatives were elected in 1979 they adopted a monetarist *macro-economic policy* in the belief that one of the UK's main problems was the persistence of high inflation which, they believed, was caused by having too much money in circulation. In order to reduce the money supply, the government raised interest rates. As the impact of this was to raise the value of sterling, British manufacturers were faced with a situation of higher borrowing costs and an exchange rate which simultaneously made their exports more expensive while reducing the cost of imports. The first two years of monetarist policy saw output fall by 6 per cent – almost exclusively in manufacturing – while the supply of money actually rose by 60 per cent in the first three years (Cairncross 1994). Although formally, monetarism was spatially neutral, the heavy concentration of manufacturing industry in northern and western regions led to disproportionate job loss in those areas.

In 1982 monetarist economics were quietly dropped and, during the 1980s, Conservative economic policy was more concerned with reducing the government's role in economic life through policies of tax cuts and the privatisation of publicly owned companies, yielding strongly disproportionate benefits to the South East (Hamnett 1997; Tickell 1998). Consumer spending mushroomed as tax cuts and the proceeds from privatisation share issues found their way into luxury goods and booming house prices. The impact was, yet again, spatially uneven: the booming economy of the mid- and late 1980s was almost exclusively concentrated in the southern regions of England, but the boom was of such a magnitude that it placed inflationary pressures on the economy as a whole. The government responded by progressively raising interest rates. While it did succeed in slowing the southern economy, it did so by precipitating a deep recession for the UK as a whole, destroying the late and fragile economic recovery of the North and West.

The recovery from the early 1990s recession then followed what had become a somewhat familiar pattern: growth in the South was fastest and strongest whilst regions elsewhere were slow to emerge from their structural economic malaise. The election of a Labour government in 1997 proved something of a false dawn for those in the peripheral regions who had felt that Conservative policies had – deliberately or not – disadvantaged them. The new Labour government handed responsibility for setting interest rates to the Bank of England, entrusting the Bank with the maintenance of persistently low levels of inflation – a policy better suited to the inflation-prone South than the deflated North. The re-emergence of inflationary pressures in the south of England during the late 1990s led the Bank of England to raise interest rates numerous times from May 1997, such that the manufacturing sector had again slid in recession by 1998. While there may be a common perception that the peripheral parts of the UK are a drag on the national economy, an alternative perspective would highlight the destructive nature of macro-economic policies which create unsustainable booms in core regions of the country while barely ameliorating long-run decline elsewhere.

Turning to *unemployment and welfare*, the Conservatives' approach was to gradually tighten the benefits regime so as to create stronger 'incentives to work'. The real value of benefits relative to average wages was allowed to fall, while a range of measures were introduced to restrict eligibility for benefits and to push those deemed 'employable' into work. For example, benefit eligibility for all 16- and 17-year-olds was removed in 1988 (with the effect of making the Youth Training Scheme (YT) effectively compulsory), while unemployed adults were – under the post-1986 Restart programme – required to take specific steps to find work (including attending work and social skills courses) or risk losing benefits. During the 1980s and early 1990s, a series of programmes for school-leavers and unemployed adults, such as YT, Employment Training and Training for Work, came to assume considerable importance, providing as they did an 'institutionalised labour market' in those depressed regions and inner-city areas where 'real' jobs remained in short supply. But these programmes differed from the job creation measures of the 1970s in that they were explicitly 'designed to work with the grain of the market, so as to encourage more realistic wage levels and more flexible working patterns' (Treasury 1984: 3). Allowances on all such schemes were set deliberately below market levels, in order to encourage more 'realistic' (i.e. lower) wages at the bottom end of the labour market. The pressure to enter low-wage work was intensified with the introduction of the JSA regime in 1996, which renders unemployment benefits conditional on *active* jobsearch, thereby compelling the unemployed to move into whatever jobs may be available locally.

In the final years of John Major's administration, the Conservatives also began to experiment with local 'workfare' programmes, under which the unemployed are required to carry out environmental and other work in order to 'earn' their benefits. The general effect of such measures, alongside the stricter benefits regime and the expansion of 'active' labour market programmes, was to drive down both the 'reservation wages' of the unemployed (the pay level at which claimants were prepared to accept a job) and the overall level of registered unemployment. Meanwhile, various forms of 'hidden' unemployment – in the form of inactivity, early retirement, long-term sickness – increased significantly, especially in depressed areas. One way of evaluating the impact of such labour market reforms is to look at the changing components of household income. In 1978/9, 72 per cent of household incomes in the UK came from salaries and wages. After seventeen years of labour market reform this had fallen to 64 per cent – with falls of similar magnitude in every region except Northern Ireland. Although part of the explanation for such changes is that Britain has an ageing population relying on investment incomes and government pensions in retirement, people without jobs rely on the social security system to pay for their basic human needs – and it is no coincidence that the regions with the highest reliance on social security are also those with the highest levels of visible and hidden unemployment.

A significant feature of Conservative labour market policy, particularly from the late 1980s onwards, was the way in which benefit reforms and provision for the unemployed were combined together with restructured systems of *training and workplace preparation*. The key moment here was the introduction of Training and Enterprise Councils (TECs) in 1988 in the context of a profound deregulation of the system of industrial training. The TECs are locally based, business-led bodies which operate on contract to central government under a wide remit embracing increasing the level of training activity, delivering market-relevant programmes for unemployed people and school-leavers, and stimulating entrepreneurship and business development in local economies. The network of localised TECs took over the key functions of the Manpower Services Commission (MSC), a tripartite body established in the 1970s to involve trade unions and business representatives, together with government, in the process of strategic labour market planning. Deeply suspicious of such corporatist organisations, the Conservatives had nevertheless used the MSC during the early 1980s to deliver large-scale training and 'make-work' for the unemployed. In fact, the MSC became a key agency in the implementation of the government's programme of neoliberal labour market reform (Peck and Jones 1995). According to King, the Conservatives used the TECs to effect

> a shift from a national tripartite regime to a local, employer-dominated neoliberal training regime . . . [During the 1980s, the British government] ended the legacies of the postwar Keynesian framework, particularly tripartism, which had informed training policy, and substituted it with a neoliberal one, dominated by employers' interests and measures to counteract labour market disincentives.
>
> (King 1993: 214–15)

In fact, while the TECs came to epitomise the Thatcherite approach to labour market reform, they were also quick to demonstrate its shortcomings and contradictions. Designed in anticipation of a tightening labour market, the TECs were rolled out into the hostile climate of the early 1990s recession, and as a result were forced to take on the role of managers of local programmes for the unemployed. They drove down the costs of such programmes and tailored them, where possible, to meet local labour market needs, but in the process revealed that they lacked the means to bring about significant increases in the level of *private* investment in training. Prior to the TECs, employers in most key sectors of the economy had been required by law to support the training of new recruits through a levy-grant system, administered by a network of Industry Training Boards (ITBs). This system was wound up on the introduction of TECs, whose remit was to encourage employers *voluntarily* to invest in training. In fact, training levels declined under this free-market regime, leaving behind it only the rump of low-level training programmes for young people and the unemployed, combined with the patchy and generally inadequate provisions of private employers (see Jones 1999). No longer underwritten by the levy-grant scheme, which through a regime of employer contributions and incentives guaranteed minimum levels of training provision, the British training system has degenerated into an uneven, low-skills, low-investment equilibrium.

Finally, perhaps the most far-reaching and controversial of the Conservatives' labour market reforms have been concerned with *industrial relations and trade unions*. The dominant pattern of economic restructuring in the UK had been undermining the bases of trade union strength for some time: the unions' organisational strength lay in manufacturing industry, especially in large plants, and membership was skewed towards male, full-time workers. This meant that the heartlands of traditional manufacturing industry were also the heartlands of British trade unionism, and while union support in these northern and western parts of the country has remained in many ways resilient, it has undeniably been eroded by the shift towards services and contingent work (Martin *et al.* 1996). Equally significant, however, was the frontal attack on the unions launched by the Conservatives in the early 1980s, when a legislative programme designed specifically to curb the political and bargaining power of the unions, coupled with a series of major confrontations with public-sector unions, served to rewrite the rules of the British industrial relations system. For the unions, this reached its nadir with the defeat of the miners in the strike of 1984–5, during which a divide-and-rule strategy had exposed damaging cleavages between the relatively conservative Nottinghamshire miners and their rather more militant fellow workers in Yorkshire, Scotland, Wales and elsewhere. The Conservatives' organised and uncompromising approach to the strike led to the effective destruction of the vanguard union, the National Union of Mineworkers (see Sunley 1990), and was to play a key role in transforming the industrial relations climate across the economy as a whole. Levels of industrial action fell dramatically, and have remained low, while single-union and no-strike deals became increasingly commonplace, perhaps most symbolically in many of the Japanese-owned manufacturing plants which began to locate in the old union heartlands of Scotland, Wales and the North of England (Munday *et al.* 1995; see also Durand and Stewart 1998).

As the extensive policy interventions of the 1980s illustrate, 'deregulated' labour markets are not free from government involvement. It follows that the labour market trends described earlier in this chapter are not necessarily 'natural' or inevitable, but instead are variously moulded and channelled by government policy. The 'flexible' labour markets of the late 1990s bear the imprint of government policy and regulation in many ways just as vividly as did their 'rigid' predecessors of the early 1970s (see Rubery 1994). What has changed – and undeniably there *has* been significant change – is the nature, goals and premises of policy itself, as quite different political choices are now being made about appropriate and desirable forms of labour market development. Flexible labour markets do not just 'happen'. Rather, they have been actively cultivated by successive governments intent on extending the scope of managerial prerogative, disorganising the labour movement, retrenching welfare provisions and individualising employment relations.

Conclusion: New Labour and the labour market

When Labour came to power in 1997 it inherited a legacy of flexibly deregulated labour markets, wide social and economic inequalities, and high but falling unemployment. While Labour criticised some of the more pernicious effects of Conservative policies, it would soon be clear that the Blair government's intention was to work within, rather than against, the neoliberal regulatory framework established in the 1980s and 1990s. The Conservatives' trade union and social security legislation would not be repealed, the privatisation programme would not be reversed, market-oriented policies in areas like training would be retained, as would both the deflationary macro-economic posture and tough stance on public expenditure. The Labour government is building on the foundations laid by previous Conservative administrations, but has also strongly advocated more activist and interventionist policies in a number of areas. Chief amongst these are welfare-to-work, where the £5 billion New Deal programme represents the Blair government's largest new public-spending commitment; social protection, where individual workplace and social rights have been extended; and wage-setting, where a minimum wage has been introduced for the first time in the nation's history. As Peter Mandelson explained on the occasion of the launch of Labour's Social Exclusion Unit:

> flexibility in its own right is not enough to promote economic competitiveness. It is the job of government to play its part in guaranteeing 'flexibility plus' – plus higher skills and higher standards in our schools and colleges; plus partnership with business to raise investment in infrastructure, science and research and to back small firms; plus an imaginative welfare-to-work programme to put the long-term unemployed back to work; plus minimum standards of fair treatment at the workplace; plus new leadership in Europe in place of drift and disengagement from our largest markets. This is the heart of where New Labour differs from both the limitations of new right economics and the Old Labour agenda of crude state intervention in industry and indiscriminate 'tax and spend'.
>
> (Mandelson 1997: 17)

It would be inaccurate, however, to portray Labour's approach to the labour market as no more than 'business as usual'. In comparison to the Conservative years, there have been changes in substance as well as in emphasis. Chancellor Gordon Brown's early Budgets were gently but persistently redistributional, while active steps have been taken – through the reform of taxes and benefits and through the introduction of a national minimum wage – to ensure that 'work pays' for welfare recipients and those in the lower reaches of the job market. A series of spatially targeted policies, such as the Employment Zones introduced in 1998, are providing focused help with local employment initiatives in areas of high unemployment. Couched in the language of 'rights and responsibilities', Labour's approach to tackling unemployment marries the carrot with the stick: the unemployed are offered new opportunities for education and training, subsidised employment and community work but in turn are *obliged* to take these up; under strict new rules, those refusing to take up places on the New Deal programme for the unemployed have their benefits withheld. Labour has proved more willing than the Conservatives to utilise 'active' labour market policies, particularly in the area of measures to tackle social exclusion. Active policies seek to work with the grain of the market to bring about higher and more sustainable levels of waged employment, and to reduce 'dependency' on welfare. Their adoption represents a marked change from the Conservative approach, which privileged narrow, market-based approaches to labour market policy, with little or no explicit role for progressive social redistribution. It also has to be said, however, that Labour's *simultaneous* embrace of the goals of labour market flexibility and social inclusion – 'flexibility plus' – exposes in a more explicit way than before the tensions and contradictions between these two quite distinctive policy objectives.

New Labour regard Britain's deregulated, flexible labour market as a major source of competitive strength. As the Prime Minister has (favourably) observed, the UK now has the most 'lightly regulated labour market of any leading economy in the world' (Tony Blair, quoted in *Guardian*, 25 May 1998: 17). Yet how far it will be possible to reconcile flexible deregulation with Labour's wider social and economic goals – such as tackling poverty and raising levels of investment in human capital – must remain to be seen. Certainly, a range of new policy measures are likely to bring about modest forms of redistribution in favour of poorer parts of the country, but it is equally clear that there will be no return to the kind of investment-driven, reflationary programme that many would argue is necessary for lagging regions to catch up with the south-eastern core. Patterns of spatial inequality in the labour market may therefore be ameliorated, but the basic rules of the game – premised as they are on short-term competitive advantage – are likely to remain essentially the same. Given the UK economy's established proneness to deep inequality, cyclical volatility and uneven development, the challenge of marrying social justice with flexible labour markets is likely to be a daunting one.

References

Allen, J., Massey, D., Cochrane, A., Charlesworth, J., Court, G., Henry, N. and Sarre, P. (1998) *Rethinking the Region*, London: Routledge.

Cairncross, A. (1994) 'Economic policy and performance, 1964–1990', in R. Floud and D. McCloskey (eds) *The Economic History of Britain since 1700, Volume 3: 1939–1992* (2nd edn), Cambridge: Cambridge University Press.

Dunford, M. (1997) 'Divergence, instability and exclusion: regional dynamics in Great Britain', in R. Lee and J. Wills (eds) *Geographies of Economies*, London: Arnold.

Durand, J.P. and Stewart, P. (1998) 'Manufacturing dissent? Burawoy in a Franco-Japanese workshop', *Work, Employment and Society* 12(1): 145–59.

Gamble, A. (1988) *The Free Economy and the Strong State*, London: Macmillan.

Green, A. (1996) 'Aspects of the changing geography of poverty and wealth', in J. Hills (ed.) *New Inequalities*, Cambridge: Cambridge University Press.

Green, A.E. and Hasluck, C. (1998) '(Non)Participation in the labour market: alternative indicators and estimates of labour reserve in UK regions', *Environment and Planning A* 30(3): 543–58.

Hamnett, C. (1997) 'A stroke of the Chancellor's pen: the social and regional impact of the Conservative's 1988 higher rate tax cuts', *Environment and Planning A* 29(1): 129–47.

Jessop, B. (1994) 'Post-Fordism and the state', in A. Amin (ed.) *Post-Fordism: a Reader*, Oxford: Blackwell.

Jones, M.R. (1999) *New Institutional Spaces: TECs and the Remaking of Economic Governance*, London: Jessica Kingsley.

Kavanagh, D. (1987) *Thatcherism and British Politics: the End of Consensus?*, Oxford: Oxford University Press.

King, D.S. (1993) 'The Conservatives and training policy 1979–1992: from a tripartite to a neoliberal regime', *Political Studies* 41: 214–35.

McDowell, L. (1991) 'Life without father and Ford: the new gender order of postfordism', *Transactions of the Institute of British Geographers* 16: 400–19.

Mandelson, P. (1997) 'A lifeline for youth', *Guardian*, 15 August: 17.

Marshall, J.N., Wood, P., Daniels, P.W., McKinnon, A., Bachtler, J., Damesick, P., Thrift, N.J., Gillespie, A., Green, A.E. and Leyshon, A. (1988) *Services and Uneven Development*, Oxford: Oxford University Press.

Martin, R. and Rowthorn, B. (eds) (1986) *The Geography of Deindustrialisation*, London: Macmillan.

Martin, R. and Sunley, P. (1997) 'The post-Keynesian state and the space economy', in R. Lee and J. Wills (eds) *Geographies of Economies*, London: Arnold.

Martin, R., Sunley, P. and Wills, J. (1996) *Union Retreat and the Regions: the Shrinking Landscape of Organised Labour*, London: Jessica Kingsley.

Massey, D. (1987) 'The shape of things to come', in R. Peet (ed.) *International Capitalism and Industrial Restructuring: a Critical Analysis*, London: Allen and Unwin.

Munday, M., Morris, J. and Wilkinson, B. (1995) 'Factories or warehouses? A Welsh perspective on Japanese transplants', *Regional Studies* 29: 1–17.

Peck, J. (1996) *Work-Place: the Social Regulation of Labor Markets*, New York: Guilford.

Peck, J. and Jones, M. (1995) 'Training and Enterprise Councils: Schumpeterian workfare state, or what?', *Environment and Planning A* 27(9): 1361–96.

Peck, J. and Tickell, A. (1994) 'Searching for a new institutional fix: the *after* Fordist crisis and global–local disorder', in A. Amin (ed.) *Post-Fordism: a Reader*, Oxford: Blackwell.

Peck, J. and Tickell, A. (1995) 'The social regulation of uneven development: "regulatory deficit", England's South East and the collapse of Thatcherism', *Environment and Planning A* 27(1): 15–40.

Peck, J. and Tickell, A. (1997) 'Manchester's job gap', *Manchester Economy Group Working Paper* 1, Manchester Economy Group, University of Manchester.

Robertson, D.B. (1986) 'Mrs Thatcher's employment prescription: an active neoliberal labour market policy', *Journal of Public Policy* 6: 275–96.

Rubery, J. (1994) 'The British production regime: a societal-specific system?', *Economy and Society* 23: 335–54.

Sunley, P. (1990) 'Striking parallels: a comparison of the geographies of the 1926 and 1984–85 coalmining disputes', *Environment and Planning D: Society and Space* 8: 35–53.

Tickell, A. (1998) 'A tax on success? Privatisation, employment and windfall tax', *Area* 30: 83–90.

Treasury (1984) 'Helping markets work better', *Economic Progress Report* 173: 1–5.

Further reading

Particularly useful general accounts of economic restructuring in the UK, which locate the processes of labour-market change in their wider social, political and institutional contexts, can be found in: R.L. Martin (1988) 'The political economy of Britain's North–South divide', *Transactions of the Institute of British Geographers* 13: 389–418; R. Hudson (1989) 'Labour-market changes and new forms of work in old industrial regions: maybe flexibility for some but not flexible accumulation', *Environment and Planning D: Society and Space* 7: 5–30; and M. Dunford (1997) 'Divergence, instability and exclusion: regional dynamics in Great Britain', in R. Lee and J. Wills (eds) *Geographies of Economies* (London: Arnold).

For considerations of the theoretical and conceptual issues in labour-market analysis, with reference to developments in the UK, see J. Rubery (1996) 'The labour market outlook and the outlook for labour market analysis', in R. Crompton, D. Gallie and K. Purcell (eds) *Changing Forms of Employment: Organisations, Skills and Gender* (London: Routledge); J. Peck (1996) *Work-Place: the*

Social Regulation of Labor Markets (New York: Guilford); and Labour Studies Group (1985) 'Economic, social and political factors in the operation of the labour market', in B. Roberts, R. Finnegan and D. Gallie (eds) *New Approaches to Economic Life* (Manchester: Manchester University Press).

Discussion of contemporary policy issues around the labour market can be found in journals such as *Renewal*, *Work, Employment & Society* and *Regional Studies*. Examples of work which explore the policy and politics of labour-market restructuring in a spatially sensitive fashion are: J. Peck (1999) 'New Labourers? Making a New Deal for the "workless class"', *Environment and Planning C* 17(3): 345–72; M.R. Jones (1999) *New Institutional Spaces: TECs and the Remaking of Economic Governance* (London: Jessica Kingsley); and R. Martin, P. Sunley and J. Wills (1996) *Union Retreat and the Regions: the Shrinking Landscape of Organised Labour* (London: Jessica Kingsley).

Chapter 9

Demography

Tony Champion

Introduction

Demography provides an important backcloth to the developments taking place in the UK, but it is neither unchanging in its nature nor neutral in its effects. The term is used here to refer to the size, distribution and composition of the population and the patterns of life-course events which maintain or alter these, notably the three basic components of change in population numbers (births, deaths and migration), but also the many factors that affect household formation and family building. The demography of the UK has undergone some major changes over the past three decades, including greater longevity, lower fertility, higher divorce rate, accelerating cohabitation, later marriage and childbearing, increasing lone parenthood, larger numbers living alone, a fluctuating exodus from large cities, a switch from net emigration to substantial international migration gain and the rise of the UK's non-white population. Each of these raises policy issues for society, most of them immediate and obvious in their implications but some with major long-term impacts such as the continuing passage of the 1960s/1970s baby boom and bust through the age structure. The significance of these trends is evident in many other parts of this book, including changes in labour supply (Chapter 8), increasing ethnic diversity of urban areas (Chapters 10, 14), growing pressures on the countryside (Chapter 11), altered lifestyles and patterns of consumption (Chapters 12, 13) and the greying of the electorate (Chapter 16). At the same time, a wide range of factors influences the course of population change itself, prompting demographers to stress that their attempts at looking into the future are essentially projections and not predictions. Some examples of uncertainty will become apparent in highlighting developments since Compton's (1991) review for a previous edition of this book.

National trends

The estimated population of the United Kingdom in mid-1996 was 58.8 million. Amongst the rest of the Developed World, this is considerably smaller than the USA's 265 million, the Russian Federation's 148 million, Japan's 125 million and unified Germany's 81.7 million, while on the worldwide scale it comprises barely 1 per cent of the planet's population. It is, however, on a par with France's 58.4 million and Italy's 57.4 million, is significantly larger than Ukraine's 51.4 million, Spain's 39.3 million, Poland's 38.4 million and Canada's 30 million, and is at least five times the size of all other European countries besides Romania and the Netherlands. Moreover, within the fifteen-country European Union as comprised in 1996, it accounted for 15.8 per cent, or almost one-sixth, of the population.

Between 1961 and 1996 the population of the UK increased by 6 million, up by 11 per cent on the 52.8 million level of just over a third of a century ago (Table 9.1). The rate of growth, however, has not been uniform across this period. It was at its highest in the early 1960s, when the annual increment averaged 367,000 or 0.69 per cent, and fell progressively until the second half of the 1970s, when the average gain was only 27,000 or 0.05 per cent. Subsequently, the growth rate has recovered somewhat, with the annual average for both the

TABLE 9.1 Population of the United Kingdom, 1961–2021

		Annual components of change (thousands)						
Mid-year to mid-year	*Population at start of period (millions)*	*Births*	*Deaths*	*Natural change*	*Net civilian migration*	*Other changes*	*Total change*	*Anual growth rate (%)*
Estimated								
1961–66	52.8	988	633	355	–8	21	367	0.69
1966–71	54.6	937	644	293	–56	20	257	0.47
1971–76	55.9	766	670	96	–55	16	58	0.10
1976–81	56.2	705	662	42	–33	18	27	0.05
1981–86	56.4	732	662	70	21	9	100	0.18
1986–91	56.9	782	647	135	60	–4	191	0.34
1991–96	58.8	756	641	116	76	7	199	0.34
Projected								
1996–2001	58.8	723	634	89	74	—	163	0.28
2001–06	59.6	696	627	69	65	—	134	0.22
2006–11	60.3	684	620	64	65	—	129	0.21
2011–16	60.9	691	621	70	65	—	135	0.22
2016–21	61.6	697	635	63	65	—	128	0.21

Sources: *Population Trends* 92, Table 5; Shaw (1998) Table 3, p. 46; data supplied directly by the Office for National Statistics.
Note: 'Other changes' refers to changes in number of armed forces and adjustments to reconcile population change between mid-year estimates with estimates of natural change and net civilian migration. Numbers may not sum because of rounding.

late 1980s and the first half of the 1990s running at 0.34 per cent, not quite half the rate of 1961–6. According to the 1996-based projections, the future trend is once again towards lower rates of increase, stabilising at around 0.22 per cent a year from the turn of the century (Table 9.1).

These fluctuations in overall numbers, however, mask a longer-term shift in the components of population change (Table 9.1). The strong growth in the early 1960s was almost entirely due to natural change, with a large surplus of births over deaths at that time. The reduction in the rate of growth over the next fifteen years was principally the result of a sharp reduction in the number of births but was reinforced by a rise in the number of deaths and the acceleration of net emigration. Since the late 1970s the number of births has risen again, but managed to recover less than a third of its previous decline by the late 1980s before falling back again somewhat in the period 1991–6 and is projected to drop below its 1976–81 low point after 2001. Equally important in the upswing in overall growth after 1976–81, however, was the shift in the UK's international migration balance from strong net loss in the early 1970s to equally strong net gain in the late 1980s – a boost to population numbers which, unlike for births, accelerated further in the 1990s. By this time, natural change was responsible for less than three-fifths of overall growth and appears destined to become even less important, with the projections showing natural change and net immigration being rather evenly matched in their contributions to national growth after the year 2001 (Table 9.1).

Fertility rate

Fluctuation over time in the number of births has played a crucial role in influencing the national population growth rate. The annual pattern for the past half-century (Figure 9.1) shows clearly the twin peaks of just over 1 million births in 1947 and 1965, separated by a trough of around 0.8 million births in the early 1950s. After the 1965 peak, there was an even more sustained fall in births, lasting twelve years and involving a reduction by one-third in numbers. Since 1977 the number of births has rebounded again, but to nowhere near the level of the early 1960s, indeed barely reaching that of the early 1950s trough.

Because of the uncertainty which this scale of fluctuation causes for projecting future population trends, these trends have been the focus of a great deal of research. The patterning of births over time is a function of three direct factors: the number of women of childbearing age (which is itself largely determined by births patterns a generation previously); the fertility rate of each birth cohort of women (referring to the number of children ultimately produced by women born in a particular year), and the timing of these births within their childbearing years.

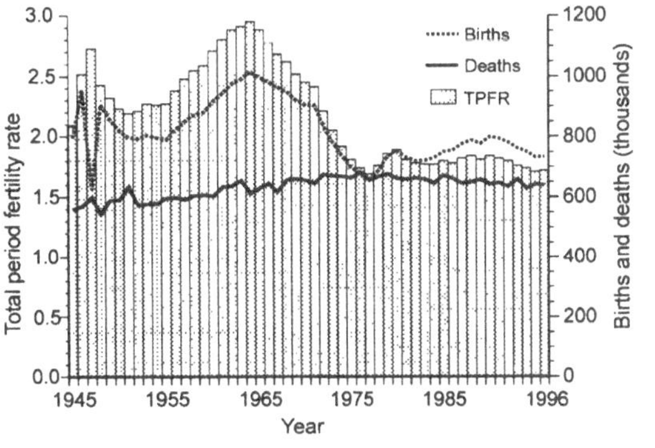

FIGURE 9.1 Deaths, births and fertility, 1945–96, United Kingdom

Sources: *Annual Abstract of Statistics*, various years; Armitage and Babb (1996).

Note: Total period fertility rate (TPFR) before 1960 is estimated.

The 1947 peak is readily explainable in terms of the effect of demobilisation of the armed forces after the Second World War, with the subsequent trough being interpreted as a return to the lower fertility rates reached at the end of the UK's passage through the demographic transition in the 1930s. It was the subsequent 'baby boom' and 'baby bust' that took the experts by surprise. With the benefit of hindsight, the rise in births in the late 1950s and early 1960s is seen to have arisen partly from an increase in desired family size, prompted largely by improved economic conditions and expectations, and partly from a bunching of births, caused both by the overlapping of births delayed by the poorer economic conditions of the early 1950s and a shift towards earlier marriage and childbearing of the next birth cohort of mothers (the latter mainly due to the improved job and promotion prospects of male breadwinners).

The 'baby bust' of 1965–77, however, has proved to have been more than the effects of birth timing and economic downturn, instead introducing a significantly different – and seemingly relatively stable – regime which has been hailed variously as a 'second demographic transition' and 'the demographic revolution'. This is illustrated most clearly by reference to the total period fertility rate (TPFR), which indicates the average number of live-born children which women would have if they experienced the age-specific fertility rate of that year throughout their childbearing lives and is unaffected by changes over time in the age distribution of fertile women. As shown in Figure 9.1, the TPFR fell by over 40 per cent from its peak of 2.95 in 1964 to 1.69 in 1977. Various factors have been put forward for this precipitous fall, including an increase in the proportion of women receiving higher education, greater female participation in the workforce, and greater control over fertility made possible by the contraceptive pill. These are long-term societal developments and their power is reflected in the fact that the TPFR has remained consistently at these lower levels for around twenty years, moving only little around the 1.80 mark during the 1980s. This is well below the rate of around 2.1 needed for the long-term replacement of one generation by another and explains why the rise in actual numbers of births in the 1980s was only a pale shadow of the 'echo effect' that would otherwise have been expected from the large numbers of mothers born in the 'baby boom' years.

On the other hand, the unprecedented stable level of the TPFR in recent years hides some important changes that would appear to be consolidating the low fertility regime and may indeed herald further reductions in the future. Fitting in with the main causes of fertility rate decline mentioned above has been an increase in the mean age of mother at live childbirth. The fertility rate of women aged 20–24 fell by over a quarter between 1981 and 1996, while that of those aged 30–34 rose by nearly 30 per cent and that of those aged 35–39 rose by some 70 per cent. The mean age of mothers for all live births in England and Wales was 28.6 years in 1996, 1.8 years above that in 1981 and the highest since records of mothers' ages were first compiled in 1938. Only teenagers have stood out against this trend of older motherhood, with their fertility rate even rising somewhat in the late 1980s. Accompanying this trend, the proportion of larger families has fallen and there has been an increase in childlessness, with the proportion of women aged 35 who have never had a child rising from 12 to 23 per cent between 1980 and 1994.

Other changes relating to fertility rates concern live births outside marriage and abortions. Despite the later occurrence of motherhood, the proportion of out-of-wedlock births has soared to over one-third (35.5 per cent in 1996), up from 12.5 per cent in 1981. As only one-fifth of these births was registered solely by the mother, this growth primarily reflects changing patterns of marriage and cohabitation. As regards abortions, the proportion of conceptions not leading to live births has almost doubled in the quarter of a century to 1996, when records for England and Wales indicate 166,400 abortions, or 1 in 5 of all conceptions that year. Further details about fertility-rate trends can be found most readily in the review paper by Armitage and Babb (1996).

Mortality rate

By comparison with births, the number of deaths in the UK has been running at a remarkably consistent level of around 650,000 for several decades, rising somewhat above this in the 1970s and early 1980s and falling below it since (Table 9.1 and Figure 9.1). The recent reduction has come about because the growth in the size of the population, particularly in the numbers of older people with their higher chances of dying, has been more than offset by improvements in people's survival chances at virtually all ages. Projections indicate a continuation of this decline, with the anticipation of further gains in life expectancy combined with a slowdown in the growth of the elderly population because of the low birth rate of the inter-war period, but the number is expected to rise again substantially from 2016 onwards as the first members of the baby boom generation reach retirement age (Table 9.1).

Life expectancy has improved consistently ever since deaths began to be recorded centrally in 1836, rising from barely 40 years then to around 50 in 1911, 60 in 1939 and 70 in 1960, and with the margin between men and women widening progressively over time. Life expectancy at birth has continued to increase over the past quarter of a century: for males, up from 68.8 to 74.6 between 1971 and 1996 and, for females, up from 75.0 to 79.5. Even at the age 70, life expectancy has increased by two years over this period. Thus men aged 70 now expect to live a further 11.5 years on average, compared with only 9.5 in 1971, and women a further 14.6 compared with 12.5 in 1971. With mortality rates currently falling by about 0.5 per cent a year, the 1996-based population projections assume that by 2021 life expectancy at birth will have gained a further three years for both sexes, putting men at 77.8 and women at 82.5 years (Shaw 1998, and see Chapter 12).

Mortality rate reductions have been occurring for most age and sex groups in recent years, but the improvements have not been evenly spread. According to an analysis of changes between 1984 and 1994 by Dunnell (1995), improvements in mortality rates were greatest for infants under one year old, with the infant mortality rate in 1994 being 65 per cent of that ten years before. Reductions of a quarter or more were also recorded by all five-year age groups of men under 20 years old and between 45 and 64 and by most groups of women in these age spans. Among the retirement-age groups, reductions of between 10 and 20 per cent were found. The weakest performance was in the range 20–44 years old, especially for men, for whom mortality rates for 30–34 year olds increased by 7 per cent and fell by less than 10 per cent for other groups.

Causes of death have altered greatly over the century, shifting away from infectious diseases towards degenerative illnesses. Even now, the picture varies greatly between age and sex. Three-fifths of childhood deaths result from accidents, neoplasms and congenital anomalies, while injuries and suicides are among the leading causes of death for young adults. For adults as a whole, approximately a quarter of all deaths is now due to cancer, a quarter to heart disease and a further one-eighth to cerebro-vascular disease (mainly strokes). While deaths from circulatory diseases have been falling, rates of cancer deaths have shown little reduction. Lung cancer accounts for almost one-third of all cancer deaths for males, though this proportion is now falling, and after breast cancer it is the second most important cause of cancer deaths for women and has been getting more important until recently (Dunnell 1995).

International migration

As noted above, the significance of international migration has altered dramatically over the past quarter of a century, with its overall balance switching from loss to gain in the early 1980s and its contribution to national population growth rising to well over one-third in the first half of the 1990s (Champion 1996). More detailed information for 1986–95, however, reveals how greatly the migration balance can vary from year to year, with the net inflow

TABLE 9.2 Net international migration, by citizenship, UK, 1986–95 (thousands)

			Non-British				
Year	*All*	*British*	*All*	*European Union*	*Old Commonwealth*	*New Commonwealth*	*Other foreign*
1986	58.2	–11.8	70.0	29.8	0.3	29.8	10.1
1987	30.1	–31.6	61.8	18.4	2.0	31.0	10.4
1988	17.8	–54.0	71.8	26.2	7.3	20.1	18.2
1989	90.9	–17.7	108.6	28.5	16.4	41.5	22.2
1990	88.3	–29.8	118.2	15.6	15.2	46.3	41.2
1991	73.3	–19.7	93.1	–1.9	9.7	45.3	40.0
1992	35.0	–33.9	68.9	5.0	3.1	42.8	18.0
1993	35.3	–35.1	70.5	2.3	7.3	34.4	26.4
1994	108.9	9.6	99.3	9.5	7.3	42.7	39.8
1995	108.8	–26.8	135.5	22.5	11.3	52.0	49.8
Annual averages							
1986–90	57.1	–29.0	86.1	23.7	8.2	33.7	20.4
1991–95	72.3	–21.2	93.5	7.5	7.7	43.4	34.8
IPS-based only							
1986–90	19.6	–29.0	48.6	9.4	8.2	21.5	9.5
1991–95	26.1	–21.2	47.3	7.7	7.7	19.9	12.0

Source: *International Migration: Migrants entering or leaving the United Kingdom and England and Wales, 1995*, Office for National Statistics Series MN 22, Table A, p. ix, and Table 2.1, p. 4 (London: The Stationery Office).

Notes: IPS = International Passenger Survey. All but the IPS-based data include adjustments for persons admitted as short-term visitors who are subsequently granted an extension, and estimates of migration between the UK and the Irish Republic. It is assumed that most of the latter are Irish and they have been treated as EU citizens. Numbers may not sum because of rounding.

for 1992 and 1993 being less than half that for the two previous years but then tripling for the next two years (Table 9.2). Part of the explanation for this lies in the fact that the net flow is the difference between two much larger gross flows which themselves can fluctuate substantially over time; for instance, between 1991 and 1995 annual recorded departures from the UK ranged between 191,000 and 239,000 and estimated arrivals between 251,000 and 300,000. At the same time, it must be recognised that, since there is no compulsory registration system of the type used for recording births and deaths, these migration statistics are based on estimates compiled from several different sources, the main one of which – the International Passenger Survey (IPS) – has a 95 per cent confidence interval of ±30,000 in its net migration figure.

The overall net figures hide some important differences in flows for separate groups of people (Table 9.2). The established pattern is of a net exodus of British passport holders and a net gain of non-British citizens. Over the ten years to 1995 the net loss of the former averaged 25,000 a year and the net gain of non-British citizens averaged some 90,000. Between the later 1980s and the first half of the 1990s, the overall upward shift in the annual average level of net immigration was produced fairly equally by a reduction in the net emigration of British citizens and a rise in the net arrivals of foreigners. The more detailed breakdown of foreigners reveals a marked slippage in the contribution of European Union passport holders but strong growth for the two other main contributors, with people from the New Commonwealth up by around 10,000 a year between the two five-year periods shown and 'other foreign' up by 14,000 (Table 9.2).

Considerable information is available about the characteristics of people interviewed by the IPS, allowing some insight into their impacts on the UK population. For instance, in 1995 four out of every five immigrants were aged 15–44, indicating a strong loading on the economically active and household-heading age groups. Emigrants also had a 15 to 44-year old peak, while the proportion of 15 to 24-year-olds was significantly lower than for immigrants and those for children and the elderly were larger, leading to net emigration for these two groups. Employed people made up around 55 per cent of both inflow and outflow, with professional/managerial personnel dominating both but being relatively more significant among the immigrants. Students were considerably more strongly represented among the inflows, while a higher proportion of the outflow was work-related. Over half the emigrants were UK-born, but only just over a quarter of immigrants were returners, with the largest net gains being of those born in New Commonwealth and 'other foreign' countries. While the IPS collects no data on ethnicity, these figures suggest a sizeable net boost to the UK's non-white population (see pp. 183–5).

Unfortunately, however, the IPS covers barely one-third of the UK's estimated net migration gains, as can be seen from comparing the IPS-based and overall five-year averages in Table 9.2. The biggest single change for the UK's migration balance since the early 1980s has been the growth in the number of asylum-seekers and 'visitor switchers', the latter being people who enter as temporary visitors but subsequently apply for permanent residence. These account for the majority of net gains that are not recorded by the IPS and, as can be deduced from Table 9.2, come almost entirely from the New Commonwealth and 'other foreign' countries.

Population distribution

Despite the UK having one of the highest population densities in the world, its population is very unevenly distributed and only quite a small proportion of its land surface is heavily settled. The degree of concentration can be measured at a variety of scales. At the broadest level, exactly half of the UK's population lives in the four standard regions which comprise the 'South' and make up barely a quarter of national territory. In terms of urban regions defined in terms of commutersheds around employment centres, just fourteen of the 283 Local Labour Market Areas (LLMAs) contain one-third of Britain's population, with London alone accounting for 1 in 7 of all residents and being larger than the next eight largest LLMAs combined (Champion and Dorling 1994). In terms of physically defined settlements, in 1991 2,307 'urban areas' (known by the census authorities as 'localities' in Scotland) accounted for almost 90 per cent of Great Britain's population despite occupying only 6 per cent of the land area (Denham and White 1998, and see Chapter 10).

This uneven distribution is a legacy of the UK's pioneering role in the industrial era and the associated rapid growth of large factory-based settlements at a time when personal mobility was very restricted. Fifty years of land-use planning policies designed to limit the outward spread of the largest built-up areas have reinforced this patterning. Even so, recent decades have seen some important changes in population distribution, in part a reflection of the changing nature and location of economic activities but also an outcome of changes in residential preferences and the importance attached to 'quality of life'. In general, population has become more concentrated in southern Britain, while at sub-regional and local scales deconcentration has been the prevailing tendency.

North–South drift

The principal feature of regional population change throughout the twentieth century has been the drift of population from North to South. As shown in Table 9.3, this tendency has continued into the 1990s, increasing the degree of population concentration in the four southernmost regions. The South East, the South West, East Anglia and the East Midlands combined accounted for over three-quarters of the 1 million national gain in 1991–96. All four regions saw their populations grow at rates of around 5 per 1,000 a year, well over twice the rate of any northern region apart from Northern Ireland. In aggregate, the South was growing by 5.4 per cent a year compared to only 1.5 per cent for the North.

The pace of this southward shift has varied over time, weakening after the first two decades of the post-war period but accelerating again more recently (Champion 1989). Dorling and Atkins (1995) have neatly summarised this shift in terms of Britain's mean centre of population, which moved southwards at a rate of 363 metres a year in the 1950s, 375 metres in the 1960s, down to 125 metres in the 1970s and back up to 369 metres in the 1980s. In 1991 it was located close to the village of Overseal between Swadlincote and the Derbyshire/Leicestershire border, some 27 kilometres south of its position in 1901. The 1990s experience represents a slowdown in this progression, but not by very much. The overall annual growth rate of the North was 0.4 per 1,000 between 1981 and 1991 and that of the South 4.9, a difference only marginally wider than the 3.9 points of 1991–6. That annual growth rates rose between the 1980s and 1990s in most northern regions appears more to do

TABLE 9.3 Population change, by country and region, 1981–96

Country and region	*Population 1996 (thousands)*	*Change 1991–96 (thousands)*	*Annual change rate 1991–96 (per thousand)*			*Annual change rate 1981–91 (per thousand)*		
			Overall change	*Natural change*	*Migration*	*Overall change*	*Natural change*	*Migration*
Northern Ireland	1,663	62	7.8	6.0	1.9	4.1	7.5	–3.3
Scotland	5,128	21	0.1	0.0	0.1	–1.4	0.5	–1.9
Wales	2,921	30	2.0	0.5	1.5	2.8	0.9	1.9
North	3,091	–1	–0.0	0.3	–0.4	–0.8	0.6	–1.4
North West	6,401	5	0.1	1.3	–1.2	–1.0	1.4	–2.3
Yorkshire and Humberside	5,036	53	2.1	1.7	0.4	1.3	1.3	–0.0
West Midlands	5,316	51	1.9	2.4	–0.5	1.5	2.7	–1.1
East Midlands	4,141	106	5.3	1.8	3.4	4.7	2.0	2.8
East Anglia	2,142	60	5.8	1.3	4.5	9.9	1.5	8.3
South East	18,120	484	5.5	3.5	1.9	3.7	2.7	1.0
South West	4,842	124	5.2	–0.1	5.3	7.7	–0.1	7.8
United Kingdom	58,801	993	3.4	2.0	1.4	2.6	1.8	0.8
North	29,556	219	1.5	1.5	0.0	0.4	1.6	–1.2
South	29,245	774	5.4	2.5	2.9	4.9	2.0	2.9

Source: Data supplied by Office for National Statistics. Crown copyright.
Note: Migration includes adjustments to reconcile population between mid-year estimates with estimates of natural change and net civilian migration, together with the effects of any boundary changes. Numbers may not sum because of rounding.

with their paralleling of the shift in the UK's growth rate than with redressing the imbalance between the two halves of the country (Table 9.3).

Table 9.3 shows the separate contributions made to these patterns and trends by natural change and migration. Natural change has been playing the lesser role, though its significance increased between the 1980s and the 1990s as the South's rate rose from 2.0 to 2.5 per 1,000 and that for the rest of the UK fell somewhat, leading to a widening of the South's premium from 0.4 to 1.0. Particularly notable in this context is the acceleration in the rate of natural increase in the South East, which in 1991–6 was responsible for more than half the UK's surplus of births over deaths. While Northern Ireland remains distinctive in this respect, owing to its much higher fertility rate than elsewhere in the UK, this province – along with most other regions in the northern half of the UK, except Yorkshire and Humberside – has seen a lowering of the rate of natural change between the two decades, against the UK trend.

Migration was responsible for the larger proportion of the North–South differential in the 1990s, though the width of the gap then was narrower than in the previous decade. It can be seen from Table 9.3 that the South's rate of migratory gain remained unaltered at 2.9 per

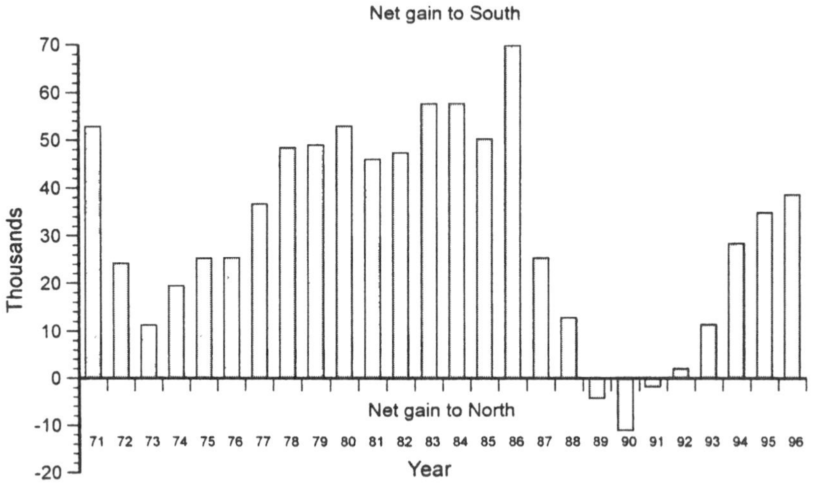

FIGURE 9.2 Net migration between the South and the rest of the United Kingdom, 1971–96
Source: Calculated from NHSCR data. Crown copyright.
Note: The South comprises the standard regions of South East, South West, East Anglia and East Midlands.

1,000, whereas the rate for the rest of the UK rose from –1.2 to a zero net balance. While the West Midlands, North West and North regions continued to experience net migration loss in the 1990s, all the northern parts of the UK recorded an upward shift in their migration balances from the 1980s. Wales showed a slight reduction in its positive migration rate. Among the four regions of the South, it is notable that, while the South West and East Anglia registered marked slowdowns from their high migratory growth rates of the 1980s, the South East and, to a lesser extent, the East Midlands saw their migration gains accelerate in the first half of the 1990s.

These rather complex changes in regional migration patterns arise from the fact that at least three different processes have been operating simultaneously and possibly interacting with each other. In the first place, the late 1980s saw a marked reduction in the drift of people from North to South within the UK, to the extent that in the three years 1989–91 the net flow switched in favour of the North (Figure 9.2). Second, around the same time, there was a cutback in the scale of net exodus from the major cities, notably that from London into East Anglia and the South West (see the next section). Third, a surge in net immigration in the late 1980s and early 1990s had a positive impact on the migration balances of most parts of the UK. However, by far the largest effect was felt within the South East which absorbed over two-thirds of the UK's net gains. The first two processes are mainly linked into changes in the economy, notably the severe recession in the South East towards the end of the 1980s which reduced its attractiveness as a destination of migrants from more northerly parts of the UK and brought a marked downturn in housebuilding more generally. It is also likely, though not as yet clearly proved, that over this period the immigration gains in the South East reduced the availability of both housing and job opportunities for potential in-migrants from the rest of the UK.

Urban–rural shift

Urban decentralisation, in the sense of suburban housebuilding, dates back to the nineteenth century, but in more recent decades it has been operating on a much broader geographical canvas. Not only is population moving out from the inner areas of individual cities but it is also shifting out of the largest urban concentrations to towns and more rural regions – a process sometimes referred to as 'counterurbanisation' and known to the media as the 'urban exodus'. As detailed in the second edition of this book (Compton 1991: 47), these changes are better studied using specially designed representations of urban regions and their internal structure as opposed to the standard administrative areas for which mid-year estimates are available, so the account below focuses on change between Census years.

Figure 9.3, based on the Functional Regions framework (Champion and Dorling 1994), clearly illustrates the dual nature of the urban dispersal process, albeit in a highly summary form. In the latest intercensal period 1981–91, a regular gradient of growth rate is evident across the four LLMA size groups in both northern and southern halves of Britain. The 'large dominants' at the top of the urban hierarchy register the lowest growth. At each subsequent level (that is, cities, towns and rural areas) rates in the South become progressively stronger. The pattern is almost as consistent for the four types of zone within the Functional Regions, with their main built-up areas (cores) losing out to their primary commuting areas (rings), and especially to the outer and rural areas beyond. More detailed analyses (Champion and Dorling 1994) show that these two dimensions of deconcentration have been occurring side by side, with the more localised movement of population out of the cores affecting not only the large dominants but also the cities and towns, which at the same time are gaining people from the former. Such is the prevalence of these shifts out of heavily built-up areas and down the settlement hierarchy that the whole urban dispersal process has been dubbed the 'counterurbanisation cascade'.

Even so, the pace of deconcentration has not been uniform over time, with the 1980s patterns just described representing a considerable convergence from those of the previous decade at both scales. Figure 9.3 shows how the large dominants in both North and South plunged from modest growth to massive decline between the 1950s and the 1970s, while the rural areas experienced a strong upward shift in growth rates, but the trend goes into reverse between the 1970s and the 1980s with a considerable narrowing of the growth gradient. The latter was principally the result of the recovery of the large dominants, notably in the South where London exerts a major influence, and a cutback in the growth of rural areas and towns. At the more localised scale of zone types a similar pattern is evident, with cores weakening through to the 1970s and then beginning to recover in the 1980s as the growth of the outer and rural areas fell back.

The partial recovery of the largest LLMAs and their main built-up areas in the 1980s is also reflected at the most detailed spatial scale – that of their inner-city areas. Special tabulations based on the small-area statistics of the last three Censuses reveal that, for London and five of the six principal metropolitan cities (Sheffield being the exception), inner areas saw a marked reduction in their level of population loss between the 1970s and the 1980s. Most impressively of all, inner London switched from being amongst the greatest losers into a position of overall gain (Figure 9.4).

In terms of explanation, recent research has focused largely on the slowdown in deconcentration since the 1970s. Particular attention has been given to the effects of

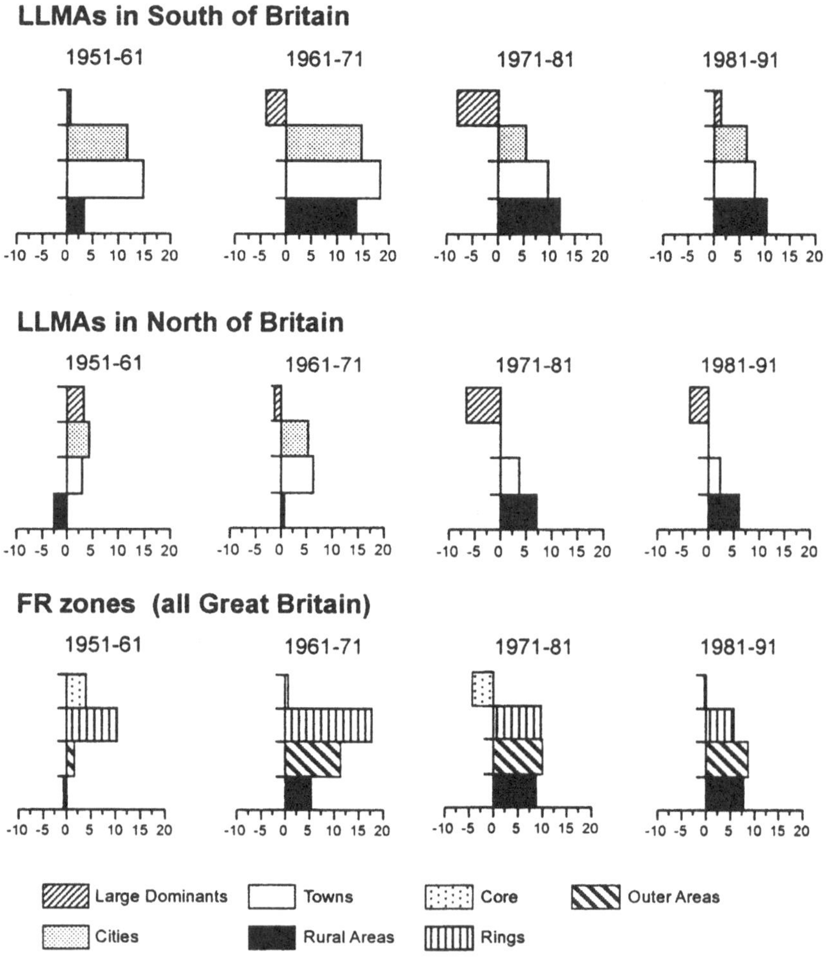

FIGURE 9.3 Population change, by Local Labour Market Area types and Functional Region zones, 1951–91, Great Britain

Sources: Calculated from Census, ESRC/JISC data purchase. Crown copyright. See Champion and Dorling (1994).

economic restructuring in the 1980s, notably the exporting to cheap-labour countries of the industries which led the job decentralisation of the 1970s and the growth of the financial and business services sectors in the hearts of the largest cities, especially London. The latter, combined with the 'baby boomers' reaching adulthood, is associated with the rise in the 'yuppy' phenomenon and the spur to the gentrification of inner-city areas (see Chapter 10). In addition, the attractiveness of more rural areas fell as increasing pressures led to rising house prices and traffic congestion, and as the quest for greater efficiencies prompted some withdrawal of local shopping, transport and public services from smaller settelements (see Chapter 11). Even so, in demographic terms, the principal reason for the urban recovery can

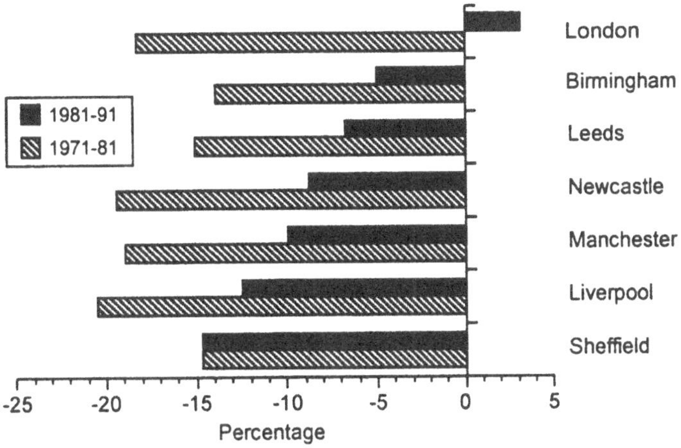

FIGURE 9.4 Population change for inner areas of London and England's six Principal Metropolitan Cities

Source: Calculated from Census, ESRC/JISC data purchase. Crown copyright.

be found in the post-1980 surge in net immigration, which has impacted most on the inner areas of London and certain other cities. Net out-migration from metropolitan England to the rest of the UK, while fluctuating in the short term, has been averaging 90,000 people a year since the mid-1970s.

In summary, population redistribution has for decades been dominated by the twin processes of North–South drift and urban– rural shift, both of which have fluctuated in pace but look set to continue into the new century. At the same time, it is important to recognise the diversity of population trends arising from the peculiar circumstances of individual places and the particular combinations of migration streams that affect them. Smaller towns can be especially affected by the opening or closure of factories and coal-mines, while alterations in the migration behaviour of specific groups like students and retirees can have significant effects even on larger towns and cities. Something of this diversity is apparent from Table 9.4, which shows not only the range of growth rates between different types of local authority districts but also the widespread occurrence of population losses, even for those types which on average are the most prosperous and fastest growing.

Population composition

While the UK's population is growing again in overall size and continuing to redistribute itself geographically, the most remarkable changes of the past quarter of a century have been those relating to demographic structure. In particular, the population has been ageing, it has been getting more diverse in ethnic terms and it has been dividing itself up into smaller and more varied types of household. These three trends have already had important implications for many aspects of national life and well-being and they look set to continue into the future.

TABLE 9.4 Population change by families and groups of local authority districts, England and Wales, 1991–5

Families and groups of local authority districts	*Population 1991 (thousands)*	*Population change 1991–5 (thousands)*	*%*	*Number of districts*		
				Total	*Increasing*	*Decreasing*
Rural areas	9,215	194	2.1	114	97	17
Upland coast and country	4,970	129	2.6	70	60	10
Mixed urban and rural	4,245	65	1.5	44	37	7
Prospering areas	12,111	266	2.2	111	92	19
Growth areas	9,549	203	2.1	86	73	13
Most prosperous	2,562	63	2.5	25	19	6
Maturer areas	5,785	116	2.0	41	30	11
Service and education centres	3,292	62	1.9	18	13	5
Resort and retirement	2,493	54	2.2	23	17	6
Urban centres	10,639	104	1.0	60	36	24
Mixed economies	5,109	49	1.0	37	22	15
Manufacturing	5,530	55	1.0	23	14	9
Mining and industrial areas	9,761	–17	–0.2	58	23	35
Ports and industry	3,899	–7	–0.2	15	5	10
Coalfields	5,862	–10	–0.2	43	18	25
Inner London	3,589	58	1.6	18	13	5
England and Wales	51,101	721	1.4	402	291	111

Source: *Monitor Population and Health* PP1 96/2, 1996, Table C and Table 4 (London: Office for National Statistics). Data are crown copyright.
Note: Trends after 1995 cannot be studied on this basis because of local government reorganisation.

Moreover, while there is a clear geography of age, ethnicity and household type inherited from earlier decades, most parts of the country have found that the recent changes in their population profiles are due mainly to their tracking of the national trends. These and other aspects of socio-demographic restructuring are examined in more detail in Champion *et al.* (1996), and some of the consequences for policy at national and local scales are explored in Champion (1993).

Age structure

The rise in the elderly population has been the most consistent theme of the past century (Grundy 1996). The proportion of people at or over the current pensionable age (65 for men, 60 for women) more than tripled from its 1901 level of around 5 per cent to 18 per cent in

1996. In recent years the most marked change has been in the numbers of the very elderly, with those 85 and over more than doubling between 1971 and 1996 (up by 120 per cent) and those aged 75–84 increasing by 45 per cent. This growth derives primarily from the very large birth cohorts of the first years of this century, which avoided the carnage of the First World War and then lived through successive periods of improving survival chances. For some years now, this has been posing a great challenge in terms of health-care and social support, not least because of the relatively large proportion who do not have children to support them owing to the low fertility of the inter-war period. Unfortunately, the continuing increases in life expectancy appear to be providing extra years of disability rather than of healthy life (Dunnell 1995).

The overall process of population ageing is also greatly affected by trends in fertility, with the most notable features currently being the low birth rate and the progress of the 1960s baby boom through the age groups. As shown in Figure 9.5, already in the past twenty-five years the proportion of under-16-year-olds has fallen by one-sixth and is projected to reduce further to under 18 per cent by 2021 (see Botting 1996 for further details of the child population). The passage of the baby boom is reflected in the big increase in 30 to 44-year-olds between 1971 and 1996 and the even larger projected rise in the proportion of between 45 and pensionable age over the next twenty-five years. Note, however, that the latter is inflated by the effects of the government's decision to raise the official pensionable age for women from 60 to 65 in its effort to curb the escalation of the pensions bill thereafter. Overall, the mean age of the population is expected to rise from 38.4 to 41.9 years between 1996 and 2021.

At the same time, there is a marked geography of age structure that means that certain parts of the UK have already been wrestling for some time with these challenges. Seaside resorts, spa towns and rural areas have for decades been characterised by older than average age structures, produced by a combination of retirement in-migration and the exodus of young adults. The extreme districts at the 1991 Census were Christchurch (Dorset) and Rother (East Sussex), both with pensioners comprising over one-third of their populations. At the other extreme, with elderly people making up less than one-eighth of their residents, were new and expanded towns like Tamworth (Staffordshire) and Milton Keynes (Buckinghamshire), which along with the earlier new towns have already had to cope with sizeable reductions in school rolls and will find their numbers of older people growing very rapidly over the next two or three decades.

Ethnic minority populations

While Britain has a long history of immigration and emigration, the growth of its non-white population is essentially a phenomenon of the last fifty years. It began with the intensive recruitment of West Indians to combat early post-war labour shortages. Initially measured in terms of numbers born in the New Commonwealth (see Haskey 1997), the non-white population is estimated to have risen from around 218,000 in 1951 to 541,000 in 1961 and to 1.15 million in 1971. From then on, unfortunately, this measure has become increasingly inaccurate, as more children were born to immigrants in the UK and as the numbers of non-white arrivals from other foreign countries mounted (see p. 175). In 1991, however, for the first time a question on ethnicity was asked in the Census of Great Britain (but not in

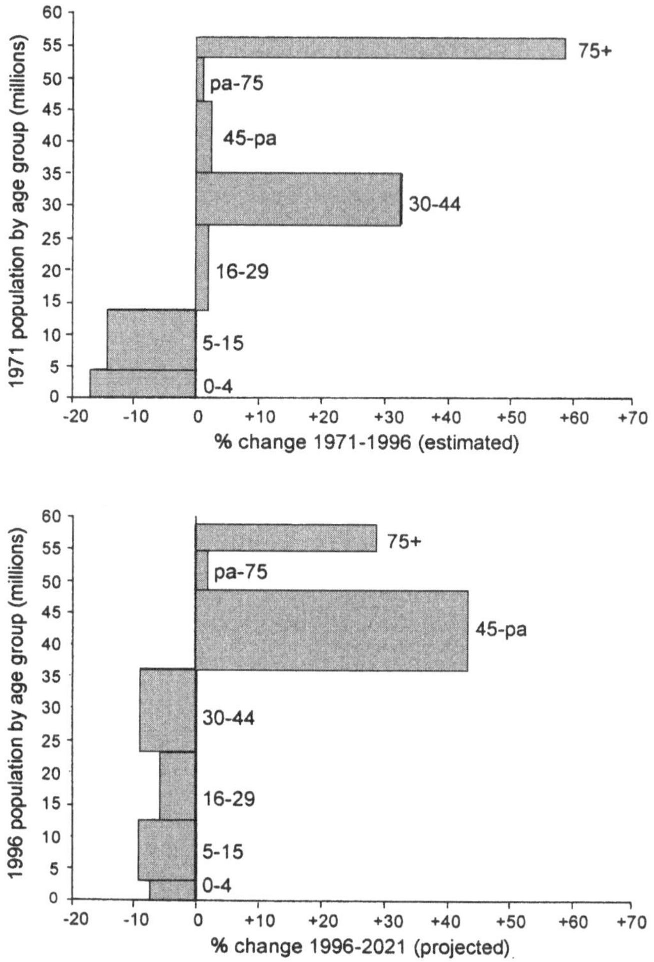

FIGURE 9.5 Population change by age group, 1971–2021, United Kingdom

Source: Calculated from *Annual Abstract of Statistics* 134, Table 2.3 (London: The Stationery Office, 1998).

Note: 'pa' refers to pensionable age, which in 1971 and 1996 is 65 for men and 60 for women but in 2021 is 65 for both sexes, see text.

Northern Ireland), eliciting a figure of 3.01 million for the non-white population. The latest estimate, derived from the Labour Force Surveys of 1995/6, is 3.25 million, or 5.8 per cent of Britain's total population. Of these, almost exactly half (49 per cent) were UK-born and almost one-third (31 per cent) were aged under 15, a far higher share than the rest of the population's 18 per cent. Perhaps most impressively, over the period 1971–96 the non-white population contributed almost three-quarters of the UK's population growth.

At the same time, the non-white population, far from being a homogeneous entity, is extremely varied in its ethnic background, geographical origins and characteristics, as shown in Table 9.5. The Black-Caribbean group makes up 15 per cent of Britain's non-white

TABLE 9.5 The population of Great Britain 1995/6, by ethnic group

Ethnic group	*Persons (thousands)*	*Share of GB's non-whites (%)*	*Proportion of ethnic group (%)*		
			Born in UK	*Aged 0–14 years*	*In Greater London 1991*
Black – Caribbean	484	15	60	23	58
Black – African	289	9	32	30	77
Black – Other (n-m)	96	3	87	47	45
Black – Mixed	147	5	91	60	n/a
Indian	868	27	44	25	41
Pakistani	554	17	51	39	18
Bangladeshi	184	6	40	40	53
Chinese	123	4	29	19	36
Other – Asian (n-m)	171	5	19	23	57
Other – Other (n-m)	135	4	32	27	42
Other – Mixed	200	6	75	51	n/a
All ethnic minority groups	3,251	100	49	31	45
White	52,903	—	96	18	10
All ethnic groups	56,154	—	93	20	12

Source: Haskey (1997) Tables 2b, 4, 5 and 8a, calculated from 1995/6 Labour Force Surveys and 1991 Census. Crown copyright.
Notes: n-m = non-mixed; n/a = not applicable, as the 1991 Census did not distinguish Mixed separately.

population and, being in the vanguard of the post-war surge of immigration, contains the highest proportion born in the UK (apart from the mixed-race groups) and has one of the smallest proportions of children. By contrast, Bangladeshis, as would be expected of the most recent arrivals among the large national groups, have a relatively low proportion born in the UK and a youthful age structure. Indians and Pakistanis form the two largest national groups and have a similar timing of arrival in the UK, but they differ in terms of their age structure, with Indians having fewer children.

The ethnic minority groups vary greatly in their geographical distribution, too. According to the 1991 Census, 45 per cent of all non-whites in Great Britain live in London, but the proportion ranges from over three-quarters for Black-Africans to under one-fifth for Pakistanis (Table 9.5). At the district level, there are some notable concentrations of individual groups. For example, Black groups make up 22 per cent of Hackney's population, Indians 22 per cent of Leicester's, Bangladeshis 23 per cent of Tower Hamlet's and Pakistanis 10 per cent of Bradford's (Champion *et al.* 1996). Nevertheless, apart from the Chinese, who are much more evenly distributed around the country, the New Commonwealth and Pakistani ethnic minority groups have in common a highly urban lifestyle, living mainly in the inner parts of the larger cities. Moreover, though they are showing some signs of moving into more suburban areas, their existing patterns of concentration are being reinforced by natural increase and new arrivals from overseas (see Chapter 10).

Household size and composition

A third theme is the decline in average household size and the growing diversity of household types (Haskey 1996). Average household size has almost halved over the past century, falling from 4.6 persons per household in 1901 to 2.43 in 1996. The most recent (1992-based) household projections for England assume that the figure will be down to 2.17 by 2016. As a corollary, the number of households has been growing much more rapidly than the population. Between 1961 and 1991, for instance, in Great Britain the former increased by 38 per cent, the latter by only 9 per cent. Similarly, the projections for the period 1991–2016 suggest that, alongside population growth of around 8 per cent, the number of households in England will rise by 23 per cent – an absolute increase of 4.4 million households. Such an increase is generating a great deal of planning concern and attempts are being made to minimise the impact on the countryside (see Breheny and Hall 1996).

The direct reasons for the fall in average household size are the rapid increase in households comprising only one person (up from 17 to 26 per cent of all households between 1961 and 1991) and the falling number of large households (the share of households with six or more people down from 6 to 2 per cent). The latter is due partly to the fall in family size since the early 1960s (see pp. 171–3) and partly to a cutting by two-thirds of the number of families sharing accommodation with each other. The growth in individual persons living alone arises from a number of factors which indicate clearly the growing diversity of household types: more elderly loners due to greater longevity, absence of (nearby) children and ability to maintain separate households; more young adults moving away from home for higher education or first job or to gain independence; and more middle-aged people divorcing or separating, particularly men leaving behind partners with children. Related both to the latter and to the growth of single motherhood is the rise in lone-parent families with dependent children, up from 2 to 6 per cent of all households between 1961 and 1991.

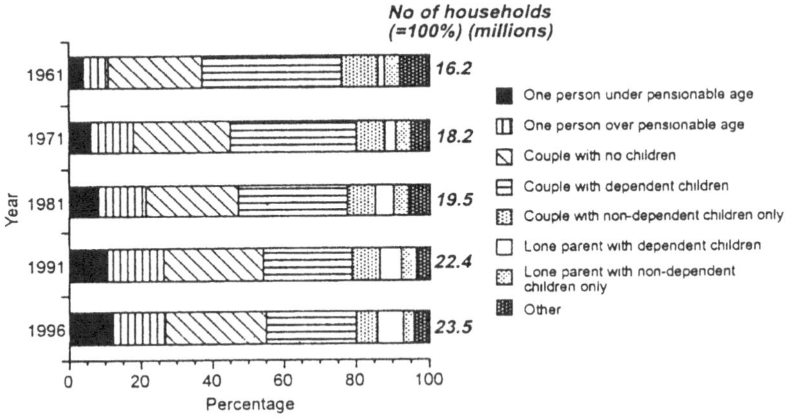

FIGURE 9.6 Composition of households, by type of household and family, 1961–96, Great Britain

Sources: Censuses, and 1996 General Household Survey. Crown copyright. After Haskey (1996).

The main casualty of these changes has been couples with dependent children, down from 38 to 25 per cent over this thirty-year period (Figure 9.6).

There is a clear geography to these household patterns, partly related to the two other aspects of population composition just described (Champion *et al.* 1996; Haskey 1996). Besides Northern Ireland where the proportion of large (6+ person) households is three times that of the next largest region (West Midlands), the main concentration of high average household size in 1991 was central southern England, arising from the strong in-migration of families with children. Below-average size is found in more remote rural areas and other areas which have gained retirement migrants, the extreme case being East Sussex at 2.24. The main conurbations are distinctive in having above-average shares of both very small and very large households, mainly as a result of their attractiveness for both young adults and ethnic minorities.

Conclusion

The processes underlying national population change have altered greatly over the past half century and are continuing to exhibit new developments. Craig (1997: 12) has summarised the key trends of the ten-year period since 1984 as follows: 'Population growth has resumed for, although fertility is still low, the expectation of life is rising and there is now a net gain of people from migration. Life styles have altered in that there has been a steady increase in cohabitation and a sharp rise in births outside marriage.' Also impressive are the steady growth of the ethnic minority population and the fluctuations in the pace of the two main dimensions of population redistribution, the North–South drift and the urban–rural shift, especially the stronger demographic basis of the major cities in recent years.

Since such important changes can occur within a decade or so, it is perhaps not surprising that those involved in anticipating the future stress the degree of uncertainty that is involved in projections. In the latest (1996-based) population projections for the UK, Shaw (1998) demonstrates the effect of adopting what he reckons to be the likely range between high and low variants for the three main components. Looking ahead to the year 2036, the mortality variants produced a range of 0.75 million around the principal projection, migration a range of around 2 million and fertility the largest range of some 3.5 million. It is certain, however, that, whatever the changes that do occur, they will – like those of the past few decades – have extremely important ramifications for the economy, society and geography of the UK.

References

Armitage, B. and Babb, P. (1996) 'Population review: (4) Trends in fertility', *Population Trends* 84: 7–13.

Botting, B. (1996) 'Population review: (7) Review of children', *Population Trends* 85: 25–31.

Breheny, M. and Hall, P. (eds) (1996) *The People – Where Will They Go?* London: Town and Country Planning Association.

Champion, T. (1989) 'Internal migration and the spatial distribution of population', in H. Joshi (ed.) *The Changing Population of Britain*, Oxford: Basil Blackwell.

Champion, T. (1996) 'Population Review: (3) Migration to, from and within the United Kingdom', *Population Trends* 83: 5–16.

Champion, T. (ed.) (1993) *Population Matters: The Local Dimension*, London: Paul Chapman Publishing.

Champion, T. and Dorling, D. (1994) 'Population change for Britain's functional regions, 1951–91', *Population Trends* 77: 14–23.

Champion, T., Wong, C., Rooke, A., Dorling, D., Coombes, M. and Brunsdon, C. (1996) *The Population of Britain in the 1990s*, Oxford: Clarendon Press.

Compton, P.A. (1991) 'The changing population', in R.J. Johnston and V. Gardiner (eds) *The Changing Geography of the United Kingdom*, London: Routledge.

Craig, J. (1997) 'Population Review: (9) Summary of issues', *Population Trends* 88: 5–12.

Denham, C. and White, I. (1998) 'Differences in urban and rural Britain', *Population Trends* 91: 23–34.

Dorling, D. and Atkins, D. (1995) *Population Density, Change and Concentration in Great Britain 1971, 1981 and 1991*, Studies on Medical and Population Subjects 58, London: HMSO.

Dunnell, K. (1995) 'Population Review: (2) Are we healthier?', *Population Trends* 82: 12–18.

Grundy, E.M.D. (1996) 'Population Review: (5) The population aged 60 and over', *Population Trends* 84: 14–20.

Haskey, J. (1996) 'Population Review: (6) Families and households in Great Britain', *Population Trends* 85: 7–24.

Haskey, J. (1997) 'Population Review: (8) The ethnic minority and overseas-born populations of Great Britain', *Population Trends* 88: 13–30.

Shaw, C. (1998) '1996-based national population projections for the United Kingdom and constituent countries', *Population Trends* 91: 43–9.

Further reading

A summary view of the main changes since the early 1970s, forming the first of a series of nine articles in the Office for National Statistics' ten-yearly review of population trends is provided by B. Armitage (1995) 'Population review. Structure and distribution of the population', *Population Trends* 81: 7–16. The topics covered in more detail in the subsequent articles are mortality and health (Dunnell 1995), migration (Champion 1996), fertility (Armitage and Babb 1996), the elderly (Grundy 1996), families and households (Haskey 1996), children (Botting 1996) and ethnic minority populations (Haskey 1997), with Craig (1997) providing a summary round-up of issues. Note that *Population Trends* appears quarterly and includes a standard set of Tables updating the key features of population change, mainly at national level. T. Champion (ed.) (1993) *Population Matters: The Local Dimension* (London: Paul Chapman Publishing) offers a collection of ten essays on local population characteristics and trends and on their policy implications, including those for schools, the labour market, housebuilding, health-care and local government finance. T. Champion, C. Wong, A. Rooke, D. Dorling, M. Coombes and C. Brunsdon (1996) *The Population of Britain in the 1990s* (Oxford: Clarendon Press) is an atlas with accompanying text, focusing on the demographic and social features which have been changing most rapidly in the past decade and which have had the most significant policy impacts. A 680-page volume, forming the principal work of reference in this area, with detailed coverage including historical background, changing distribution, components of population change, ethnic minority populations and policy issues is D. Coleman and J. Salt (1992) *The British Population* (Oxford: Oxford University Press). An impressive atlas drawing primarily on Census data and providing a highly original perspective on the British population using population-based cartograms at district and ward level, with sections on population change, demography, labour market, housing, health, society and politics is D. Dorling (1995) *A New Social Atlas of Britain* (Chichester: John Wiley and Sons). S. Jackson (1998) *Britain's Population* (London: Routledge) is a readable introductory text which includes chapters on population data sources and demographic methods as well as on historical trends, fertility and mortality, migration and distribution, and current policy issues.

Chapter 10

Towns and cities

David Herbert

Introduction

The forces which transformed Britain's economy from an agricultural to an industrial base during the eighteenth and nineteenth centuries also brought about a major redistribution of population and established the bases for an urban nation. Already by the mid-nineteenth century, over half the British population could be classed as urban dwellers, and by the end of the twentieth century this figure was of the order of 80 per cent and mirrored the overall situation in the European Union. The shape of urban Britain was influenced by pre-existing settlements, but many new towns and cities were added. As British cities grew, so changes in transport technology allowed them to widen their spheres of influence and create city regions. The description of transport as the 'maker and breaker' of cities summarised its effects. Throughout the nineteenth century and the early twentieth, transport as the maker of cities focused on their centres and underpinned the dominant processes of concentration and centralisation. From about the 1920s, more flexible transport systems, principally private cars, led to deconcentration and decentralisation, the dispersal of people and activities; transport became the 'breaker' of cities.

Accompanying the dominant flow outwards from city centre towards the peripheries were regional-scale shifts prompted by planning strategies to contain urban growth and divert some activities to the north and west of Britain. Containment was effected principally through planning controls and green-belt policies, aided by major initiatives such as new towns and town expansion schemes. The redistribution of activities proved more difficult to achieve, as a succession of failed regional policies testified (Prestwich and Taylor 1990). At the same time, the urban system had become more complex. Concepts such as the city region, metropolitan area, and local labour market area, reflected attempts to capture the essence of new urban forms resting on flows and linkages rather than physical contiguity. Globalisation added a further dimension and, whereas the idea of external, global impacts on local urban systems was not new, the rising intensity of transnational flows of goods, capital

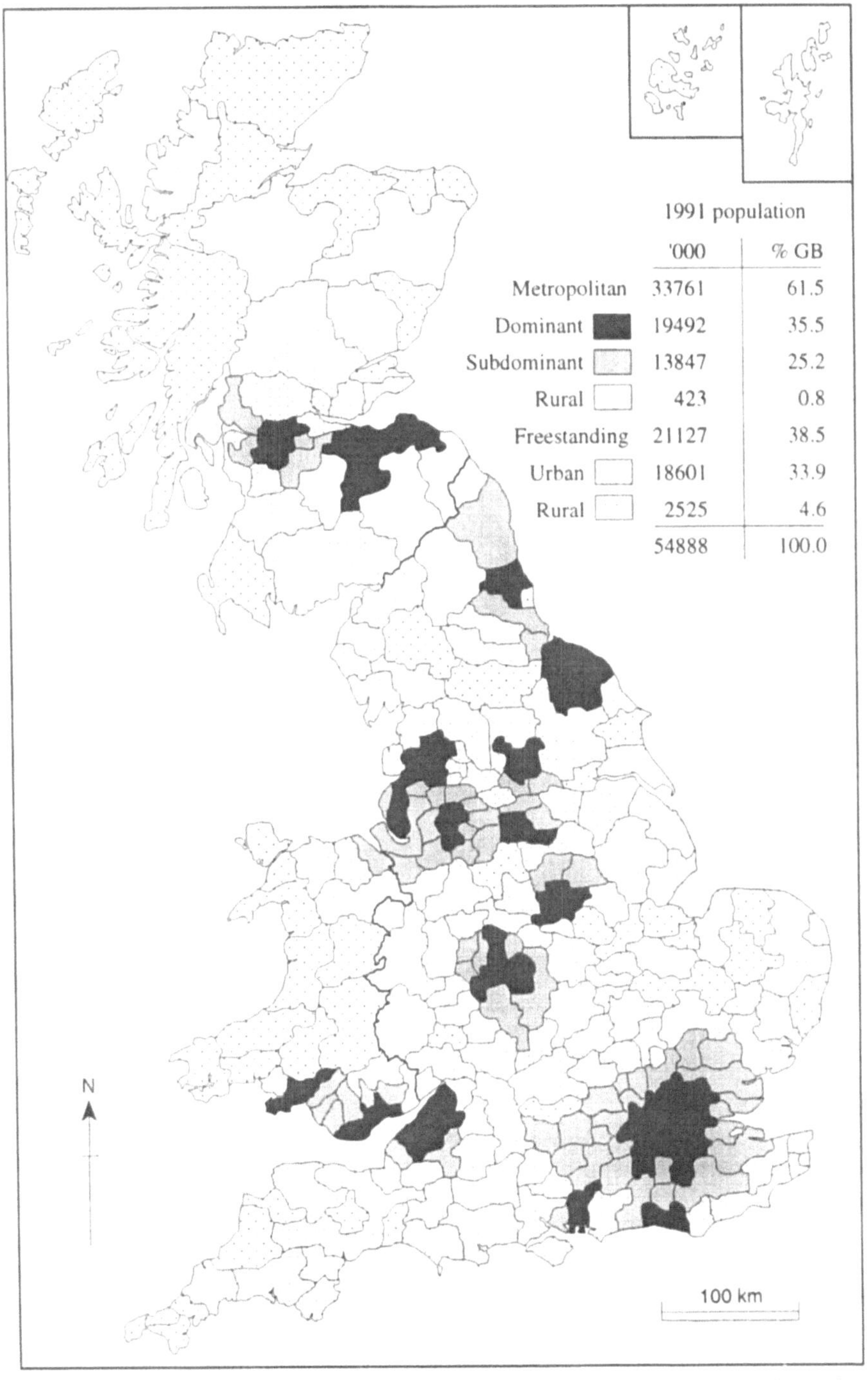

FIGURE 10.1 Framework of local labour-market areas (LLMAs) in the United Kingdom
Source: After Champion (1994).

and information over the last two decades of the twentieth-century has given this process a new meaning. The nature of global–local interaction is a major issue in a debate which both recognises the power of wider national and international forces and the relevance of local socio-political formations. The argument that cities have vital roles as creative control and cultural centres within the broad dynamics of globalisation (Ash and Graham 1997) approaches the right balance between local and global effects.

Monitoring urban change in the United Kingdom has become a focused research activity most closely associated with the Centre for Urban and Regional Development Studies (CURDS). The Daily Urban System consists of a Core and its primary commuting field or Ring; the Local Labour Market Area (LLMA) adds the Outer Areas from which commuters are drawn (Figure 10.1). In 1981, it was estimated that 61.6 per cent of the British population lived in cores, 26.6 per cent in rings, 6.6 per cent in outer areas, and 5.2 per cent in rural areas. Comparisons up to 1991 are difficult because of redefinitions of areas, but the main divisions of LLMAs are metropolitan areas, divided into dominant, subdominant and rural, and free-standing areas, divided into urban and rural. Metropolitan LLMAs held 61.5 per cent of total population in 1991 compared with 62.8 per cent in 1981, but free-standing LLMAs had increased from 37.2 per cent to 38.5 per cent.

This chapter will consider both macro processes of economic and social change which affect cities and patterns of change and interaction within urban areas. Initially, there is some focus on processes such as decentralisation and counter-urbanisation that have regional impacts within the United Kingdom. Within the mosaic of the city, there is discussion of specific sectors such as housing, inner city, commercial centres and employment and also of key issues such as problem areas, ethnic districts and the trends towards gentrification. Later sections are concerned with management and the application of urban policies.

Processes of urban change

Economic decentralisation has been a continuing feature of British cities, and its impacts on the manufacturing and industrial bases were its main manifestations. As Hall (1985) observed, it took over a hundred years, from 1851 to 1951, for technology and foreign competition to halve the numbers employed in British agriculture, but it took only thirteen years from 1971 to 1983 to cut manufacturing jobs by one-third.

> There are striking parallels between the 1880s and the 1980s. Then rural England was in the process of losing much of its traditional economic base. Now it is urban England's turn. The root cause in both was structural transformations arising from the new technologies and the changing balance of geographical disadvantage.
>
> (Hall 1985: 10)

Deindustrialisation ran strongly during the 1970s and 1980s, and between the early 1970s and the early 1990s manufacturing jobs in Britain declined from 7.5 to 4.3 million. The job losses in the inner cities could be most strongly tied to the closure of factories unable to cope with cramped sites, inadequate accommodation and intense market competition. There were transfers of firms, forced moves because of urban renewal, and 'deaths' of uneconomic units. Outside the inner cities, firms were more successful and the expansion of information technology and services employment was marked on the crescent from Cambridge to the M4 corridor; numbers of employees in the south of the country increased by 1.35 million between 1979 and 1990. The North–South divide, despite some apparent reversals, shows a remarkable persistence, and Champion and Green (1991) concluded that most of the best performing LLMAs on economic indicators were in a halo around London whereas the worst were from northern Britain. The divide had widened, but there was also greater diversity within regions that made a simple dichotomy misleading. Green *et al.* (1994) noted that when recession hit the British economy in the early 1990s large northern cities were among the last to feel the impact. In this latest recession it was the construction, financial and services sectors which experienced retrenchment and losses, halting the rapid rise of the producer services sector which had fuelled growth in London and the South East. The fortunes of London were also adversely affected by the collapse of the commercial property market in 1991 that left the capital with a huge surplus of office space. What were described as the 'first green shoots' of recovery appeared in the North, Midlands and Wales, rather than in the South. Older regions such as South Wales had shown some success in attracting investment and there were some forty Japanese firms there by the mid-1990s, with total employment rising from 940,000 in 1981 to 964,000 in 1991.

Decentralisation of people

All parts of metropolitan Britain, the large free-standing cities and the older industrial regions lost population through migration between 1981 and 1991. Overall population loss was greatest from the central parts of main cities and there was little to suggest that Britain is not experiencing a continuing drift away from metropolitan areas. The drift was slower in the 1980s than in the previous decade, but smaller towns and rural areas continued to show relative growth. Migration was reflected in increased suburbanisation, and counter-urbanisation was recognised as an additional trend. In some ways counter-urbanisation and suburbanisation were indistinguishable and formed parts of an ongoing dispersal process. There was also, however, a form of product cycle dispersal whereby some economic activities were relocating to small towns in rural settings. Allied to this were indications that increased use of the Internet and 'Tele-working' were having some effect on the distribution of workplaces. Such changes were most evident in outer Metropolitan area corridors formed around motorways in parts of southern England such as Berkshire, Hampshire, Surrey, Bedfordshire and Buckinghamshire. Bibby and Shepherd (1997) spoke of the 'Golden Belt' in England from Dorset, through Wiltshire, Oxfordshire, Berkshire and Buckinghamshire, to Cambridge. It was here that they predicted that the pressure for more land for housing to accommodate the 4.38 million extra households expected in England by the year 2016 would come, thus threatening green-belt land. Remoter, mainly rural districts, showed significant decennial increases in population in all standard regions of the United Kingdom and reached 115 per cent in the South West and 84 per cent in East Anglia. In *Social Trends* (1997), it was predicted that the population of the non-metropolitan counties and Greater London would increase by between 3 per cent and 4 per cent between 1993 and 2001. Greater London would show the greatest increase in numbers of children and East Anglia of those of working or retirement ages. Counter-urbanisation remains a credible but not unambiguous British trend and the clearest indicators are those of a resurgence of non-metropolitan growth.

Given these impacts upon the countryside, there are questions on the survival of rurality in Britain. Is there an 'urban' and a 'rural' in traditional senses of the terms? Clearly in terms of lifestyles and values the distinctions have become blurred, but differences persist. Cloke *et al.* (1997), in their studies of rural England, found that people, particularly long-term rural residents, acted as if rurality still had a geographical dimension. They had an imagined rural community to which they felt they belonged, and attitudes towards in-migrants who either tried to take over or did not participate in the local community were antagonistic.

London, with its status as a 'world city', has been a special case. Inner London has shown a significant improvement in its population change since the low point of the early 1970s and its ability to achieve positive change may be symptomatic of new forces at work in the centres of really major urban areas. The London phenomenon remains perplexing. Whilst there were clear new forces at work in the 1980s – renewed interest in parts of the inner city, new trends in the housing market, and the swing towards producer services – the net longer-term impacts of these remain difficult to judge. Gordon (1988) argued that there were no fundamental changes in the process of population deconcentration but fluctuations stimulated by the vagaries of the housing market and demographic characteristics of new inner-city populations. Although London, and some other large British cities, continues to attract large numbers of in-migrants, often from overseas, it is not of the scale experienced by large American cities such as New York and Los Angeles where such in-migration creates overall increases in inner-city populations. Finally, it must be noted that London is a large area and contains a range of urban places with often separate profiles and experiences.

Urban housing markets

Economic restructuring and the changing geography of employment opportunities are critical factors in understanding the changing form of British cities. They were not the only factors, however, and in some ways they followed rather than led the outward movement of people. To understand this outward movement, there are social as well as economic factors to consider: residential preferences, the housing market, the infrastructures including transport, and changing government policies. Urban housing development was strongly influenced by the fiscal advantages of owner-occupancy, the rising costs of land and buildings, and an interventionist planning system.

The advantages of owner-occupancy were related to advantageous tax positions on mortgage repayments and rapid increases of house prices that made purchase a very worthwhile investment. This has remained true though *Social Trends* (1997) estimated that on average owner-occupiers with mortgages had expenditures on housing three times as much as either outright owners or social sector tenants. Successive governments encouraged the growth of owner-occupancy, and the only real deviations from this policy came in the later 1980s with restrictions on the amounts of mortgage funds qualifying for tax relief. By the early 1990s there was also a recession in house prices, affecting London in particular, and widespread negative equity (people holding mortgage-loans which were greater than the value of their property), especially among buyers in south-east England who had purchased between 1988 and 1991. This recession was short-lived and Nationwide Building Society estimated a halving of negative equity between late 1996 and mid-1997, with falls most pronounced in the South East although it was still the worst affected area. In 1997, Nationwide recorded a 12.5 per cent rise of house prices over the past year.

The state has intervened in the housing market as both a builder and landlord of 'affordable housing'. The public sector share of total housing stock rose most rapidly between 1947 and 1961 when it changed from 12 per cent to 25 per cent in England and Wales. By 1995, this public sector share had fallen to 18 per cent in England and Wales, though it remained higher in Scotland (31 per cent) and Northern Ireland (24 per cent). Housing associations, which also have a brief to provide affordable housing, held 4 per cent of housing stock in 1995 (*Social Trends* 1997). The major changes in the 1980s and 1990s were linked with the slowdown in housing construction in the public sector and with legislation, such as the Housing Act of 1980, which gave tenants the right to buy (Figure 10.2). Between 1979 and 1995, about 2.2 million public sector dwellings were sold under a scheme which gave generous incentives to sitting tenants; sales have now stabilised at around 100,000 per year. By 1991, 14 per cent of these properties had been re-sold and had become part of the private housing market. The sales from the public sector have been uneven in their impact. Whereas high-rise flats made up 25 per cent of public sector stock, only 4 per cent of these had been purchased. Sales have proceeded most briskly in favoured estates and the overall effect has been to widen the disparities between the private and public sectors. Public sector housing has become more closely aligned with the most disadvantaged sections of society, and through the 1990s over 25 per cent of all public sector tenancies went to homeless households and most of the rest to those classed as priority need. Also in the 1990s there have been increases in larger-scale disposals of public sector housing, usually to housing associations.

Owner-occupancy has continued to grow and moved from 54 per cent in 1981 to 68 per cent in 1996. Highest levels of owner-occupancy (80 per cent) were among the higher

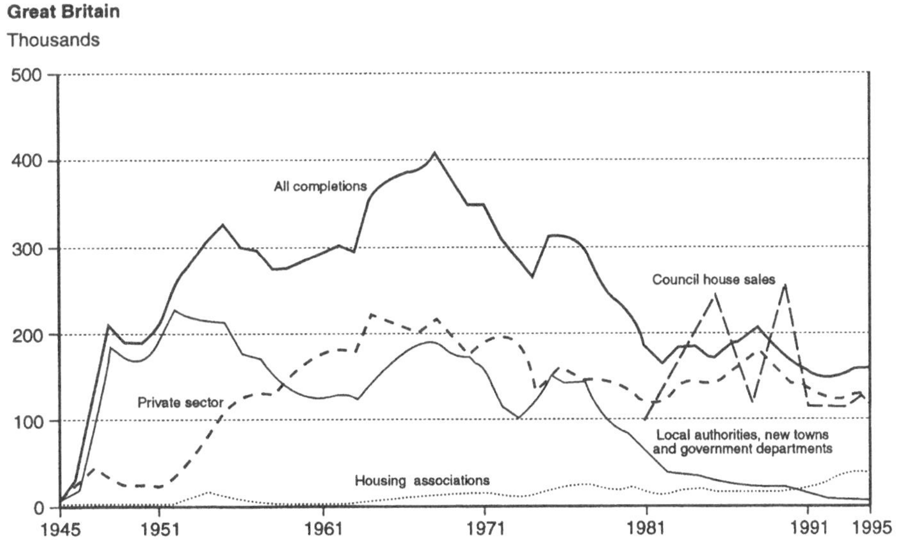

FIGURE 10.2 Housebuilding completions by sector and council house sales
Sources: Department of Environment, Welsh Office, Scottish Office, *Social Trends* (1997).

social class groups, but greatest change occurred among manual and lower-skilled workers. Owner-occupancy rates among skilled manual workers changed from 52 per cent in 1981 to 76 per cent in 1996, and among unskilled manual workers from 27 per cent to 43 per cent. The policies of council house sales and an emphasis upon housing improvement rather than renewal have underpinned these trends. The private rented sector showed the greatest decline over the longer period with a fall from 61 per cent in England and Wales in 1947 to 15 per cent in 1970. The advent of housing associations added a dimension to the private rented sector, but in 1996 the share was still only of the order of 14 per cent. Private landlords remain important in central parts of major cities, especially London, and the development of controls to give a wider protected housing market has been a feature. One estimate suggested that over 90 per cent of British urban housing enjoyed a significant measure of protection.

The housing market is often cited as a significant factor in understanding the ongoing process of urban change. As already mentioned, negative equity became a feature during the early 1990s and marked, if not the end, at least a set-back to the long process of rising value of residential property. Up to the 1960s, land value was a key consideration with an index change from 100 in 1939 to 1615 in 1963. After 1974, there was far more stability in the value of land. Regional variations in property values were marked, with least signs of improvement and investment in the older industrial cities. Inner-city dereliction signified the lack of confidence in the urban housing market. By 1994, the United Kingdom had the lowest annual number of housebuilding completions in the European Union, apart from Sweden and Denmark, with about 150,000 new dwellings. This compared with over 350,000 per year in the 1960s and 250,000 in the 1970s. Specialised new build for the elderly, the disabled and the chronically sick formed 38.8 per cent of all new completions in 1981 and 43.5 per cent in 1991, falling back to 27 per cent in 1995. The major shift was the diminished role of the public sector. In 1981 local authorities built 14,375 new units for these three groups, but only 2,312 in 1991 and 219 in 1995; between the same dates the private sector contributions changed from 162 to 1,994 to 445 and Housing Associations from 2,425 to 2,660 to 1,948. In this critical area of housing provision for groups at risk the changing responsibility from public to private sectors is very clear.

During the 1980s there was an 'access crisis' in urban housing with growing numbers of people qualifying for help under the Homeless Persons legislation. Since 1990, the DETR has committed over £180 million to the Rough Sleepers Initiative and has built 3,300 new accommodation units for the homeless in London alone. Estimates of the numbers of homeless people are notoriously difficult to form, but the number is growing and the problem is becoming more acute. The homeless comprise more variety than the stereotype single male, and although the root causes may be economic and lack of jobs, they are exacerbated in major cities by housing shortages and an influx of refugees. Deinstitutionalisation has forced many mentally ill people out onto the streets and alcohol problems and drug abuse often exacerbate their condition. Homeless people form an extreme example of the process of social exclusion, being denied access to the most basic of human needs. The introduction of an enterprise culture and the market ethic in housing, as in other areas, has made the plight of the 'losers' more visible and extreme. There is also a more telling human dimension, with research now showing that men living rough are almost forty times more likely to die young than their contemporaries in secure homes (Hawkes 1998). As in the United States (Wilson 1987), a 'truly disadvantaged' class of urban dwellers has emerged.

The central cities

Commercial functions

City-centre development has been an important facet of urban change in Britain. Initially stimulated by the legacy of war damage, it has transformed city centres. The continuing viability of city-centre retailing has led to a sustained flow of investment into shopping facilities, and between 1965 and 1989 8.9 million square metres of retail floor-space had been provided in 604 city centres of over 4,650 square metres (the Department of Environment inspects all developments above this threshold). Between 1984 and 1994, 95 per cent of new retail space had been developed in existing city centres.

Allied with new shopping developments have been positive moves towards traffic control and management and in-town shopping-centre projects, such as Nottingham's Victoria Centre and Broadmarsh and Newcastle upon Tyne's Eldon Square, which have been designed to add to or upgrade more traditional retail provision. The perceptible shift to policies with a stronger commitment to public transport and more constraints on the use of private cars has been matched by the appointment of some eighty-nine city-centre managers. Funding remains an issue as the state looks for sponsorship from partners such as Chambers of Trade and Commerce (representing local retail businesses) and commercial interest groups such as property investors and developers (which have considerable political influence). Zoning laws control out-of-town development and it is often argued that there is no other country in the world which exercises such stringent planning controls over the retail system, particularly by resisting the market pressures for a greater amount of decentralisation.

Given this support, large department stores continue to invest in the central city and their presence is essential for the success of in-town schemes. Local authorities have also improved central-city environments by landscaping and traffic management schemes, and many office functions that have a high level of direct contact with consumers continue to locate centrally. For financial and commercial offices, face-to-face contact, especially for higher-level management, remains important. London, with its large share of the head-quarter offices of the largest UK companies, is dominant and its employment in producer services increased from 13 per cent in 1971 to 23 per cent in 1989. Large-scale office developments, especially in London, have been part of the cycles of property speculation and rising land values that have brought employment and new consumers to the central city.

British inner-city policies have tended to understate the job-creation potential of retailing despite its labour-intensive nature. The long-term decentralisation of jobs and people has clearly affected retailing and an intra-urban hierarchy has emerged. In the inter-war years a specialised central-city shopping area extended outwards along main traffic arteries, with local clusters and individual shops in the surrounding residential areas. With time a more clear-cut set of shopping areas developed, classified as:

- a central area serving a population of at least 150,000;
- regional shopping centres which have developed from the smaller central areas in conurbations;
- district centres serving local catchments of around 30,000;
- neighbourhood centres selling convenience goods in catchments of 10,000;
- local or sub-centres with small clusters of stores serving 500 to 5,000 people.

Much of this framework already existed but was consolidated in the 1950s when planned shopping centres were added to several levels of the hierarchy. It has been estimated that in larger cities (over 250,000) 36.8 per cent of retail provision is found outside the central area; this proportion is inversely related to city size and is 55.3 per cent for cities in the 40,000 to 49,000 range.

Retail trade has experienced considerable structural change since 1945. There have been upheavals in the methods and organisation of retailing, a blurring of the retailing/wholesaling distinction, escalation of multiples, an increase in store size, and greater bulk buying by consumers. These changes adversely affected small stores and between 1950 and 1966 the number of general stores fell by 56.2 per cent and of grocery retailers by 16.3 per cent. In part this trend was related to out-migration and the subsequent decline of the 'corner-shops', but it can also be tied to economies of scale and the changing organisation of retailing. Already by the early 1970s, four organisations together operated nearly 4,500 grocery shops and accounted for nearly 22 per cent of sales. There is some evidence for a renewed role for small convenience stores in suburban locations. Large-scale retail organisations have sought to develop large out-of-town sites, although in the late 1970s only Brent Cross in suburban north London could be described as 'out-of-town'.

By the late 1980s, three 'waves' of retail decentralisation had been recognised. The first wave involved the emergence of superstores and hypermarkets during the period 1964 to 1975. A second wave between 1975 and 1985 was composed of retail warehouses, retail warehouse parks and retail parks; the Enterprise Zones at Swansea and Dudley were typical of this type of development. The initial sales emphasis was on DIY products, furniture, carpets and electrical goods but expanded to include clothing, footwear, toys and car accessories. The third wave dates from 1984 and the proposal by Marks and Spencer to open out-of-town stores. This type of decentralisation affected the outlets for quality goods and involved firms such as Habitat, Laura Ashley and World of Leather. Gateshead's Metro Centre, opened in an Enterprise Zone in 1986, was the first major example of an integrated regional shopping and leisure complex. Following Metro Centre with its 136,430 square metres of floor space, were Merry Hill, Dudley (143,000 square metres), Meadowhall, Sheffield (116,250 square metres) and the Lakeside Centre at Thurrock (116,250 square metres), with other planned developments in London, Leeds, Manchester and Glasgow (Figure 10.3). The idea of a fourth wave in the 1990s focuses on the conflation of retailing with leisure tourism, with the emergence of outlet malls or shopping villages, such as the Clarks' Village at Street in Somerset, the shopping villages at Bicester, Oxford and at Swindon. Other additions are the informal car-boot sales, new convenience chains, and the emergence of tele-shopping which may have wider implications for the geography of retailing. There is now much greater diversity in the provision of retail sales and consumer choices are changing accordingly.

Objections to out-of-town centres stemmed from fears of their impact on city-centre trade but the earlier hypermarkets affected smaller branches of multiples rather than independent corner stores and many retailing firms retained city-centre stores. There is some force to the concept of the disadvantaged consumer but empirical evidence is equivocal. Mobility is unevenly available but there has been an expansion of convenience stores, discount stores and shopping transport, to offset the effects of out-of-town centres.

Overall, retail provision has responded to the two basic needs of redeveloping outworn parts of the central city and adding new facilities to rapidly growing suburbs. Both of these

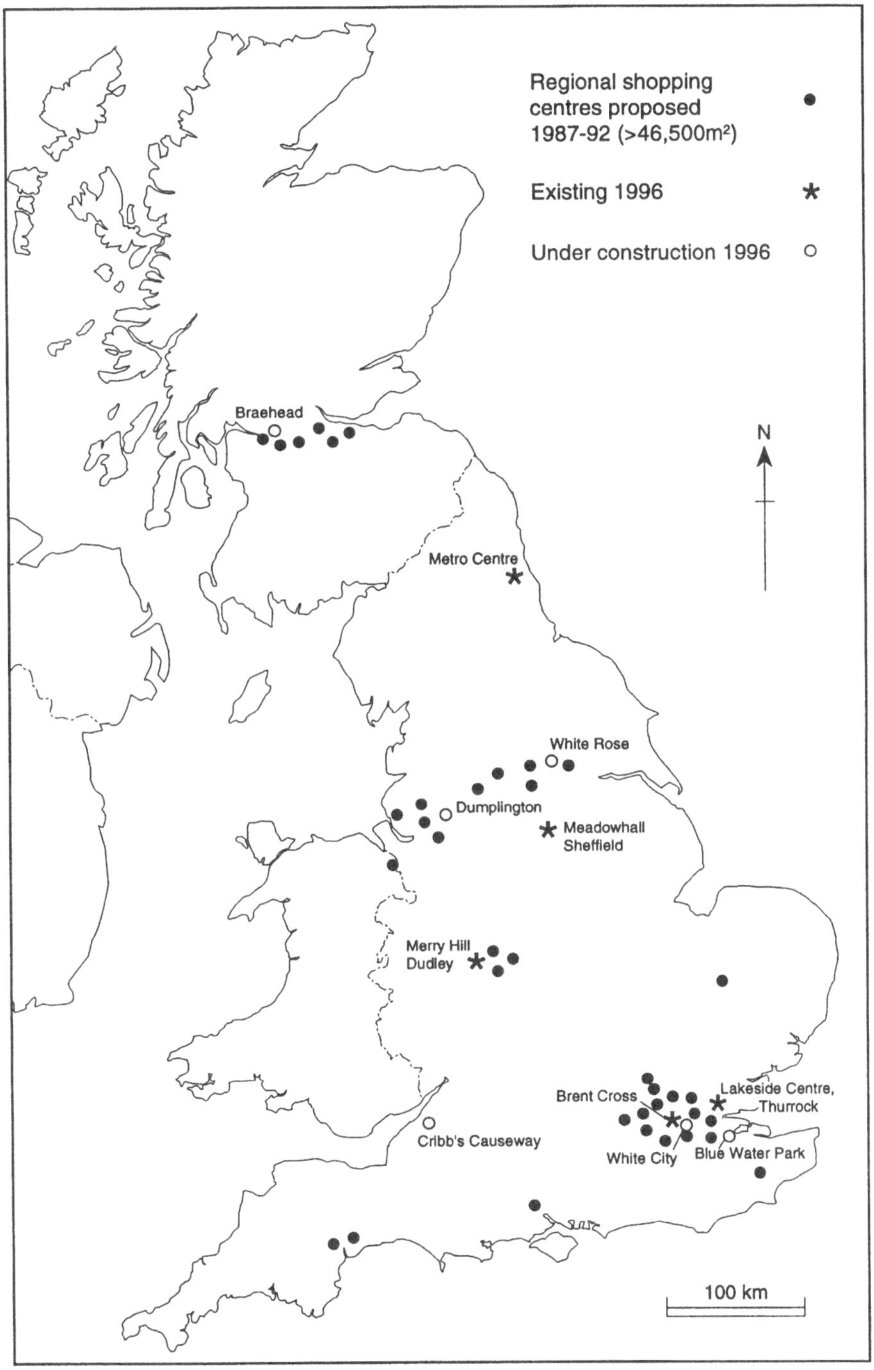

FIGURE 10.3 Regional shopping centres in the United Kingdom, 1996
Source: After Herbert and Thomas (1997).

have been paralleled by changes in the organisation of retailing, shifts in consumer behaviour and a fairly consistent planning attitude towards decentralisation. The protected central city remains a strong and viable location for commercial functions, but there still remains a need for a comprehensive planning strategy for retail development in and around the city which reconciles the role of the centre with the continuing pressures for commercial decentralisation. During the 1990s there are signs that city centres are being adversely affected.

Jobs and the inner city

The industrial and manufacturing base of the inner cities virtually collapsed in the second half of the twentieth century with devastating effects on the resident population. Between 1951 and 1981, inner areas of the conurbations lost 45 per cent of their employment; over a million manufacturing jobs were lost in the same period; Hall (1985) cited a figure of 2 million lost factory jobs between 1971 and 1981 in the UK alone. London, Liverpool, Manchester, Birmingham all suffered serious losses mainly through factory closures and the disappearance of traditional employers. The resident populations were deskilled, and for the new kinds of jobs entering the inner city and London in particular the demands were for professional skills and IT in keeping with the new wave of producer services and finance. A continuing expansion of retail trade, office employment and other services ensured a high demand for women's employment but these did little for unemployed industrial or dock workers. Again, there was an impact on lower-skilled white-collar jobs as banks, for example, displaced their traditional counter staff and tellers in an age of automated transactions. Commuting flows increased in importance and resident inner-city populations, apart from those in the new gentrified areas, had little stake in the restructured employment market. In 1996, inner areas of cities such as London and Glasgow continued to have unemployment rates of around 15 per cent, which were well above those of surrounding areas. Green (1996) showed that during the 1980s inner London and parts of other metropolitan areas emerged as the main losers from the processes of social and economic change. Yet there was evidence that employment was expanding during the 1990s in places where contraction had dominated over the past thirty years. Cities with business and electronic networks placed to act as centres of skill and knowledge could offer milieux favourable to innovation and change. Large areas of derelict land and pools of labour might be sources of opportunities rather than of despair (McLennan 1998). The optimism is real but has to be tempered by the continuing low employment opportunities for black youth and less-qualified sections of society.

Quality of life

Quality of life, mirrored in indicators such as lack of jobs, substandard housing, educational disadvantage, crime, vandalism, drugs, and deprivation, has always been a concern for the inner city. There are pockets of deprivation, and generalisations should not obscure the considerable diversity which exists. It will be shown that the problem estates in the outer urban rings can be at least as great a problem as the older parts of the central city. Matthews (1991) estimated that with 7 per cent of the total population, British inner cities contained

14 per cent of the unskilled workers, twice the average number of single-parent families, three times the national rate of long-termed unemployed, ten times those below the 'poverty' line, and most of the schools classed as exceptionally difficult. Although many individual households do not suffer these specific disadvantages, they are still afflicted by deteriorating environments, vandalism, petty crime and traffic congestion.

Inner cities also tend to have disproportionate vulnerability to hazards in the urban environment. The 1992 Earth Summit agreed a policy titled 'Agenda 21' aimed at the improvement of urban environments. This has proved most difficult to implement in the inner cities where the key challenges of controlling infectious or parasitic disease, reducing chemical and physical hazards, achieving high-quality environments, minimising transfers of environment costs, and progressing towards sustainable consumption (Satterthwaite 1997) can all be easily identified. As Gibbs (1997) noted, cities are the key economic units, producing 60 per cent of global GNP – and there are inevitable environmental impacts. Clean production and consumption needs to be accompanied by greater equity and democratic involvement. There are many working examples. Leeds and Southampton have environmental strategies; Cardiff, Manchester and Kirklees have environmental programmes set by planning or environment departments.

Problem residential areas

For much of the nineteenth century, poverty areas were recognisable parts of the British inner city. They were characterised by high levels of substandard housing and many indicators of deprivation. The early waves of urban renewal and slum clearance schemes had a major impact on these areas during the early twentieth century and many thousands of households were transferred to social housing outside the traditional inner-city terraced areas. It became clear during the 1970s and 1980s that the problems of the inner city remained and were once again becoming acute. The pockets of deprivation became highly visible, with outbreaks of rioting and social disorder, often with racial connotations. Inner-city areas had become repositories of people such as the unskilled, some ethnic minorities, the elderly and the disadvantaged. Unemployment was high, housing conditions were often poor, and tensions were high. Social indicators consistently showed wide disparities between inner-city areas and suburbs, and Green (1996) confirmed the persistence of these disparities and the poverty areas that still typified the inner city. Such areas in particular have been adversely affected by the collapse of urban funding and the withdrawal of inner-city strategies in the 1990s.

Such problem residential areas, however, are not restricted to the inner city. The problem estate, variously referred to as the difficult-to-let or the sink estate, has become a feature of public sector housing. Often the products of past policies of 'dumping' problem families in the least-sought-after housing, these estates are typified by high levels of crime, vandalism, drugs and forms of social disorder. Estates may go through community careers, with the worst problems evident when demographic cycles endow them with higher numbers of young people. Studies of incivilities, or signs of disorder in the urban environment, often pick out estates of this kind at the earlier stages of a downward spiral (Figure 10.4). Such estates have been described as disgraceful and degrading, adding poor states of repair and lack of amenities to the high-unemployment and accompanying social

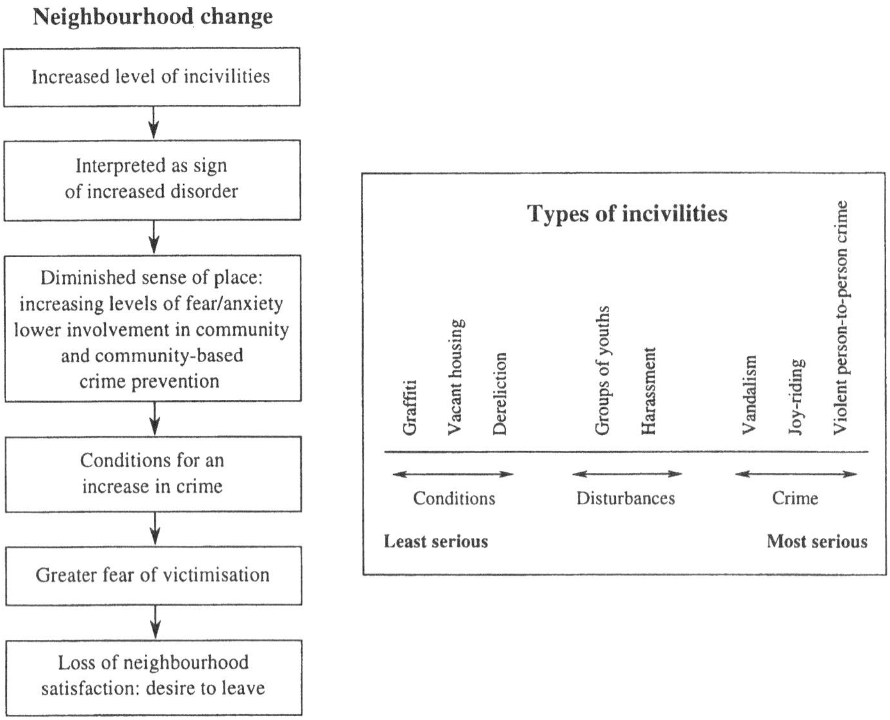

FIGURE 10.4 Incivilities and neighbourhood change

problems. At Meadowell on Tyneside, Barke and Turnbull (1992) documented a high unemployment estate where a drug culture, endemic crime and a sense of hopelessness prevailed. The worst estates become characterised by high vacancy rates, and the City of Swansea in 1998 was reputedly seeking to allocate unlet properties to refugees in an attempt to generate income from government subsidies. Low demand for housing on a specific estate is often the starting point for a cycle that leads to pockets of extreme deprivation (Hall 1997). Hall identified six reasons for the deepening problems of the worst estates. First, right-to-buy legislation had removed better housing to the private sector; second, allocations of decreasing stock were made almost exclusively to marginalised groups such as the homeless and single-parent families; third, gentrification was displacing poorer tenants to edge estates; fourth, high rents without subsidies were excluding the working poor; fifth, cut-backs in state spending reduced maintenance; and sixth, the gap between council tenants and jobs was widening. One of the problems for the people on problem estates is their relative remoteness from jobs, opportunities and housing managers.

Gentrification and the return to the central city

The process of gentrification involves the upgrading of specific inner-city districts to attract higher-income renters or owners. Such districts are often those with Georgian or Edwardian terraces, most commonly three-storey, capable of modernisation and rehabilitation. Places such as parts of Islington in London have experienced a considerable transformation as a result of this process. Housing improvement grant legislation during the late 1960s and 1970s enabled gentrification. One intention was to improve conditions for sitting tenants, but speculative owners, developers and the housing market saw the potential to change the whole character and marketability of these areas. Large numbers of dwellings were improved and sold at much inflated prices and this had the effect of displacing former low-income tenants. These negative impacts of gentrification involve displacement, loss of traditional communities, and overcrowding elsewhere in the city. Positive impacts revolve around the physical and social revitalisation of older areas at little public cost, new demands for goods and services and positive spillovers.

The return of investment to the central city gathered impetus in the 1980s with major redevelopments, funded by a mix of public and private funds, which transformed old, derelict docklands in many cities into attractive and expensive residential areas. This maritime quarter boom fuelled an inflow of investment, with the construction of many new residential properties in imaginative mixes of dwelling types. The transformation of London's Docklands has had far-reaching effects linked both with the expansion of producer services and the attractions of central water-fronted sites. The Isle of Dogs was designated as an Enterprise Zone in 1982 and land values rose from £100,000 to £7 million a hectare in five years. The Docklands Light Railway linked Canary Wharf to the City and this single site could offer direct employment to 40,000 people. There were caveats to this story of growth. First, the new 'yuppie' communities were being constructed in close proximity to the residual docklands communities and presented huge disparities in wealth and quality of life. There were inevitably social conflict problems. Second, the investment into Docklands was overdone and a recession was evident in the early 1990s, leading to a loss of confidence and falling values of property. Some of the major property investors were badly affected and recovery has been slow.

Urban conservation has been a feature of urban planning in Britain since the formalisation of the listed building procedures, covering buildings of architectural or historic interest, in the Town and Country Planning Acts of 1944 and 1947. The conservation area concept arose most clearly from a court case involving two terraced houses in St James's Square, London in 1964. Conservation areas are subjective judgements, but the criteria include special architectural and/or historical interest and buildings with a character or appearance worth enhancing or preserving. With the growth of heritage tourism in its various forms (Herbert 1995), historic cities such as Chester and Bath have become significant attractions for visitors. Many of the major tourist attractions, including truly historic sites such as the Tower of London and Edinburgh's Holyrood Palace, and more recent constructions such as the Jorvik Centre at York, are located in cities. Many older cities, faced with a loss of traditional economic activities, have turned to tourism as a panacea and have striven to make use of whatever heritage, medieval or industrial, old or reconstructed, that they might possess. For the really historic cities there are pressures from weight of numbers. There is a 'local fatigue' (Strange 1997) arising from numbers of visitors, demands on local infrastructures and conflicts with resident populations. Such pressures can threaten the very qualities that make these places attractive and new regulatory mechanisms may be needed.

Immigrants and ethnic areas

Although the return to the central city of higher-income groups is not an insignificant trend, the role of the inner city as a destination for immigrants and expanding ethnic minority communities has far more telling effects on urban demography. None of the major ethnic minority groups, Afro-Caribbean, Indian, Pakistani and Bangladeshi, had less than two-thirds of their populations in the conurbations in 1991 (Figure 10.5). All these groups had shown growth in their conurbation populations over the previous decade, ranging from 1.6 per cent for Afro-Caribbean to 5.7 per cent for Bangladeshis. The Greater London and West Midlands conurbations tend to dominate with, for example, almost two-thirds of all the Afro-Caribbean and Bangladeshis resident in one or the other. The lesson from the 1991 Census was one of little change; ethnic areas in British cities have emerged and are consolidating, but conditions do not resemble the ghettos typical of many cities in the United States (Peach 1996; Robinson 1993). There is variety in ethnic housing and there is some evidence that Afro-Caribbean households were decentralising from the innermost parts of London and, to a greater extent than the other groups, had a significant presence in the public sector of housing. The large majority of South Asians remained rooted to owner-occupancy, low-cost areas of cities with a limited amount of professional suburbanisation.

Ethnic areas have been features of British cities for many years in areas such as London's East End and comparable districts in other major port cities. The modern wave of immigrants from New Commonwealth countries began in the 1950s stimulated by poverty and lack of opportunities in their own countries and the promise of better jobs and prospects in the United Kingdom. In 1963/4 legislation to control the flow of migrants stopped the main movements, but certain categories such as dependants and those with special skills have continued to arrive. The earliest migrants were Afro-Caribbean, recruited at a time of labour shortage by London organisations such as the large teaching hospitals, London Transport and the Hotels and Restaurants Association. Other Caribbean immigration followed the expanding car plants and engineering works in the London area and the West Midlands. Indians followed a similar pattern (Robinson 1993), but Pakistanis moved further afield to towns in the north of England such as Manchester, Oldham, Blackburn, Leeds and Bradford. There were later Asian concentrations in the East Midlands (such as at Leicester), amplified by forced migrations out of East Africa.

Estimates put the total size of the UK ethnic minority population at 2.6 million in 1990, forming 4.8 per cent of the total population, with the largest groups being Indian (827,000), Afro-Caribbean (496,000) and Pakistani (455,000). Annual rates of increase peaked with 98,000 in 1981, but were around 55,000 by the early 1990s. There is diversity within the groups, with divisions by islands of origin among the Afro-Caribbeans and sub-groups such as Sikhs and Gujeratis within the South Asians. Although the ethnic minorities tend to occupy the same broad types of housing within cities, there is some evidence for segregation within these sub-groups. The 'ethclass' had emerged, with clear social gradations reflected in housing areas.

The Race Relations Act of 1976 was a key piece of legislation, but there has been persistent discrimination in employment and housing markets against non-white immigrants. The whole thrust of public policy since the 1970s has been to improve employment chances, but always in difficult situations. There are greater numbers of ethnic minorities progressing further in the educational system, and Asians in particular have proved very

FIGURE 10.5 Numerical distribution in 1991 of (a) Afro-Caribbeans, (b) Indians
Source: After Robinson (1994).

successful in several kinds of business enterprises. Ethnic school-leavers may be vital to local economies and in 1987 45 per cent of all births in Tower Hamlets were to New Commonwealth or Pakistani parents.

The continuing disadvantages of black youths, among whom levels of unemployment have continued to be high, combined with displays of prejudice, have led to conflicts in many inner-city areas. Urban riots, which have occurred in cities like London, Bristol, Nottingham and Liverpool all have racial overtones. Relationships between the police and black youths have yet to develop properly at acceptable levels of trust. These forms of urban conflict pale into insignificance when compared to the situations in the towns and cities of Northern Ireland. Roman Catholic and Protestant communities have retreated into their own 'fortress areas' in Belfast and Londonderry, with uneasy lines of demarcation. Differences are ethnic as much as they are political or religious, but political initiatives from the British and Irish governments have led to a positive vote for a representative assembly in Northern Ireland and the prospect at least of a more stable and peaceful future.

Managing the city

Between 1974 and 1996, the functions of local government were commonly divided between county and city authorities and this led to duplication and conflicts of interest. Since 1996, the wider introduction of unitary authorities has given greater autonomy to free-standing urban areas. For the larger metropolitan areas, strategic planning and policy has been difficult, and since the early 1980s the powers of local government have progressively been undermined by the appointment of special bodies and agencies to manage critical areas of change. Invariably during this period the agencies, such as Urban Development Corporations (UDCs), were charged to introduce market criteria and an enterprise culture that focused on job and wealth creation and pushed welfare into subsidiary roles. The record was variable. Thomas and Imrie (1997) noted that whereas in Sheffield and Teesside, the UDCs represented no radical change in policy and worked closely with local government, the Cardiff Bay Development Association consistently turned to international consultancies and refrained from recruiting personnel from local agencies. More telling for local government of the cities was fiscal intervention such as 'capping' and the freezing of housing funds. One outcome of the latter was the virtual removal of local authorities as housebuilders and suppliers. At one key level of city management, therefore, the changing roles of central and local government and the creation of powerful local agencies were considerable forces for change. Another dimension was added by greater interest in place promotion and various forms of city 'boosterism': the importance of attracting events or labels, such as Glasgow's designation as European Capital of Culture in 1990 and European City of Architecture in 1999.

Managers in the urban housing market

Market forces and government policy have similarly determined the extent and form of urban residential development, but there are many managers interposed between the producers and consumers of housing who have key entrepreneurial roles in the allocation of scarce resources.

The powers of such managers or gatekeepers have varied over time. State interventions, planning controls and the welfare imperatives, for example, have diminished the supremacy of landowners. Many of the macro changes, such as the growth of suburbs and the return to the central city, can be attributed to flows of capital and the willingness of investors to consider particular parts of the city. The state as a developer of both housing and commercial areas has been a key figure in British cities, but during the 1980s and 1990s there was a stronger focus on balanced private and public urban investments.

Planners have significant powers and key interventionist roles in British cities. On a wider stage there are the visionaries, and Peter Hall (1988) talked of the cities of the imagination which were products of the minds of influential planners such as Howard with his new towns and Le Corbusier with his towers. In their less prosaic roles, planners control land use and development, provide utilities and highways and chart the directions of urban growth. Although there is clearly scope for innovation and initiative, much planning has to work within the natural thrusts of urban change, modifying and accommodating rather than revolutionising. Containment remains a priority and, faced by predictions of an additional 4 million plus households by the year 2016, planners have a strong preference for 'brown field' and high-density solutions. However, the nature of the anticipated housing demand, much of which will come from single-person households and one-parent families, is not well understood.

Particularly since the Housing Acts of 1969 and 1974, planners have favoured rehabilitation and the improvement of better-quality older property rather than demolition. General Improvement Areas (GIAs) and Housing Action Areas (HAAs) are area policies designed to achieve positive change in both environment and housing. The HAAs in particular were targeted at areas of multiple deprivation and remain in place in the late 1990s. They can be declared where the majority of dwellings are substandard and allow 90 per cent grants to householders. In 1984, the peak year for HAAs, half a million dwellings were renovated, about half of which were in the public sector. Bailey and Robertson (1997) studied HAAs in Glasgow and Edinburgh and noted that since 1974 there had been 1,700 HAA schemes declared in Scotland affecting 70,000 dwellings. They noted that HAAs could involve a *social approach*, with original welfare ideals, but also and increasingly, a *private approach* where they were much more closely tied with improvement as part of a gentrification process. Rehabilitation is likely to become of greater significance as ageing and deteriorating housing from the early part of the twentieth century reaches the end of its 'working life', though for much of this housing, rehabilitation may not be a viable option.

Once housing is constructed and becomes part of the housing market, new managers or gatekeepers become active. Landlords in the private sector control housing units to rent, and for house-buyers the key managers are those representing building societies and banks that fund over 90 per cent of housing purchases. Estate agents or realtors are the entrepreneurs of the housing market who have the ability to form as well as to follow the market. Local authority housing managers determine who qualifies for local authority accommodation and of what type, and when and where accommodation is offered.

Urban policies

Urban policies have shown major shifts since the early 1980s. These shifts relate primarily to varying government initiatives but also to the tensions of priorities between regions, cities and more general issues such as unemployment and the need to promote growth. By the later 1980s, some of the main commentators were painting pessimistic scenarios on the effectiveness of urban policies, and Stewart (1987) described inner-city policy as a charade. Similarly, Robson (1988) argued that the weaknesses of urban policy included lack of

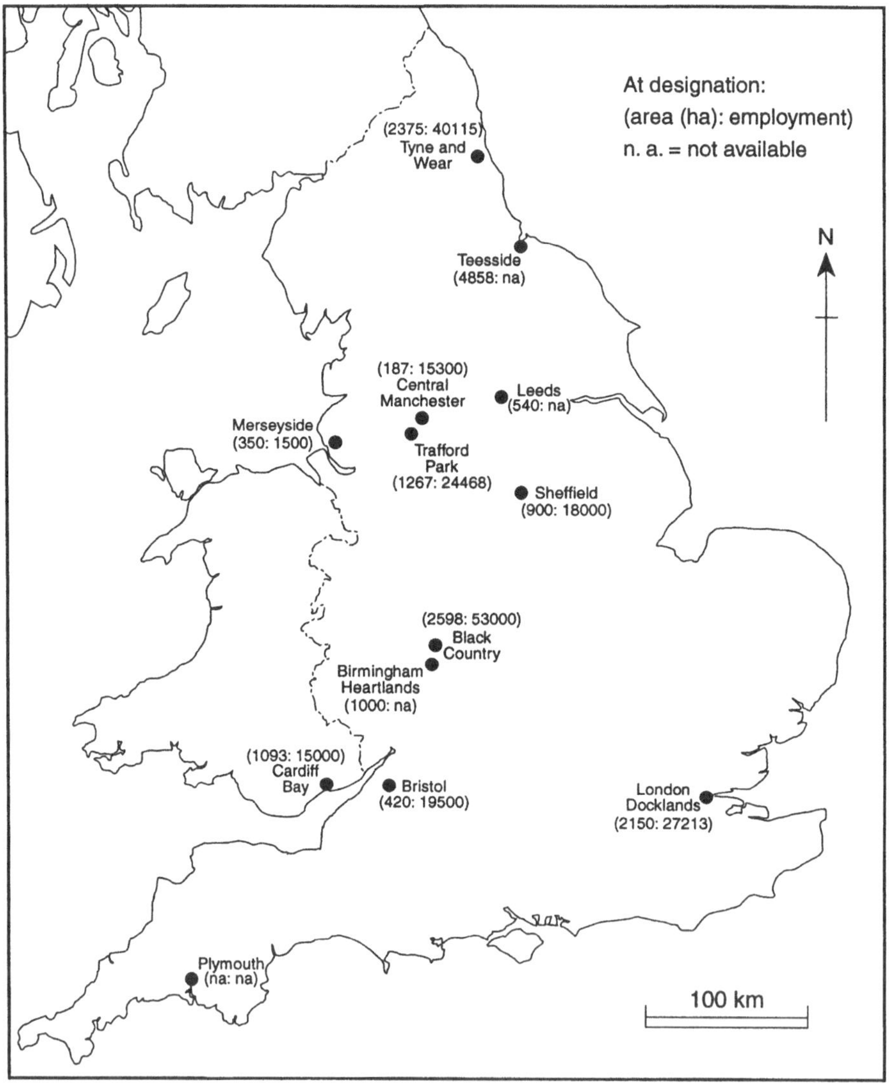

FIGURE 10.6 Location of Urban Development Corporations (UDCs) in England and Wales
Source: After Thomas and Imrie (1997).

national co-ordination, as shown by the withdrawal of rate support funds with the effect of counteracting urban aid policies; conflict between central and local government; and a serious imbalance between economic and social objectives. Undoubtedly, these have been the major issues as elected local government has been usurped by nominated bodies with executive powers and as an enterprise culture has submerged any welfare or social considerations beneath the priorities of economic growth and wealth creation.

Since the early 1980s, the government has based its range of initiatives on the belief that competitive and market economies could deliver equitable and efficient solutions to urban problems (Nevin *et al.* 1997). Major agencies, such as the Urban Development Corporations (Figure 10.6), were set up with nominated members, often drawn from business backgrounds, to drive change. There was a plethora of urban policies, and Robson (1994) reported that Manchester had six Urban Programmes, two Enterprise Zones, two Task Forces, two Urban Development Corporations, two Safer City projects and a City Action Team. There was the criticism of lack of co-ordination, but also of the fact that funds were simultaneously being pulled out of other parts of city budgets. The DoE housing budget fell from £4.5 to £1.9 billion between 1981 and 1987; seven inner-city partnerships lost almost £850 million in revenue support funds; and expenditure on social and community programmes fell from 34 per cent of Urban Group budget in 1979 to 16.6 per cent in 1988. Robson (1994) concluded that the impacts of the urban policies of the 1980s were at best modest. Although there was some relative improvement in unemployment indicators across the fifty-seven targeted Urban Programme Authorities, conditions in the worse areas had not improved; the old industrial conurbations still dominated the ranks of the most deprived places. This situation was exacerbated in the first half of the 1990s by the effective withdrawal of inner-city policies. Between 1990 and 1996, urban funding was cut by 40 per cent and successor schemes such as City Challenges did not amount to a clear inner-city strategy.

The Single Regeneration Budget introduced in 1993 merged twenty programmes from five different government departments and rendered irrelevant the 1980s policy programmes (Table 10.1). At a stroke this move brought the promise of co-ordination and more strategic thinking and also made openings to involve the local authorities more closely in the whole process of urban policy. The commentators remain sceptical. Robson (1994) feared that the swing back to local authorities and the Training and Enterprise Councils (TECs) would go too far and saw little evidence of strategic vision. Nevin *et al.* (1997) argued that because of the downgrading of need as a criterion, urban regeneration strategies that brought benefits to the disadvantaged groups would be hard to achieve. The social dimension was still a minor part of the equation. An analysis of successful bids on the first round showed that whereas 34 per cent targeted ethnic minorities and poor housing, 95 per cent had employment as a main objective. Whereas a business perspective would recognise the virtues of this set of priorities, there is also a compelling need to add a community-led dimension and to recognise the centrality of social integration as a goal and social exclusion as a danger.

Conclusion

Britain remains an urban nation with a diversity of towns and cities that reflects both its long history of urbanisation and the rapid changes of the twentieth century. Cities contain the

TABLE 10.1 Budgets subsumed within the Single Regeneration Budget (SRB), 1994/5

Programme	*£ million*
Urban Development Corporations*	286
Housing Action Trusts*	88
English Partnerships*	181
Estate Action	373
City Challenge	213†
Urban Programme	83
Task Force	16
City Action Teams	1
Section 11 (part)	60
Ethnic Minority Grant/Business Initiative	6
Safer Cities	4
Programme Development Fund	3
TEC Challenge	4
Local Initiative Fund	29
Business Start-up Scheme	70
Education Business Partnerships	2
Compacts	6
Teacher Placement Service	3
Grants for Education Support and Training	5
Regional Enterprise Grants	9
Total	1,442

Source: *Hansard*, 21 March 1994, col. 918.
Notes: *Denotes programmes which have been ring-fenced in the SRB. †Programme supported by £19 million from the Housing Corporation.

most obvious problems and pose the greatest challenges but they also serve as the location for the main institutions of society and economy in the United Kingdom. London in particular is a world city and has roles which are global and international as well as national, regional and local. Many other cities possess the same qualities, but to a lesser degree. It is one of the fascinations of the great city that within short distances it accommodates the needs of major players on a world stage and those of ordinary families in their neighbourhoods. Many of the long-recognised problems of cities remain – the ailing traditional economies, the deteriorating urban fabric, the traffic strangulation and the great disparities of wealth and poverty. Yet cities and their landscapes contain the icons and symbols to which people relate. Major investments in new projects, or individual buildings such as the Millennium Dome or the new British Museum Reading Room, are always contentious yet add to the identity of a city as place. The task for cities is to achieve gradual transformations, to raise the quality of life and reduce the sources of conflict but yet to inject where one can the spectacular and the visionary into their landscapes.

References

Ash, A. and Graham, S. (1997) 'The ordinary city', *Transactions, Institute of British Geographers* 22: 411–29.

Bailey, N. and Robertson, D. (1997) 'Housing renewal, urban policy and gentrification', *Urban Studies* 34: 561–78.

Barke, M. and Turnbull, J. (1992) *Meadowell: the Biography of an Estate with Problems*, Aldershot: Avebury Press.

Bibby, P. and Shepherd, J. (1997), 'Projecting rates of urbanisation in England 1991–2016', *Town Planning Review* 68: 93–124.

Champion, A.G. (1994) 'Population change and migration in Britain since 1981: evidence for continuing concentration', *Environment and Planning A* 26: 1501–20.

Champion, A.G. and Green, A. (1991) 'British economic recovery and the North–South divide', *Geography* 2: 249–54.

Cloke, P., Milbourne, P. and Thomas, C. (1997) 'Living lives in different ways? Deprivation, marginalization, and changing life-styles in rural England', *Transactions, Institute of British Geographers* 22: 210–30.

Gibbs, D. (1997) 'Urban sustainability and economic development in the United Kingdom: exploring the contradictions', *Cities* 14: 203–8.

Gordon, I. (1988) 'Resurrecting counterurbanisation: housing market influences on migrations from London', Paper presented at the Institute of British Geographers Conference, Loughborough, January.

Green, A. (1996) 'Changing local concentrations of poverty and affluence in Britain', *Geography* 8: 15–25.

Green, A.E., Owen, D.W. and Winnett, C.M. (1994) 'The changing geography of recession: an analysis of local unemployment time series', *Tijdschrift voor Economische en Sociale Geografie* 19: 142–62.

Hall, P. (1985) 'The people: where will they go?', *The Planner* 71: 3–12.

Hall, P. (1988) *Cities of Tomorrow: An Intellectual History of Planning and Design in the Twentieth Century*, Oxford: Basil Blackwell.

Hall, P. (1997) 'Regeneration policies for peripheral housing estates: inward and outward looking approaches', *Urban Studies* 34: 873–90.

Hawkes, N. (1998) 'Living rough boosts death rate 40 times', *The Times*, January.

Herbert, D.T. (ed.) (1995) *Heritage, Tourism and Society*, London: Cassell.

Herbert, D.T. and Thomas, C.J. (1997) *Cities in Space: City as Place*, London: Fulton.

McLennan, D. (1998) 'Better cities for Britain', *ESRC* 37: 5.

Matthews, M.H. (1991) *British Inner Cities*, Oxford: Oxford University Press.

Nevin, B., Loftman, P. and Beazley, M. (1997) 'Cities in crisis: is growth the answer?', *Town Planning Review* 68: 145–64.

Peach, G.C.K. (1996) 'Does Britain have ghettos?', *Transactions, Institute of British Geographers* 21: 216–35.

Prestwich, R. and Taylor, P. (1990) *Introduction to Regional and Urban Policy in the United Kingdom*, London: Longman.

Robinson, V.R. (1993) 'Making waves? The contribution of ethnic minorities to local demography', in A.G. Champion (ed.) *Population Matters: The Local Dimension*, 150–69, London: Paul Chapman.

Robinson, V.R. (1994) 'The geography of ethnic minorities', *Geography Review* 7: 10–15.

Robson, B.T. (1988) *Those Inner Cities*, Oxford: Clarendon Press.

Robson, B.T. (1994) 'Urban policy at the crossroads', *Local Economy* 9: 216–23.

Satterthwaite, D. (1997) 'Sustainable cities or cities that contribute to sustainable development', *Urban Studies* 34: 1667–91.

Social Trends 27 (1997) Office for National Statistics, London: HMSO.

Stewart, M. (1987) 'Ten years of inner city policy', *Town Planning Review* 58: 129–45.

Strange, I. (1997) 'Planning for change, conserving the past: towards sustainable development policy in historic cities?', *Cities* 14: 227–33.

Thomas, H. and Imrie, R. (1997) 'Urban development corporations and local governance in the UK', *Tijdschrift voor Economische en Sociale Geografie* 88: 53–61.

Wilson, W.J. (1987) *The Truly Disadvantaged*, Chicago: University of Chicago Press.

Further reading

A short paper which gives an up-to-date picture of disparities within the UK is A. Green (1996) 'Changing local concentrations of poverty and affluence in Britain', *Geography* 8: 15–25. P. Hall (1988) *Cities of Tomorrow* (Oxford: Blackwell), provides a broad sweep of the city past, present and future. D.T. Herbert (1996) 'Western cities and their problems', in I. Douglas, R. Huggett and M. Robinson (eds) *Companion Encyclopaedia of Geography*, 730–51, (London: Routledge), looks at cities in a wider global context and reveals common trends and issues. Other chapters in the book refer to the 'Saviour City' (Robson) and Third World Cities (Drakakis-Smith). A more general text in urban geography that provides the ideas and theories behind urban change and a large number of case studies is D.T. Herbert and C.J. Thomas (1997) *Cities in Space* (London: Fulton). For an accessible and well-illustrated account of urban problems and policies, read M.H. Matthews (1991) *British Inner Cities* (Oxford: Oxford University Press). A short paper which summarises the changing distribution of ethnic minority groups up to the early 1990s is V. Robinson (1994) 'The geography of ethnic minorities', *Geography Review* 7: 10–15.

Chapter 11

Rural change and development

Malcolm Moseley

Introduction

The first problem with rural Britain is to define it. Indeed, the first task is to question its existence altogether! That statement is not an excuse for bemoaning the so-called fact that our leafy shires are rapidly disappearing under a tidal wave of concrete – a much exaggerated bit of scaremongering – but merely a recognition that although nine-tenths of our island is still visually 'countryside', in social and economic terms the distinctiveness of 'rural Britain' is fast disappearing. Our island is shrinking, as much because of advances in telecommunications as because of those in transport availability, and the argument that most of midland and southern England is now one vast dispersed metropolis, whatever the view from the window, is now a compelling one.

But 'rural areas' – however defined – do have certain characteristics, notably low population density and a much sought after green environment, which give national social and economic trends and problems a peculiar twist when they surface in that milieu. It is those 'twists' to which this chapter will be devoted. Look in other chapters if your interest is predominantly in that minority but necessarily rural industry called agriculture (Chapter 5), in land use *per se* (Chapter 20), or in the physical environment of the countryside (Chapters 17 and 20).

As for a definition, the criteria most commonly used are population density, the proportion of built-upon land, remoteness from urban centres, and degree of reliance upon 'land extensive' economic activities (notably agriculture, horticulture, forestry, quarrying, fishing, environmental management, green tourism and outdoor recreation). But every researcher applies such criteria in a different way (how remote is 'remote'?; what density is 'low density'?) and data availability imposes its own complications. In short, there is no unambiguous domain called 'rural Britain'.

One recent piece of research rested upon designating every tiny census tract of Britain as 'rural' or 'urban', based on two criteria: population density and land use. In this way, some

TABLE 11.1 The 'rural' and 'urban' population of Britain, 1991

'Rural' people tend to:		*Rural (%)*	*Urban (%)*
Be older	Over age 45	22	19
Be healthier	With limiting long-term illness	10	13
Be less mixed ethnically	'Non white'	1	7
Be more likely to live in detached houses	In detached residences	51	16
Be owner-occupiers	Buying or owning	72	65
Be car owners	Households with at least one car	85	63
Be *two*-car owners	Households with two or more cars	40	20
Be 'middle class'	Households in the professional and managerial socio-economic groups	32	19
Be engaged in agriculture, forestry or fishing	Heads of households thus engaged (self-employed or employed)	11	1
Travel to work by car	Employed or self-employed travelling by car	67	59
Work at home	Employed or self-employed working at home	14	3

Source: C. Denham and I. White (1998) 'Differences in urban and rural Britain', *Population Trends* 91, Spring: 23–34.
Note: See text for the definition of 'urban' and 'rural'.

2,300 'urban areas' emerged, over half of them with fewer than 5,000 residents. The remainder, termed 'rural', were seen to have about 10 per cent of the national population; some data are presented above on their characteristics. But note that *this* definition of 'rural Britain' is an exacting one and that most small towns and the larger villages are excluded from these data.

That said, and on the basis of the 1991 Census, the population of Britain's rural areas clearly tends to differ in some significant respects from that of our towns and cities (Table 11.1).

But if the statistics shown in Table 11.1 suggest a rather 'rosy' view of rural Britain, the reader should remember two salutary points. First, it is an aggregate view and ignores considerable local variation. Second, it conceals the real possibility that those without cars, who are not middle class, who are in an ethnic minority, who are obliged to seek rented accommodation, or whatever, may well be *worse* off than their urban counterparts through living in the seemingly idyllic world of rural Britain in the late twentieth century. We will return to this theme.

Demographic and social change

Looking at a key indicator – population size – depopulation has *not* been the norm in post-war rural Britain, as it was hitherto and as it still is across large swathes of continental

Europe. Indeed the converse is true. In 1951 there were 10.4 million people in the administrative rural districts of England, Wales and Northern Ireland and the county districts of Scotland; this was 20 per cent of the total. By 1971 the corresponding figures were 12.8 million and 23 per cent.

Moving on to the next twenty-year period, 1971 to 1991, and focusing on rural England only, the population of its 150 'most rural' local authority districts was as in Table 11.2. The reasons for the 'rural renaissance' shown in Table 11.2 are complex, but two sets of factors are important. The first are '*permissive*' – they have enabled but not caused the migration movements that underlie these statistics. Principal amongst these have been the rise in car ownership, the improvement in transport infrastructure and the penetration into the rural areas of sophisticated telecommunications benefiting both businesses and, more recently, private households alike. In short, our already small island has got progressively smaller. To this list of key 'permissive factors' we should add higher disposable incomes (especially pensions) as well as ubiquitous television, because they too have made possible an urban lifestyle without an urban residence.

The second set of factors underlying the post-war growth of the rural population are of a more *causal* nature. They relate in part to business owners and managers weighing up the pros and cons of conurbation locations compared with those of smaller settlements – and increasingly opting for the latter. Such decisions have served to draw the people, their labour force, out of the big cities and old industrial areas to join them (see the discussion of the so-called 'rural economy', pp. 219–21). And in part they relate to private individuals and households similarly preferring a village or small to medium-sized town location but this time for 'lifestyle' reasons of their own – i.e. not just to 'follow the jobs'. These 'lifestyle' reasons have been varied but relate to both rural 'pros' and urban 'cons' – with factors such as lower crime levels, more space in and around the home, a greener environment and a perceived greater sense of community proving decisive.

Of course, and this point is fundamental, not everyone has been able to indulge these residential preferences, and a measure of 'social class imbalance' and of 'age imbalance' has come about as a result. As for age imbalance, we should not underestimate the capacity of occupational pensions – still a rarity in the early post-war years – to permit a growing number of retired people, no longer needing to be near the main employment concentrations, in effect to live where they please; and clearly a lot of them of them are pleased by the coast and the more tranquil areas inland.

TABLE 11.2 Population change in England, 1971–91

		1971	*1991*	*Change 1971 to 1991*
Rural England*	Total	11.07 million	12.94 million	+ 1.86 million
	Percentage	23.9	26.8	+ 16.9
Non rural or 'urban' England	Total	35.34 million	34.76 million	–0.58 million
	Percentage	76.1	72.9	–0.2

Note: *Defined here as the most rural 150 districts using a range of criteria.

Of course different parts of the country have been affected to different degrees by this 'counter-urbanisation'.

The geography of population change in post-war Britain (see also Chapter 9) is largely the product of a complex web of interrelated migration flows including:

- planned population movement from the conurbations to the new and expanded towns;
- job-led out-movement from the cities as described above;
- retirement migration, sometimes linked to earlier decisions to buy second homes;
- localised 'commuter' relocation out into the villages and small and medium-sized towns;
- continued depopulation in some smaller and remoter parishes, especially where the loss of agricultural employment has been pronounced or a large employer has closed.

And different rural areas, though each exhibiting some degree of population growth, may be very different in character. Compare for instance some remoter rural areas with an ageing population (the product of younger people moving out and older people moving in) – with all that that implies for the local community, birth rate and service provision – with prosperous commuter areas having a more balanced age structure, thriving services and more generally the characteristics of suburbs surrounded by fields rather than suburbs surrounded by other suburbs.

Rogers' review (1993) of population and social change in England's rural communities up to the early 1990s concluded with these key observations:

- households are getting ever smaller so that the same number of people in an area will over time make greater claims on the housing stock;
- on average people move every seven years and 'an ongoing and complex re-sorting' is at work as people in different circumstances periodically reassess their options;
- in this regard, the preferences and behaviour of the growing 'service class' – broadly those in business and management occupations – tend to be the driving force in the whole process, with other groups adjusting their residential behaviour accordingly;
- regional patterns conceal a great deal of very local variation with adjacent parishes often faring very differently.

A survey in 1990 of the perceptions of some 3 000 people living in a range of English rural areas (Cloke *et al.* 1994) revealed some of the tensions generated by this constant 're-sorting' of the local population. Some longer-established residents clearly felt strangers in their own communities. One Shropshire respondent observed: 'The town types living here are not true country people; they commute to the towns and don't adapt to country life. They should respect the countryside and not try to take over social life', while a kindred spirit in rural Warwickshire complained that 'they think they own the place . . . they come in and take over everything . . . they have put the price of homes up so that the "natives" have to move out'. Of course, however, it is a fine line between 'taking over' and 'breathing new life' into the community.

BOX 11.1 The Forest of Dean: rural *and* industrial

Administratively part of Gloucestershire, this former coal-mining area has more in common visually, culturally and historically with industrial South Wales to its west than with the picturesque Cotswolds to its east. Sandwiched between the Wye and the Severn, more forested than farmed and with a scattering of former colliery villages, restored spoil tips and redundant tramways to remind the visitor of its industrial heritage, this densely populated but still rural area is seeking a new vocation, with too large a proportion of its workforce still wedded to manufacturing and land-based industries with limited potential for long-term prosperity. Only 56 per cent of the workforce is in the service sector compared with over 70 per cent nationwide.

To gauge local opinion, several village or town appraisals were conducted in the early 1990s, and taken together they revealed a strong popular attachment to the area and a love of its wooded environment and sense of community, but six issues emerged as causes of real concern:

- a need for affordable housing for local low-income people – 'the youngsters are unable to stay in the area';
- a need for more jobs locally and better access to paid employment;
- a fear of crime and the perceived insufficiency of the police presence – vandalism, petty theft and a fear of physical abuse were frequently mentioned;
- the very limited mobility of households without a car (5,800 of them in 1991, or 21 per cent of the total), with the difficulties of access to health-care of most concern;
- traffic-related problems such as poor road maintenance and street lighting, various road hazards and heavy vehicles trundling through the villages;
- a general sense of ignorance and alienation regarding the local council and 'the powers that be'.

In response to such concerns, the Forest of Dean Rural Development Programme has placed emphasis not just on attracting new firms from outside the area – the mainstay of development programmes of the 1960s and 1970s – but on forging a spirit of partnership working between the various agencies in the public, private and voluntary sectors, and on encouraging development *of* and *by* the local community, not just *for* it. But this is a long-term process; tangible results take time to come through.

Welfare, deprivation and exclusion

The fact is that the demographic and social changes briefly summarised above – as well as trends in the economy and deficiencies in the availability of affordable housing and of essential services – have affected different rural residents in different ways, and the circumstances of a substantial minority are increasingly a cause for some concern. In this regard four concepts recur in the literature, with subtle differences in meaning. They are worth setting out, even though they are not always employed consistently, if only to highlight the fact that the nature of the problem and its possible causes and remedies are not as simple as they might at first seem:

- *Disadvantage* is the inability of individuals or households to share in styles of life open to the majority.
- *Poverty* is the inability of individuals or households to share in styles of life open to the majority because of a lack of financial resources (the term is usually made operational by using as a yardstick some measure of departure from an accepted 'income norm').
- *Deprivation* is a less precise concept but denotes something more than just the lack of material resources; it tends to involve value judgements about what is or is not morally acceptable.
- *Social exclusion* is the *process* whereby the various systems that should guarantee the social integration of individuals or households fail to do so. Classic examples are the employment and housing markets and the way that we deliver education and training to those in need.

With those interlocking concepts in mind – some of which focus on the sufferer and some on the sufferer's context – all that can be attempted here is to list a number of key conclusions from an in-depth study in 1990 of a random sample of 3,000 households scattered across a dozen representative rural areas of England (Cloke *et al.* 1994):

1. About 20 per cent of all rural households are on the margins of poverty (defined as equivalent to 140 per cent of the income support entitlement of the household in question).
2. Elderly people comprise the biggest single group susceptible to poverty.
3. Poverty and deprivation are to be found in all twelve study areas, though more in some (e.g. the Nottinghamshire coalfield) than others (e.g. the commuter belts of Cheshire and West Sussex).
4. For those of working age without their own transport, poor access to job opportunities is a problem because of the limited supply of well-paid employment and poor or non-existent public transport.
5. There is a significant variation of access to private transport *within* car-owning households, with women typically disadvantaged.
6. The difficulty of access to shops and to health-care is keenly felt in no-car households.
7. The limited availability of affordable rented accommodation – especially for younger low income people – is a serious problem.

And, shifting the focus to people's *attitudes and feelings*:

8. Very different perceptions are held of the benefits and costs of 'rurality'. One person's splendid isolation is another person's loneliness. One person's 'close knit community' is another person's world of prying, gossiping and intrusion.
9. Many people who have lived in the area for most or all of their lives now feel left

out and marginalised as other people move in who are affluent, influential, have different social and political ambitions and even a different view of what 'rurality' is all about.

10 Most people living in rural areas do not admit to the existence of poverty and deprivation in their midst. Those with such problems tend to be very largely 'out of sight out of mind'; this simply is not the case in our big cities and run-down industrial areas.

Two important questions flow from this scenario. First, how far is 'rurality' itself at fault in 'causing' or accentuating these problems? And how far does their resolution lie in the hands of local or explicitly 'rural' agencies? The answer to both questions is a tantalising 'to a certain extent'.

The 'rural economy'

It is arguable whether a distinctive 'rural economy' really exists any more. It seems more valid to say that Britain has an 'economy' with somewhat different characteristics in different parts of the country and with the traditional rural/urban distinction becoming less and less pronounced. In large part this reflects the steady erosion of 'traditional rural jobs'.

Thus, as far as England is concerned, the total farm labour force in 1950 was about 1 million, including around 700,000 paid agricultural workers of different kinds, while by 1994 the total was 431,000, including just under 200,000 paid workers. Extractive industry is no longer a major employer in the countryside; indeed as far as coal-mining is concerned there are only about 5,500 people employed in rural collieries (1994 figures), compared with over 60,000 in the early 1980s. As for quarrying, the number of jobs in that sector fell by over 30 per cent between 1981 and 1991. And, with the ending of the Cold War, defence-related jobs – since the Second World War a major staple of many rural areas – have similarly fallen significantly (Rural Development Commission 1995).

However, the evidence suggests that the economy of at least the more accessible rural areas has adapted well to the decline of employment in the traditional industries. This is largely because the post-war growth of manufacturing industries, and more particularly of service activities, has been freed from a need to locate in the traditional urban and industrial areas by new technologies, especially in communications. As explained earlier this outflow has been inextricably linked to the residential aspirations of entrepreneurs, managers and skilled staff in a complex of cause-and-effect relationships. In the remoter areas, however, the traditional land-based elements of the economy retain a distinct significance, though even this distinction – between 'accessible' and 'remote' rural areas – seems to be melting as telematics (the fusion of telecommunication and computer technology) removes a major disadvantage of physical remoteness.

The birth of small firms has been particularly important in this respect – and of Britain's 2.7 million businesses, 96 per cent employ fewer than twenty people. Analysis at county level shows that it is rural counties in central or southern England such as Devon, Hereford and Worcester and Cambridgeshire that have the greatest propensity to generate new, almost always small, firms while the conurbations have the lowest rates of new firm growth. And research also shows that the majority of new business founders set up their

BOX 11.2 The Western Isles, Skye and Lochalsh: does remoteness matter?

The most north-westerly of the European Union's 850 'LEADER'* Rural Development Programmes covers nearly 6,000 square kilometres of islands, moorlands and mountains and just 43,000 inhabitants in this, the remotest region of Scotland. The main economic activities of the Western Isles, Skye and the neighbouring mainland are crofting, fishing, fish farming, fish processing, quarrying, construction, forestry, tourism and the servicing of the local population.

The area's deep-seated problems include high unemployment and the steady out-migration of its young people (initially for further and higher education and then for the better employment opportunities of the urban areas), plus, of course, the difficulty and cost of transport within the area as well as to and from the major markets of Britain and Europe. Its assets include the Gaelic language, which has moulded every facet of the cultural heritage of the area and helped give an underlying cohesion, and a wild and beautiful environment which brings over half a million tourists every year.

More prosaically, its assets also include 'Objective 1 status', meaning recognition in Brussels that this is one of the European Union's poorer regions with a Gross Domestic Product less than 75 per cent of the EU average. With that recognition has come substantial financial aid. That part of the aid coming as part of the 'LEADER' programme has had to be spent on projects suggested and championed by local people themselves – the 'bottom-up approach' which develops 'the human resource' as much as the economy in a conventional sense.

Of the projects supported over the last six years by the local LEADER partnership of private, public and voluntary bodies (which must put in 'matching funds' as a measure of their own resolve), most have sought to 'add value' to the local resource – be it physical, cultural or human – and/or to grapple with the challenge of overcoming geographical remoteness. They have therefore included:

- establishing a network of '*community animateurs*' to help local people develop their own business ideas;
- *distance learning courses*, often using telematics, to foster training and advanced education without the trainee or student having to leave the area;
- *interpretative centres* to celebrate and explain the history, culture and life of the area for locals and tourists alike;
- various *marketing initiatives*, including the setting up and support of craft associations and the opening of dedicated retail outlets.

The future should be bright for remote and rugged regions like north-west Scotland as discerning tourists and entrepreneurs are increasingly attracted to places of tranquillity, natural beauty and cultural distinctiveness, and as the Internet promises to shrink intra- and inter-regional distances still further.

*LEADER (Liaison Entre Actions de Developpement de l'Economie Rurale) is a EU programme to promote innovative rural development via locally focused partnerships and community involvement.

business in the locality in which they are living – though most new firm founders in rural areas had moved to the countryside prior to setting up the firm.

This 'geography of entrepreneurialism', coupled with the role of environment and quality of life factors in explaining the changing geography of potential entrepreneurs, is central to our understanding of the future and character of rural Britain in the years to come. The really interesting question, in the age of the Internet, is whether in the early twenty-first century the growth zone will be not so much the Herefordshires and Cambridgeshires but the remoter parts of rural Wales, northern England and the Scottish Highlands and Islands. To the extent that the answer is 'yes' then the challenge of conserving the natural and human-made heritage and beauty of those areas will be correspondingly greater.

Another important element of the renaissance of rural areas is the growth of 'countryside tourism', given the growing interest in the environment and the potential for the countryside to attract short-break and overseas tourists. Rural tourism has been boosted by the growth of countryside accommodation and other tourist facilities, assisted by farm diversification and by new products such as timeshare, all-weather holiday villages, and what some term the 'commodification of the countryside'. But often the economic challenge is to get visitors to stay overnight (picnicking day-trippers and coach parties pump little money into the local economy) while the environmental challenge is to channel and accommodate rural tourism and recreation so that it does not degrade the very environment that attracted it.

Services and accessibility

A key issue in rural Britain is the steady decline of many services traditionally found in our villages and small towns and the isolation that this implies for many rural people in no-car or even one-car households. Two factors in particular are at work, whether the service be health-care, education, retailing or whatever. The first is 'economies of scale', meaning the lower average costs *for the supplier* of delivering services in fewer and larger units. The second is the growing mobility of the car-borne majority which *permits* service suppliers – be they retailers, health authorities or breweries – to benefit from those economies of scale secure in the knowledge that most, but not all, of their clients will be able to reach the fewer but better endowed larger 'units' (superstores, large health centres, etc.).

The slow but steady withdrawal of rural services matters for four reasons:

- people with limited mobility suffer – especially elderly people, households too poor to run a car, young people seeking work or a richer social life, and mothers in one-car families at home with young children;
- the 'balanced community' is eroded if schools, surgeries and post offices are not there to attract and retain the sort of people just referred to;
- the village loses 'venues' – school gates, post-office counters and the like – where people can rub shoulders and undertake the exchange of information which is the stuff of a caring community;
- very local jobs are lost (jobs for school cleaners, surgery receptionists, shopkeepers and bar staff), which also sustain the balanced community as well as circulate money in the local economy.

TABLE 11.3 Parishes without certain services

Worsening situation compared with 1991	*Apparently stable situation*	*Improving situation compared with 1991*
Permanent shop (42%)	*Pub (29%)	Village hall or similar (28%)
Post office (43%)	Petrol station (56%)	Bottle bank or recycling scheme (60%)
Seven-day bus service (75%)	*GP surgery (83%)	Community minibus or social car scheme (79%)
Bank or building society (91%)		Public or private nursery (86%)
Police station (92%)		Day care for elderly people (91%)

Notes: Percentages are of parishes *without* certain services. *Other evidence suggests a decline in the number of pubs and surgeries in the 1990s.

Survey evidence provided every three years by England's 8,000 or so parish clerks at the behest of the Rural Development Commission charts the sorry tale (Rural Development Commission 1998). The picture in 1997 is shown in Table 11.3.

Not surprisingly, population size plays an important part in explaining variations from parish to parish in service provision. For example, of parishes with 200 to 299 inhabitants (i.e. mainly 'very small villages'), the proportions *without* certain services in 1997 were: permanent shop (82 per cent); pub (61 per cent); seven-day bus service (89 per cent); school (97 per cent). Corresponding figures for parishes with 1,000 to 2,999 inhabitants (i.e. mainly 'large villages'), were 4 per cent, 4 per cent, 43 per cent and 8 per cent, respectively.

Not all, however, is doom and gloom. Many public and voluntary agencies are trying to arrest the decline or to come up with compensatory measures. For example, the Rural Development Commission offers advice and support to village shopkeepers, and the county-based Rural Community Councils often help run 'Use it or Lose it' campaigns. And some enterprising local or regional entrepreneurs are using lateral thinking, with the post bus, the post office inside the pub and the mini library inside-the-shop all demonstrating ways of 'doubling up' on service provision by running two or more services from one building or vehicle. Growing community involvement – for example, in community transport schemes or in supporting small schools – is also helping. So too may 'telematics' – the linking of computer and communications technology with, for example, small post offices becoming able to offer a much wider range of services to their clientele.

Annual surveys of what is happening on the ground, as recorded by the field officers of the county-based Rural Community Councils and collated by the present author for the Rural Development Commission, reveal some of the forces at work. Table 11.4 sets out some 'good news' and some 'bad news' for several local services, as revealed in these surveys . But what is 'good' news and what is 'bad'? Is a petrol filling-station selling groceries on a bypass half a mile from a village centre a 'good' or 'bad' thing for car-less residents of the village? Better than nothing? Or the last nail in the coffin of the village shop . . . which may of course be destined to close soon anyway?

TABLE 11.4 Some factors affecting the viability of village services

	Good news	*Bad news*
Food shops	Relief from Uniform Business Rate for some small shops in small villages	Superstores opening in quite small towns and extending their opening hours into the evenings and weekends
Post offices	New deals negotiated by Post Office Counters to help small post offices broaden their product range (e.g. travel insurance and 'bureau de change')	Small post offices downgraded to part-time 'community office' status and becoming unattractive as going concerns
Public transport	Increase in community-provided transport schemes which can now be subsidised by parish councils	County councils which lost 'their' larger towns to 'unitary council' status having reduced scope for cross-subsidy of rural transport
Pubs	Increasing multi-use of pub premises – especially restaurant meals for the 'food and family' market, but also the pub hosting a post office, elderly persons' luncheon club, etc.	Increased import of cheap liquor from Continent; stronger anti-drink drive culture; supermarket liquor consumed at home
'Village schools'	The clustering of schools located in neighbouring villages so that they share resources (e.g. head teacher management)	Growing trend for children not to attend their most local school as schools increasingly compete
Primary health-care	Growth of 'primary care led purchasing' is strengthening the GP focus in health-care provision. Many surgeries are now effectively 'mini health centres'	Closure of GP 'branch surgeries'; NHS dentistry unavailable over quite large rural areas

Housing and planning

The nature of 'the rural housing problem' has changed enormously in the post-war period. No longer is there a serious *physical* housing problem – the absence of piped water, electricity or adequate sewerage, gross overcrowding, dampness or squalor. The problem now is one of *access* to housing, as well-heeled commuters and retired couples compete for what was traditionally working-class housing. In commuter villages (detached suburbia) and second home villages (seasonal suburbia) the conflict is between those who *can buy* and those who *must rent*, with house prices, particularly in rural areas which are close to the cities and/or environmentally attractive, precluding purchase by lower-income households.

Well-intentioned policy initiatives have often made the situation worse. The 'right to buy' legislation of 1980 which compelled local authorities to sell their 'council housing' with favourable discounts to those sitting tenants wishing to buy, coupled with a virtual embargo on new housebuilding by those same local authorities, has eroded the pool of affordable rented housing stock in the countryside. Housing associations were intended to take up the slack but, for various reasons set out below, their contribution, especially in the smaller settlements away from the market towns, proved to be insufficient right through the 1980s and 1990s.

All of this matters for two reasons. First, social justice. Most people – indeed most politicians regardless of party – feel that young people raised in a certain area, especially if they are working there as well, have some sort of implied right to find reasonable accommodation locally when the time comes for them to set up their own homes. Similar arguments tend to be made for other social groups of local origin – retired people wanting to move back to be near their grown-up children, lone parents setting up a second household, etc. Yet frequently such people are obliged to move away to the more plentiful stock of affordable housing in the towns – even if this means 'reverse commuting' to jobs back in the countryside.

The second reason relates to 'social balance'. To the extent that our villages become 'havens of the middle classes' then, the argument runs, society as a whole is the loser – *including* the middle-class people living in those villages. This is part of a bigger argument about the dangers of 'social polarisation', but at a more prosaic level it is about a reduced capacity for 'community care' in the wider sense of the expression. Where have all the care assistants, school secretaries and for that matter single adult sons and daughters gone? Few of them on low incomes are to be found living in the villages and thereby in a position to support their neighbours and the wider community outside working hours.

Restrictive land-use planning policies, while serving to protect the precious countryside of this small island, have not helped in this respect. Strict development control, linked to a variety of green belt, 'village envelope', 'key village', conservation and aesthetic objectives, have served to curtail the supply of new affordable housing by pushing up land prices. And there is a sort of vicious circle at work here. As the rural areas become more populated by middle- or upper-income people who have chosen to move there for environmental reasons, so the NIMBY cry of 'Not In My Back Yard' gets correspondingly louder when the suggestion is made that a few new houses need to be built for local people.

'Housing need' surveys often reveal that just half a dozen new dwellings in a given village would make a real contribution to both the 'social justice' and 'social balance' objectives set out above – *so long as the rents are kept low and so long as 'nomination rights' rest with a benign local agency* which will ensure that low-income local households get first refusal when a property falls vacant. But even this modest target is often beyond the housing associations' capacity to deliver – for a combination of reasons:

- the prohibitive cost of land at open market prices;
- insufficient money earmarked by the government's Housing Corporation to fund such rural schemes;

- a lack of 'scale economies' – it generally being cheaper to build thirty-six houses in a neighbouring town than six in each of six villages;
- the 'hassle' of NIMBY opposition, especially if the parish council is only lukewarm in its enthusiasm for low-cost housing for local people.

A selection of statistics relating to different years in the early 1990s provides a crude quantification of the situation:

- at least 16,000 new affordable houses are needed each year;
- but only about 2,000 are being built;
- an estimated 17,000 homeless people are to be found in rural England;
- about 70 per cent of all rural residents in a national survey perceived the lack of affordable housing as a serious social problem in their area;
- young married people, followed by young single people, are perceived as those in greatest need by the same sample of residents.

Looking to the future, only the availability of moderately priced land earmarked for social housing will permit progress on this issue. In this regard the creation of 'land banks' – i.e. files of landowner promises to sell land cheaply for affordable housing in the right circumstances – and the operation of what has come to be known as 'exceptions planning policy' allow some reason for modest optimism. The latter allows housing development on sites within or adjoining existing villages *which would not otherwise get planning permission* – so long as the existence of local housing need is proven and legally binding mechanisms are in place to ensure the allocation and reallocation of tenancies to low-income local people.

But in the end much depends on local attitudes. If local communities are adamantly opposed to even modest housing developments intended for their poorer neighbours, then developers will tend to go elsewhere. This in turn implies a long-term programme of 'community development' if some measure of social balance is to be maintained in rural England into the next century.

Power and decision-making

In recent years geographers with an interest in the sort of issues reviewed in this chapter have paid increasing attention to the decision-making processes that underlie them. Many relevant decisions are of course taken by individual people – and decisions to move house, to take holidays in one place rather than another or to shop here rather than there, have been much researched. But it is increasingly realised that decisions taken by powerful public and private sector 'actors' are at least as influential as the choices made by millions of individuals.

In this connection, it should be stated at the outset that 'the planner' – meaning the statutory land-use planning authority at county, district or unitary authority level – is but one actor on the rural stage, and a relatively weak one at that. This is for two reasons. First, by and large, the 'planner' can only steer development not cause it to happen or prevent it completely. Second, there are so many more important decision-makers. For example, the 'accessibility problem' experienced by so many rural households with limited mobility is the product not just of their own personal circumstances but of a web of ill co-ordinated decisions taken by the bus and rail operators, the health authorities and trusts, the big retail chains, Post Office Counters, etc., as well as by those in the local authorities who try to co-ordinate transport provision and 'outlet location' decisions in a variety of strategic and local planning exercises.

Indeed a new word is entering the lexicon of those who study the changing rural scene. It is 'governance':

> The concept of 'governance' is broader than that of 'government' because it encapsulates not just the formal agencies of elected local political institutions, but also central government, a range of non elected organisations of the state at both central and local levels, as well as institutional and individual actors from outside the formal political arena, such as voluntary organisations, private businesses and corporations, the mass media and increasingly, supra national institutions such as the European Union.
>
> (Ward and McNicholas 1997: 1)

The point is twofold. First, only by trying to untangle how all these agencies interrelate can we really get a handle on how rural Britain is changing. Second, given that the decision-making arena is so crowded, attempts are increasingly being made to pull them together at the local or area level into formal or informal partnership arrangements. Two of this chapter's three brief case-studies of Britain's rural areas – those of the Scottish Western Isles and of England's Forest of Dean – demonstrate examples of local partnerships, or, to put it differently, attempts at 'governance'.

One other key development in 'rural decision-making' warrants mention, touching as it does all of the issues raised in this chapter – that of increased 'local community involvement'. More and more ordinary people living in our small towns and villages are demanding a say in the decisions that affect them and often have a role in actually delivering services at the very local level. Equally striking, local and national government is increasingly encouraging this trend, troublesome though on occasions it might prove to be.

There are various reasons for this official encouragement of community action. First, better decision-making: rural people are an increasingly sophisticated source of information and of ideas for addressing local concerns that it would be folly to ignore. Second, if a local authority policy – relating, for example, to a school reorganisation programme or to the designation of land for different kinds of development – can be firmly based upon a local consensus then it is less likely to be 'ambushed' when it comes to implementation. Third, self-help can obviously save money – especially in those smaller or remoter settlements where public and private sector agencies are increasingly concerned at the costs of service delivery. And local people are frequently ready to give freely of their time, expertise and money, not to mention spare seats in their cars and underused space in their buildings if it is for the good of their immediate local community. Fourth, involving the community can enhance what is now called 'capacity building', meaning the building up of the human resource, the social networks and the informal institutions of an area which can all bear fruit in the longer term.

One increasingly popular way of stimulating community involvement is the 'village appraisal' – over 2,000 of which have been carried out in British parishes, villages and small towns over the past fifteen years. These are social surveys 'of the people, by the people, for

BOX 11.3 Goring and Streatley: a slice of 'middle England'

Goring and Streatley lie on the Oxfordshire/Berkshire border, astride the River Thames, ten miles upstream from Reading, forty miles from London, and well and truly in the relatively prosperous 'South'. In 1991 a group of volunteers, helped by the two parish councils and the local amenity association, decided to carry out a 'village appraisal', delivering a long questionnaire to all 1,700 households and achieving a 70 per cent response. The idea was to 'provide an accurate picture of the villages as they are now . . . for information and interest . . . and to help villages and organisations interested in the future'.

The attractive 62-page report that emerged, *Goring and Streatley: A Portrait*, is one of a thousand or more produced up and down the country in the last decade and it is the very 'ordinariness' of the picture revealed which is of interest. This is not a former coal-mining area with run-down colliery villages or a collection of crofting communities on remote islands 600 miles from London (see Boxes and 11.1 and 11.2). It is a slice of middle class, middle aged, south midland, 'middle England' beloved by millions of semi-rural/semi-urban (does it matter which?) people spread across broad swathes of our country.

Car ownership in the two villages is high. Even in 1991 43 per cent of households had one car, 33 per cent had two and 9 per cent had three or more. By the same token, 15 per cent did not have a car at all. And this is 'commuter land', with half of those employed travelling daily to Reading or London and with most of the rest working in a broad swathe of Oxfordshire, Berkshire and neighbouring counties. Unemployment hardly gets a mention in the report, but there is concern that strikingly few adults aged 20–40 live in the two parishes, and thereby few young children. Housing is expensive and most lower-income families cannot afford to buy in this part of the Thames valley.

The concerns expressed by a majority of the households surveyed were far from being narrowly self-interested. Most households are themselves well housed and car owning, but a need for affordable housing for rent, and for better public transport, was widely appreciated. High on the list of concerns were two almost universal *bêtes noires* of middle-class, semi-rural, outer-metropolitan England. First, traffic – congestion, road safety, the long mooted but still awaited bypass. Second, 'planning' – the sense of having to be constantly vigilant lest 'the planners' foist on the community 'unwanted development' – in this case a shopping and office development on a vacant site in the village centre.

And the responses of the people of Goring and Streatley also neatly exemplify the widely held sentiments of village and small town communities across the country: an affection for 'the village atmosphere and environment' and a concern for very local environmental nuisances – in this case traffic intrusion and car parking, a few eyesores and derelict buildings, litter and dog dirt in the park.

the people', with the words 'by' and 'for' being the crucial ones. Data amassed at local level on people's hopes and fears, likes and dislikes, pulled together by volunteer teams into reports on 'our village and where it's going' and debated in village hall and parish council meetings, have proved to be a powerful spur to action. Such action may be by outside agencies who feel empowered or cajoled into re-routing the bus service or part-funding a play scheme; and/or by local people themselves who feel encouraged to set up 'good neighbour schemes' or community shops, say, by the local mandate for action and the freshly replenished reservoir of community spirit.

The danger, of course, is that 'do it yourself' will come to be seen as the panacea for the ills of our rural communities. Certainly there was a whiff of this in the Rural White Paper (DoE and MAFF 1995), which set as a key objective 'encouraging active communities which are keen to take the initiative to improve their quality of life'. It would be a tragedy if the growth of community involvement were to become the state's mandate to abdicate as far as rural Britain is concerned. Rather, the fortunes of Britain's rural communities in the twenty-first century may better lie in the notion of 'partnership' and in ever seeking a judicious balance between the benefits of orchestrating integrated strategies and programmes at area or regional level on the one hand, and on the other of fostering grass roots initiative and entrepreneurship. Certainly the old notion of 'state decides, state provides' seems gone for ever.

Conclusion

At the end of the twentieth century, rural Britain continues to undergo rapid social and economic change, driven by forces emanating largely from outside the rural areas themselves. The next century will see a further erosion in the economic distinctiveness of rural areas, although socially they may increasingly be home to the more privileged members of our society who will strive to preserve, indeed to 'manicure', the idyllic character of their environment even if that imposes hardship on their less fortunate neighbours.

Listing a few overarching concepts provides a convenient way of summarising some of the challenges that lie ahead:

- *sustainability* – managing the inevitable change so that it minimises the erosion of the inherited 'capital' of rural Britain – the word 'capital' embracing the natural and human-made environment as well as traditional culture and social support mechanisms in the widest sense;
- *social balance and social justice* – ensuring that we do not drift towards a polarised society in which our countryside, villages and attractive small towns become just a haven or playground for the middle classes while the cities deteriorate in quality and social acceptability;
- *governance, partnership and community involvement* – all terms that imply a need to galvanise the full range of local people and of relevant agencies, from parish councils to Regional Development Agencies, in resolving these key issues.

References

Cloke, P., Milbourne, P. and Thomas, C. (1994) *Lifestyles in Rural England*, Salisbury: Rural Development Commission.

Department of Environment and Ministry of Agriculture, Fisheries and Food (1995) *Rural England, a Nation Committed to a Living Countryside* ('The Rural White Paper'), CM3016, London: HMSO. (*Note*: Parallel documents were produced simultaneously by the Scottish and Welsh Offices.)

Moseley, M.J. (1997) 'Parish Appraisals as a tool of rural community development: an assessment of the British Experience', *Planning, Practice and Research* 12(3): 197–212.

Rogers, A. (1993) *English Rural Communities: an Assessment and Prospect for the 1990s*, Salisbury: Rural Development Commission.

Rural Development Commission (1995) *Rural Economic Activity*, Salisbury: Rural Development Commission.

Rural Development Commission (1998) *1997 Survey of Rural Services*, Salisbury: Rural Development Commission.

Ward, N. and McNicholas, K. (1997) *Reconfiguring Rural Development in the UK: Objective 5b and the New Rural Governance*, Newcastle upon Tyne: Centre for Rural Economy, University of Newcastle upon Tyne.

Further reading

LEADER Magazine (LEADER Observatory AEIDL, Chaussee St Pierre 260, B1040, Brussels). This tri-annual magazine is available free of charge from the Observatory.

Three excellent ongoing series of reports on aspects of rural development in the UK are worth noting, namely those by:

1 the Scottish National Rural Partnership on 'Good Practice in Rural Development' (contact the Scottish Office);
2 the Centre for Rural Economy, University of Newcastle upon Tyne;
3 the Rural Development Commission, London and Salisbury (now subsumed into the Countryside Agency).

Chapter 12

Life chances and lifestyles

Daniel Dorling and Mary Shaw

Introduction

This chapter considers some of the most simple of life chances in Britain. It shows how they are affected by social influences on lifestyle at the individual and geographical level. People's chances of dying young in the 1990s are used as an indicator of misfortune. These are compared with their chances in the 1950s and the two geographies are mapped for different age groups to show the changing regional patterns of mortality. Normally medical explanations are made for premature mortality, but this chapter focuses solely on social and behavioural factors as explanations for mortality rates for particular groups of people in Britain. The chapter begins by illustrating the changing regional geography of mortality in Britain through a series of maps for different age groups at two points in time. We then show how geographical inequalities in mortality can be measured. Since the 1950s, and particularly during the 1980s, geographical inequalities in mortality and life chances in general have grown in Britain. The factors which may underlie this rise are then discussed. The chapter then shows the reader how they can estimate their life expectancy, or, rather, the life expectancy of someone of their age, sex and characteristics. The categories of factors considered are basic demographics (age and sex), social class background (father's social class, own social class, height and unemployment), geography (country of birth, region of residence, housing tenure), behavioural factors (smoking, alcohol, diet, exercise, weight and injecting illegal drug use) and relationships (sexual activity and marital status). The purpose of this exercise is to show how life chances in contemporary Britain have both social and spatial dimensions. The geography of society influences the behaviour and opportunities of its members which, in turn, alters the geography of the outcomes of their actions – in this case, death.

The changing geography of death in Britain

There is a geography to premature death in Britain. This geography is largely the mirror image of the geography of good health. The concept of health is very difficult to define, and even more difficult to measure. For example, the World Health Organisation (1948) defined health as 'a state of complete physical, mental and social well-being and not merely the absence of disease'. Because good health is so difficult to define and measure, geographers often use the map of premature death – a much more easily definable and measurable outcome – as a proxy for a map of ill health. In this chapter we shall introduce you to the geography of good and poor health and how health can be counted and the factors behind health measured. We shall show you maps of areas of good and poor health for different groups of the population and describe how and why these maps have changed over the last fifty years. We shall also explain how the geography of health in Britain has been changing over recent decades. These patterns and changes will be illustrated by showing you how you can estimate your own life expectancy.

We start our description of the geography of mortality in Britain by looking at the differences in people's chances of dying fifty years ago. Here, for simplicity, we concentrate on the ten places for males and females which had the highest and lowest mortality rates for each of six age groups. Males and females are considered separately because different factors have differing influences on their lives and they often die from different causes of death. The obvious differences are biological; for example, no men die giving birth to a child or of ovarian cancer. More importantly men and women, in general, tend to suffer differently as a result of the social changes which underlie the geography of health. For instance, when the shipyards and mines were closed it was almost all men who lost their jobs and who were subsequently more likely to suffer from ill health. Women are, in general, more likely to suffer from the effects of insecurity of employment which have been shown to be health damaging, but are less likely to suffer, for example, from accidents in adolescence as they are brought up to be more careful than boys. We separate the six different age groups described here for much the same reasons, both biological and social.

Our data are based on the Registrar General's Decennial Report on Mortality in England and Wales for 1950–3 and the Scottish General Registrar Office Reports of Mortality for these same years. The geographical areas we use are the 292 London Boroughs, Metropolitan Boroughs and Urban and Rural Remainders of Counties used in these reports. We have also used a Geographical Information System to produce equivalent statistics for mortality in the period 1990–2 so that changes in the patterns can be evaluated for these same areas. This was the most recent period which could be compared with the past as we are reliant upon the 1991 Census to provide the population estimates to calculate mortality rates for these areas.

Figure 12.1 shows the geography of infant mortality across Britain between 1950 and 1953 and can be compared with Figure 12.2 which shows the same distribution forty years later. Figure 12.2 highlights the ten areas with the highest and lowest rates of infant mortality for both boys and girls. Usually these areas coincide, areas with high rates for boys also have high rates for girls and in these cases the figure splits the two symbols in half and the combined symbol is placed over the area being indicated. There was a clear regional divide in infant mortality in the 1950s with the highest rates found in large cities in the North of England, Scotland, South Wales and the Potteries while the lowest rates were all south of a line between the Severn and the Wash, with the exception of Edinburgh for infant girls. This divide reflected the living standards of the parents of infants across Britain at that time, their levels of nutrition, whether their homes were damp, levels of overcrowding, and so on. Forty years later, as Figure 12.2 shows, the map has not altered greatly, despite huge reductions in the national rates of infant mortality and great improvements in general levels of nutrition, housing conditions and overcrowding. This is because the relative differences between the standards of living in the North and South have not narrowed. There has been a concentration of the areas with the highest rates of infant mortality into northern English cities, and two affluent places in the north (small towns in Cheshire and parts of Rural Derbyshire) now have some of the lowest rates.

Figure 12.3 shows mortality rates for boys and girls aged 1 to 4 between 1950 and 1953 with Figure 12.4 showing these same rates forty years later. Again the North–South divide is evident, although slightly less clear cut. Part of the reason for this is that, particularly in the later period, this is the age group in which mortality is least common and so a few extra deaths occurring in any one year can alter the overall picture. The congenital conditions which lead to the majority of infant deaths are no longer so vital for this age group and low rates can now be seen west of the Penines as well as in the South and in Edinburgh, as before. However, by 1990–2 rates in two of the poorer London Boroughs come to be within the top ten areas for boys aged 1 to 4, while in the North rates are generally only low outside of the larger towns. The relative deterioration of infant health in London will be seen later to be reflected by a new concentration of ill health for adults in the capital city by 1992.

Figures 12.5 and 12.6 show the rates for the next age group of children aged 5 to 14 in 1950–3 and 1990–2. Again the North–South divide is evident, although rates are low in the 1950s in small towns in Durham, the West Riding, Lancashire and Cheshire. All high rates are again north of a line between the Severn and the Wash. Fifty years on the pattern is similar, although rates in the small towns of Wiltshire (principally Swindon) and in the old London Borough of Deptford are now in the top ten. Accidental death is a major factor for this age group, often including road traffic accidents. But again this is not a random toll. A child's chances of being killed by a car are strongly related to their social advantages and disadvantages, as we discuss later.

For young adults Figures 12.7 and 12.8 also show an extremely strong North–South divide for both the 1950s and 1990s. Initially, only in the rural districts of the West Riding and Lancashire were young adult mortality rates low in the North. Forty years later the greatest change has been the sudden concentration of areas of high mortality into six London Boroughs. Of all the age–sex groups we are examining here, men aged 15–44 have seen the worst improvement in their overall chances in recent years, with their death rates now actually rising in many areas. This rise has followed the pattern of high unemployment in Britain – the relatively new concentration of mass unemployment and poverty within London has led to this geographical concentration.

For older adults aged 45–64 the story is quite different, as Figures 12.9 and 12.10 show. The North–South divide is almost as clear as for infants in the 1950s, with only the rural districts of the West Riding bucking this trend. However, by the early 1990s this anomaly has disappeared and the country is clearly divided between northern cities where the old die young and the generally rural areas of the south where they are more likely to live to healthy old age. Compare Figures 12.1 and 12.10, infant mortality in the 1950s and older adult mortality in the early 1990s. Notice any similarities? In general, the children who were

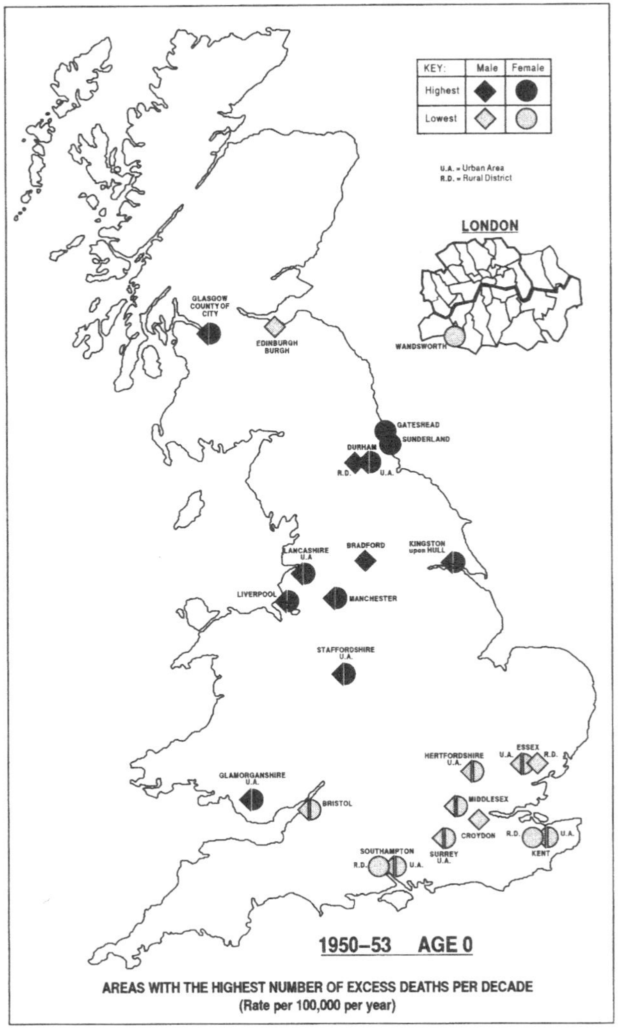

FIGURE 12.1 Infant mortality rates for males and females, 1950–3

FIGURE 12.2 Infant mortality rates for males and females, 1990–2

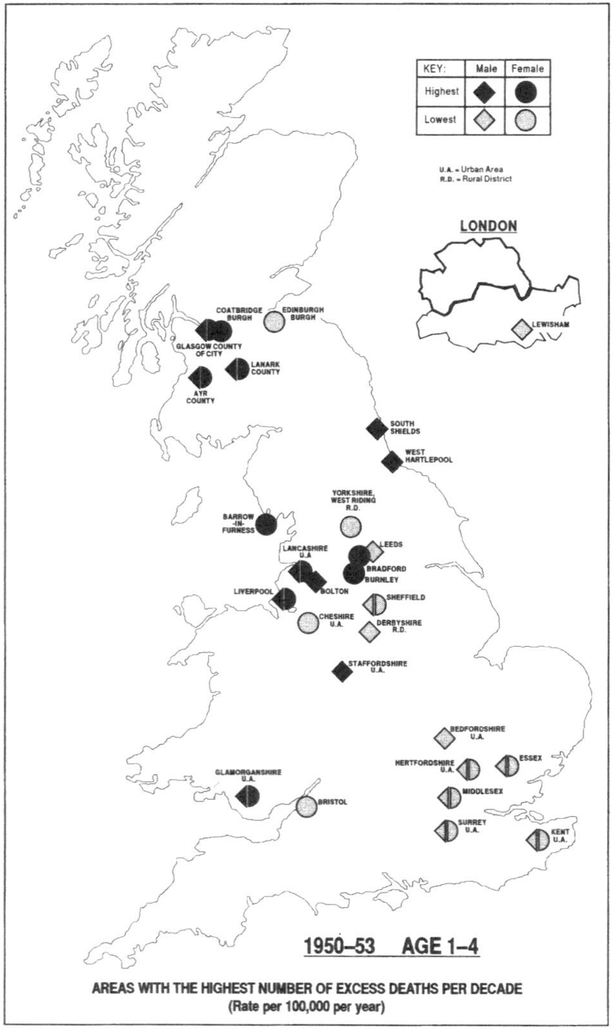

FIGURE 12.3 Mortality rates for males and females, aged 1–4, 1950–3

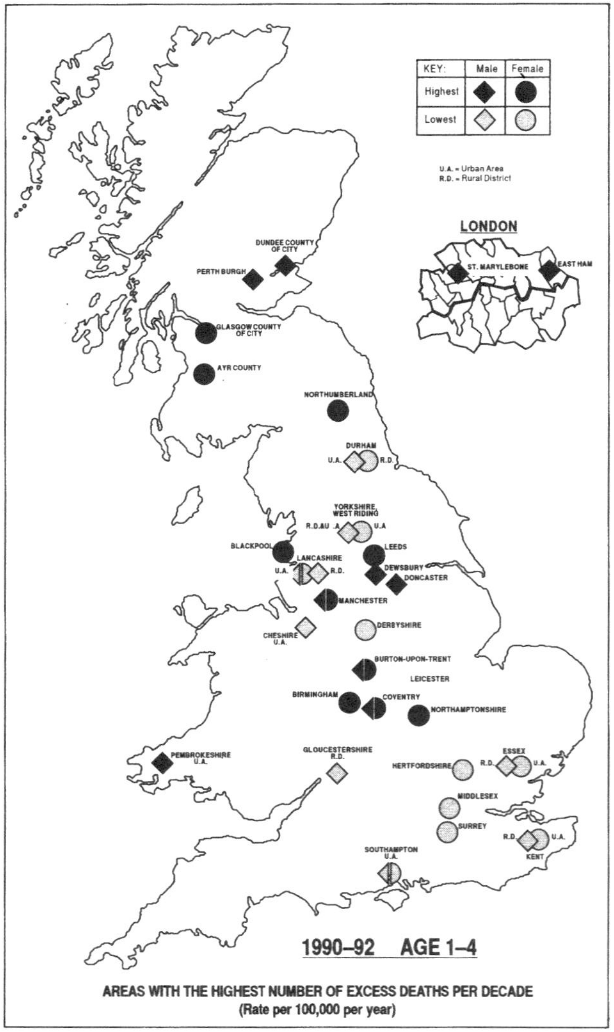

FIGURE 12.4 Mortality rates for males and females, aged 1–4, 1990–2

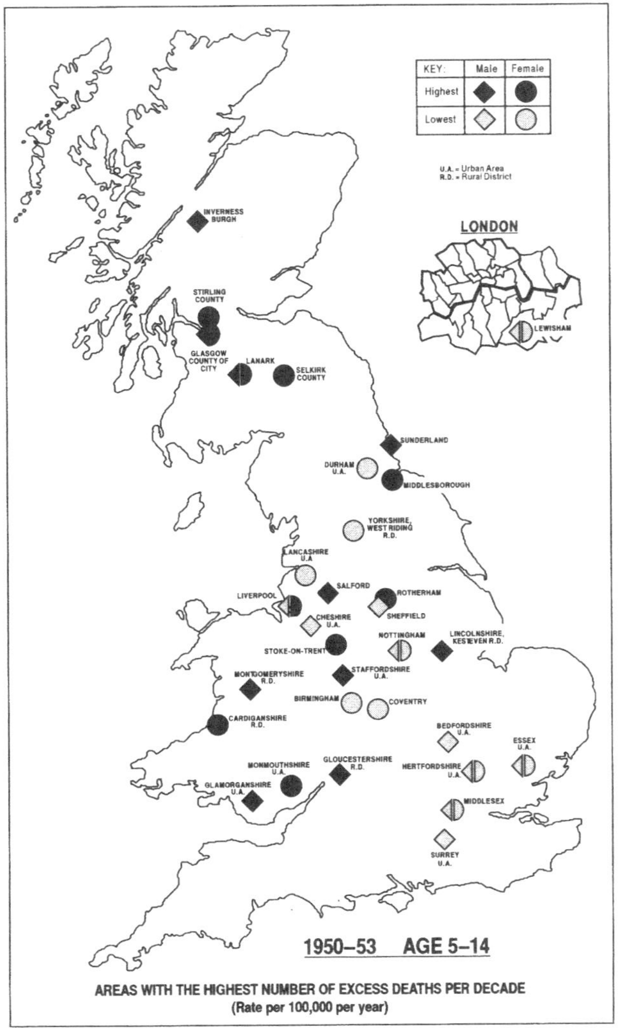

FIGURE 12.5 Mortality rates for males and females, aged 5–14, 1950–3

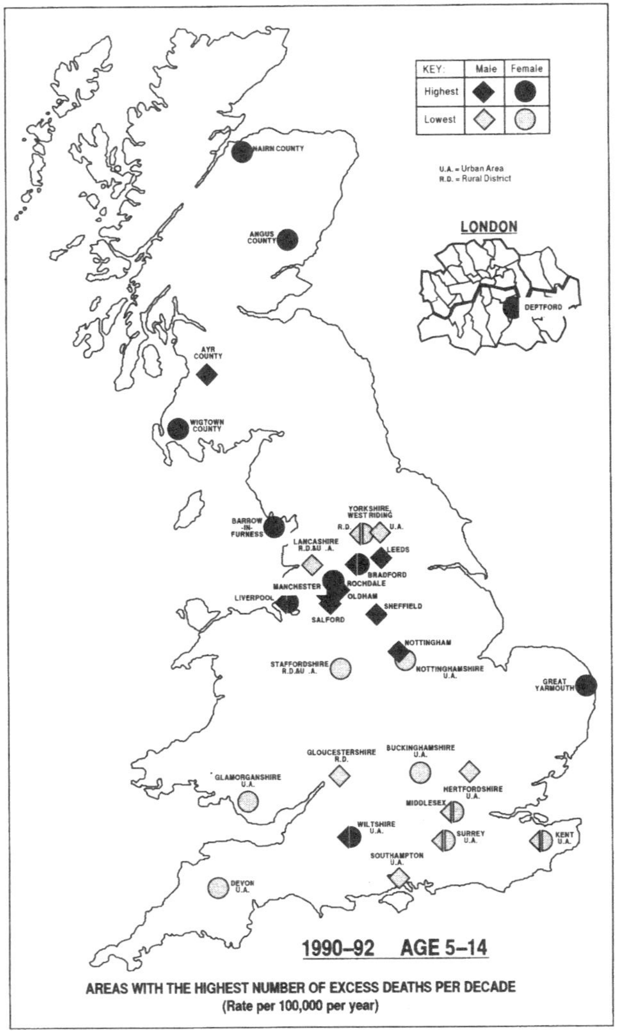

FIGURE 12.6 Mortality rates for males and females, aged 5–14, 1990–2

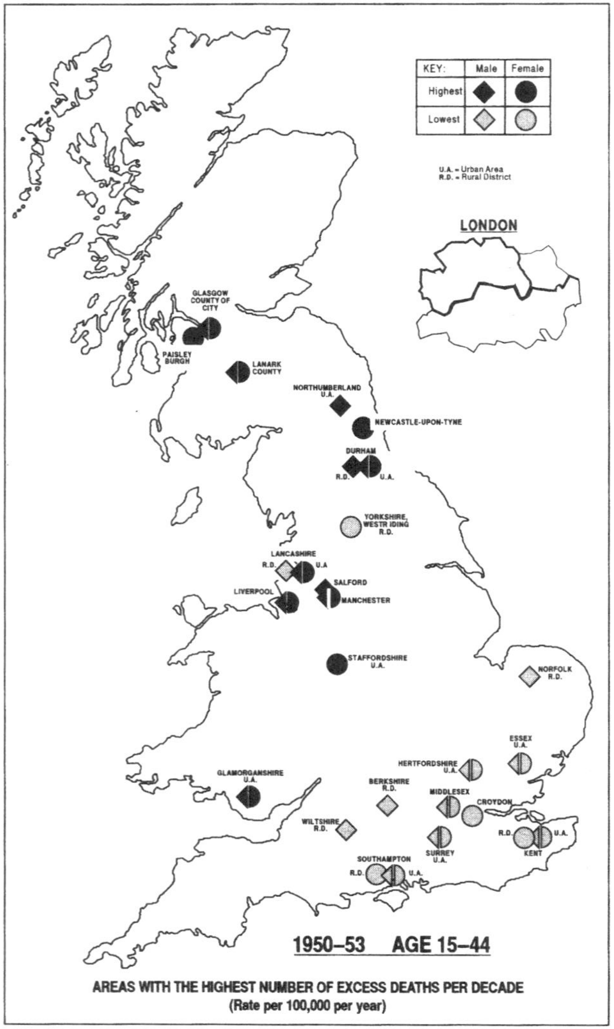

FIGURE 12.7 Mortality rates for males and females, aged 15–44, 1950–3

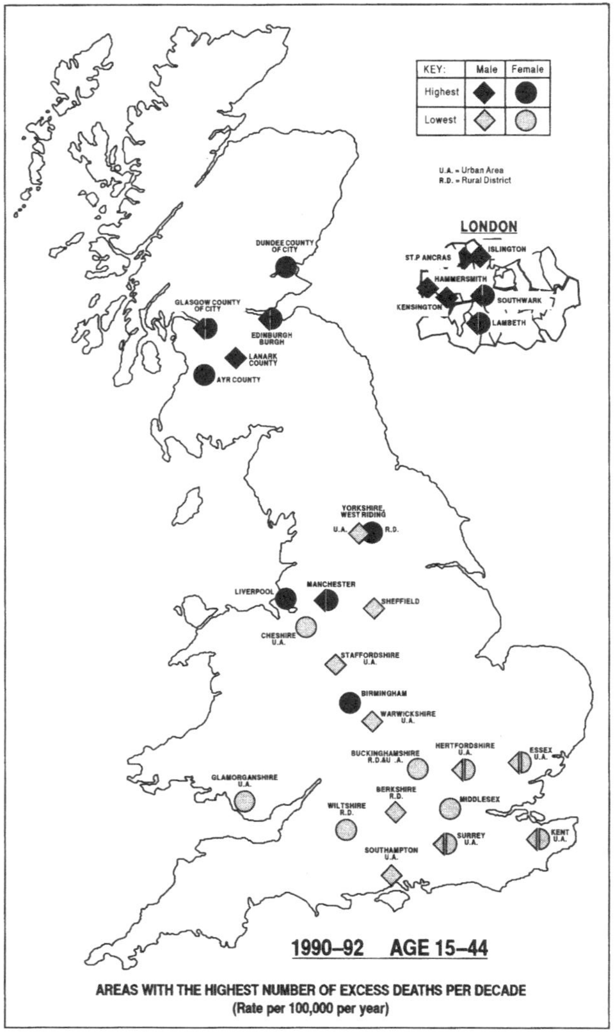

FIGURE 12.8 Mortality rates for males and females, aged 15–44, 1990–2

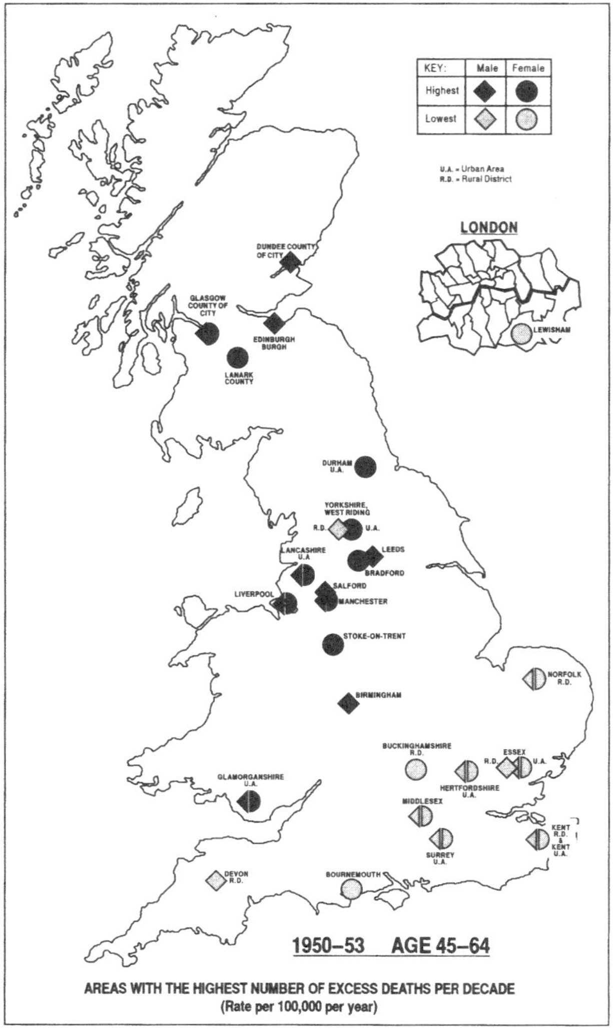

FIGURE 12.9 Mortality rates for males and females, aged 45–64, 1950–3

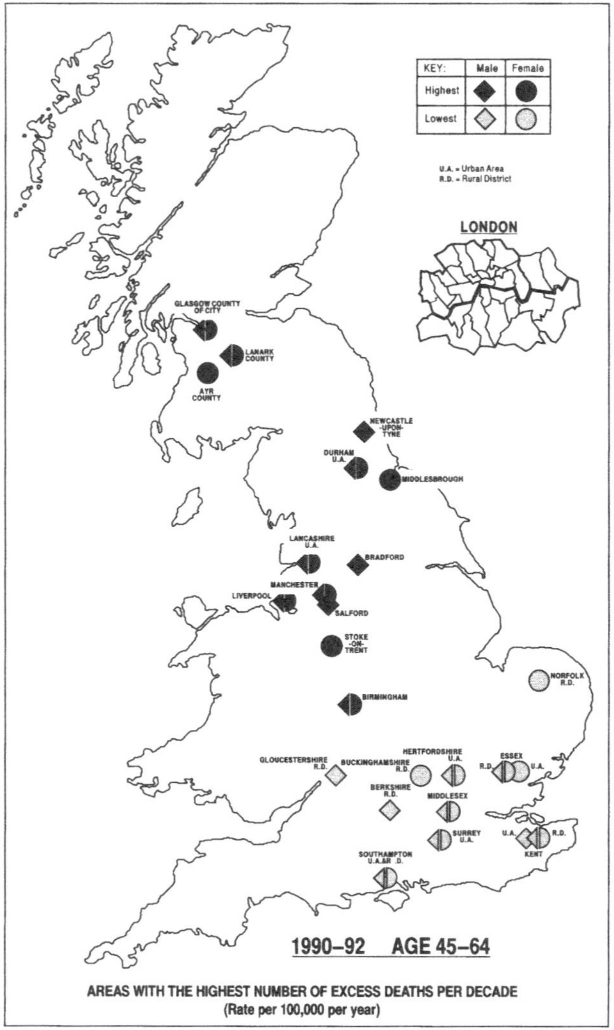

FIGURE 12.10 Mortality rates for males and females, aged 45–64, 1990–2

least likely to die in the 1950s became the adults least likely to die before retirement. This is because of the effects of the living conditions in the places in which they grew up upon their subsequent lives and health. Those children who survived the first years of life in areas with the highest rates of mortality were still more likely to die young as adults.

To conclude this description of the geography of mortality in Britain Figures 12.11 and 12.12 show the age standardised mortality ratios in each area at both dates. We describe the process of age standardisation below, but basically these are the areas in which mortality in general is most and least likely for men and women, irrespective of their ages. In the 1950s the overall pattern of mortality is made most clear by combining the six age groups. Mortality rates were very high in Scottish towns, in South Wales and for men in the north-west of England. Rates were low across the south of England. The two aberrant areas of Nairn county in Scotland and the City in London are both very small districts containing only a few thousand people. By the early 1990s the overall pattern had changed little. The highest rates of mortality still clustered in Scottish towns and in the north-west of England, and rates were still low across the South. The most significant change over this second half of the twentieth century is that three London Boroughs now rank amongst the areas with the highest standardised mortality ratios and we know – from having looked at the individual age groups – that this is due largely to the concentration of relatively higher rates of mortality among young adults in particular London Boroughs. To sum up, the North–South divide in Britain epitomises the geography of mortality over the period 1950–92. In recent years there has been a renewed concentration of mortality within large urban areas and this has led to parts of London now joining the North in the top ten of the mortality league table of Britain.

The rise of inequality

One way of measuring geographical changes in health is to calculate what proportion of the population are living in places where Standardised Mortality Ratios (SMRs) are either very high or very low. SMRs are the chances of dying for people living in a particular area, having taken account of the age and sex structure of the population of that area. They are presented in terms of a ratio as compared to a national average of 100 (in this case we use the contemporary mortality rates at each period in England and Wales as these are used in official statistics). Thus people living in an area with an SMR of 125 are 25 per cent more likely to die at any given time than is the population more generally. Table 12.1 on page 240 shows those changing proportions of the population for each of the time periods being considered here for the 292 areas in Britain, used in the maps described previously (which were referring to 1950–3 and 1990–2). The time periods shown in this table are those for which Registrar General's Reports were available or for which comparable statistics could be calculated.

Table 12.1 can be interpreted in different ways. On the one hand it shows a process of convergence. In the first period, 57.8 per cent of the population lived in areas with close to average SMRs of between 95 and 114. This proportion has risen slowly but steadily every decade such that it has reached 65.5 per cent of the population of Britain by the latest period. More and more people are living in areas where their aggregate chances of dying are near the national average. The proportion of people living in areas with low ratios (under 95) rose to almost a third of the population (32.8 per cent) over the first three periods but has now fallen to almost a quarter (26.4 per cent).

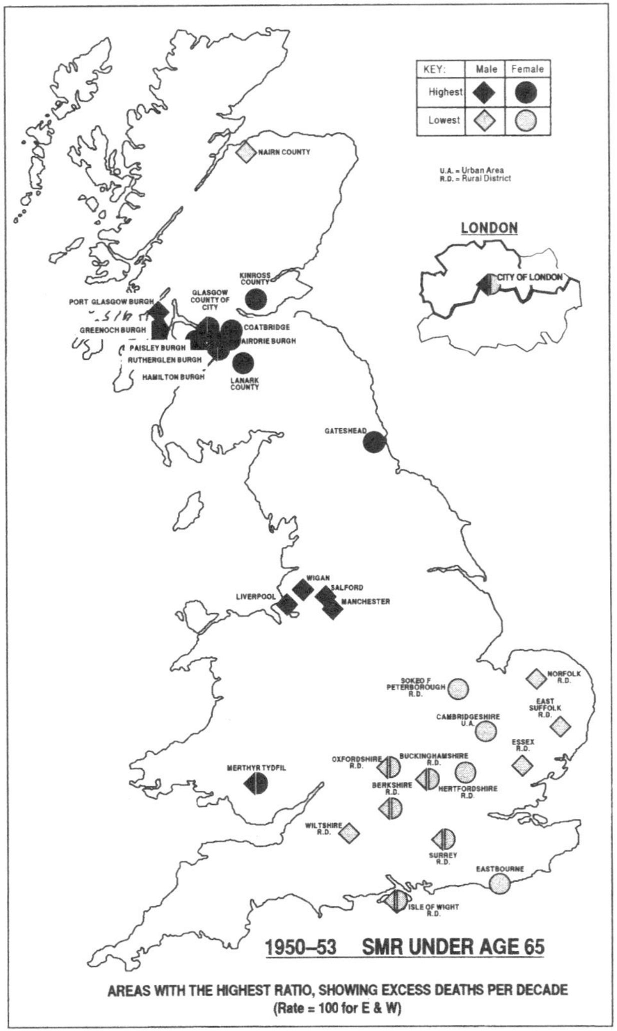

FIGURE 12.11 Age standardised mortality ratios (under 65) for males and females, 1950–3

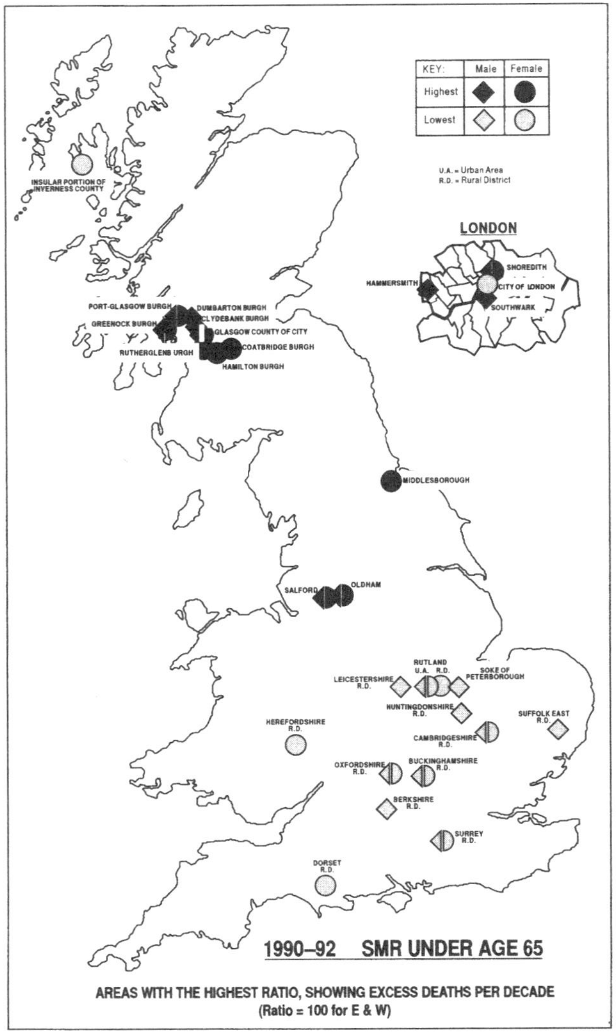

FIGURE 12.12 Age standardised mortality ratios (under 65) males and females, 1990–2

TABLE 12.1 People in Britain living in local authority areas by SMR (figures are per thousand people)

	Contemporary SMR of people aged under 65 (England and Wales = 100)				
Period	*125+*	*115+*	*95 to 114*	*Under 95*	*Under 87*
1950–53	23	103	578	319	25
1959–63	21	77	598	325	11
1969–73	8	65	607	328	37
1981–85	15	66	638	296	29
1986–89	14	80	650	270	42
1990–92	17	82	655	264	11

However, the opposite trend is seen for the proportion of the population living in areas with high SMRs which fell from 10.3 per cent to 6.5 per cent over the first three periods, but which has risen steadily since then. And, as the table shows, there is even more fluctuation at the extremes. The problem with measuring changes in geographical inequality using this technique is that the results can depend too much on what SMR threshold is used to calculate them.

An alternative and more robust measure is to consider the number of deaths which occur in each area above the number that would be expected given the age and sex of the population of each area. This statistic, when expressed as a proportion of all deaths, is labelled excess mortality and fluctuates far less than do figures based on simple thresholds. Table 12.2 below presents the figures for excess mortality for the whole population and for the selected age and sex groups which contain the majority of early deaths in Britain.

Table 12.2 shows that just after the Second World War 4.3 per cent of deaths in Britain were accounted for by excess mortality (i.e. deaths above those expected in areas where the SMR was greater than 100). This proportion fell slightly during the 1950s, then rose again to 4.3 per cent in the late 1980s to settle in the contemporary period again at 4.1 per cent. As Table 12.2 shows, the figures are more dramatic when smaller groups of the population are considered. For men aged 15 to 44 and 45 to 64, respectively 5.2 per cent and 9.6 per cent of all deaths are now accounted for by excess mortality. While for women of the same ages excess mortality now accounts for 3.9 per cent and 8.9 per cent of all deaths. Most premature deaths occur to people in the older of these two age groups, and excesses there were highest in the late 1980s for men and early 1990s for women.

Geographical inequality in mortality in Britain is rising for older adults and has been fairly stable for the population as a whole. It is thus likely that this country will fail to meet its commitment to the reduction of inequality as laid out in Target One of the World Health Organisation (WHO 1985). Indeed, for particular groups of the population there will probably be a rise in inequality, rather than a reduction of the specified 25 per cent. For the population aged under 65 there is a significant rise in area inequality over the period. Table 12.3 shows the SMRs for the population aged under 65 when Britain is divided into ten areas each containing approximately the same number of people. In the early 1950s the worse off 10 per cent of the population had a mortality rate below the age of 65 which was 31 per cent higher than the average. By the early 1990s this had risen to 42 per cent – the highest recorded.

TABLE 12.2 Excess mortality in Britain by local authority areas (figures are percentages of all deaths in the age group) of the whole population, of men aged 15–44 and 45–64, and of women aged 15–44 and 45–64

Period	*Whole population*	*Men 15–44*	*Men 45–64*	*Women 15–44*	*Women 45–64*
1950–53	4.3	5.6	7.5	7.4	5.6
1959–63	4.1	5.0	6.9	5.4	5.4
1969–73	4.2	6.0	7.1	5.2	6.8
1981–85	4.2	5.2	8.8	4.1	7.8
1986–89	4.3	5.2	9.8	2.6	8.4
1990–92	4.1	5.2	9.6	3.9	8.9

TABLE 12.3 SMRs under 65 for population deciles of equal size

Decile	*1950–53*	*1959–63*	*1969–73*	*1981–85*	*1986–89*	*1990–92*
1	131	135	131	135	139	142
2	118	123	116	119	121	121
3	112	117	112	114	114	111
4	107	111	108	110	107	105
5	103	105	103	102	102	99
6	99	97	97	96	96	94
7	93	91	92	92	92	91
8	89	88	89	89	89	86
9	86	83	87	84	83	80
10	82	77	83	79	78	76

Table 12.3 provides the most robust single summary we have of how geographical inequalities in mortality in Britain have risen over time. This rise is not dependent on the decile divisions we have used, as both the first and second decile of the population have seen their position deteriorate since 1981 (in both cases SMRs have risen since 1973, rising quickly during the 1980s). Similarly there has been a steady improvement in the mortality rates of the two deciles of the population which had the best life chances after 1973. The best off 10 per cent of the population have seen their mortality ratio fall from being 83 per cent of the national average between 1969–73 to 79 per cent, 78 per cent and now 76 per cent of the national average. If we compare this with the worse off 10 per cent of the population's current standardised mortality ratio of 142 we see that people living in these areas are almost twice as likely to die at any given time under the age of sixty-five when compared to the best-off decile of the population. The gap between the best and worse off in Britain has never been wider.

Underlying this rise in geographical inequality is the changing geographical concentration of people with different individual chances of dying in different parts of Britain. This could have happened in two ways and both probably occurred to some extent. First, the populations within an area can change in terms of their aggregate behavioural patterns and social circumstances (for instance, people in Surrey may start smoking less). Second, different groups of people may have different likelihoods of moving to different parts of the country at different times (for instance, Liverpool may have seen an exodus of its more qualified population). It is difficult to unravel the varying importance of these two types of change, but first we need to know how the relative importance of the different choices people make and the constraints under which they operate in determining their life chances. Only after that can we begin to explain the changing geography of health in Britain.

Life expectancy: demographic basics

Age and sex

There are many factors which influence how long you will live. One of the most important of these is chance, and chance by its very nature is unpredictable. We cannot predict, for example, the likelihood of an individual being killed in a car accident, by a bolt of lightning or by a crazed madman. An individual may also have a specific medical condition which dramatically alters their life chances, but this is also very unpredictable. Although different people have different chances some factors are more certain than others and some affect much larger groups of people than others.

Here, we concentrate on those factors which have the most predictable effect on your life expectancy and which do not require a doctor to diagnose. These include five things which you cannot change: your age, your sex, your father's occupation when you were 14 years old, your country of birth and your height. Also included are five factors which you may be able to influence to some degree: your occupation, whether you are unemployed, where you now live, whether you own or rent your home and whether you have a partner. Finally there are seven factors over which you have the most control: smoking, alcohol consumption, injecting drug use, diet, your weight, exercise, and sexual activity.

All of the factors we consider here have a geography to them, some more than others, and just living in a particular place can have an effect on your health after all other factors have been taken into account. You have no control over some of the factors and do not have total control over any of them. You can only choose to smoke, for instance, if someone is prepared to sell you cigarettes; although it is far easier to smoke than it is to inject illegal drugs – and it could be argued that it is far easier to drink alcohol in excess than either of these. It is society as a whole that proclaims what behaviour is acceptable, and thus society – as much as the individual – is responsible for people's behaviour.

The geographical aspects to health will be discussed as we explain the impact, in terms of increasing or reducing years of total life expectancy, of the various factors. In order to estimate your life expectancy, you will need a starting point, which refers to the average life expectancy for your age and sex group. Choose your starting point from Table 12.4. For instance, if you are a 20-year-old female you start at 80.

The first thing to see from Table 12.4 is that the younger you are the lower is your life expectancy. All else being equal your parents and grandparents are likely to live longer than you. At first this may appear counterintuitive, but there are two reasons why this is so. Most importantly this is a table for survivors. The life expectancies refer only to people who have already managed to live to a certain age. Someone who is currently 85 years old has to have a

TABLE 12.4 Life expectancy starting points: men and women in Britain

Men		*Women*	
Age	*Life expectancy*	*Age*	*Life expectancy*
0	74	0	79
1–35	75	1–40	80
35–49	76	41–54	81
50–59	77	55–64	82
60, 61	78	65	83
62, 63	79	66–69	84
64, 65	80	70–74	85
66–70	81	75	86
71, 72	82	76, 77	87
73, 74	83	78, 79	88
75	84	80–82	89
76, 77	85	83, 84	90
78, 79	86	85	91
80	87		
81	88		
82, 83	89		
84, 85	90		

life expectancy greater than 85 (our Table stops at 85 as we envisage few readers will be older than this). Second, Table 12.4 is based on the life expectancies of the population today. It is possible that there will be improvements in general nutrition, conditions of work and in medicine that will improve life chances in the future, as they have improved in the past. So the life expectancies in Table 12.4 may well be an underestimate. It is also possible that there could be general catastrophes in the future, as there have been in the past. In this case these figures could be an overestimate. If, for instance, we were producing this table a hundred years ago we would not have known of the future toll of the Great War of 1914–18, or of the influenza epidemic of 1918–19 (which killed more people world-wide than the Great War).

From Table 12.4 note how, on average, women begin with an expectation of five more years of life than men, but that this advantage diminishes with age. It is still not known with any degree of certainty why women now tend to live longer than men (they have not always had this 'advantage'). Several factors are likely to matter. First, it is well known that there are gender differences in socialisation – in the way that men and women are taught and expected to behave in society. Little boys, for example, are expected to be boisterous and adventurous and they learn to be ambitious and assertive. Girls, on the other hand, are encouraged to be more passive and co-operative, careful and gentle. It may be that behaviour that is seen as appropriate for males and females results in very different risks for those two groups in society. For example, women have much lower rates of death from accidents than do men. Second, biological explanations may also play a part in the sex differences in death rates. Fewer women die of heart and circulatory diseases in middle age – hormones seem to protect women from heart attacks until the menopause around the age of 50. After the menopause, female rates of heart disease gradually rise and by the age of 75 more women than men die from this cause of death. Medical advances may also be a factor – until the last century childbirth was a very dangerous activity; death as a result of childbirth is now extremely rare.

From the influences of age and sex alone we can see that there will be a geography of life expectancy in Britain because in some parts of the country there are more elderly people (for example, coastal retirement towns such as Bournemouth and Worthing). The ratio of men to women also varies across Britain. It is important that this distribution of age and gender is not seen as a map of health in Britain. In fact it is the opposite of that, as elderly people are much more likely to be ill than are younger people and there is some evidence that women are more likely to be ill than men (Miles 1991). Most studies of the medical geography of Britain first standardise mortality rates to 'take into account' the differing age–sex profiles of different areas, but it is important to understand that age and sex alone produce the largest variations in our risks of mortality.

As you calculate your own life expectancy we will use two imaginary students throughout this chapter as examples to illustrate some extremes. These fictitious students are Gary and Victoria. Gary is a 24-year-old mature student at Teesside University and Victoria is 21-year-old undergraduate at Bristol University. From Table 12.4 their respective starting life expectancies are 75 and 80.

Lifecourse factors

Father's social class

A map of life expectancies based only on age and sex would not show the full range of geographical differences in life chances in Britain. In recent years there has been a rapid divergence in the life chances of people both living in different social groups (Drever and Whitehead 1997) and between people living in different parts of the country (Dorling 1997). This growing division of mortality within Britain is largely accounted for by the polarising social, cultural and economic geographies of Britain. In the 1980s and 1990s where you were brought up and where you live had a greater influence on your chances of success in the labour market and in other areas of life than was the case at any time since the 1930s.

Evidence has recently been emerging that health and the risk of premature death are strongly effected by socio-economic factors acting throughout life (Bartley *et al.* 1997). There is evidence that widening inequalities in mortality in adulthood reflect the accumulating influence of poor socio-economic circumstances throughout life. Studies have shown that there is an association between cardiovascular risk and socio-economic position during adulthood and childhood, and that both past and present socio-economic circumstances affect mortality.

Davey Smith *et al.* (1997a) reported an analysis referring to respondents' fathers' social class, first social class and current social class. The effect of fathers' social class (taking other factors including your own social class into account) is shown below. You should modify your life expectancy according to the rule in the Question 1 box.

Question 1: What was your Father's occupational class when you were 14 years old?

If your father was in a manual occupation or not working when you were 14 years old subtract one year from your life expectancy, otherwise add two years.

To return to our hypothetical examples, Gary's father was a shipbuilder when he was a child and so his life expectancy falls to 74, whereas Victoria's father was an accountant and so her life expectancy rises to 82. We use the father's social class as the best simple indicator of material conditions in childhood.

Why should your father's social class matter and what are the geographical implications of differing material conditions in childhood on health? One reason for these differing chances is that people whose father was in a manual occupation or unemployed when they were born are less likely to have been well nourished when they were children. Food makes up a much higher proportion of the expenditure of the poor than of the rich and access to good food is worse in poorer areas of Britain (Blackburn 1991; Sooman and MacIntyre 1995). Another reason is that children whose father is in a manual occupation are many times more likely to suffer from accidents and illness in childhood. Today children of parents in the lowest social classes are five times more likely to be killed by a car than those of parents in the highest classes. The cumulative effect of all the disadvantages of poor income and status result in the average three-year gap in life expectancies given in the question above. Geographically this means that parts of the country where adults were more likely to be in manual occupations in the past will have higher mortality rates today. Although here we have only referred to one previous generation, the geography of the great economic depression of the 1930s can still be seen reflected in the pattern of mortality in Britain in the 1990s.

Current social class

Next we refer to your current social class as this also has a large effect on your life expectancy. This relationship has been known since the Registrar General first devised social class categorisations based on occupation in 1911. The differences observed are due to the material and social advantages of people in particular classes. The effects are greater for men than women because, traditionally, the gap in incomes has been greater between men than between women. However, there has been a rapid growth in the proportions of people in higher social classes, particularly for women, and this growth has not been evenly spread across the country. Part of the explanation of the diverging trends in mortality are diverging geographies of class. It is also possible that occupational status has a greater effect on the health of men than on women because men spend a greater proportion of their lives in paid work. It may also be that the classification groups men more effectively than it does women – that an occupational stratification is not as sensitive to the range of socio-economic differences between women as it is for men. The Question 2 box shows how your life expectancy should be adjusted given your current social class.

Question 2: What is your occupational class?

To which of the following occupational categories do you belong? (If you are not currently working refer to your last full-time occupation; if you have never worked use your partner's or parent's class.)

	Men	*Women*
A or B	+3	+2
C1	+1	+1
C2	0	0
D	–2	–1
E	–4	–2

The classification of social classes shown in Question box 2 refers to thousands of occupations which cannot be listed here (for more information see OPCS 1992). Table 12.5 indicates the type of occupations in each category.

Once current class is added as a factor our two students' chances diverge even further. Neither have a current occupation and so both have to use their parent's. Gary's dad has been a caretaker in the local school since the shipyard closed and so his life expectancy falls to 70, whereas Victoria's father now manages a firm of accountants and so hers rises to 84.

Why should current social class matter so much and why can we use a relative's or partner's class as a proxy? To answer the second part of that question first, most people when they get an occupation tend to have a similarly graded occupation to their parents and, as men were more likely to work than women in the past, we use father's occupation as the best proxy.

Different explanations have been put forward to explain social class patterns in health and mortality (these are outlined in the Black Report, which is usually taken as the starting

TABLE 12.5 Occupational classification

*Class**	*Occupation type*
A	Professional (e.g. accountants, doctors, electronic engineers)
B	Managerial and technical/intermediate (e.g. proprietors and managers – sales, production, works and maintenance managers, nurses, teachers)
C1	Skilled non-manual (e.g. clerks, secretaries, cashiers – not retail)
C2	Skilled manual (e.g. drivers of road goods vehicles, metalworking production fitters)
D	Partly skilled (e.g. storekeepers and warehousemen, machine tool operators)
E	Unskilled (e.g. building and civil engineering labourers, cleaners, etc.)

Source: Drever and Whitehead (1997).
Note: *These classes are also referred to as I, II, III non-manual, III manual, IV and V.

point for the analysis of such differentials: DHSS 1980). The first explanation is artefact: that problems of measurement and the classification of social classes produce the pattern observed. While there are indeed some problems of this nature, these do not account for social class differences.

The second possible explanation is natural or social selection, a theory which suggests that those who are most fit will rise to the top of the social strata and those who are ill or weak will drop to the bottom. So here the argument is that health differences produce social class differences, rather than the other way around. While this may be the case for a small minority, particularly those who are most marginalised, there is evidence that selection only accounts for a small proportion of differences by social class. For example, most social class mobility occurs early in working life (Blane 1985).

A third explanation refers to material and structural factors. This concerns differences between individuals which are the result of socio-economic structural factors, rather than individual lifestyle factors (discussed next). This explanation places emphasis on an individual's position in society in terms of their class background affecting their education, which then affects their labour market position and their income, which consequently affects their health through their material living conditions such as housing.

A fourth explanation of social class differences in health focuses on the cultural, behavioural and lifestyle factors that are associated with poor health outcomes. This explanation says that people in lower social classes are more likely to smoke, to have a poor diet, and so on, and that these behavioural differences lead to different health outcomes. There is some evidence to support this theory (for example, that a greater proportion of those in lower social classes smoke) but again, this only accounts for a minority of the class difference observed.

All of these different types of explanations are likely to play a role, to a greater or lesser degree, in explaining social class differences in health. We cannot easily isolate one explanatory factor, as they are interconnected. For example, behaviours are a social as well as individual phenomenon – what may seem like an individual choice may be framed by constraints imposed by class. Victoria, for example, may like to spend her time going to polo matches at Cowdray Park and Gary may prefer playing darts in the local pub – these are individual behaviours which are shaped by class.

Class introduces an important factor which complicates the estimation of life expectancy here – that of confounding. Confounding means that the different factors we are considering are interconnected, and we need to take this into account. For example, unemployment is concentrated geographically, lower social classes smoke more, men and women have different dietary habits. In the statistics shown here confounding is taken into account, where possible, by using studies which control for such factors.

The Question 2 box referring to class differentials in life expectancy (Drever and Whitehead 1997) gives a difference of seven years of life expectancy between occupational classes I and V for men, and a four-year difference for women.

Height

A good example of the effect of confounding is height. A number of studies have reported that height and mortality are related, such that taller people live longer. For example, findings from the Whitehall study (Davey Smith *et al.* 1990) show that height is inversely associated to all cause mortality, coronary heart disease (CHD) and non-CHD mortality. While height was also related to car ownership and employment grade and other risk factors, controlling for these factors did not abolish the association between height and mortality. Thus although height is a confounded variable it is also a valuable piece of independent evidence. Why should this be? There are two main reasons why height may matter. First, height can be a proxy for factors such as class. Simple occupational grading does not measure social advantage as well as does occupational grading in conjunction with height. Second, height may have simple medical advantages in its own right.

Question 3: How tall are you?

Alter your life expectancy by adding or subtracting the following number of years:

	Men		*Women*
Men less than 5'8"	–1	Less than 5'3"	–1
5'8"–5'10"	0	5'3"–5'5"	0
More than 5'10"	+1	More than 5'5"	+1

In case you were wondering, Gary, at 5'7" is an inch taller than Victoria, but because he is under average and she is over – he loses a year and she gains one. Gary is taller than either of his parents but still one of the shortest when it comes to getting attention at the bar to buy a drink. Most young people today are taller than their parents were when they reach adulthood (Usher 1996), but this change has also not been evenly spread across the country, with height rising fastest in the South where nutrition has been better and in the past people have been more affluent.

Unemployment

Question 4: Have you been unemployed for the past three months?

If so deduct two years.

Unemployment carries a risk of premature mortality (Drever and Whitehead 1997). However, this is not because the unemployed are more likely to be ill or because they are more likely to be from a lower social class (they are likely to be both of these things, but when we take these into account the effect is still apparent). Data from the British Regional Heart Study has shown that differences in mortality between employed and unemployed men remains after adjustment for factors including smoking and alcohol consumption (Morris *et al.* 1994). Why should this be? There are many reasons. Unemployment decreases a person's income and status, and so the same factors that apply in the case of social class apply here. But in addition to this unemployment carries an extra social stigma of being rejected from the world of work for which most adults have been trained and conditioned. Work is central to our identity – one of the first things we want to know about a person when we meet them is what they do for a living. Unemployment can lead to depression and even despair and hence can cause psychological ill health which in turn affects physical health and life expectancy. Most directly, unemployed people are twice as likely to kill themselves as compared to the employed population.

The changing geography of unemployment is one of the most important possible recent influences on the changing map of mortality in Britain. Soon after 1979 unemployment in Britain rose to unprecedented levels which, for families, was higher than in the 1930s. This rise was very much concentrated first in the Celtic periphery then in Scotland and the north of England and Wales, only spreading to the south of the country by the late 1980s. The coincidence between the geographies of unemployment and the geography of mortality a decade later is very high and it is likely that rising inequalities in unemployment have had a large part to play in rising inequalities in premature mortality.

Let's return to Gary and Victoria. Victoria has never been unemployed, but Gary has been unemployed many times since leaving school. Most recently he has just signed on as unemployed, as his course at university has finished and he hasn't found a job. So his life expectancy (or more correctly the life expectancy of people in a similar situation to Gary) falls to 67, eighteen years below Victoria's. The main reason for Gary being unemployed and still finding it difficult to find work is that as a young man growing up on Teesside in the 1990s his chances of finding a job, regardless of his individual characteristics, are poor. Unemployment rates vary dramatically geographically. While Gary was struggling to find work in Teesside, Victoria has found work during the summer in a tea shop in Chichester to finance her skiing holiday the following winter.

Geographical influences on health

So far we have only considered demographic and social factors which have indirect geographical causes and consequences, but there are also direct geographical influences on health.

Country of birth

There is well-established evidence that people who live in Britain but were born in different countries have different life expectancies (see Question 5 box). This is a result of both their experiences and circumstances when they were young, the selective effects of the migration that brought them to where they are now, and the effects of the way they are treated once they have arrived. Britain has welcomed continuous waves of immigrants and at different times they have tended to settle in different areas in Britain. This may have had a minor influence on the geography of life chances across the country. Note that in recent years governments have put laws in place which are implicitly designed to limit the immigration of black people to Britain.

Question 5: Where were you born?

Alter your life expectancy by adding or subtracting the following number of years:

Caribbean	+2
Africa	–3
England or Wales	0
Indian Subcontinent (Indian, Pakistan, Bangladesh, Sri Lanka)	–2
Ireland (all parts)	–3
Scotland	–3
Other	0

The addition of two years for people born in the Caribbean is likely to be due partly to the selective effect of the pattern of migration to Britain that occurred mainly in the 1950s and 1960s. In general it was the most mobile and most healthy Caribbean citizens who came to Britain and so their life expectancies are now higher than the average in this country. The opposite is true for the more recent migrants from Africa and Asia and also for people who were born in Ireland. This may be due partly to the 'push' factors which brought many of these people to Britain (as opposed to the 'pull' factors for those from the Caribbean) and also to the effects of living in the parts of the country in which they tended to settle. However, the country of birth effect that influences the life expectancy of the largest group of people in Britain is reserved for people born in Scotland. Again, there is no generally agreed reason why this should be so, but a mixture of factors, including the fact that Scotland suffered economic recession long before mass unemployment reached England and Wales in the early 1980s, may well be a factor. These figures were calculated from data published in Drever and Whitehead (1997).

Both Gary and Victoria were born in England, and so their life expectancies are still 67 and 85. When we looked at social class we referred to the previous as well as the current generation. There is very little research available about the relative mortality rates of second generation migrants. Raftery *et al.* (1990) found that Irish immigrants and second-generation Irish had raised mortality, and that the effect was not solely a result of social class. However, they do not say whether they think such findings can be inferred for other ethnic groups. Thus, although Gary's mother was born in Scotland we will not adjust his life expectancy as a consequence. Victoria's family can trace their origins to the 'Domesday Book' and have lived in southern England ever since.

Region of residence

Question 6: Do you live in England, south of, or in, Gloucestershire, Warwickshire, Leicestershire or Lincolnshire?

If so add a year, if not subtract a year.

Many researchers have reported regional inequalities in mortality within the UK. Patterns have been documented for over a century in Britain, and it is consistently found that mortality rates are highest in the North and in Scotland and lower in the South. Similar evidence of a North–South divide is presented by Britton (1990) who, looking at data up to 1983, argued that there was a continuation, and if anything a worsening, of the regional gradient in mortality, from high in the North and West to low in the South and East for both men and women. This is the case for almost all of the main causes of death. In reference to particular causes of mortality, Strachan *et al.* (1995) report regional variations in cardiovascular disease and stroke with a South–East to North–West gradient in mortality, the North–West having the higher mortality. Similarly, Howe (1986) found regional differences in heart disease and lung cancer for males; for females the number of deaths overall from these conditions were less, but the pattern of regional differences was similar to that for males.

As observed in the Black Report, Britain can be divided into two zones of relatively high and low mortality (DHSS 1980). Howe (1986) proposes an imaginary line reaching from the Bristol Channel to the estuary of the Humber separating those experiencing favourable and unfavourable life chances, whereas Britton (1990) suggests a divide from the Severn to the Wash separating areas of low and high mortality. The former dividing line is used here. Again the changing geography of the population may well have a part to play in the strengthening of this dividing line in mortality in Britain. The population who live below this line has been growing for the last century, partly due to migration from the North. If the migrants were less likely to, say, suffer unemployment than those who stayed, then this changing human geography may have also altered the medical geography of the country.

Using this geographical division, Gary lives in the northern region and thus loses a year of life expectancy, taking him to 66, whereas Victoria adds a year as she lives in southern England. This takes her to 86. These geographical differences in life expectancy are not merely the effect of the aggregation of particular groups of people in certain places – it is not just that there are more working-class and unemployed people in the North, for instance. The differences between areas are a result of the context of a place as well as the concentration of people who live there. Areas can affect health in a number of ways – there may be environmental pollution, for example. Also living in a deprived area as opposed to an advantaged area may mean that you have less access to services which promote good health – for example, sporting, leisure and community services, and of course health services. Victoria has many social, cultural and sporting opportunities close to both her home and university. Where Gary lives there are fewer things to do and young people spend most of their time hanging around on the streets.

Housing

Housing tenure is also a spatial factor as certain types of housing, for instance council housing, is often geographically concentrated. Where you live in terms of the type of home you live in also affects health. The Question 7 box shows the years of life expectancy to add or subtract according to housing tenure. Tenure patterns have changed particularly rapidly in the last twenty years – there has been a dramatic rise in the number of owner-occupiers in the last two decades, rising from 10 million owner-occupiers in 1971 to 16 million in 1993

(Dorling 1995). Changes have been partly due to the policy of giving social tenants the opportunity to buy their property and partly due to the Conservative government's refusal to build new social housing. This has led to a residualisation of the social rented sector to serve particular groups in particular places, which in the extreme concentrates mortality into particular places. For instance, having a serious medical condition increases your chances of being allocated a council house whilst decreasing your chances of being granted a mortgage.

Question 7: What is your housing status?

Alter your life expectancy by adding or subtracting the following number of years:

Own your own home/have a mortgage	0
Rent from a private landlord	–1
Rent from a local authority or housing association	–1
Live in a hostel for the homeless	–12
Have no home – sleep rough	–25

The figures given in the Question 7 box for the traditional housing tenures (owner-occupation, social housing and private renting) were calculated from data provided in Drever and Whitehead (1997). These have taken social class and car access into account, so the effect we observe is not due to these factors. Owning your own home may be beneficial to health because of the feelings of security and pride it creates. Alternatively, living on a council estate may adversely affect health if people are confronted with day-to-day problems such as poor levels of maintenance, and higher than average vandalism and crime. While the privately rented sector is very mixed the lower life expectancy of this group is probably due to the fact that on aggregate this type of accommodation is likely to be of poorer quality.

Two groups stand out from the Question 7 box, however – hostel residents and the street homeless. There is much data available on the health problems encountered by homeless people. Bines' survey (1994), the most comprehensive study of the extent of health problems in the homeless to date, found that people using hostels, living in temporary bed-and-breakfast accommodation and sleeping rough had many physical and mental health problems (including musculoskeletal problems, chronic chest or other breathing problems, and depression and anxiety) at a much greater rate than the general population. It is those who sleep rough, however, who experience the worst health problems, due to exposure to the elements, lack of good nutrition, lack of access to washing facilities, and so on. While we have some data on the health problems of the homeless, there is much less evidence, however, about their mortality. The source used, referring to mortality rates of hostel residents and rough sleepers, is Shaw (1998a). Homeless people are more likely than the general population to die from causes of deaths such as suicide and homicide, pneumonia, bronchitis and TB, and drug- and alcohol-related causes of death.

Victoria's parents bought her flat for her when she moved to Bristol. This not only provides her with a home during her studies but is also a sound investment for her parents. Gary rents privately with other students. His life expectancy consequently falls to 65 and the gap between them is twenty-one years. Tenure thus represents many things other than current income, family inheritance and wealth being among them.

Behavioural influences on health

In this section we consider actions which affect health which, while having social determinants (as discussed above) and geographical dimensions, are those factors over which the individual has most control. The factors we include here are the four main health-related behaviours: smoking, alcohol consumption, diet and exercise. We also refer to weight, which is related to these, and briefly cover a less common but far more dangerous activity, injecting illegal drug use.

Smoking

The adverse effects of smoking on health are perhaps the most well-known of all health effects. Approximately 120,000 people die prematurely in the UK each year due to smoking. The question below shows the average life expectancy lost by a smoker – there is a seven-year difference between a smoker and someone who has never smoked. There are also differences according to the amount and type of cigarettes smoked, but for the sake of simplicity we have used three categories in the Question 8 box. These rates were calculated from findings published by Doll *et al.* (1994).

Question 8: Are you a smoker?

Alter your life expectancy by adding or subtracting the following number of years:

A smoker	–4
A former smoker	0
A never smoker	+3

Since the age of 13 Gary has smoked on average twenty cigarettes a day (his parents both smoke and his grandfather died of lung cancer). Victoria, on the other hand, has never smoked (taking a few drags of a cigarette at a summer ball at the age of 15 doesn't count). While undoubtedly an individual exerts a choice over whether to smoke or not, there are also social and cultural aspects to this behaviour, as those who are brought up in a household where their parents smoke are more likely to smoke themselves. People in Gary's situation are most likely to die at 61 whereas Victoria's contemporaries – those who act like she acts and have the advantages that she has – will live, on average, to 89.

Just as there is a geographical dimension to every other factor affecting health and lifestyles there is also a spatial dimension to smoking. It is easier and more acceptable to smoke in areas where it is more usual to smoke and thus the geography of smoking is self-reinforcing. For instance, in those parts of Britain where it is less usual to smoke, fewer cafés and bars will allow smoking and it is more difficult to buy cigarettes. Where smoking is more usual it is more difficult, particularly as a child, to avoid smoking. The effects can clearly be seen when particular causes of death related to smoking, such as lung cancer, are mapped. What is not so clear is whether changes in the prevalence of smoking have had a great effect on the geography of mortality in Britain. This is because smoking rates have fallen fairly uniformly across the country.

Alcohol consumption

Alcohol consumption also affects health and life expectancy, but the nature of the relationship is rather different to that for smoking. Also, very little is known about the changing geography of alcohol consumption.

Question 9: How much alcohol do you drink?

Alter your life expectancy by adding or subtracting the following number of years:

Abstainer	–1
Moderate*	+1
Heavy†	–3

* 14 units or less per week for women, 21 or less for men, spread over the week.

† More than moderate (i.e. recommended levels).

A moderate intake of alcohol has a protective effect. Those who do not drink alcohol at all have a year of life expectancy less than those who drink moderate amounts. Those who drink over the recommended amounts lose three years.

Gary qualifies as a heavy drinker, as he regularly meets his mates in the pub for more than a few beers. Victoria is a moderate drinker, tending to have a glass of wine with her evening meal. When she goes out with her friends she drinks sparkling mineral water and the occasional glass of Pimms. His life expectancy is thus 58 and hers is now 90. Again these are individual behaviours which have a clear social component, as drinking habits are as much learnt from one's cultural context as they are an individual creation. The map of deaths from cirrhosis of the liver is a good proxy for the map of excess drinking in Britain.

Diet

Eating a healthy diet is also a factor which affects health and life chances. Diet is particularly associated with cardiovascular disease, and also bowel and stomach cancer. Eating habits are very difficult to measure, however (Shaw 1998b). In particular, changes in the geography of eating habits are very under-researched. The rising importance of supermarkets means that those without a car who do not live near a supermarket may now have to spend more on a more restricted choice of foodstuffs. Here we use a simple question which refers to just one aspect of a healthy diet.

Question 10: Do you usually eat fruit every day? (do not include juice)

Alter your life expectancy by adding or subtracting the following number of years:

Yes	+1
No	–1

Gary was always told by his mother to eat his greens, but basically he's a pie and chips kind of man. He eats fruit now and then, but not usually every day. Victoria on the other hand is a vegetarian (although she eats fish) and loves to eat fresh salads and vegetables. She makes it a part of her health routine (which also includes exercise and skin-care) to eat fresh fruit every day. Gary's life expectancy is now 57 and Victoria's is 91.

Exercise

Physical fitness has an independent impact on health – those with sedentary lifestyles have lower life expectancies compared to those who regularly exercise. The changing geography of exercise is unknown but it is likely that two factors play a large part. First, the decline in walking and manual labour has reduced exercise. Second, exercise is now seen as a recreational activity in some places more than others. Both these factors are likely to have changed the map of exercise and hence have had an impact on the map of mortality.

Question 11: How much exercise do you do?

Alter your life expectancy by adding or subtracting the following number of years:

Regular vigorous (e.g. aerobics or running twice a week.)	+2
Regular moderate (e.g. a brisk walk at least twice a week)	+1
Less	–1

Victoria plays tennis and also goes to aerobics, so she adds two years. Gary is a bit of a couch potato (he plays a bit of footie every other Sunday with his mates, but that is not twice a week and they almost always end up in the pub afterwards). Gary's life expectancy (and that of many of his mates) is now 56 and Victoria's is now 93.

Weight

Related to diet and exercise, of course, is whether or not a person is overweight. Here we are referring not to weight *per se*, but to weight-for-height. Because of this the same factors that have led to a changing geography of height will have affected the geography of weight, the influence of which we are reporting here. The changing geographies of both diet and exercise are also important factors.

Question 12: What is your weight?

Alter your life expectancy by adding or subtracting the following number of years:

Underweight	–2
Normal weight	+1
Overweight	–1
Very overweight	–2

The data in Question box 12 indicates that it is bad for health to be very overweight or underweight, as people in these categories lose, on average, two years. Those who are overweight also lose one year, whereas those who are in the 'normal' weight range add one year to their life expectancy. Victoria is underweight (as is usual for her cohort of students), so she subtracts two years. Gary is overweight (as is usual for his) and subtracts one year. This is the only question on which Gary has an advantage over Victoria.

Injecting drug use

Question 13: Are you an injecting illegal drug user?

If so subtract eight years.

The habit of injecting illegal drugs (we are primarily referring to heroin here) is harmful to health in a number of ways. First, there is the direct danger of overdosing (which may be deliberate or accidental). Second, there are also possible indirect effects on health, through infection with diseases such as HIV and Hepatitis C. There have only been a few studies of the death rates of injecting drug users, but these report that this group have death rates much higher than the general population, with approximately a 2 per cent chance of dying per year (Frischer *et al.* 1997; Fugelstad *et al.* 1997). Neither Gary nor Victoria have used heroin. Because only about 1 per cent of the population inject illegal drugs their geography and its changes do not have a great effect on the map of mortality.

Relationships

Sexual activity

As we have already shown our social relationships affect our lives indirectly, but they also affect our health in a direct sense. Perhaps surprisingly, few studies have looked at the health effect of sexual activity. However, there is some evidence that sexual activity is associated with longevity (Persson 1981; Davey Smith *et al.* 1997b). From the available evidence it seems that for men it is the frequency of sex that is important, but for women it is not merely the frequency but the enjoyment of sex that is important.

Question 14: For men: how often do you have sex? For women: how often do you enjoy sex?

Alter your life expectancy by adding the following number of years:

Twice a week or more	+1

So far Gary and Victoria have led very different and separate lives. However, they do share some interests, and like many young people they both enjoy attending music festivals during the summer months. During a hot and sultry day in late June (and perhaps under the influence of a variety of legal and illegal but not injected drugs) they met at the Glastonbury festival. They have been visiting each other regularly since, while waiting for their exam results to be posted. They both add a year to their life expectancy as a result. The geography of sex is not well known as the British public are particularly reticent to answer questions on sexual behaviour, and as this was most true in the past we cannot be sure of regional changes. However, we do know that sexual activity has increased and that the young are more likely to have sex than the old (although we are unsure as to how much they enjoy it).

Marital status

Question 15: What is your current marital status?

Alter your life expectancy by adding or subtracting the following number of years:

Married/cohabiting/never married	+1
Divorced/Widowed/Separated	–1

Relationship can cause anguish as well as be pleasurable, however. There is some evidence that those who experience the stress of divorce, separation or widowhood are at a greater mortality risk (Ben-Shlomo *et al.* 1993). Both Gary and Victoria have never married, and thus add a year to their life expectancy. If, however, they were to marry and subsequently to separate or divorce (or one of them dies) then this situation would change. The geography of both marriage and divorce is well known and is highly correlated with younger people living in urban areas, although rising unemployment is also known to increase individual chances of separation.

Conclusion

In this chapter we have both described the polarising geography of mortality by area in Britain in the 1990s and outlined those factors which are thought to contribute to determining mortality most strongly. We have also paid attention to how the geography of these various factors has been changing in recent years. There has been a geographical 'filtering' of the population of Britain over the last twenty years which has led to those with better life chances being less likely than before to live near those with worse chances. This process is likely to continue in the future and we have tried to indicate how you personally will contribute to it both through your own social circumstances and behaviour and in terms of how you affect other people's chances. Personal behaviour, such as encouraging someone to smoke, can have an effect, but collective behaviour can be far more damaging. For instance, university admission systems which favour students from privileged backgrounds will reinforce the status quo and hence help to build on the social inequalities in Britain which underlie so much of the geography of health. Gary's admission to university is more likely to change his life chances in the future than any other single socio-structural event.

You may think it unrealistic given their very different backgrounds that Gary and Victoria should cross paths, let alone hit it off. You are right, it would be unusual, and it is partly the rigidity and predictability of the class system in Britain which makes it possible to produce estimates of people's life expectancies given the answers to just fifteen questions and knowing their age and sex. However, the polarisation of life chances in general in Britain means that there are fewer people who are average and more people in Gary and Victoria's positions than before.

Having answered all the questions Gary's life expectancy, at 57, was thirty-six years less than Victoria's at 93, but what were the most important factors in determining this difference? Nine years of the difference was due to their differing social class backgrounds and seven years was due to their differing ages, sexes and heights. There is little therefore either of them could do about roughly half of the gap between them. However, smoking alone accounted for seven of the years of difference, and drinking, diet, exercise and weight for a further eight years. Being unemployed and living in a northern region in privately rented accommodation accounted for the final five years of difference.

Gary and Victoria are towards the extremes of the distribution of life expectancies in Britain today, but it is extreme groups like this which will determine the most prominent features of tomorrow's map of mortality and extreme groups which are growing in size. Gary and Victoria went to university in 1997 and graduate in the year 2000. People in Gary's situation will on average die in the year 2033, whereas Victoria will live until 2062. Even if she does stay with Gary she is likely to spend almost as much of her life after he has died as she spends with him.

This chapter has covered some of the factors by which life expectancy has been found to vary. Others which could also have been included, had the data been available, include dangerous activities, such as riding a motorbike or rock climbing; dangerous jobs such as deep-sea diving; family medical history; unprotected sex; mental state/depression; breast feeding and birth weight. However, we hope we have included enough to illustrate both how variable socially and geographically life chances are, how they are changing, and how this variability will have an impact on the human geography of Britain for many years to come.

References

Bartley, M., Blane, D. and S. Montgomery, (1997) 'Health and the life course: why safety nets matter', *British Medical Journal* 314: 1194–6.

Ben-Shlomo, Y., Davey Smith, G., Shipley, M. and Marmot, M.G. (1993) 'Magnitude and causes of mortality differences between married and unmarried men', *Journal of Epidemiology and Community Health* 47: 200–5.

Bines, W. (1994) *The Health of Single Homeless People*, York: Centre for Housing Policy, University of York.

Blackburn, C. (1991) *Poverty and Health: Working with Families*, Buckingham: Open University Press.

Blane, D. (1985) 'An assessment of the Black Report's explanations of health inequalities', *Sociology of Health and Illness* 7: 423–45.

Britton, M. (ed.) (1990) 'Mortality and geography: a review in the mid-1980s, England and Wales', OPCS, Series DS No. 9, London: HMSO.

Davey Smith, G, Shipley, M.J. and Rose, G. (1990) 'Magnitude and causes of socioeconomic differentials in mortality: further evidence from the Whitehall study', *Journal of Epidemiology and Community Health* 44: 265–70.

Davey Smith, G., Hart, C., Blane, D., Gillis, C. and Hawthorne, V. (1997a) 'Lifetime socioeconomic position and mortality: prospective observational study', *British Medical Journal* 314: 547–52.

Davey Smith, G., Frankel, S. and Yarnell, J. (1997b) 'Sex and death: are they related? Findings from the Caerphilly Cohort Study', *British Medical Journal* 315: 1641–4.

Department of Health and Social Security (1980) *Inequalities in Health: Report of a Research Working Group'*, London: DHSS.

Doll, R., Peto, R., Wheatley, K., Gray, R. and Sutherland, I. (1994) 'Mortality in relation to smoking: 40 years' observations on male British doctors', *British Medical Journal* 309: 901–11.

Dorling, D. (1995) *A New Social Atlas of Britain*, Chichester: Wiley.

Dorling, D. (1997) *Death in Britain: How Local Mortality Rates have Changed: 1950s–1990s*, York: Joseph Rowntree Foundation.

Drever, F. and Whitehead, M. (1997) *Health Inequalities*. ONS Series DS No.15, London: HMSO.

Frischer, M., Goldberg, D., Rahman, M. and Berney, L. (1997) 'Mortality and survival among a cohort of drug injectors in Glasgow, 1982–1994', *Addiction* 92: 419–27.

Fugelstad, A., Annell, A., Rajs, J. and Agren, G. (1997) 'Mortality and causes and manner of death among drug addicts in Stockholm during the period 1981–1992', *Acta Psychiatrica Scandinavica* 96: 169–75.

Howe, G.M. (1986) 'Does it matter where I live?', *Transactions of the Institute of British Geographers* 11: 387–414.

Miles, A. (1991) *Women, Health and Medicine*, Milton Keynes: Open University Press.

Morris, J.K., Cook, G.D. and Shaper, A.G. (1994) 'Loss of employment and mortality', *British Medical Journal* 308: 1135–9.

OPCS (1992) *1991 Census Definitions*, London: HMSO.

Persson, G. (1981) 'Five-year mortality in a 79-year-old urban population in relation to psychiatric diagnosis, personality, sexuality and early parental death', *Acta Psychiatrica Scandinavica* 64: 244–53.

Raftery, J., Jones, D.R. and Rosato, M. (1990) 'The mortality of first and second generation Irish immigrants in the U.K.', *Social Science & Medicine* 31: 577–84.

Shaw, M. (1998a) *A Place Apart: the Spatial Polarisation of Mortality in Brighton*, Bristol: University of Bristol.

Shaw, M. (1998b) 'Measuring eating habits: some problems with the National Food Survey, in D. Dorling and S. Simpson (eds) *Statistics in Society*, London: Arnold.

Sooman, A. and MacIntyre, S. (1995) 'Health and perceptions of the local environment in socially contrasting neighbourhoods in Glasgow', *Health and Place* 1: 15–26.

Strachan, D.P., Leon, D.A. and Dodgeon, B. (1995) 'Mortality from cardiovascular disease among interregional migrants in England and Wales', *British Medical Journal* 310: 423–7.

Usher, R. (1996) 'A tall story for our time', *Time*, November: 11.

World Health Organisation (1948) *The Constitution*, Geneva: WHO.

World Health Organisation (1985) *Targets for Health for All 2000*. Copenhagen: WHO.

Further reading

Textbooks on geographical and sociological aspects of health include S. Curtis, and A. Taket (1996) *Health and Societies: Changing Perspectives* (London: Arnold); K. Jones and G. Moon (1987) *Health, Disease and Society: an Introduction to Medical Geography* (London: Routledge); G. Moon and R. Gillespie (1995) *Society and Health: an Introduction to Social Science for Health Professionals* (London: Routledge); S. Nettleton (1995) *The Sociology of Health and Illness* (Cambridge: Polity Press); and S. Platt (ed.) (1993) *Locating Health: Sociological and Historical Explorations* (London: Avebury).

Empirical work on health and society can be found in D. Dorling (1995) *A New Social Atlas of Britain* (Chichester: Wiley); D. Dorling (1997) *Death in Britain: How Local Mortality Rates Have Changed 1950–1990* (York: Joseph Rowntree Foundation); D. Gordon and C. Pantazis (1995) *Breadline Britain in the 1990s* (London: Ashgate); and M. Shaw, D. Dorling, D. Gordon and G. Davey Smith (1999) *The Widening Gap* (Bristol: Policy Press).

For a more in-depth, and fascinating, discussion of possible explanations for inequalities in health see R. Wilkinson (1996) *Unhealthy Societies: the Afflictions of Inequality* (London: Routledge).

Chapter 13

Leisure and consumption

David Crouch

Introduction

In this chapter the key changes in the geography of leisure are introduced and outlined. In the last decade or so there have been significant changes in terms of the location of key types of leisure which have had dramatic impacts on city and country alike. These changes have been both physical and visual affecting the fabric and land-use of cities and countryside. Certain kinds of leisure that two decades ago were in one kind of location in the UK now occupy very different locations. In addition there have been important changes in the 'symbolic landscape' of city and country. For example, the countryside has hitherto been regarded as a place of peace and quiet, a location for gentle leisure activity that is also small scale. Recent developments have reshaped the countryside in very significant ways, changing what we imagine the countryside to be. There have also been changes in the symbolism of cities, most notably in the revamping of many inner-city districts in response to new investment and demand for dance and music clubs, with zones of decline being transformed into symbols of new youth culture.

This review of changes in the geography of leisure in the UK makes particular reference to the process called consumption, which will be more fully explained in a later section (pp. 268–71). Here, suffice it to say, leisure has become much more commercialised during recent years and this has led to new patterns and locations of leisure development and activity. The attraction of UK leisure sites in terms of global tourism is an important influence on the kinds of places and sites that have been developed. Tourism and leisure overlap as activities and many attractions are valued for both.

This chapter consists of three main sections. The first identifies key aspects of the geography of leisure and considers the shape and scale of change. The next interprets this new geography through the lens of consumption, with particular reference to the commercialisation of leisure in recent years. Last, I consider an alternative interpretation of leisure which offers an enlarged focus to more traditional perspectives.

A new geography of leisure in the UK

The new geography of leisure in the UK may be investigated under the following key headings:

- a cluster of developments that may be labelled 'themed places';
- the revamping of shopping streets as shopping malls;
- new sports stadiums;
- the extension of intensive, 'city-type' leisure developments in the countryside;
- a new geography of heritage leisure;
- city-centre cultural revival;
- surprising microgeographies of virtual leisure.

Themed places

Theme parks have become a symbol of a new wave of leisure places. Especially designed and built with leisure in mind these provide a 'complete' experience in terms of facilities for all ages and a particular storyline to the whole site. The geography of theme parks can be summarised by considering the three kinds of places where these have been developed: city fringes, holiday areas and historic estates (Urry 1995; Harrison 1991). A number of theme parks are sited on the fringes of cities and their urban motorway peripheries, exemplified by Chessington World of Adventures and Thorpe Park, in the south and west of London and just off the orbital M25 motorway. Other theme parks have been located in traditional holiday areas, sometimes as part of the revamping of these areas as the more traditional sites such as piers and sea fronts became less popular, as in the example of Pleasurewood Hills in East Anglia. In the south-west of England there are several theme parks, including Flambards, which capitalises on the TV series of that name and is located without any particular local reason other than to capture the market of people in the area on holiday. Another type of theme park location is that attached to a historic landed estate, which is the case of Alton Towers in Staffordshire (although the house has been demolished). These theme parks have a landscape that combines children's stories with history and fantasy. Sometimes they draw upon local landscape features. This is especially the case of Center Parcs. These are larger than theme parks and provide 'complete' holiday places, including residential facilities. In general these are located in attractive countryside but where they can be hidden from 'spoiling the view', although all are imaginatively designed (WWFN/Tourism Concern 1993). The first two Centre Parcs were built in Thetford Forest, Norfolk and in Sherwood Forest, Nottinghamshire during the 1980s. Further parks of this kind have been developed at Lockinge Forest in Kent, near Bath and in the southern approach to the Lake District (see Figure 13.1).

Shopping leisure

Two decades ago it would have seemed absurd to write of shopping as leisure except in the case of very specific holiday-buying or window-shopping for a very special purchase, but

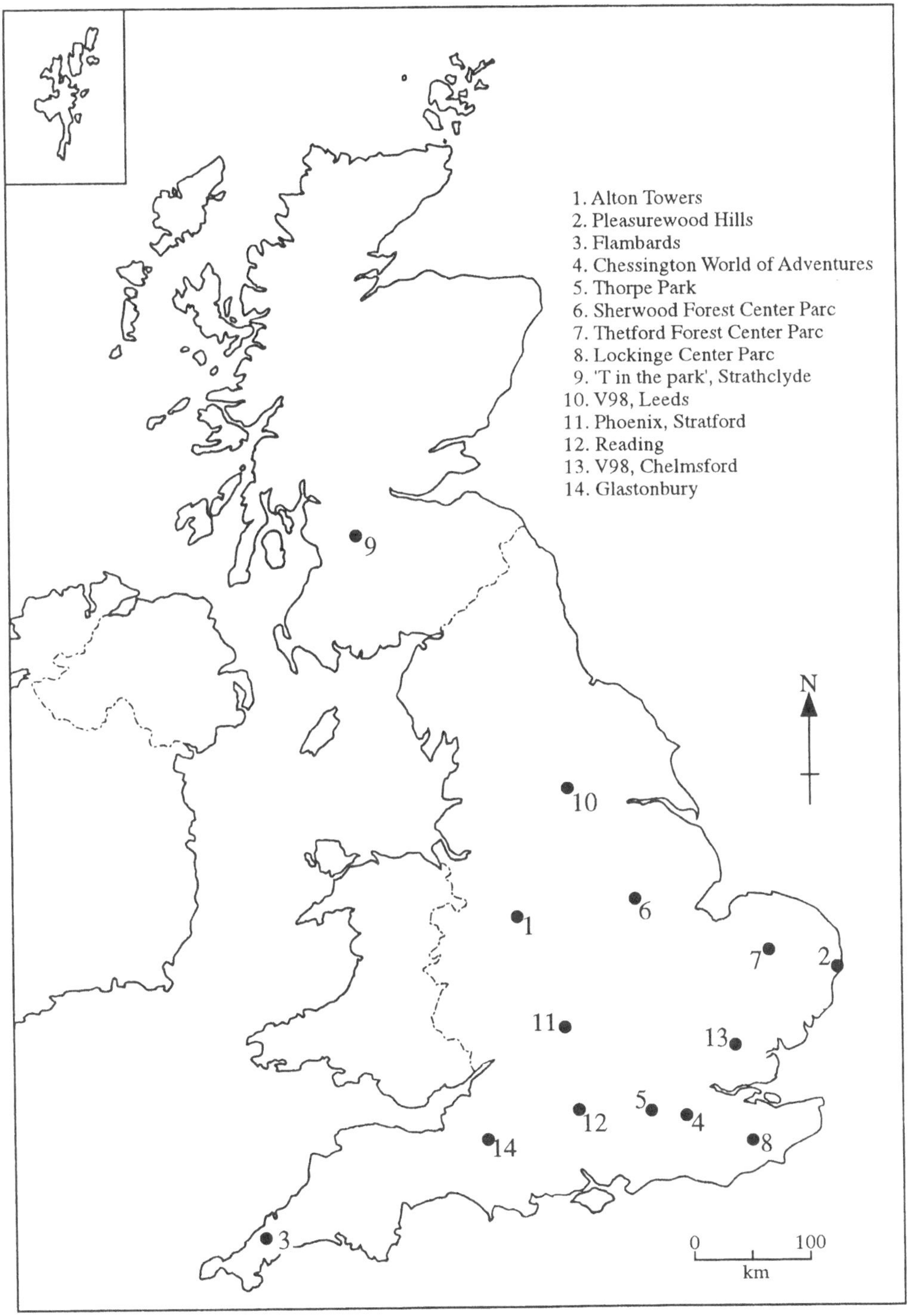

FIGURE 13.1 Theme parks, Center Parcs and music festivals (examples cited in the text)

shopping has profoundly changed. Stimulated by the financial success of the 'mall' in most American cities and small towns, there has been considerable development of indoor-style shopping places in the UK. These tend to provide more comfortable shopping areas, with a strong emphasis on attractive design dovetailed to market research evidence of what provides a good ambience in which to buy goods. Frequently these malls have associated facilities such as cafés, restaurants and children's play areas (Shields 1992). Like theme parks, these take on a different character in different places, and many are designed on one single idea or theme. Many shopping malls have been developed in the run-down areas of older cities in response to the loss of local manufacturing buildings and as part of revitalisation programmes, often with government financial support. The example of Sheffield's Don Valley is a good one. Like the earlier Gateshead Metro Centre it combines a range of stylised buildings with associated facilities and indoor space. Both are also policed for shoppers' safety. In the 1990s most large cities have their shopping mall. Another major location for these leisure-oriented malls is on the edge of major cities. Here car accessibility is important as these malls are large scale and depend upon large numbers of customers and mass shopping. The first mall around London was at Brent Cross, once near the end of the M1, now at a junction with the newer M25. Others have followed in similar locations around London. These centres are the result also of the relocation of shopping from more congested suburban areas of the city, and have capitalised on the growing attraction of shopping in contemporary cultural life (Chaney 1996; Miller and Jackson 1998). More numerous are the supermarkets of the 'big five', many of which have been relocated from city and suburb to urban fringes, for reasons mirroring the main malls. Shopping malls have accrued alongside them a number of other leisure-related buildings, often indoor, such as superbowls (Figure 13.2). A very different, more temporary geography of leisure is the weekend car-boot sale, where people get together to buy and sell in fields, school playgrounds and behind pubs (Gregson Crewe 1994).

New sports stadiums

A similar large-scale relocation of leisure has occurred in sport spectatorship. This is particularly the case in the more mass-spectator sports like football. This sport has been influenced by its increasing display on television which has increased rather than decreased the number of people wanting to attend a match (Bale 1995). In order to increase their ground capacities, a number of Premier League teams have sold their old sites, mostly located in the middle of a city or town where they were first established often a century ago, and developed larger, new sites with very stylish designs on the edges of cities. These developments, like the new out-of-town shopping malls and theme parks, have often attracted new developments nearby, a further example of the construction of larger centres of leisure.

Leisure and the countryside

Increasingly, intensive leisure sites have been developed outside the city altogether and, in consequence, the countryside has become dramatically changed in many different ways. This results from the commercialisation of countryside leisure. The range of active sports participated across rural areas includes angling (still the most popular), cycling, archery, golf, motor cycling, mountaineering, orienteering, polo, shooting and walking, four-wheel driving and many others (Clark *et al.* 1994). Many of these activities tend to be located in the zones within about fifty miles of an urban area where they are able to draw weekend and Sunday afternoon visitors, which includes green belts and other urban fringes, including Community Forests. Some of these activities are located away from urban areas but most are still accessible by the main motorway network. These sites located in 'deep country' and around the coasts depend upon particular physical attractions such as surface water, higher ground, flat land, deep ravines and river valleys.

Some pursuits have developed a more specialist geography. Golf has expanded beyond its traditional sites to the perimeters of many cities and towns. It also occupies an increasingly dispersed geography, as many people are now more prepared to travel and prefer a more rural location for the game (Cloke 1993; Crouch 1992, 1997).

Pop festivals and raves signify a new temporary rural geography of leisure. Here warehouses and large barns outside the city or large fields and abandoned aerodromes have spawned a nationwide diary of music festivals: Phoenix, Glastonbury, Reading, Chelmsford and Leeds V98, (Figures 13.1, 13.2). The environment has become a much more significant part of the geography of leisure, although this is often an adjusted, managed 'nature' (Urry 1995). This has been a powerful influence on the emergence of new kinds of places to see 'nature': butterfly farms, country zoos, wildlife parks and farm leisure centres. Many of these 'nature' sites are located in more frequented 'tourist' areas. Farms have acquired an increasing leisure role too, and farms like these are located especially in holiday districts such as the south-west of England, but also more scattered in locally popular areas. Skiing is another outdoor sport which is growing in popularity. Ironically skiing is also a cause for environmental concern as a result of the damage caused by the overuse of sites and associated development. This activity is especially limited by physical conditions, although the enthusiasm for skiing has produced a number of urban and suburban edge artificial ski slopes, for example the Beckton Alps in London's East End. Holiday homes and caravan parks continue to be popular options for many people. Their geography is guided as much by available land as anything else (Clark *et al.* 1994).

A new geography of heritage leisure

The geography of leisure has been changed significantly as a result of the development of 'heritage leisure'. The attraction of visiting heritage sites has been fuelled by the increasing interest in history. Many sites are surrounded by myth or memory. These sites are located in cities, towns and in rural areas. Many have been developed (invented) to capitalise on tourist potential, and some have been renovated by leisure enthusiasts, exemplified in the revival of old railway lines. Story-telling about domestic and working lives by previous generations has stimulated industrial heritage sites, exemplified in Ironbridge and also the Rhondda coal-mine themed site in South Wales. There are key heritage sites, the most well known including Stonehenge and Hadrian's Wall. These two in particular are examples of World Heritage Sites – that is, places of international heritage importance – and their visitors are drawn from across the world. Special investment is directed to places like these to make them accessible to large numbers of people whilst at the same time protecting them from overuse.

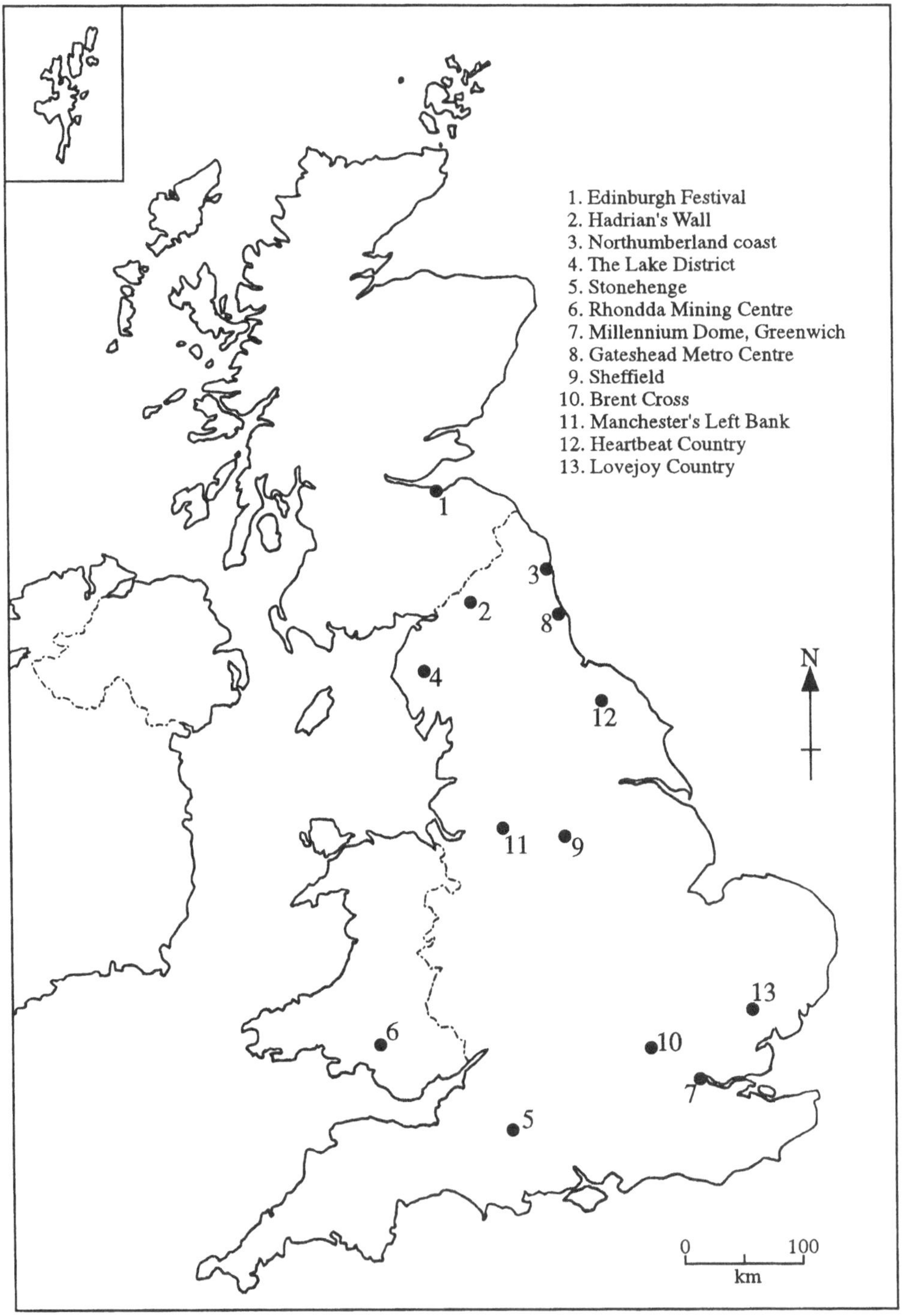

FIGURE 13.2 Heritage, shopping, club culture and 'TV Tourism' (examples cited in the text)

The wider 'discovery' of much less well-known sites has been significant too. Heritage captures the interest of local enthusiasts and casual visitors as well as tourists at large. Old canals provide new attractions and 'open up' new areas of city and countryside; for example, inner Manchester and the Kennet and Avon canal across the Wiltshire and Berkshire countryside in western England. Often the geography of these sites is shaped by their own distinctive past, but some areas have been especially developed in areas of larger visitor numbers. There are particular swathes of heritage sites extending from Hadrian's Wall at Penrith and the Lake District to the castles of the Northumberland coast. Areas of Wales, Cornwall and Scotland have been newly promoted for leisure/tourism based on their distinctive Celtic heritage.

A more temporary geography of leisure has been the creation and revival of numerous festivals of the arts across Britain. Places such as Bath, Worcester, Bury St Edmunds and Chester all offer 'a cultural experience' (Hewison 1987; Urry 1990). Many curious new 'heritage' places have been stimulated by television, literature and films set in locations that become an important part of the content. Examples are *Brideshead Revisited* (Castle Howard), *Heartbeat* in Yorkshire, Cookson Country in Northumberland and Lovejoy Country in Essex/Suffolk. These are set in real places that become the object of imagined projection of heritage (Figures 13.1 and 13.2).

Leisure as part of city-centre cultural revival

Heritage has contributed to the cultural interest of city centres for leisure use. The Globe Theatre in London is an excellent example, where there is an 'authentic' link with history and site, but the theatre itself has been completely rebuilt. Many cities have renovated their castle areas and linked these with new leisure/tourism developments. The leisure geography of many inner cities is one of dramatic restyling. Assisted by innovative urban landscaping often geared to the youth market inner urban districts have experienced enormous renewal and complete cultural change. This has often produced a new culture of clubs, music and dance venues, restaurants, multiplex cinemas and a 'café society' where people can walk and relax in a redesigned street. These include the examples of the old areas of Sheffield (subject of the film *The Full Monty*), and Manchester's 'left bank' (see Chapter 14). Such change produces a night-time geography of leisure. In the daytime many of these areas are largely quiet or the busy zone of office employees, until club-time arrives. The city has become increasingly a leisure space for strolling, looking and being looked at, for display and spectatorship (Fyfe 1998).

Another example of major city leisure that is both creating as well as celebrating heritage is the Millennium Dome. This will add a new scale of leisure to the reshaping of the inner East End (Figure 13.2). A much more temporary city geography emerges from the revival of street events, most notably in the case of the annual multicultural celebration of Notting Hill Carnival, when the whole life of the area of a square mile is joyfully disrupted for a weekend. Centrally located parks have been reinvigorated with investment to encourage the attraction of city areas for tourism, and leisure (Comedia 1995).

Micro-leisure and virtual leisure sites

A significant contrast to many of these key shifts in the changing geography of the UK are the highly dispersed, individualised and domestic sites of virtual leisure provided by computer games and the Internet. Instead of requiring large developments in particular parts of the country or the city this leisure requires only a terminal and connections. In an ironic combination, heritage and micro-leisure are connected in cyberspace. It is possible to 'visit' Stonehenge, 'see it' from all angles, 'walk round' it, move inside it with a cursor and mouse in cyberspace at the computer screen (Rojek 1995). There are also new 'transit' leisure sites epitomised by the airport lounge and the motorway service station. These are places where people punctuate journeys, make encounters and anticipate being somewhere else.

Comprehending the new geography of leisure

Central to these shifts in the geography of leisure in the UK are aspects of what is called 'consumption'. This is a term which is frequently applied to the consumer buying from a producer. However it can mean more than this. It is important to note that the term 'consumption' also applies to the 'making-sense' and 'making-value' of things we use, places we visit, things we do, in a way that may or may not involve purchase or payment of any kind. However, it is the case that in the last decade or so many more leisure activities and sites have become part of the commercialised world. The more obvious geographies have been noted already in this chapter. There are several other key aspects which are to be described in this section; for example, the commercialisation of leisure, lifestylisation, leisure as an individual pursuit, landscape style as part of the commercialisation of leisure, the leisure 'gaze', and new symbolic geographies.

'Street cred'

During the last decade there has been a massive growth in the commercial promotion of goods as objects of making lifestyle. Since manufacturing investment fell during the 1980s significant interest has focused on using land and investment for leisure developments. To make this new geography work it became necessary for the consequent commercial development to attract consumers. This coincided with a new profile of demand for leisure. For example, a new youth culture sought more varied leisure than its parents. Amongst older people a larger proportion had acquired resources through private pensions and investments, and others from redundancy. An ageing population means more older people making leisure, too.

A key feature of the new commercialisation of leisure is its offer of a route to a new lifestyle, a promotion of 'product' directed at both young and old. Investing in new clothes, cars, kinds of food, visiting new places, taking a longer holiday and taking up a new pastime provide a new style that offers the prospect of real change in one's life. Leisure plays a very central part in this, and has become a major object of contemporary lifestylisation, making oneself anew, 'freshening one's image', gaining status amongst one's peers, gaining 'street cred' (Chaney 1996). The presentations of leisure sites with commercial 'wrapping' becomes evermore important to promote this new leisure and connect with desires for lifestyle.

'Shopping' became a metaphor for 'making lifestyle'. This has been interpreted as part of the break from neighbourhood communities, workplace friendships and traditional social classes; for example, working classes with their traditional forms of leisure kept very much within the community (Harrison 1991; Chaney 1996). Part of this diversity and extended choice in leisure places and styles is the possibility of detachment, of getting away from familiar, perhaps claustrophobic, social moorings. At the extreme, people 'buy' leisure rather than 'make it' amongst themselves. Sharing time becomes less important. We enjoy leisure more on our own, choice is for the individual and we only need ourselves to buy something, perhaps as long as the result is seen by 'someone that matters'. People, as long as they have enough resources, can take part in anything; this provides freedom. They can shop, visit theme parks, heritage sites, football matches and new snooker complexes and film multiplexes. As we have seen in this chapter there has been an increased diversity and more fragmented association between a kind of leisure and its location as users change. The geography of socio-economic groups has also become more fragmented in recent years. The geography of the leisure activity of any individual and any social group has become more fragmented and complicated, more finely tuned to local differences. This contrasts with the time when broad areas of similar social class might have been expected to make leisure near home. Participation in leisure is much more diverse, and its geography is too (Clark *et al.* 1994).

Moreover, the shifts in the economy over the last decade or so have assisted this geographical adjustment in other ways. These shifts have produced places wanting development (such as redundant manufacturing sites and less profitable farmland). There have been shifts of investment into new areas of production, notably leisure, providing chances to exploit the increased wealth available to part of the population and to provide new local employment. In order to attract visitor use and investment there has been commercialisation of places that had previously not been regarded as interesting for commercial leisure development. The countryside is an example of this, where previous leisure pursuits had been connected with the open countryside (for example, for walking, riversides for fishing). This geography has been changed, as noted earlier, with the advent of heritage developments, theme parks, holiday-home parks, and so on.

'McDonaldisation'

Across a number of these new zones of leisure there is a new landscape, epitomised by the labels 'McDonaldisation', or 'Disneyfication'. This signifies the imprint of new, very stylised commercial developments which are oriented around food, eating out and new retailing. These stylised developments are customised and have repeated and universal designs for customer-product recognition. The sites are individually 'themed'. Making places commercial and attracting people to use or visit them depends heavily on making new images, relabelling, and using these images and labels in advertising these places. Advertising is often very visual and can make strong use of landscapes. The idea of 'theming' is that the development is designed as one concept, integral, and landscaped together, as typified in theme parks. This creates a certain repetition of landscape in urban and rural areas. The stylised copying of 'rural' barns and churches by the major supermarket stores, increasingly located on the edges of cities, makes a curious contrast. In all these cases the design of places is seen to be all-important (Urry 1990). This connects with the shift towards heritage fascination.

Heritage leisure, themed sites and 'the gaze'

The increased interest in 'heritage' has contributed new geographical commodities and new places of interest to visit. There has been a burgeoning interest in heritage, old buildings and sites. Sites associated with the past have become invoked with new value and this has been converted, with commercialisation, into new commodities in new sites (Hewison 1987). Where new sites are produced as heritage sites there is arguably a false geography of 'unauthentic', pretend, pastiche, unreal, places. These places become objects of fascination and the 'gaze'.

Observers of leisure activity have noted the changing interest towards watching events. The increasing size of sports stadiums; the development of heritage as displays and spectacles of past culture, countryside visiting to see the 'sights' and the attraction of street strolling amongst the young are all part of a wider change. In all these there is an emphasis on watching, gazing detached from the world around us, taking in the view rather than perhaps participating in an event oneself. This has been labelled 'the age of the spectacle' (Urry 1990). This also produces a new symbolic geography of leisure.

The new symbolic geography of leisure

In terms of symbolic geography changing patterns of leisure are the result of new ways in which we 'think' of places. Again, the commercialisation of places has changed meanings previously held. The new themed landscapes can convey new meanings, and can assist in making a place previously unpopular become very popular. Hence, places like the run-down canalsides of Birmingham and Manchester two decades ago would not have been attractive for leisure and instead were frequently regarded as areas of decay and danger. These have now been reinvested for leisure, and given new labels, new names and become new landscapes. The promotion of place has become an important feature in the promotion of leisure, making the remaking of these places a reality. For example, the theme park Pleasurewood Hills combines in its name key features that make possible the complete re-imaging of a traditional seaside resort (Yarmouth on the coast of East Anglia) and make Yarmouth marketable to a new generation. Geography as landscape has become very important in the promotion of new leisure places. In a similar way, the symbolic meaning of 'countryside' has changed too, as noted at the beginning of this chapter. Once a symbol of peace and quiet, the countryside now provides a picture of activity, but activity not associated with traditional farming so much as playtime, holidays and weekends, Sunday walks (Clark *et al.* 1994). More than this, it means a mixture of often highly developed and intensively used sites alongside 'reserves' of wildlife (Crouch 1992, 1997; Urry 1995).

The 'new symbolic geography of leisure' reaches an ultimate in the micro-spaces of virtual reality, although these provide simultaneously a global and, moreover, fantastic supra-spatial geography that can be created by turning on on-line. The imaginary landscape can become almost tangible, providing new dimensions to much older representations of painting, literature and the imagination alone (Rojek 1995).

Cybercafés

The strange geography of the 'Net and virtual games' epitomises one pervasive feature of this new geography of leisure; that is, the softening of the boundaries, and differences between many of the leisure activities and the sites where they occur. In particular, the old distinctions between city and countryside, peace and quiet on the one hand and excitement and pleasure on the other, have been significantly, though not completely, loosened and diluted in recent years. There is much more diversity and complexity in the kind of leisure sites located across the city and across the country. In others places there is increased distinction and specialisation, such as the new café quarters in city downtowns that also include cybercafés for swapping Net stories, and in large-scale specialised watersports areas like Holme Pierrepoint in the East Midlands. In some ways this produces a very mixed, diverse and fragmented geography of leisure that has been called 'post-modern'. The extreme dimensions of this geography are captured in this almost exotic narrative from the cultural theorist Iain Chambers: 'the migrant landscapes of contemporary metropolitan cultures, deterritorialised and de-colonised, re-situating and re-presenting the common signs . . . It is perhaps something that we can hear when Youssou N'Dour, from Dakar, sings in Wolof, a Senegalese dialect, in a tent pitched in the suburbs of Naples' (Chambers 1995: 14–15).

Enlarging this changing geography leisure

Perhaps Chambers is being ironic. There are further dimensions to the geography of leisure so far considered in this chapter, and these are considered in this section (Crouch 1994, 1997, 1998). At the core of 'consumption' is an appreciation of 'culture'. 'Culture' can be encapsulated as sets of beliefs or values that give meaning to ways of life and produce (and are reproduced through) material and symbolic forms (Crang 1998). In terms of leisure we make and reproduce culture through what we do and use and the sense we make of places. The way we 'make sense' of leisure 'products', events and places is influenced by numerous contexts: advertising and other media, family, friendships and schooling. All of this helps us to make sense of places and our lives. My assertion is that this amounts to influences and 'doing' much more than can be 'shaped' by the commercial world; that would only provide a certain amount of structure and attraction – we make the rest. We also make leisure in places that have nothing, or very little, to do with the products of market leisure.

Enjoyment

However, it may be argued that we still 'consume' these places and activities. This is so if we extend the word 'consumption' to include experience, enjoyment, and just making sense of things and places. It is important in mapping the changing geography of leisure in the UK that the degree of emphasis that is given to the significant physical products of commercialised leisure, such as big theme parks and renovated districts of the city, is not exaggerated. Leisure happens in many less dramatically developed places, in everyday familiar districts of people's lives. Interpretation depends upon the sources of data considered. Much of the data available to interpret the geography of leisure is based on commercial leisure, measures what people spend money on more than what they do, and thus tends to be self-fulfilling (Crouch and Tomlinson 1994).

Friendship

Many people do take part in leisure away from their familiar social worlds of everyday life, and many people no longer do exactly what their parents did (and most likely their parents did not do what *their* parents did, and so on). Yet friendship and social relationships remain very important in the way people take part in leisure (Finnegan 1997). Who has been to a theme park alone? – perhaps only geography researchers! Leisure can provide a shared orientation around which identity is forged, through shared participation rather than through buying a lifestyle. Instead of traditional, planned residential areas being the focus of 'community', the new geography of community is likely to be found around networks of leisure sites, dance halls, 'banger' racing, caravan sites, historic enactment sites (Urry 1995). The new commercialised leisure is usually consumed socially.

Embodiment

Alongside the argument of increasing detachment and 'gazing' it is evident that in many of these events people do 'join in' (theme parks, historic enactments, snooker, dancing, shopping). When people attend more visual displays and performance they are often very much involved in preparation, walking around the site, eating out and using many different senses. People may 'gaze', but the gazing is part of doing other things. A more personal, and shared, symbolic geography is produced from places where people make leisure. The symbolic geography of this activity consists of parks, everyday streets, clubs and bars, walks in fields and woods as well as the commercialised sites of consumer leisure with their particular symbolism (Crouch 1998). This means that leisure geography is 'all around us', in familiar places and 'down the road'. This is an everyday geography of leisure. It is also a changing geography, because the changes already discussed in this chapter influence and are influenced by the enjoyment of these everyday places. Moreover there are adjustments in the social groups attracted to participate in different kinds of leisure.

Social composition

Although there has been a large social change in Britain that has reduced the 'working class' and their traditional patterns of leisure, such a change is incomplete. Even in mining areas, brass bands continue, as attested by *Brassed Off*; and allotment holding remains very popular too. Evidence of social trends suggests there is a continuing link between socio-economic groups and kinds of leisure, and thus the social geography of leisure activities may be less radically changed than some interpretations assert (*Social Trends* 1995). There does seem to be a number of significant differences between social groupings. This has important consequences for the popular, 'lay' geography of leisure; that is, what people do, and where. This may be due to the continuing influences of social class and other social categories on leisure preference, or because social access to leisure is still very much constrained according to wealth, or both. Thus 46 per cent of social class A/B visit the cinema and only 24 per cent of D/E do; 6 per cent of C1 go to bingo, and 17 per cent of D/E do. Theatre visiting subdivides thus: A/B 33 per cent, C1 24 per cent, C2 16 per cent and D/E 10 per cent. Theatre is traditionally a more 'middle class' leisure as are Henley Regatta and Cowes Week (*Social Trends* 1995). Camping and caravanning, theme park visiting and rock concerts figure almost the same in each grouping and there is little difference in terms of disco and night club attendance. However, the aggregations of these categories probably disguise differences in the kind and cost of participation. There is an age dimension to this, and 'going to the pub' peaks at 25–34 across the aggregated groups, fast food at 25–34, but going to restaurants remains high from 25 years old to just before retiring age. Many of these activities are of course very social activities, and most visits from home are made to visit friends/relatives and to drink/eat (17 per cent and 18 per cent each) (*Social Trends* 1995).

Mixing

There is also an extension of traditional leisure activities to new social groups, a 'discovery' of 'new' leisure activities. Allotment holding is participated by an increasingly broad social spectrum, which exemplifies a movement in the opposite direction to the assertion of the gaze. A young middle class has been attracted to a 'green' lifestyle and is reconnecting with ideas of locality, place, and partnership (Crouch 1997). Public parks in ordinary residential areas in every city, town and village remain highly valued everyday 'lay' geographies amongst many different social groups (Comedia 1995; Crouch 1994).

This series of notes on an alternative reading points towards a larger geography of leisure in the UK. It has been possible to enlarge the geography of leisure beyond a focus on production (the location of new developments in leisure) in order to embrace more dimensions of leisure activity and their geographies.

Conclusion

Whatever emphasis is given, the new consumption geography of leisure, in its many components and places, is influenced by an increasing commercial leisure economy. Simply describing the most recent large-scale and highly invested leisure developments and their location does not explain, however, how and where people are making leisure, nor does it adequately account for the shape of the symbolic geography of leisure.

It has become much more difficult to be dogmatic about the location of leisure. With increasing levels of leisure activity so its geography has become more dispersed. The commercialisation of leisure has contributed to its relocation in purpose-built and larger, more profitable sites. There remain, however, real profound differences in the availability of leisure sites and in how different groups access leisure opportunities. However, the changing boundaries between different activities and their location, the complexity of small districts of the city, and the diversity of the country make the complexity and the small-scale variations in the geography of leisure more important to observe and understand.

References

Bale, J. (1994) *Landscapes of Modern Sport*, Leicester: Leicester University Press.

Chambers, I. (1995) Popular Culture: the Metropolitan Experience, London: Methuen.

Chaney, D. (1996) *Lifestyles*, London: Routledge.

Clark, G., Darrall, J., Grove-White, R., Macnaughten, P. and Urry, J. (1994) *Leisure Landscapes*, London: CPRE.

Cloke, P. (1993) 'The commodification of the countryside', in S. Glyptis (ed.) *Leisure and the Environment*, London: Wiley.

Comedia (1995) *Parklife*, London, Comedia-Demos.

Crang, M. (1998) *Cultural Geography*, London: Routledge.

Crouch, D. (1992) 'Popular culture and what we make of the rural', *Journal of Rural Studies* 8(3): 229–40.

Crouch, D. (1994) *The Popular Culture of City Parks*, London: Comedia-Demos.

Crouch, D. (1997) 'Others in the rural: leisure practices and geographical knowledge', in P. Milbourne (ed.) *Revealing Rural Others*, London: Cassell.

Crouch, D. (1998) 'The street and geographical knowledge', in N. Fyfe (ed.) *Images of the Street*, London, Routledge.

Crouch, D. and Tomlinson, A. (1994) 'Leisure, space and lifestyle: modernity, postmodernity and identity in self-generated leisure', in I. Henry (ed.) *Modernity, Postmodernity and Identity*, Brighton: Leisure Studies Association.

Crouch, D. and Ward, C. (1997) *The Allotment: its Landscape and Culture*, Nottingham: Five Leaves Press.

Finnegan, R. (1997) 'Music, performance and enactment', in G. Mackay (ed.) *Consumption and Everyday Life*, London: Sage.

Fyfe, N. (ed.) (1998) *Images of the Street: Planning, Identity and Control in Public Space*, London: Routledge.

Gregson, N. and Crewe, L. (1994) 'Beyond the high street and the mall: car boot sales as an enterprise culture out of control?', *Area* 26: 261–7.

Harrison, C. (1991) *Countryside Recreation in a Changing Society*, London: TMS Partnership.

Hewison, R. (1987) *The Heritage Industry*, London: Methuen.

Miller, D. and Jackson, P. (1998) *Shopping, Place, Identity*, London: Routledge.

Rojek, C. (1995) *Decentring Leisure*, London: Routledge.

Shields, R. (1992) *Lifestyle Shopping*, London: Routledge.

Social Trends (1995) London: OPCS.

Urry, J. (1990) *The Tourist Gaze*, London: Sage.

Urry, J. (1995) *Consuming Places*, London: Routledge.

World Wide Fund for Nature/Tourism Concern (1993) *Beyond the Green Horizon*, London: WWFN/Tourism Concern.

Further reading

Connections between consumption and place, as well as the idea of symbolic landscapes and McDonaldisation, are summarised neatly by M. Crang (1998) *Cultural Geography* (London: Routledge). On aspects of consumption a useful introduction is D. Chaney (1996) *Lifestyles* (London: Routledge). Key works on leisure and 'the gaze' include J. Urry (1990) *The Tourist Gaze* (London: Sage) and C. Rojek (1995) *Decentring Leisure* (London: Routledge). The final chapter of J. Urry (1995) *Consuming Places* (London: Routledge) discusses issues of shared leisure and the new environmentalisation in leisure. *Leisure Landscapes* (Clarke *et al.* 1994) examines data on consumers and includes a variety of valuable insights on leisure outside the cities. Further statistical evidence is to be discovered in *Social Trends* and *Regional Trends*, both published by HMSO annually. D. Crouch (1997) 'Others in the rural: leisure practices and geographical knowledge', in P. Milbourne (ed.) *Revealing Rural Others* (London: Cassell), explores the newest work in the field, and redefines the limits of consumption. The journal *Leisure Studies* is valuable in exploring the interfaces between geography and leisure. A new text of relevance is D. Crouch (ed.) (1999) *Leisure/Tourism Practices, Knowledge and Geographies* (London: Routledge).

Chapter 14

Cultures of difference

Peter Jackson

Introduction

During the 1980s, geography (along with most of the other social sciences) underwent a 'cultural turn', involving a widening of its academic horizons and an increasing openness to a range of intellectual movements including feminism and post-colonial studies. As a result of the growing dialogue between social geography and cultural theory, the study of cultural geography was transformed to emphasise a *plurality of cultures* and the *multiple geographies* (landscapes, spaces and places) with which those cultures are asscociated (Jackson 1989). Rather than approaching culture in narrowly aesthetic terms, cultural geographers embraced a much wider agenda, exploring the many ways in which people attach meaning to places, investing the material environment with cultural significance. As meanings and values vary over time, from place to place, and between groups and individuals, geographers quickly became immersed in complex issues of *cultural politics*, whereby dominant ideologies are negotiated and contested by those in less powerful positions.

This chapter applies these ideas to the changing geography of the United Kingdom during the 1980s and 1990s. It suggests that these decades have been characterised by a growing sense of multiculturalism, when what it means to be 'British' has been undergoing rapid change. While some have welcomed these changes, acknowledging a growing tolerance of cultural difference as a positive feature of contemporary British society, others have seen them as undermining what they regard as the core values of 'Britishness' or as a threat to older forms of solidarity that appear to be fragmenting into ever more complex and plural identities. This chapter will argue that cultures cannot be contained within narrowly defined national boundaries. Notions of 'Britishness' are unstable and unbounded, a hybrid blend of diverse cultural influences that originate from (and extend) well beyond the national boundaries. An openness to cultural difference therefore implies a shifting sense of national identity which many find threatening. Some of those who align themselves politically on the Left have seen the rise of 'identity politics' as a threat to the allegedly 'universal' appeal of class-based politics, while many on the Right have sought a return to an apparently more stable sense of national identity. This chapter aims to chart this dynamic cultural geography through an exploration of the UK's increasingly complex 'cultures of difference'.

The chapter begins with an example of those who would deny or seek to limit the extent of cultural difference. It goes on to map various forms of multiculturalism that characterise the UK's changing geography, exploring the extent to which cultural differences have been subsumed within the market through the process of commodification. The chapter ends with a discussion of the limits of tolerance, arguing that 'living with difference' will be a major social and political challenge as we enter the new millennium.

The denial of difference?

According to Lord Tebbit, speaking at a fringe meeting of the Conservative Party conference in October 1997, multiculturalism is a divisive force in British society. No one, Norman Tebbit argued, can be loyal to two nations. Immigrants should be taught that the Battle of Britain is part of their history. The alternative, Tebbit warned, is the kind of ethnic and cultural division that led to the break-up of the former Yugoslavia. His words recalled earlier interventions from right-wing politicians, including Enoch Powell's 'rivers of blood' speech in 1968 which warned of the inevitability of violent conflict if immigration was permitted to continue unchecked. Lord Tebbit had himself made several earlier speeches on the subject of multiculturalism, including his infamous 'cricket test' speech in 1990 when he argued that the loyalty of ethnic minorities should be judged by whether they supported the English cricket team: 'Which side do they cheer for?' he asked. ' Are you still harking back to where you came from or where you are? I think we've got real problems in that regard.' Suggesting that many British Asians continued to search for husbands and wives in the family's native country, he concluded that 'you can't have two homes. Where you have a clash of history, a clash of religion, a clash or race, then it's all too easy for there to be an actual clash of violence' (*The Times*, 21 April 1990).

Unlike these earlier speeches, however, Tebbit's comments at the 1997 Conservative Party conference received very little public support. The Tory leadership rapidly distanced itself from his remarks, claiming that Lord Tebbit was out of touch with the mood of the country, while the Queen spoke out in praise of ethnic minorities. The recognition of cultural difference is no longer the unique preserve of left-leaning cultural critics. Even the right-wing press has come to accept a more 'hybrid' view of British national identity. While the *Daily Mail* printed Tebbit's speech in full, the *Daily Telegraph* took a very different tack, arguing that 'A child with a Welsh father and a mother from Ulster can eat Indian food, listen to reggae, and watch Italian football without experiencing cultural confusion and political alienation' (9 October 1997). As Stuart Hall argued in a *Guardian* interview, the denial of difference is no longer tenable: 'You can't go on, generation after generation, denying immigration' (11 October 1997).

While it may no longer be a popular public position, Lord Tebbit's denial of difference has deep roots within British society. For a nation that prides itself on its tolerance and sense of fair play, British history is marred with recurrent outbreaks of anti-Semitism, racism and xenophobia (Holmes 1991). Given this persistent ambivalence about multiculturalism, defining 'the nation' has often been problematic. When T.S. Eliot (himself an American émigré) sought to define the 'national culture' shortly after the Second World War, he reeled off a bizarre list of ingredients: 'Derby Day, Henley Regatta, Cowes, the twelfth of August, a cup final, the dog races, the pin table, the dartboard, Wensleydale cheese, boiled cabbage cut into sections, beetroot in vinegar, nineteenth-century Gothic churches and the music of Elgar' (1948: 31). This exclusionary vision of Britain was later parodied by Hanif Kureishi, reflecting Britain's increasing multiculturalism. For Kureishi, British culture would now include: 'yoga exercises, going to Indian restaurants, the music of Bob Marley, the novels of Salman Rushdie, Zen Buddhism, the Hare Krishna Temple, as well as the films of Sylvester Stallone, therapy, hamburgers, visits to gay bars, the dole office and the taking of drugs' (1986: 168–9).

It is not simply that 'British culture' has been enriched by 'ethnic diversity'. The very existence of separate (national or ethnic) cultures is now in doubt. Popular music is a particularly good instance of this kind of 'hybridity' with *bhangra* frequently cited as a key example of such intercultural fusion. Performed by artists such as Apache Indian, *bhangra* blends a range of sounds from Punjabi folk styles to African-American house music and is listened to by a variety of audiences from Bombay to Brixton. Acording to one observer:

> Bhangra created an over-arching reference point cutting across cleavages of nationality (Indian, Pakistani, Bangladeshi and other), religion (Sikh, Muslim and Hindu) and caste and class . . . [Its emergence] in the 1980s signalled the development of a self-conscious and distinctively British Asian youth culture.
>
> (Back 1996: 219–20)

Or in the words of one young person from Bradford:

> We've got bhangra happening. It's amazing. I love listening to people like Apache Indian because that's literature and music and dance all together and it's a mixture, it's all in fusion. He's saying Urdu, Punjabi, English, everything thrown together. He's rapping away . . . and he's going on about arranged marriages, and there's Indian instruments in the background, there's western instruments. And when you see him on TV, there's Indian dancers and there's Afro-Caribbean men doing his stuff, Asian kids doing their stuff, and there's white kids doing their bit. I just think, 'Wow!' I just think all this fusion will throw up new art forms. New things will come, they're already happening and it'll be really nice to see where we end up in a few years to come.
>
> (Bradford Heritage Recording Unit 1994: 164–5)

It is possible, of course, from a position of cultural privilege, to romanticise the blending of cultures and to champion the emergence of 'hybrid identities' with too little concern for the material conditions that enable a positive fusion of cultures to be achieved. Picking and mixing is certainly easier for those with sufficient economic and cultural resources. But, at least in some instances, it seems clear that academic understanding of 'multiculturalism' is being outstripped by people's everyday practices. Producing and consuming such hybrid cultural forms transcends established boundaries of taste and nation, compelling social scientists to coin new terms such as 'inter-being' and 'multiculture' in an effort to map the contours of the UK's changing cultural geography.

As these examples suggest, the United Kingdom is now a thoroughly multicultural place, though one might question how deeply the tolerance of cultural diversity extends. Even a cursory reading of the national newspapers shows plenty of evidence of continued intolerance towards minorities of every kind, underpinned by legal and other forms of institutional exclusion. Examples include recent debates about lowering the age of consent for homosexual sex, persistent evidence of sexual and racial harassment in the police and armed forces, and the orchestration of recurrent 'moral panics' against single mothers (McRobbie 1994). Differences of 'race', religion, sexuality and gender still provoke prejudice and evoke widespread hostility. Mapping the extent of 'British' multiculturalism allows us to reflect on the limits of tolerance as well as on the possibilities of mutual recognition and understanding.

Mapping multiculturalism

'Britishness' can no longer be confined (if it ever could be) within the boundaries of the nation-state. Britain's imperial past established a series of transnational connections that continue to be reproduced through the economic networks of multinational corporations, amplified by the flow of capital, the migration of labour and the transmission of ideas and other kinds of information. While some movements are controlled by nationality and immigration laws, other boundaries are more permeable, such as the relatively free movement of information and ideas across the Internet. As a result, 'British culture' has, for many years, exceeded the political boundaries of the United Kingdom.

In terms of international migration, for example, Britain has been a net exporter of labour for most of the post-war years. The single most important source of immigration has been the Irish Republic. Yet it has been immigration from the New Commonwealth and Pakistan that has generated the most intense public debate, fuelled by concerns over the 'assimilation' of ethnic minorities. As overt racism has become less publicly acceptable it has been transformed into a more subtle cultural form. A polite concern for cultural difference may then disguise much more sinister racialised fears. Margaret Thatcher's comment in 1979 about the country being 'swamped by people with a different culture' has often been interpreted in this light as has the racialised subtext of concerns for the 'inner city', for 'law and order' and other apparently 'non-racial' subjects (Jackson 1988).

From this perspective, debates about 'race' and racism can be seen as central to British political debate rather than of only marginal concern. This helps explain why public anxiety about immigration often seems so out of proportion to the actual numbers of 'immigrants' (and their descendants) who have settled in the UK. According to the 1991 Census, less than 5 per cent of Britain's population were from 'non-white' ethnic groups, with the Muslim population generally estimated at around 2 per cent.[1] The ethnic minority population is, however, far from evenly distributed across the country, with the heaviest concentrations in urban areas, particularly in London, the West Midlands and the older industrial cities of the North and North West (see Figure 14.1). Localised concentration in areas of declining economic opportunity has given rise to the potential for conflict, creating in Lord Scarman's analysis of the Brixton riots 'a predisposition towards violent protest' (1981: 2.38).

The migration of people has also led to a flow of commodities and cultures. British multiculturalism is often referred to in these terms, with references to 'samosas and saris' or a blending of 'roast beef and reggae music' (Sarup 1986; Jeater 1992). This may, however,

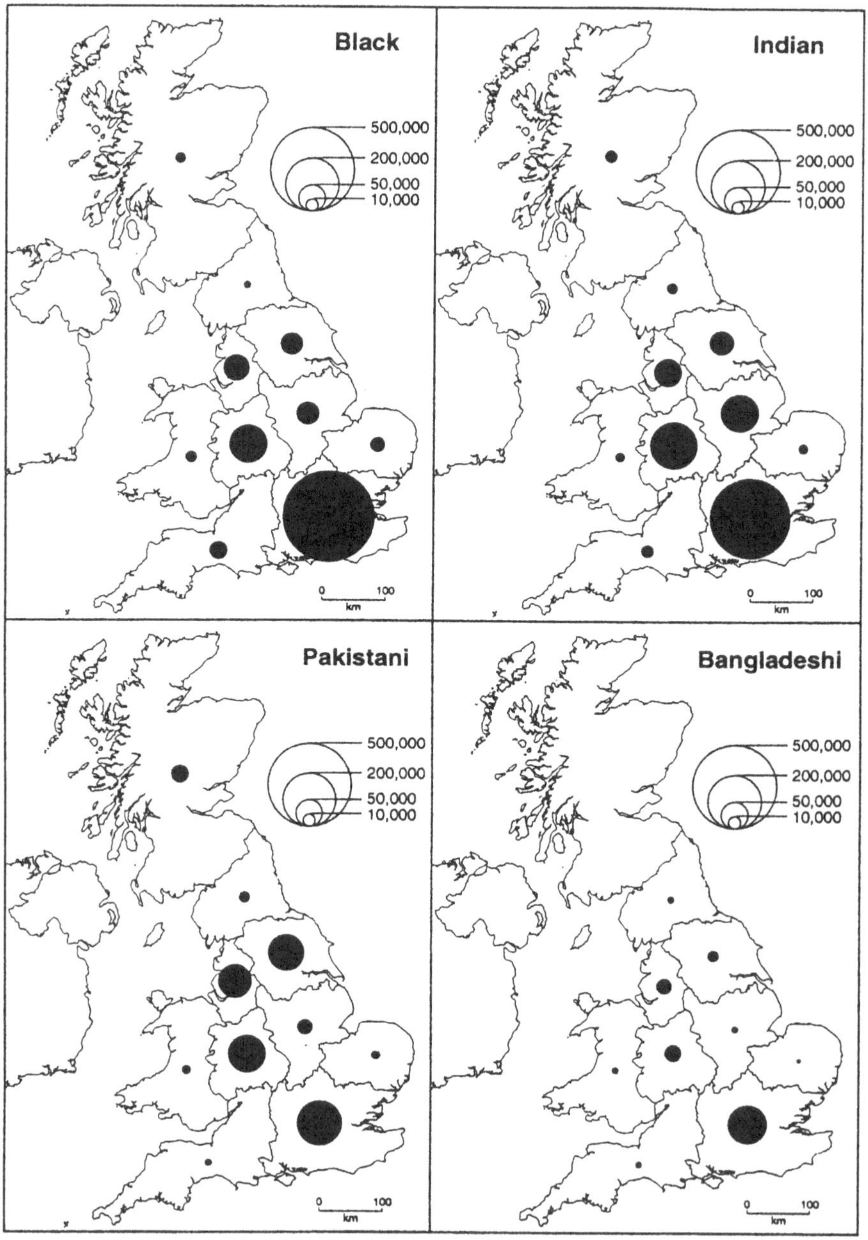

FIGURE 14.1 Distribution of ethnic groups, 1991
Source: OPCS, Census of Population, 1991.
Note: The Census category 'Black' includes Black-Caribbean, Black-African and Black-Other.

involve only a superficial involvement with other cultures as opposed to a more fundamental acknowledgement of the political rights of minority groups and the provision of equal economic opportunities. An understanding of the 'culture of commodities' allows some further reflections on the nature and limitations of British multiculturalism.

The tendency to engage with other cultures on a purely superficial level is nowhere clearer than in relation to culinary culture. Researching the consumption of 'ethnic' food in North London, two geographers provide this wonderful illustration from the listings magazine, *Time Out* (16 August 1995):

> The world on a plate. From Afghan ashak to Zimbabwean zaza, London offers an unrivalled selection of foreign flavours and cuisines. Give your tongue a holiday and treat yourself to the best meals in the world – all without setting foot outside the fair capital.
>
> (Cook and Crang 1996: 131)

But what exactly are we consuming when we indulge our tastes for such 'exotic' food and what other cultural meanings are at stake?

Research in the gentrifying inner-city neighbourhood of Stoke Newington in North London suggests that food consumption is related to a whole range of social issues, including the formation of class-based identities. Residents in this study were shown to be using their economic and cultural capital (money and educational resources) to indulge in 'a little taste of something more exotic', including a preference for 'exotic' dishes such as Thai food, pizza or curry rather than an 'ordinary sandwich' (May 1996). But the cultural politics of food is far more complex than we might at first think. The British taste for curry, for example, can be traced back through generations of colonial exchange between Britain and India. Taken from a Tamil word (*kari*), 'curry' originally referred to a range of local *masalas*, prepared for consumers 'back home' via the British invention of curry powder (Crang and Jackson in press). Some of the most basic signifiers of British culinary taste, such as the good old 'English' cup of tea, also have more complex cultural roots. As Stuart Hall has demonstrated, the 'English cuppa' may be an accepted symbol of national identity but its origins are anything but British:

> Because they don't grow it in Lancashire, you know. Not a single tea plantation exists within the United Kingdom. . . . Where does it come from? Ceylon-Sri Lanka, India. That is the outside history that is inside the history of the English. There is no English history without that other history. . . . People like me who came to England in the 1950s [from the West Indies] have been there for centuries; symbolically, we have been there for centuries. . . . I am the sugar at the bottom of the English cup of tea. I am the sweet tooth, the sugar plantations that rotted generations of English children's teeth. There are thousands of others besides me that are . . . the cup of tea itself.
>
> (Hall 1991: 48–9)

The quintessentially 'English cuppa' is, then, the product of globally extended commodity networks. Yet these complex cultural origins are conveniently forgotten in countless daily acts of domestic consumpton. A similar process of selective forgetfulness applies to the foreign (immigrant Jewish) origins of high street favourites such as Marks & Spencer, now widely regarded as purely and simply 'British'. Even the 'national dish' of fish and chips conceals a hybrid cultural history involving French styles of preparing fried potatoes and an East European Jewish tradition of frying fish (Back 1996: 15).

A similar argument can be applied to our understanding of 'English' literature, where even the most revered works, such as Jane Austen's novels, can be shown to exhibit complex multicultural geographies. Though now regarded as a staple feature of our national culture (with film and television adaptations attracting millions of viewers) and as an outstanding representation of 'British' culture abroad (via exports worldwide), the geography depicted in the novels stretches well beyond the UK. As Edward Said's work on *Culture and Imperialism* (1993) demonstrates, the plot of Jane Austen's *Mansfield Park* (1814) revolves around Sir Thomas Bartram's absence from Britain, attending to his plantation in Antigua. His departure allows for a temporary absence of moral restraint which is only restored by his timely return at the end of the novel. Said charts a complex cultural geography of overlapping territories and intertwined histories where, even in the apparently tranquil world of Jane Austen's novels, events 'over here' are crucially connected to distant events 'over there'.

The flow of people, information and goods has, of course, increased dramatically since the nineteenth century with the process of 'time–space compression' now characteristic of our post-modern world (Harvey 1989). In such circumstances of change and instability it should be no surprise that representing the 'national culture' is becoming evermore problematic. The world of advertising is a particularly good illustration of this tendency, with its need to boil down complex ideas into brief, concentrated messages. Long-established symbols of national identity no longer have unambiguously positive meanings. The Union flag, for example, has been endowed with right-wing associations, including some of the most virulent forms of exclusionary racism such as those associated with the National Front and the British National Party (Gilroy 1987). How, then, are more recent appropriations of the national flag to be interpreted? What does it signify about our 'national culture' when the Spice Girls drape themselves in the Union Jack or when Patsy Kensit and Liam Gallagher wrap it around themselves on the cover of *Vanity Fair*? Why, too, have British Airways dropped the flag from their airline livery in a multi-million pound refit, designed to help retain its position as 'the world's favourite airline' via a more outward-looking multicultural image, while the Labour Party included the flag in its *New Labour, New Britain, New Vision* electoral campaign in 1997, along with the British bulldog?

Other symbols of 'national unity' have also been refigured in recent years. The Royal Family, for example, no longer evokes an unequivocal sense of national loyalty. Despite the nationwide outpouring of grief over the death of Diana, Princess of Wales, in 1997, revelations of Prince Charles's adultery, together with the Queen's reluctance to pay income tax or to curb expenditure on 'luxuries' such as replacing the royal yacht, have all damaged the credibility of the Royal Family (though there is as yet no serious discussion within Britain of an alternative to the monarchy).

Imagining the nation is now much more complex and cannot be represented in unambiguously 'heroic' terms. One response has been the emergence of a nostalgic concern for 'national heritage', accompanied by the proliferation of industrial museums, heritage sites and commercial ventures such as the *Past Times* chain of high-street stores – a romanticisation of the past that is inevitably associated with a sense of national economic decline (Wright 1985). Another response has been the emergence of more critical 'visions of

Britain'. In the cinema, for example, such a view can be traced back at least to Steven Frears's bleak representation of the inner city in *Sammy and Rosie Get Laid* (1987). But it is now a much more widespread and acceptable way of seeing, with images of urban decline now a central motif in some of the most popular British films such as *Trainspotting*, *Twin Town* and *The Full Monty*, set respectively in Edinburgh, Swansea and Sheffield.

The sociologist Anthony Giddens (1991) refers to the weakening of long-established sources of identity as the advent of 'post-traditional' society, characterised by a plurality of new sources of identification. Traditional class-based identities, for example, rooted in stable working-class communities, with the expectation of a job-for-life, clear gender divisions and a relative lack of social and spatial mobility, have been thoroughly disrupted by the process of industrial restructuring and the move to more flexible modes of capital accumulation. New sources of identity have emerged, associated with the politics of health and the body, gender and sexuality, environmentalism, nationalism and a host of other social movements.

In many cases, these new forms of politics have come to be associated with particular local geographies or 'sites of struggle'. Think, for example, of the way that 'race' politics in Britain were shaped by the Brixton riots, or how recent developments within the women's movement came to be associated with the protests at Greenham Common. Similarly, the development of environmental politics using various forms of non-violent direct action have come to be associated with the protests over live animal exports at Brightlingsea and with opposition to the construction of the Newbury bypass or the extension of the runway at Manchester airport.

For some activists on the Left, these multiple forms of 'identity politics' are, at best, a mixed blessing. At worst, they are interpreted as a threat to traditional forms of class-based solidarity. Writing in the *New Left Review*, for example, Eric Hobsbawm (1996) contrasts the particularism of 'identity politics' and their appeal to sectional interests of gender, ethnicity and sexuality, with the alleged universalism of class-based politics. While it could be argued that class politics in the UK were far from 'universal' in their very partial incorporation of women and ethnic minorities, for example, Hobsbawm warns that identity politics will never be able to mobilise more than a minority of the population. Others have taken a more optimistic view of the 'New Times' (Hall and Jacques 1989), arguing that new social movements have the potential to include many of those who were excluded from traditional forms of labourism.

Commodifying cultural difference

One of the most significant responses to the proliferation of cultural difference has been the attempt to commodify the process, to exploit its exchange value within the marketplace. For some observers this is a cynical move within contemporary 'consumer culture', designed to capitalise on difference, using ethnicity, for example, as a kind of 'spice' to liven up the dull dish of mainstream white culture (hooks 1992). From this perspective, the commodification of difference threatens to undermine the 'authenticity' of autonomous cultural production and dull the radical edge of oppositional cultures. What happens, such critics ask, when Malcolm X's political vision is appropriated by Spike Lee and transformed into a range of merchandise, from tee shirts to baseball caps, designed to promote a major motion picture? Does it matter that the movie-maker is African-American or that the tee shirts and baseball caps are worn by teenagers across the world, many of whom have only a limited understanding of Malcolm X's politics?

It is certainly possible to think of ways in which 'the united colours of capitalism' (Mitchell 1993) lead to a diminution of cultural autonomy and involve only the most superficial commitment to multiculturalism. But notions of 'authenticity' are of questionable significance if we accept the earlier arguments about the increasingly 'hybrid' character of contemporary culture. Rather than searching nostalgically for the *roots* of a lost 'authenticity', geographers are now seeking to explore the *routes* of different forms of cultural production and their associated 'displacements' (Crang 1996). In the previous example, this would mean examining the way that Malcolm X's message was transformed as it moved through different media, being adopted and deployed in different contexts and by different groups, all of whom are situated differently in terms of the social relations of 'race', class and gender. A 'horizontal' logic of connection and differentiation comes to replace the 'vertical' logic of depth and authenticity that has dogged cultural studies for so long.

To pursue these ideas, let us examine a couple of examples in more detail: the first concerning the commodification of 'youth culture' in the Manchester club scene during the 1980s; the second concerning the commodification of contemporary masculinities through the growth of the UK's men's magazine market.

The 'Madchester scene'

Following the post-war decline of manufacturing employment, Manchester (along with many other northern industrial cities) made a concerted effort to reposition itself to take advantage of the expansion of job opportunities in the service industries (retailing, financial services, tourism, etc). Besides the attraction of external economic investment, this has also involved an attempt to raise the city's cultural profile through a series of initiatives, including a bid to host the Olympic Games, the construction of an exhibition centre (the G-Mex), an arts centre (the Cornerhouse) and a new concert hall (the Bridgewater Hall). Another aspect of this 'cultural renaissance' was the development of the 'Madchester' club scene, built on the success of local 'independent' bands like The Smiths, Inspiral Carpets, the Stone Roses, the Charlatans and Oasis. Commenting on the distinctiveness of the city's independent music scene, Halfacree and Kitchin provide this (highly contentious) chronology:

> First, in the late 1970s and early 1980s, there was the distinctive post-punk sound of Joy Division, which mutated into the slightly more cheerful tones of New Order. Bedsit blues returned in the mid 1980s with The Smiths and James, whilst the tempo and mood was revived around 1988, in the wake of 'Acid House', with the arrival of the club-and-Ecstasy sounds of 'Madchester' led by The Happy Mondays, The Stone Roses and Oldham's Inspiral Carpets . . . Madchester was pronounced 'dead' by 1991–2.
>
> (Halfacree and Kitchin 1996: 50–3)

As Ian Taylor's recent study has shown, however, the transformation in Manchester's cultural life resulted not only from 'the raw populism of the music and the creative energy of a series of local bands' but also from the efforts of the Central Manchester Development

FIGURE 14.2 Canal Street: 'before' and 'after' reinvestment in Manchester's Gay Village
Source: Peter Jackson.

Corporation to rescue city-centre warehouses and other commercial buildings through a programme of tax incentives and grant aid (Taylor *et al.* 1996: 272). Among these redevelopments, the Hacienda club was created from an old industrial building on Whitworth Street, and Affleck's Palace, a multi-storey mecca for teenage consumers, was created on the site of a former department store. Meanwhile, numerous record companies have sought to emulate the success of Factory Records, established by local entrepreneur Tony Wilson in 1978.

Equally significant (though often neglected) in Manchester's cultural revival has been the development of the city's Gay Village in another old warehouse district along the canalside (Figure 14.2). While there have been gay venues in Manchester for many years (notably the Rembrandt and the Union pubs), the liberalisation of sexual attitudes in the 1960s (following the Wolfenden Report and the Sexual Offences Act of 1967 which decriminalised same-sex relations among consenting adult men in private) provided the opportunity for more open expressions of sexual difference. The area that became known as the Gay Village had a history of prostitution which may have deterred 'mainstream' developers from reinvesting in the area. It emerged as a publicly gay space in the 1980s and even then faced opposition from a number of quarters. Chief Constable James Anderton adopted a high-profile offensive against public expressions of 'homosexuality', including much-publicised raids on gay venues such as Napoleon's and Rockie's nightclubs (leading, in one case, to a prosecution for 'licentious dancing'). Raids continued after Anderton's resignation in 1991. Meanwhile, opposition to the Conservative Party's championing of 'family values', mobilisation around 'safer sex' and the HIV-AIDS epidemic and resistance to the government's attempt to prohibit the public promotion of 'gay lifestyles' (under Section 28 of the Local Government Act, 1988) all served to increase support for gay rights.

But it was the economic potential of the 'pink pound', coupled with support from the Central Manchester Development Corporation, that provided the impetus for the area's commercial development as a Gay Village. Canal Street and the surrounding area now boasts at least fifteen gay or mixed pubs and clubs, three cafés and a range of other services aimed primarily at young gay consumers. Commercial success has had its downside, as well as its more celebratory aspect. Some have argued that Manchester's gay 'scene', targeting young, high-spending gay consumers, is a form of discrimination towards older, poorer gay men, driven by commercial motives and supported politically by narrow interest-group considerations (Taylor *et al.*, 1996: 189).

Others have also capitalised on the commercial success of the 'Madchester' phenomenon. Among these, Shami Ahmed probably best embodies the 'rags-to-riches' dream of many immigrant entrepreneurs. Arriving from Pakistan in 1964, Mr Ahmed worked on his parents' market stall in Burnley before setting up his own company to manufacture and market jeans under the very 'English' name of Joe Bloggs. His business is now estimated to be worth some £50 million, making him Britain's thirteenth richest Asian businessman. Significantly, however, his company trades not on Mr Ahmed's 'ethnic' identity but on the associations of his product with the city's club scene, emphasising youth culture, sports and leisure. 'Throughout the world, people know about Manchester,' Mr Ahmed claims, 'they know Manchester United and Joe Bloggs' (*Guardian*, 29 September 1997).

The men's magazine market

If the 'Madchester scene' represents the successful commodification of youth culture, then the growth of the UK men's magazine market represents an equally successful commodification of contemporary gender anxieties. Until the 1980s, the only men's magazines were so-called soft pornographic titles like *Penthouse* and *Playboy*, or 'special interest' magazines that addressed their readers in terms of their specific interests in football or cars, fishing or photography. As recently as 1986, the conventional wisdom in the UK publishing industry was that 'men don't buy magazines' or, at least, that 'men don't define themselves as men in what they read, they define themselves as people who are into cars, who play golf, or fish . . . Successfully launching a general interest men's magazine would be like finding the holy grail' (*Campaign*, 29 August 1986). The initial breakthrough came with a smaller independent publisher, Wagadon, who were persuaded by Nick Logan to launch *Arena* in 1986, capitalising on his experience working on *The Face* and *Smash Hits* (Nixon 1996). The major publishers (like IPC, EMAP and Conde Nast) still regarded the sector with suspicion. Interviewed in *The Observer* in 1987, for example, Mark Boxer from Conde Nast argued that English men simply weren't ready for such a magazine: 'The only area where there might be a gap', he concluded, ' is for a style magazine – but in that area Englishmen are uneasy; they don't admit to taking fashion seriously' (26 April 1987).

Within ten years, however, the men's magazine market had been transformed, with the trade press describing it as 'The fastest growing of all consumer magazine markets [with] currently the highest profile' (*Keynote* 1996). *Loaded* ('for men who should know better') was voted 'magazine of the year' in 1995, reaching a circulation of over 200,000 in the first half of 1996. It was shortly overtaken by *FHM* (For Him Magazine) which now reportedly outsells *Cosmopolitan* (*Independent*, 15 August 1997). Several US men's magazines launched UK editions (*GQ* in 1988, *Esquire* in 1991 and *Men's Health* in 1995). They have been joined by several more specialised magazines such as *Xtreme* (focusing on 'extreme sports'), *Escape* (on the Internet) and *Attitude* (for gay men). Several other titles have 'spun off' from their successful 'parent' magazines (*GQ Active* from *GQ*, *Stuff for Men* from *Maxim*, *XL for Men* from *FHM*).

But what does the success of the magazines tell us about the UK's changing geography? To begin with, it is clear that the magazines are written by and for a very metropolitan audience. Most of the fashion stockists and restaurants listed in the magazines are concentrated in London and the South East. While the images of masculinity portrayed in the magazines might extend to Brighton or Bristol, even to Manchester and Leeds, cities like Sheffield or Sunderland hardly ever feature in their pages. There is little sense in the magazines of how men, across the UK, are responding to changing economic and political circumstances. Rather than interpreting these changes as a generalised 'crisis of masculinity' (Connell 1995), a more complex pattern of local dislocations and reconfigurations might be anticipated. Instead of seeing changing masculinities in terms of a comprehensive shift from 'new man' to 'new lad', as implied by current media stereotypes, most academic studies actually show that gender relations are changing quite slowly (Segal 1990).

Our own research on the production, content and readership of these magazines used focus groups to examine how different groups of men were attempting to 'make sense' of the changes that are taking place in contemporary gender relations and 'consumer culture'.[2] The research shows that men are currently facing a number of contradictions, being called on to be more 'natural', 'open' and 'honest' while simultaneously seeking reassurance for their insecurities and using the magazines as a form of cultural approval for ways of behaving that are sometimes regarded as offensive. Some readers spoke of the magazines as a response or 'backlash' to feminism, helping men feel easier about themselves. Others thought that there had been an opening up of the very narrow definitions of acceptable roles for men, but that the gap had soon narrowed down again as the market settled on more 'laddish' versions of masculinity.

While some readers regarded the magazines as a reaction against 'political correctness' or a response to what they called 'extreme' forms of feminism, others saw the magazines in purely commercial terms. Our research (Jackson *et al.* 1999) suggests that the men's magazines have succeeded commercially by commodifying men's gender anxieties. As men are encouraged to 'open up' previously repressed aspects of their masculinity (including health, fashion and relationships) they face an increasing sense of risk and anxiety about their gender and sexual identities. By providing advice and information in an entertaining format, often using an 'ironic' sense of humour, the magazines have succeeded in reassuring men, responding to their anxieties about job insecurity, health worries or the problems of establishing and maintaining a relationship. A recent MINTEL report on *British Lifestyles* showed that men's anxieties focused on money and health problems, with a third of 20 to 54-year-olds worried about losing their jobs within the next five years and nearly half anxious about their health. The same report showed that expenditure on treats and rewards, such as magazines, chocolates and jewellery, had increased significantly (*Guardian*, 12 February 1997). While the magazines signify the possibility of significant change in men's identities and gender relations, they are simultaneously reinscribing more traditional forms of masculinity (celebrating 'laddishness' as a legitimate response to the perceived excesses of feminism and political correctness). As with many of the examples in this chapter, a complex cultural politics is at work here that defies easy explanation.

The limits of cultural tolerance?

This chapter began with Lord Tebbit's recent speech on multiculturalism as an example of the 'denial of difference'. Before concluding, I want to consider some other indications of what might be called the limits of cultural tolerance. The 'Rushdie affair' provides a striking example where an individual's right to free speech came into conflict with the demands of a minority community (or of some groups within it) to respect its religious values. Following publication of *The Satanic Verses* in 1988, a *fatwa* was issued by the Ayatollah Khomeini calling for the author's execution as a religious duty because of the novel's alleged blaspheming of Islam. The situation was immediately complicated by the British legal system which only recognises blasphemy in relation to Christianity. Forced into hiding, Salman Rushdie has frequently criticised the Foreign Office's reluctance to pursue his case with the Iranian authorities, suggesting that his earlier public involvement in anti-racist politics had done little to endear him to the British establishment.[3]

Some observers saw the *fatwa* as evidence of the growing strength of Islamic 'fundamentalism', tapping into deep veins of Orientalist prejudice (Said 1978) that had been reinvigorated by Britain's recent involvement in the Gulf War. Reactions to the 'Rushdie affair' were further complicated by the complex local politics of Britain's Muslim communities in Bradford, Tower Hamlets and elsewhere, providing activists with an opportunity to articulate a variety of grievances. In Bradford, for example, earlier events, such as the provocative remarks about multicultural education made by a local head teacher, Ray Honeyford, in 1984 had heightened tensions with many Muslim residents which were compounded when the two-year dispute ended in the teacher accepting a generous early retirement deal. In Bradford and elsewhere, the 'Rushdie affair' also provided an opportunity for 'traditionalists' within the Muslim community to reassert their authority over those (including many Muslim women) who had begun to challenge their power.

A second example of an apparently intractable form of cultural difference is provided by the (contested) right of the Orangemen in Northern Ireland to parade along their 'traditional' routes, even where these pass through predominantly Catholic neighbourhoods and where they are vehemently opposed by local residents. Unionists trace the history of parading back for 300 years, celebrating the victory of William, Prince of Orange, at the Battle of the Boyne in 1690 which guaranteed the Protestant ascendancy to the British throne. As with all such 'invented traditions' (Hobsbawm and Ranger 1983), its significance and continuity can be disputed. Though parading in Northern Ireland certainly has a long history, evidence of its declining popularity during the eighteenth and nineteenth centuries is often conveniently ignored. Following the signing of the Anglo-Irish Agreement in 1985, interpreted by many Unionists as a weakening of their constitutional position and a threat to their sense of national identity, parading experienced a revival in popularity.

In cities where residential segregation is intense (signified by the painting of curbstones and murals, the establishment of 'peace walls' and formal and informal means of surveillance), parading has become 'the most prominent means of asserting collective identities and claiming political dominance over territory' (Jarman 1997: 79). While Nationalist parades are represented as a threat to public order, Loyalist parades (which are nearly ten times more numerous) are seen as a key element of Unionist culture and tradition, to be defended as an inalienable civil right. The parading season lasts from Easter to the end of August. Within this period, the Twelfth of July parades in Belfast have become a pivotal event, attracting an estimated 100,000 people. The parades are noisy and spectacular events dominated by aggressive 'blood and thunder' bands (Figure 14.3). While some participants wear the 'respectable' regalia of an Orange sash and bowler hat, others carry the insignia of Loyalist paramilitary groups such as the Ulster Defence Association and the Ulster Volunteer Force, as well as more traditional banners depicting William of Orange and other Protestant heroes. The route of Belfast's Twelfth of July parade extends for six miles from Carlisle Circus, up the Lisburn Road (following the route that 'King Billy' is supposed to have taken from Belfast in 1690), down to the Lagan valley and along country lanes out to the Field. By the end of the day, as Neil Jarman argues in his recent study of parades and visual displays in Northern Ireland, 'the city is claimed by the Orangemen as theirs and theirs alone' (1997: 102).

Nationalists regard the parades as a triumphalist display of Loyalist superiority, symbolising their resistance to the changing social geography of Northern Ireland by their insistence on marching through areas that have become solidly Roman Catholic (such as the Apprentice Boys' parading around the walls of Derry). Major disputes occurred in 1995 and 1996 along the Garvaghy Road in Portadown and in Belfast's Ormeau Road. Attempts to modify the parade routes or to request the consent of residents to traverse their neighbourhoods, have been greeted with hostility by Loyalists, leading to weeks of

FIGURE 14.3a Showing faith, Belfast, July 1991
Source: Neil Jarman (1997) (reproduced with permission from *Material Conflicts*, Oxford: Berg).

FIGURE 14.3b On the way home, Sandy Row, Belfast, the Twelfth, 1994
Source: Neil Jarman (1997) (reproduced with permission from *Material Conflicts*, Oxford: Berg).

confrontation, road-blocks and rioting. Some Unionists have interpreted any curtailment of their traditional rights as a capitulation to the IRA, an infringement of their civil liberties and a threat to the Union.

Violent confrontation recurred in July 1998 following a ruling by the Northern Ireland Parades Commission, banning the Orangemen from marching down the Garvaghy Road. As armed police and soldiers dug trenches and erected barbed-wire fences, the 'standoff at Drumcree' threatened to undermine the entire peace process inaugurated earlier in the year by the Good Friday Agreement on constitutional change

One final example provides further evidence that 'multiculturalism' is rarely an easy option. Living with difference demands mutual respect and a tolerance for other cultures that may sometimes present quite daunting challenges. The example concerns the recurrent battle within the UK over the state-funding of separate Muslim schools. Several thousand Church of England, Roman Catholic and Jewish schools have received financial support from the state while attempts to secure state funding for separate Muslim schools have consistently been thwarted. Some state schools have provided Islamic teaching as part of their religious studies programme and *halal* meat is available in some school canteens. Compromises have also been reached over the provision of places in single-sex schools for Muslim pupils and over the development of appropriate dress codes (though these are occasionally contested).

While Muslim schools could, in theory, be established under existing legislation regarding the funding of denominational voluntary-aided schools, efforts to establish such schools in Batley (West Yorkshire), Brent (West London) and elsewhere faced constant opposition. As Claire Dwyer's (1993) research has shown, the Islamia school in Brent became a national symbol for these struggles. An initial application for state funding in 1986, supported by Brent Council, was turned down by the Department of Education and Science (DES) because the school was considered too small to be viable. Subsequent attempts to expand the school were blocked by the Brent Planning Sub-Committee. An appeal to the DES in 1988 was rejected because of the availability of surplus places in neighbouring schools and because of the withdrawal of Council support, this now being seen to contravene their policy on multicultural education. The application was formally rejected by the Secretary of State for Education in 1990. Following a judicial review, however, the High Court ruled that there was 'manifest unfairness' in the DES's decision. Meanwhile, a new Jewish voluntary-aided school was opened in Redbridge and grant-maintained status was awarded to a neighbouring (non-Muslim) school in Brent. Eventually, in January 1998, two Muslim primary schools (the Islamia school in Brent and the Al Furqan school in Sparkhill, Birmingham) were awarded grant-maintained status, the Education Secretary David Blunkett conceding that 'This is an issue of fairness and ensuring good community relations' (*Guardian*, 10 January 1998).

Conclusion

The evidence reviewed in this chapter suggests that, as we look towards the new millennium, there are several positive signs of growing tolerance towards cultural difference, informed by new ways of understanding 'multiculturalism'. In the case of food and music, for example, 'hybrid' identities are now common features of 'British' culture. Reflecting an increasingly complex cultural politics, representations of the nation have also become more complex whether one takes the example of the national flag or the Royal Family, and whether one looks at British cinema or television advertising.

Examples of intolerance remain all too frequent, however, and serve as a reminder that religious, racial and sexual tolerance have to be struggled for. Approaching the UK's changing cultural geography in a critical way, via concepts of hybridity and difference, we have argued that the notion of 'multiculturalism' applies as much to dominant groups as to minorities. Radically destabilising our understanding of 'Britishness' and unsettling established notions of the 'national culture' provides one way forward in meeting the difficult challenges of 'living with difference'.

Notes

1 Media estimates place the size of the Muslim population in Britain at between one and two million. The social geographer Ceri Peach has contested this view, arguing that the actual figure is between 550,000 and 750,000 (Peach 1990).

2 The research was funded by ESRC under grant number R000221838 and was undertaken in collaboration with Nick Stevenson and Kate Brooks at Sheffield University.

3 Among other interventions, Rushdie had argued that 'racism is not a side-issue in contemporary Britain; . . . it's not a peripheral minority affair. . . . It's a crisis of the whole culture' (Rushdie 1991: 129).

References

Back, L. (1996) *New Ethnicities and Youth Culture: Racisms and Multiculture in Young Lives*, London: UCL Press.

Bradford Heritage Recording Unit (1994) *Here to Stay: Bradford's South Asian Communities*, Bradford: Bradford Heritage Recording Unit.

Connell, R.W. (1995) *Masculinities*, Cambridge: Polity Press.

Cook, I. and Crang, P. (1996) 'The world on a plate: culinary culture, displacement and geographical knowledges', *Journal of Material Culture* 1: 131–53.

Crang, P. (1996) 'Displacement, consumption, and identity', *Environment and Planning A* 28: 47–67.

Crang, P. and Jackson, P. (in press) 'Consuming geographies', in K. Robins and D. Morley (eds) *British Cultural Studies*, Oxford: Oxford University Press.

Dwyer, C. (1993) 'Constructions of Muslim identity and the contesting of power: the debate over Muslim schools in the United Kingdom', in P. Jackson and J. Penrose (eds) *Constructions of Race, Place and Nation*, 143–59, London: UCL Press.

Eliot, T.S. (1948) *Notes Towards the Definition of Culture*, London: Faber & Faber.

Giddens, A. (1991) *Modernity and Self-identity*, Cambridge: Polity Press.

Gilroy, P. (1987) *There Ain't No Black in the Union Jack*, London: Hutchinson.

Halfacree, K.H. and Kitchin, R.M. (1996) '"Madchester Rave On": placing the fragments of popular music', *Area* 28: 47–55.

Hall, S. (1991) 'Old and new identities, old and new ethnicities', in A.D. King (ed.) *Culture, Globalisation and the World System*, 41–68, London: Macmillan.

Hall, S. and Jacques, M. (eds) (1989) *New Times: the Changing Face of Politics in the 1990s*, London: Lawrence and Wishart.

Harvey, D. (1989) *The Condition of Postmodernity*, Oxford: Basil Blackwell.

Hobsbawm, E. (1996) 'Identity politics and the Left', *New Left Review* 217, 38–47.

Hobsbawm, E. and Ranger, T. (eds) (1983) *The Invention of Tradition*, Cambridge: Cambridge University Press.

Holmes, C. (1991) *A Tolerant Country?*, London: Faber and Faber.

hooks, b. (1992) 'Eating the other', in *Black Looks: Race and Representation*, 21–39, London: Turnaround.

Jackson, P. (1988) 'Beneath the headlines: racism and reaction in contemporary Britain', *Geography* 73, 202–7.

Jackson, P. (1989) *Maps of Meaning*, London: Unwin Hyman.

Jackson, P., Stevenson, N. and Brooks, K. (in press) 'Making sense of men's lifestyle magazines', *Environment and Planning D: Society and Space*.

Jarman, N. (1997) *Material Conflicts: Parades and Visual Displays in Northern Ireland*, Oxford: Berg.

Jeater, D. (1992) 'Roast beef and reggae music: the passing of whiteness', *New Formations* 18, 107–21.

Kureishi, H. (1986) 'Bradford', *Granta* 20: 149–70.

McRobbie, A. (1994) 'Folk devils fight back', *New Left Review* 203: 107–16.

May, J. (1996) '"A little taste of something more exotic": the imaginative geographies of everyday life', *Geography* 81: 57–64.

Mitchell, K. (1993) 'Multiculturalism, or the united colours of capitalism?', *Antipode* 25, 263–94.

Nixon, S. (1996) *Hard Looks: Masculinities, Spectatorship and Contemporary Consumption*, London: UCL Press.

Peach, C. (1990) 'The Muslim population of Great Britain', *Ethnic and Racial Studies* 13: 414–19.

Rushdie, S. (1991) *Imaginary Homelands*, London: Granta.

Scarman, Lord (1981) *The Brixton Disorders of 10–12 April 1981*, Cmnd 8427, London: HMSO.

Said, E. (1978) *Orientalism*, New York: Pantheon Books.

Said, E. (1993) *Culture and Imperialism*, London: Chatto and Windus.

Sarup, M. (1986) *The Politics of Multi-racial Education*, London: Routledge and Kegan Paul.

Segal, L. (1990) *Slow Motion: Changing Masculinities, Changing Men*, London: Virago.

Taylor, I., Evans, K. and Fraser, P. (1996) *A Tale of Two Cities: Global Change, Local Feeling and Everyday Life in the North of England: a Study in Manchester and Sheffield*, London: Routledge.

Wright, P. (1985) *On Living in an Old Country*, London: Verso.

Further reading

The best introductions to the 'cultural turn' in human geography include P. Shurmer Smith and K. Hannam (1994) *Worlds of Desire, Realms of Power* (London: Arnold); M. Crang (1998) *Cultural Geography* (London: Routledge); and P. Jackson (1989) *Maps of Meaning* (London: Unwin Hyman (reprinted in 1992 by Routledge)). Up-to-date statistics on cultural issues are hard to find, but *Social Trends* and *Regional Trends* (HMSO) are valuable sources, along with trade reports by marketing organisations such as *Mintel* and *Keynote*. C. Smith (1998) *Creative Britain* (London: Faber), provides a critique of the recent media hype about 'Cool Britannia'. For place-based case studies informed by current social theory, see D. Sibley (1995) *Geographies of Exclusion* (London: Routledge) and T. Cresswell (1996) *In Place, Out of Place* (Minnesota: University of Minnesota Press). The changing geography of contemporary youth culture is charted in T. Skelton and G. Valentine (1998) *Cool Places: Geographies of Youth Cultures* (London: Routledge). For an introduction to issues of 'race' and racism see J. Solomos (1993) *Race and Racism in Britain*, 2nd edition (London: Macmillan). Recent studies of gender and sexual difference include the Women and Geography Study Group (1997) *Feminist Geographies* (London: Longman) and D. Bell and G. Valentine (1995) *Mapping Desire* (London: Routledge). There are fewer studies of the geography of religion, but see L. Kong (1990) 'Geography and religion: trends and prospects', *Progress in Human Geography* 14: 355–71. Finally, for an introduction to the geography of consumption, see D. Bell and G. Valentine (1997) *Consuming Geographies: You Are Where You Eat* (London: Routledge) and the material reviewed in P. Jackson and N. Thrift (1995) 'Geographies of consumption', in D. Miller (ed.) *Acknowledging Consumption* (London: Routledge).

Chapter 15

Local government and governance

Joe Painter

Introduction

All modern societies are governed and over the past 300 years (and especially since the nineteenth century) the *territorial nation-state* has developed as the dominant form of government to the point where almost the entire land surface of the globe is divided politically into a mosaic of states. The dominance of the nation-state has recently been challenged by processes of global economic and political transformation, and some commentators predict that the conventional nation-state will lose its pre-eminence as the basic building block of the world political map. Nevertheless, for the time being, the state remains as the most important structure of government for most purposes. On the other hand, only the tiniest of nation-states are able to carry out the numerous functions and processes of government through a single, national, set of institutions. The territories and populations of almost all countries are large enough to justify, or even to require, one or more additional tiers of government and administration organised at spatial scales below that of the nation-state. The form and function of these different tiers and the relationships between them can vary widely from country to country. In the United Kingdom a more or less uniform two-tier structure of formal local government was in place from the mid-1970s to the mid-1980s. Since then a much more complex picture has emerged, with a number of places acquiring a new single-tier system of local government. In addition new devolved regional governments are being established in Scotland, Wales and Northern Ireland (albeit different in each case), while the English regions have seen a degree of administrative decentralisation with the setting-up of government offices for the regions with responsibility for the organisation of a range of central government services at a regional scale (Figure 15.1).

Further complexity is added if we broaden the focus to consider local governance and politics as a whole rather than limiting discussion to the system of formal, elected local councils. Here we need to consider a range of unelected public agencies, voluntary

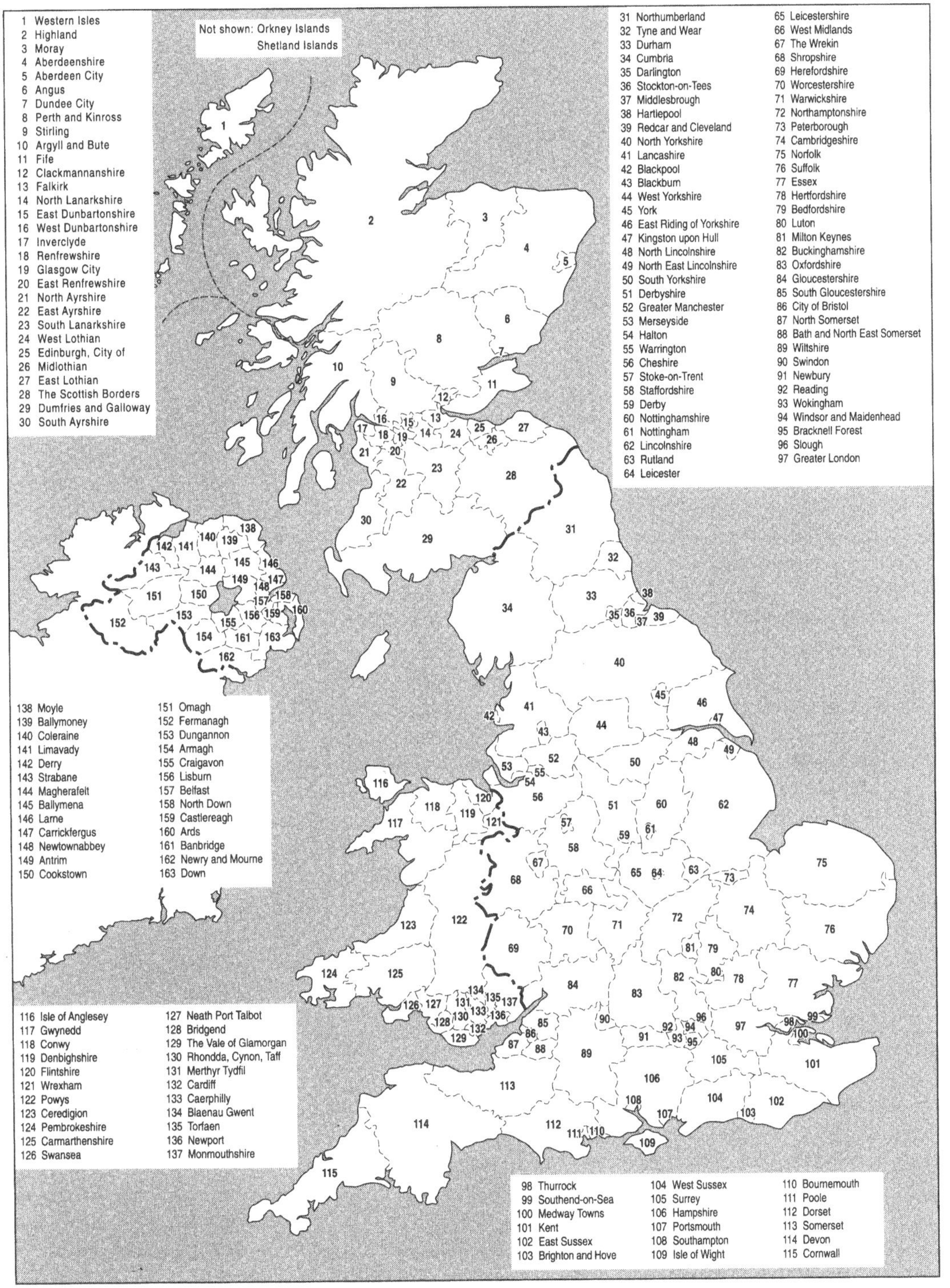

FIGURE 15.1 The 'top tier' of local government in the UK from April 1998: counties (including former metropolitan counties and Greater London) and new unitary authorities

organisations, businesses, and individuals and groups of people who have a role in the formulation and implementation of local policies and the organisation and delivery of local services, as well as the ongoing processes of political conflict, co-operation, campaigning and decision-making. In this chapter I will use the following definitions for these various terms:

- *Local government* refers to elected local councils comprising county and district councils and the new unitary (single-tier) councils where those exist. Local councils are also known as local authorities.
- *Local governance* is the process of the formation and implementation of public policy at the local level involving both elected and non-elected organisations.
- *Regional government* refers to the newly developing Scottish and Northern Irish governments and Welsh Assembly.
- *Regional governance* is the process of the formation and implementation of public policy at the regional level involving both elected and non-elected organisations.
- *Local politics* and *regional politics* refer to processes of conflict, co-operation, agenda-setting and decision-making at the local and regional scales respectively, over the nature, scope and content of public policy, particularly, though not necessarily exclusively, as these affect the locality or region concerned.

The focus of this chapter is on governance at the local, rather than the regional, scale.

Elected local government

Although local governance today involves many agencies and organisations in addition to elected local councils, and while local councils have lost many of their former powers, the formal system of elected local government remains the single most important element in the geography of local governance in the United Kingdom. The scale of the system can be seen from some simple statistics (Central Office of Information 1996). Between 1984 and 1994, local authorities together were responsible for, on average, 25.39 per cent of general government expenditure in Britain. In the early 1990s English local authorities were spending about £45 billion a year on providing services such as education, policing, social services and transport, and employing almost 2 million people (1.4 million full-time equivalents) to deliver them. Local government, in other words, is big business. At the same time, the discretion of local councils over the way in which their resources are allocated has been significantly eroded since the mid-1970s. An extensive programme of national legislation has been enacted to regulate and constrain the ability of councils to decide for themselves how to respond to local needs and demands. There has also been a decline in public interest in local government and local elections that may be related to this decline in local autonomy. According to the government's recent Local Government White Paper, on average, in recent elections only about 40 per cent of local electors voted in local council elections compared with 60 per cent in Portugal, 72 per cent in Germany and 80 per cent in Denmark; in the elections of May 1988, turnout reached record low levels in many areas. New government proposals in the 'White Paper' of July 1998 aim to redress some of these problems (Department of Environment, Transport and the Regions 1998).

The development of elected local government

A brief history

The origins of local self-government in Britain are ancient. From the twelfth century cities and certain towns (called boroughs or burghs) were given the right to manage their own affairs by royal charter. The system failed to keep pace with social change however, and by the early nineteenth century some ancient boroughs were virtually depopulated but still retained political privileges, such as the right to elect members of parliament (these became known as the 'rotten boroughs'), while other settlements had expanded substantially in the wake of the Industrial Revolution but did not have the privileges of borough status. The shire counties are also of ancient origin and provided a framework for military defence and the justice system. The smallest scale of local government was the parish, which had responsibility for the relief of poverty (albeit on a very limited scale by modern standards).

The nineteenth century saw a series of major reforms to the local government system to do away with the 'rotten boroughs', to harmonise structures and to introduce new local services. In 1835 the Municipal Corporations Act established directly elected corporate boroughs in place of the self-electing medieval ones. A series of new specialist local boards was established. The Poor Law Amendment Act 1834 set up Boards of Guardians to oversee local poor relief. The Public Health Act 1848 established local health boards, while highways boards and elementary school boards were established by the Highways Acts of 1835 and 1862 and the Education Act of 1870, respectively. The latter, together with the Scottish Education Act of 1872, introduced universal elementary education into Britain for the first time.

Then at the end of the nineteenth century, three Acts introduced a new system of local government to England and Wales that was to serve for eighty years. The Local Government Act 1888 set up sixty-two elected County Councils, including the London County Council, and sixty-one all-purpose County Boroughs. The Local Government Act 1894 revived parish councils and established a middle tier (between parish and county) of 535 Urban District Councils, 472 Rural District Councils and 270 non-county Borough Councils. The London Government Act 1899 set up twenty-eight Metropolitan Borough Councils in London and the Corporation of London in the tiny City of London itself. A parallel series of reforms in Scotland produced a similar but not identical system. Following the Local Government (Scotland) Act 1929 local government consisted of two tiers. The upper tier had thirty-three county councils and four 'counties of cities' (like the English County Boroughs) in Edinburgh, Glasgow, Dundee and Aberdeen. The lower tier comprised twenty-one Large Burghs, 176 Small Burghs and, in the rural areas, 196 District Councils. Parish councils were abolished as civil authorities.

By the 1960s this system too had become unwieldy and was seen as too complex and inefficient. In 1974 in England and Wales and 1975 in Scotland a new, and somewhat simplified, system of local government was put in place. Everywhere (with the sole exception of the Scottish Islands) was now covered by two main tiers. In Scotland, in addition to the three island councils, there were nine large regional councils subdivided into fifty-three districts. In England and Wales there were six metropolitan county councils, covering the major conurbations, divided into thirty-six metropolitan districts; a Greater London Council (which had been introduced in 1965) and thirty-two London boroughs plus the City

TABLE 15.1 The structure of UK local government, mid-1970s to mid-1980s

	London	*Major English conurbations*	*Rest of England and Wales*	*Islands of Scotland*	*Rest of Scotland*	*Northern Ireland*
First tier	Greater London Council	6 Metropolitan County Councils (West Midlands, Merseyside, Greater Manchester, Tyne and Wear, South Yorkshire, West Yorkshire)	47 Non-Metropolitan County Councils	3 Island Councils	9 Regional Councils	Many 'first tier' functions undertaken by the Northern Ireland Office – a department of UK central government, or through 9 area boards accountable to government ministers
Second tier	32 London Boroughs + City of London	36 Metropolitan District Councils	333 Non-Metropolitan District Councils	n/a	53 District Councils	26 District Councils
Third tier	n/a	Parish Meetings or Councils	Parish or Community Meetings or Councils	Community Councils	Community Councils (optional)	n/a

of London Corporation in the 'square mile'; and forty-seven non-metropolitan county councils divided into 333 non-metropolitan district councils (see Table 15.1).

Current characteristics

Commenting on the key themes of the UK local government system as it stood in the early 1990s, David Wilson and Chris Game (1994) identified seven defining characteristics. First, it is a system of local government, not local administration. In other words, the system involves decentralisation of not only administrative authority, but also of political authority:

> Local authorities are far more than simply outposts or agents of central government, delivering services in ways and to standards laid down in detail at national ministerial level. Their role, as representative bodies elected by their fellow citizens, is to take such decisions themselves, in accordance with their own policy priorities: to govern their locality.
>
> (Wilson and Game 1994: 20)

Second, and on the other hand, the UK does not have local self-government. In contrast with community governments in some other countries in Europe (such as Switzerland), UK local authorities do not possess a 'power of general competence'; that is, the ability to act in any way that they feel is necessary to promote local interests or to serve local populations, unless the action proposed is specifically forbidden. In the UK local councils can only do those things they are explicitly permitted to do under law. Other actions are considered *ultra vires* (beyond the powers). Third, and related to this, UK local government is wholly subordinate in law to the national government and parliament in London. In so far as local authorities have any decentralised authority at all, they have it because parliament has allocated it to them. Local authorities are created by parliament and can be abolished by it. Fourth, this situation leads Wilson and Game to identify another characteristic: partial autonomy. Although central government has the theoretical right to govern all aspects of local authority activity in detail, if it did this in practice the purpose of the system of local government would have disappeared: it would have become local administration. The actual history of central–local relations in the UK, therefore, has been one of a (sometimes uneasy and disputed) balance of powers between local authorities and the national government, giving rise to a situation of 'partial autonomy'. Fifth, local government gains this limited autonomy mainly from the fact that it is directly elected. Elected councillors are thus accountable to their constituents, and have a degree of legitimacy not enjoyed by appointed members of health authorities or other local bodies. Sixth, local councils are multi-service organisations responsible for dozens of local services from adult education to zebra crossings. Seventh, and finally, local councils are also multi-functional organisations. Perhaps the best known is 'direct service provider' in which the council undertakes all the activities associated with providing a particular service itself. Other important roles include:

- *regulating* other organisations and individuals (such as licensing cinemas and taxi drivers);
- *facilitating* activities undertaken by others, such as economic development;

- *contracting* with other agencies or private companies to provide services not undertaken directly by council staff.

In addition to all of these characteristics, Wilson and Game note that local councils possess one further vital feature: the right to raise taxes. Although this right has been much reduced, with most local government finance now provided by central government and local authorities tightly circumscribed by law in their ability to increase local taxes above centrally set limits, the power to tax remains in principle an important element of the ability of local government to exercise its partial autonomy and to tailor its activities to local circumstances.

The functions of elected local government

Elected local government is responsible for an extremely wide range of local public services. Many of these are mundane (such as refuse collection) but undeniably essential. Others, such as education, are equally important, but rather more complex and expensive. Indeed school education is the single biggest element of local government budgets. Increasingly, elected local authorities are sharing the provision of services with a range of other private and voluntary sector organisations, using a number of different contractual and partnership arrangements to try to ensure that local needs are met. There are a number of different ways of classifying local government functions. One way is to consider the different services that are provided by the different tiers of local government. This classification has inevitably become less clear-cut as the universal two-tier system described above has given way to a hybrid system in which some areas now have single-tier ('unitary') local authorities. Nevertheless, in the many areas which still have two main tiers there is some logic to the distribution of functions between them:

> Broadly, the functions are allocated between the two main tiers on the basis of operational efficiency and cost-effectiveness. Thus it is both cheaper and more effective to have responsibilities such as major planning, police and fire services operated over fairly large areas. Similarly it was strongly argued by the Redcliffe-Maud Report and the Wheatley Report [two major 1960s government reports on local government in England and Scotland respectively] that the education service should be administered over an area containing at least 200,000–250,000 people in order that such authorities should have at their disposal 'the range and calibre of staff, and the technical and financial resources necessary for effective provision' of the service. On the other hand services such as allotments, public health and amenities can be effectively administered over smaller areas, and are thus the responsibility of district authorities.
>
> (Byrne 1992: 59)

In the light of the government reports mentioned by Byrne, with their stress on economies of scale in the provision of education, it is ironic that one of the major local government reforms of the 1979–97 Conservative governments was to introduce 'opted-out' schools in which some individual schools were given independence from local government control (see p. 312). Moreover, the two-tier system was never organised completely rationally. Factors such as tradition, democratic control and political influence also affect which services are provided at which scale (Byrne 1992: 59).

An alternative way of classifying services and functions is to recognise four categories (Wilson and Game 1994):

1. Need services which are provided for all regardless of the ability to pay. They involve redistribution of resources within the community.
2. Protective services provided for all to national standards.
3. Amenity services provided to locally determined standards to meet the needs of each community.
4. Facility services that people can use if they wish.

Table 15.2 shows the various services provided by local government between these four categories and their relative scale in financial terms for the financial year 1990/1 in England and Wales. Between them, the three biggest services in financial terms, education, social services and the police, account for very nearly three-quarters of all local government expenditure. These services are also among the most controversial and important public services in the UK, and it is therefore hardly surprising that local government has become such a political hot potato.

Another key aspect of the political controversy that has surrounded local government in recent years is its financing. A large proportion of the bill for local government is actually met by central government. There is considerable logic to this arrangement. The geography of wealth and income in the UK, as in most other countries, is very uneven. If local councils had to raise all their resources locally there would be dramatic disparities between the ability of councils to finance their activities and the demands placed on them by users of council services. Simply put, wealthier areas are better able to pay for local government than poorer areas, but poorer areas make much greater demands on local public services. Therefore, to ensure that local government in poorer areas is able to function effectively a system of redistribution is required. In the UK this is provided by central government in the form of grants to local councils financed out of general national taxation. During the 1970s central government provided 60–65 per cent of the financial resources for local government in the form of two main types of grant: a general grant called the Rate Support Grant and special grants for specific activities (Service Specific Grants). The remaining 35–40 per cent of the cost of local government was met by local taxation.

Until 1989 in Scotland and 1990 in England and Wales, the local element of local government finance was collected through a local property tax, the rates. Although not directly related to the ability of people to pay, rates were correlated to some extent with wealth, in so far as wealth was held in the form of land and property. On the other hand the rates did have a number of disadvantages (Wilson and Game 1994: 164–5) and they were particularly unpopular with the Conservative government of Margaret Thatcher. The Thatcher government was in the vanguard of a wave of radical right-wing politics that emerged in Europe and North America in the 1980s. Re-elected in 1979, the Conservatives in the UK pursued economic policies inspired by the doctrines of monetarism developed by right-wing economists such as Milton Friedman. One of the central planks of the monetarist approach was a reduction in the economic and social role of the state in general and the implementation of tight controls on public expenditure in particular. Greater controls over

TABLE 15.2 Typology of local government services, showing percentage of total net expenditure on all local government services in 1990/1, England and Wales

Need services	%	*Protective services*	%	*Amenity services*	%	*Facility services*	%
Education	48.6	Police	12.9	Highways	5.0	Libraries	1.5
Personal social services	12.4	Fire	2.7	Street cleaning	1.0	Museums and art galleries	0.3
Housing benefit	0.6	Courts	0.8	Consumer protection	0.3	Refuse collection	1.4
		Probation	0.8	Refuse disposal	0.6	Housing	0.8
				Environmental health services	1.1	Recreational centres	1.3
				Parks and open spaces	1.7	Cemeteries	0.2
				Economic development	0.3		
				Town and country planning	1.2		
				Other services	2.4		
				Administration	2.1		
Totals	61.6		17.2		15.7		5.5

Source: Welsh and Game (1994).

local government finance had been put in place in 1976 by the then Labour government and these were significantly strengthened by the Conservatives. Central government grants to local government were constrained to the point where, by the end of the 1980s, they met only 53 per cent of the costs of local government. Faced by these limits on the contributions from the centre, many local authorities tried to make up the shortfall by increasing income from rates on both residential and business properties. Because this revenue also counted as public expenditure, the Conservative government, still seeking to control such expenditure overall, began to restrict the rights of councils to raise revenue through rate increases. Political conflicts over this programme of 'rate-capping', as it was known, were among the bitterest-fought issues in local politics in recent years. However, worse was to come.

The Conservatives adopted a number of strategies to try to rein in what they saw as the excesses of local government (especially in councils controlled by the Labour Party). At the forefront of these was the abolition of the rates system and its replacement with what the government called the Community Charge, but which everyone else referred to as the 'poll tax'. Wilson and Game refer to the poll tax as 'arguably the British government's single biggest policy disaster of the past 50 years' (1994: 132). The poll tax was based on a simple principle: everyone in a given local area would pay the same, single, flat-rate charge as their contribution to the cost of providing local services. The Conservatives hoped that this would encourage voters to support parties (most notably the Conservative Party) that offered to cut the charge, and thus the cost of local government, and therefore both weaken the Labour Party's control over many local councils and reduce public expenditure overall. In the event the poll tax turned out to be a huge disaster for local government, for the Conservative Party and for Margaret Thatcher personally, as it was a key factor in her replacement as party leader and as Prime Minister by John Major.

Although most, if not all, taxes are unpopular, the main problem with the poll tax was the enormous depth and breadth of its unpopularity. To all intents there was a popular uprising against the tax, with many people refusing to pay it, national and local anti-poll tax campaigns being established, many non-payers ending up in court and even in gaol, and eventually a full-scale riot in central London. The extra cost to central government in 1991/2 alone amounted to £7.5 billion – or £125 for every man, woman and child in the country. In addition, the government also decided to introduce a national business rate for business premises and to increase the proportion of the cost of local government met by central government from 47 per cent in 1989/90 to 54 per cent in 1992/3. The effect of these measures was to reduce the locally determined proportion of local government finance to just 15 per cent, making it yet more difficult for councils to decide for themselves how much money to raise and what kinds of services to provide.

In April 1993 the poll tax was scrapped and replaced with a new tax called the council tax that abandoned the per capita system of taxation on which the poll tax was based and returned to property as the basis of local revenue generation. Now each household receives a bill related to the value of the property. The vestige of the poll tax per capita principle that remains is that single-person households gain a discount of 25 per cent. The amount that councils can raise through local taxation remains tightly controlled by central government, although the Labour government under Tony Blair, elected in May 1997, has recently relaxed the constraints to a limited extent.

The contemporary geography of elected local government

Changing territorial structures

The straightforward two-tier structure of local government established in the 1974 reorganisation lasted until the mid-1980s. In 1986, as part of its attempts to control the power of radical left-wing local councils, the Thatcher government took the dramatic step of abolishing the Greater London Council, then led by the populist Labour left-winger, Ken Livingstone, and the six metropolitan county councils of West Midlands, Merseyside, Greater Manchester, Tyne and Wear, South Yorkshire and West Yorkshire. This move left London and the other major English conurbations with only a single tier of elected local government. In many cases, and particularly in London, the lack of a strategic local authority to provide city-wide services and consider planning issues over a wide area has been keenly felt. In many cases the metropolitan district councils are simply too small to provide cost-effective services such as police and fire. In these cases a confusing pattern of 'joint boards' has been set up. Joint boards are not directly elected but are made up of representatives of the various district councils in the area covered by the board. Different boards are responsible for different services such as fire, transport and so on.

Although the imposition of single-tier local government in the big cities has been widely criticised, there are also problems with the two-tier system with its potential for waste and duplication. In the early 1990s many commentators from across the political spectrum were arguing that a universal single-tier system of all-purpose local authorities (so-called 'unitary' authorities) should be introduced. The idea of unitary authorities was supported particularly by the Conservative government of John Major, and in 1992 the government began reviewing the structure of local government with a view to moving to a unitary system. In England, a Local Government Commission was established under the chairmanship of Sir John Banham. In Scotland and Wales reviews were conducted by the government itself. Northern Ireland already had a single tier of elected councils.

The Banham Commission was instructed to undertake a review of the structure, boundaries and electoral arrangements for local government with two criteria in mind: that local councils should reflect local identities and community interests and that local government should be 'effective and convenient'. The initial assumption was that unitary authorities would be established virtually everywhere, and as Wilson and Game note, the Commission's early reports

> advocate forcefully the advantages of unitary authorities: their ability to develop a co-ordinated approach to service delivery; their claimed enhancement of local accountability; their improved efficiency and effectiveness.
>
> (Wilson and Game 1994: 303)

In the event the process of reorganisation turned out to be a much more piecemeal affair than the vision of a universal single-tier system suggested. The Commission worked through a process of local consultation (including opinion polling to discover the degree of public identification with existing and possible alternative units of local government). In most areas, the Commission recommended that the present two-tier system should be retained, but it did recommend the creation of forty-six new unitary authorities, mainly in larger towns and medium-sized cities such as Darlington, York and Bristol. All the new unitary authorities in England were in operation by April 1998.

In Scotland and Wales, where the process of reorganisation was undertaken by the government directly, the two-tier system was replaced by a universal system of unitary authorities in April 1996. In Scotland twenty-nine unitary councils have been established to replace the fifty-three districts and nine regional authorities. In Wales twenty-two new unitary councils replaced the eight counties and thirty-seven districts.

The result of the reorganisation process has been to produce a much more complex geography of local government structure in the UK. This is summarised in Table 15.3. A major difficulty with the whole process of reorganisation is identified by Wilson and Game:

> The establishment of the Local Government Commission has preceded meaningful debate about the role, constitutional position and function of local government in a democratic society. It has been compared to putting up a building without first determining its use.
>
> (Wilson and Game 1994: 300)

The Labour government has recently brought forward new proposals for reviving the role of local government and is also introducing some further structural changes, notably the proposed new strategic authority for London, with a directly elected mayor.

Electoral geography

All the discussion and debate about structure, function and finance can obscure the fact that local government also involves electoral and party politics. Although the political power of local authorities has been significantly circumscribed since the late 1970s, local elections

TABLE 15.3 The structure of UK elected local government, 1998

Areas	*Structure*
One-tier system	
England: new unitary authorities	46 unitary authorities
England: metropolitan areas	36 metropolitan boroughs
Northern Ireland	26 districts
Scotland	32 unitary authorities
Wales	22 unitary authorities
Two-tier system	
England: other non-metropolitan areas	1st tier: 35 counties 2nd tier: 241 districts
London	1st tier: proposed new London strategic authority from May 2000 2nd tier: 32 London boroughs and City of London Corporation

still provide a degree of democratic accountability for the activities of local authorities and democratic control over the distribution of resources between different local services. Moreover, the present Labour government is proposing a range of reforms to the operation of local government with the aim of reviving local democracy, improving consultation and participation and bringing local government closer to local people (Department of Environment, Transport and the Regions 1998).

Local councils are made up of councillors who are elected to represent small geographical areas (known as wards) for four years at a time. In most areas of the country the major political parties play a central role in local politics. In some places (mainly remote rural areas) many councillors serve without formal political affiliations as independents. In all there are some 25,000 local councillors in the country each representing an average electorate of 2,200 voters. District councils in England typically have around 35–60 councillors. County councils and metropolitan boroughs are somewhat larger with around 55–100 councillors. Where one political party gains a majority of seats on the council it is able to control the council and ignore, if it wishes, the views of the minority councillors. In many councils, however, no party is in a majority and the council is said to have 'no overall control'. Table15.4 illustrates something of the geography of local government electoral politics. It shows that in 1997 the Labour Party was the dominant party of local government controlling 46.7 per cent of all councils in Great Britain. The Liberal Democrats are in second place with 11.2 per cent of councils and the Conservatives a poor third with just 4.3 per cent. This picture reveals the extent to which the Conservative Party has been eradicated from swathes of local government in the UK, even in its traditional heartlands, the English shire counties. Only six of the thirty-five English counties were controlled by the Conservatives in 1997. Labour is particularly dominant in the cities, shown by its high levels of control in the English metropolitan districts and the English unitary authorities. The regional picture adds another dimension. Labour dominates much of local government in Wales and Scotland, but in the more remote parts independents have strong support. In addition the nationalist parties, Plaid Cymru and the Scottish National Party, between them control four councils. The Liberal Democrats have pockets of electoral strength, especially in south-west England. In Northern Ireland the structure of local government and the use of a system of proportional representation ensures that no single party dominates, with most councils having no overall control.

Although the geography of council control is clearly important, the British 'first past the post' electoral system tends to exaggerate the popular appeal of the biggest party (in this case Labour). A complementary way of examining the electoral strength of the different parties in different types of council is to look at the numbers of councillors elected for each (Table 15.5). Table 15.5 confirms the dominance of the Labour Party in local government nationally. Only in the English counties does another party (the Conservative Party) have more councillors than Labour. However, there is also a clear geography to party strength, with Labour being strongest of all in the metropolitan districts (large cities) where it has three-quarters of all seats. The Liberal Democrats are strongest in rural areas and medium-sized towns (non-metropolitan districts and English unitary authorities). Table 15.5 also confirms the importance of Independent councillors in the remoter parts of Wales and Scotland. The weakness of the Conservative Party is also evident; its present position compares very unfavourably with its historical position as a strong party of local government in both rural and urban areas.

TABLE 15.4 Political control of British local government, May 1997

	London boroughs		*English counties*		*English non-metropolitan districts*		*English metropolitan districts*		*English unitary authorities*		*Welsh unitary authorities*		*Scottish unitary authorities*		*Great Britain total*	
	No.	*%*	*No.*	*%*	*No.*	*%*	*No.*	*%*	*No.*	*%*	*No.*	*%*	*No.*	*%*	*No.*	*%*
Conservative	4	12	6	17	8	3			1	2					19	4.3
Labour	16	48	9	26	87	36	32	89	31	67	14	64	19	59	208	46.7
Liberal Democrat	3	9	1	3	39	16			7	15					50	11.2
Nationalist parties											1	5	3	9	4	0.9
Independent/Other	1	3	1	3	10	4					4	18	5	16	21	4.7
No overall control	9	27	18	51	97	40	4	11	7	15	3	14	5	16	143	32.1
Total	33		35		241		36		46*		22		32		445	

Source: *Municipal Yearbook*.

Notes: Percentages may not sum to 100 because of rounding.

'No overall control' means that no party held a majority of seats. It includes councils where one party had exactly half the seats and governed with effective control using the casting vote of the chair, as well as councils where minority Conservative groups govern with support from Independents.

In Northern Ireland two councils were controlled by the Ulster Unionist Party, one by the Social Democratic and Labour Party and one by Sinn Fein. The remaining twenty-two councils had no overall control.

* Figure includes nineteen district councils that did not become unitary authorities until April 1998.

TABLE 15.5 Party strength by number of councillors, May 1997

	London boroughs		*English counties*		*English non-metropolitan districts*		*English metropolitan districts*		*English unitary authorities**		*Welsh unitary authorities*		*Scottish unitary authorities*		*Great Britain total*	
	No.	*%*	*No.*	*%*	*No.*	*%*	*No.*	*%*	*No.*	*%*	*No.*	*%*	*No.*	*%*	*No.*	*%*
Conservative	516	25.0	928	37.6	2,504	21.3	194	7.8	412	16.9	39	3.1	82	6.6	4,675	19.7
Labour	1,018	49.2	875	35.4	4,747	40.3	1,871	75.5	1,381	56.8	718	56.4	623	50.1	11,233	47.3
Liberal Democrat	315	15.2	530	21.4	3,009	25.5	358	14.4	549	22.6	79	6.2	120	9.7	4,960	20.9
Nationalist parties											117	9.2	178	14.3	295	1.2
Independent, other or vacant	219	10.6	138	5.6	1,522	12.9	55	2.2	91	3.7	319	25.1	240	19.3	2,584	10.9
Total	2,068		2,471		11,782		2,478		2,433		1,272		1,243		23,747	

Source: *Municipal Yearbook*.
Notes: Percentages may not sum to 100 because of rounding.
* Category includes nineteen district councils that did not become unitary authorities until April 1998.

The electoral geography of local government affects the geography of local government service provision. Different local authorities do provide somewhat different levels and mixtures of services to local people and these differences reflect in part the political differences between councils. This seems obvious, since local elections are traditionally fought on issues of the balance between local taxation and local service provision and the mix of services to be provided. However, there are other factors at work too. As Duncan and Goodwin showed in 1988, variations in levels of service provision are not perfectly correlated with political control. In fact, their research suggested that the geography of local government activity has more to do with the pattern of uneven social and economic development in the country than with the party-political affiliations of local councils. In addition, the thrust of central government policy over the past twenty years has been to work to even-out such local variations. For example, central government grants to local councils are calculated using a complicated formula known as the 'Standard Spending Assessment' (SSA). The SSA for each local authority is calculated as that amount of local expenditure required to provide a standard package of service to the local population, taking into account demographic and socio-economic differences between local areas. If councils wanted (for political reasons) to provide a more generous, or less generous, level of services they would have to raise the money from local taxation. However, central government has also limited the ability of local authorities to increase local taxes. This has made it very difficult in practice for local councils to deviate very significantly from a national pattern of service provision. Nevertheless, council services and tax rates do vary from place to place – partly because the Standard Spending Assessment system is far from perfect as a way of assessing local needs, partly because councils vary in the efficiency with which they use the resources at their disposal and partly because even within the tight controls imposed by central government there is still some scope for local discretion in the way resources are deployed.

The relationship between central and local government, of which the financial system is one aspect, can be thought of as another key element in the changing geography of local government. In many countries, local government has much more freedom and autonomy from central control. In some cases this is guaranteed constitutionally, in others it is a matter of custom and practice. The United Kingdom is a unitary state, in which local government is constitutionally subordinate to central government, but it has also become a highly centralised state in which most of the activities of local government are closely regulated by the centre. This trend began during the 1970s, but was accelerated dramatically by the Conservative governments of Margaret Thatcher and John Major. The present Labour government has put forward proposals which will modify this centralisation to some extent, but there is still a considerable emphasis on national standards of service provision, quality control, systems of consultation and mechanisms of governance. Even if the new proposals are implemented in full, the UK will still not have anything approaching local self-government. Thus local government is both an arena of political conflict, debate and competition, but also the object of national political debate and action.

Non-elected local government

One of the most important aspect of central–local relations in the UK is the growth of non-elected local government, mainly as a result of central government action. Elected local councils remain the most important element when the local government system is considered overall (they are the main multi-functional service providers, covering the whole country and spending the largest proportion of locally budgeted resources). However, in particular fields of local government activity, such as local economic development, and education and training, a range of new institutions and agencies have grown up with considerable governmental powers but no direct electoral accountability. Some commentators argue that such organisations help to 'get things done' by bypassing the sometimes cumbersome processes of local council decision-making. Many others, however, see the growth of non-elected local government in much more negative terms as removing significant areas of service provision from democratic control by local communities through their elected representatives.

The agencies concerned are often referred to as 'quangos' (quasi-autonomous non-governmental organisations), although technically many of them are not quangos in the strict sense. In official government terminology, quangos are known as 'non-departmental public bodies' (that is, bodies set up by the government, with members appointed by the government, to undertake certain public functions outside the activities of existing government departments). Many of these organisations are national in scope and do not form part of non-elected local government.

At the start of this chapter I defined local governance as the process of the formation and implementation of public policy at the local level involving both elected and non-elected organisations. Non-elected organisations are thus central to the overall governance of localities. The range of non-elected agencies involved is considerable. Two examples will give a flavour of their importance.

Until the late 1980s, state education until the age of 18 was governed by local councils acting as local education authorities or LEAs (except in central London, where a single 'Inner London Education Authority' covered a number of boroughs). Councils provided schools, appointed teachers, determined educational policy and funded colleges of further education. From the late 1980s, parents were given the right to vote to 'opt-out' of local authority control and to see their children's school funded directly by central government (this is known as Grant Maintained Status) and governed by its own board of governors, including some elected from among the parents. Supporters of this scheme argued that it gave more power to parents to influence their children's education, increased efficiency because there were no council overheads, and allowed head teachers to run schools without interference from councils. Opponents claimed that the system would increase inequalities between schools (with Grant Maintained Status being adopted by schools in better-off neighbourhoods and poorer schools that remained with local councils being increasingly underfunded), remove democratic oversight of education from local communities and result in greater inefficiency because county-wide economies of scale would be lost.

In the field of local economic development, the Conservative government set up a series of Urban Development Corporations (UDCs) in the most run-down areas of major cities, such as Newcastle, Manchester, Merseyside and Leeds. UDCs were given sweeping powers to initiate and control urban redevelopment, based on the regeneration of the physical infrastructure in inner-urban areas. Each UDC was governed by a board whose members were appointed by the government. Although each UDC included representatives of elected local government, the UDC itself was outside the control of local government and effectively took over the functions of local government within its development area with respect to planning and local economic development. In many cases relations between the UDCs and the elected councils in their areas were frosty, or even hostile. Supporters of UDCs claim that they have managed to transform the physical environment in their areas, dramatically improving the cityscape much more quickly and comprehensively than would have been possible by normal local government means. Critics claim that they are anti-democratic, with only a limited voice for the local communities in which they work and that their much-vaunted improvements are mainly cosmetic, and have largely failed to produce employment and social regeneration to match the regeneration of the landscape.

During the 1990s the emphasis has moved away from single specialist agencies to partnership forms of local governance. A partnership consists of a number of organisations, elected local authorities, unelected bodies (such as UDCs, Training and Enterprise Councils and Health Authorities), private companies and voluntary and community organisations acting together for a common purpose, often related to community development. The main potential strength of a partnership is that it brings together expertise and resources from a number of different sectors which, it is hoped, will add up to more than the sum of the parts and represent a wider cross-section of local interests than a single agency could. Critics argue that partnerships are weak, because they can lack overall co-ordination and have no means of ensuring the participation of unwilling partners. Partnerships too are by definition unelected and their popularity further increases the size of non-elected local government.

The geography of non-elected local government is inevitably complex because of the number and range of agencies involved. Some non-elected agencies, such as Training and Enterprise Councils (which are based on local labour markets), cover the whole of the country. Others, such as UDCs, are highly localised and cover small areas of city-centres, but aim to have much wider effects. The system of Grant Maintained Status for schools is a national scheme, but its take-up by parents has been very uneven, with schools in more affluent areas tending to adopt Grant Maintained Status and those in poorer localities staying with the local authority. In other words there are multiple geographies of non-elected local government which combine with elected local government in different ways in different places to produce a highly differentiated and complex map of local govern*ance*.

The future of local governance

The system of local governance involving both elected councils and unelected agencies continues to develop and change. Over the coming years the new pattern of unitary authorities in Wales, Scotland and some parts of England will be consolidated and a judgement will be possible about their effectiveness and popularity. The recent growth of non-elected local government and the parallel decline in public interest in elected local government (reflected in very low turnout figures for local elections) are causes of concern to many. In its recent White Paper (Department of Environment, Transport and the Regions 1998), the new Labour government has brought forward proposals to reform and democratise local government to make it more responsive to the views of local people. One of the most important initiatives is the option for councils to have a directly elected mayor who would act as a kind of local president. Proponents of this scheme suggest that a high profile mayoral election would generate much greater interest in and enthusiasm for local government and politics than is currently the case. Others suggest that it would place too much power in the hands of one person and distract attention from the real social and economic problems facing many parts of the country. Other ways of involving citizens more directly in local decision-making, such as citizens' panels and focus groups, are also being tried out in many areas. It remains to be seem whether such initiatives will succeed or whether a greater degree of devolution of political power and financial resources will also be required to install genuine local democracy in the United Kingdom.

References

Byrne, T. (1992) *Local Government in Britain* (revised 5th edn), London: Penguin.

Central Office of Information (1996) *Local Government*, London: HMSO.

Department of Environment, Transport and the Regions (1998) *Modern Local Government: In Touch with the People*, Cm 4014, London: HMSO.

Duncan, S. and Goodwin, M. (1988) *The Local State and Uneven Development*, Cambridge: Polity Press.

Wilson, D. and Game, C. (with Leach, S. and Stoker, G.) (1994) *Local Government in the United Kingdom*, London: Macmillan.

Further reading

The best and liveliest introduction to UK local government is provided by D. Wilson and C. Game. (with S. Leach and G. Stoker) (1994) *Local Government in the United Kingdom* (London: Macmillan). Their book covers the whole of the subject in reasonable detail and in a very accessible way. A less comprehensive, but also lively treatment of the turbulent period of the 1980s and the early 1990s, is provided in A. Cochrane (1993) *Whatever Happened to Local Government* (Buckingham: Open University). A very thorough account of most aspects of British local government is T. Byrne (1994) *Local Government in Britain*; 6th edition (London: Penguin). The post-war history is documented in a scholarly overview by K. Young and N. Rao (1997) *Local Government Since 1945* (Oxford: Blackwell). An important discussion of the issue of local citizenship is provided by D.M. Hill (1994) *Citizens and Cities* (Hemel Hempstead: Harvester Wheatsheaf). Different theoretical approaches to local governance are covered by the contributions to D. Judge, G. Stoker and H. Wolman (1995) *Theories of Urban Politics* (London: Sage). Local elections in Britain are analysed in detail by the foremost authorities on them in the UK in C. Rallings and M. Thrasher (1997) *Local Elections in Britain* (London: Routledge). Finally, three periodicals: the annual *Municipal Yearbook* provides an encyclopaedia of UK local government on a service-by-service and council-by-council basis. Packed with facts, figures, names and addresses, it is an essential tool for any research project on local government; *Local Government Studies* carries a range of scholarly articles on the subject; and the fortnightly *Municipal Journal* provides topical coverage of local government news as well as feature articles on current developments.

Chapter 16

A (dis)United Kingdom

Charles Pattie

Introduction

National identity is an act of imagination, a social construct (Anderson 1983). Outside nationalist ideology, nations are not mono-ethnic, mono-cultural entities. The United Kingdom, no exception, has always been a multi-ethnic, multinational state which has grown by the incorporation of smaller nations. Wales was incorporated into England in 1536. Scotland joined the Union in 1707. Ireland apart (where, despite the 1801 Act of Union, nationalist sentiments continued, culminating in independence for southern Ireland in 1922), those living in the constituent parts of the United Kingdom have generally shared a common sense of 'Britishness', over and above loyalties to their own nations. That shared identity, forged in the eighteenth century from a mix of Protestantism, war and nascent Empire (Colley 1992), was sufficiently flexible to allow most citizens of the United Kingdom to hold a sense of dual nationality – both British and Scottish, for instance.

By the mid-twentieth century, regional and ethnic loyalties appeared to have been replaced by a sense of shared nationhood as economic progress enhanced communications across the 'national' territory (Agnew 1987). National politics were dominated by left–right ideological debates between the Labour and Conservative parties. Neither the unity nor the long-term future of the United Kingdom were in question.

At the end of the twentieth century, however, that judgement seems premature. The 1970s and 1980s saw social and political upheaval, as the major parties gradually abandoned the policy nostrums of the post-war period. After a prolonged period in which Britain had become a more equal society, the pendulum swung back, and by the end of the century society was arguably more unequal than at any time since the Second World War.

Meanwhile, Westminster's authority was increasingly challenged. The factors creating a sense of Britishness in the eighteenth century had lost their resonance. Nationalist movements in the Celtic Fringe were increasingly assertive, and legislation establishing separate Scottish, Welsh and Northern Irish parliaments was passed in 1997 and 1998 (in

Northern Ireland alone, this meant the reintroduction of a form of 'home rule'). At the same time, moves towards the ever-greater integration of the European Union led to greater 'pooled' sovereignty between the United Kingdom and other EU countries. For some commentators, at least, the country at the end of the century seemed a *dis*united kingdom.

From consensus politics . . . to a new consensus?

British politics has moved since 1945, from cross-party consensus, through a period in which consensus disappeared, to a new consensus (Taylor 1991). The post-1945 consensus was built on a cradle to grave welfare state, and on Keynesian policies for economic growth and full employment. The new policies, largely agreed by the wartime coalition, were enacted by the 1945 Labour government. But the Conservative Party too had to adapt to the new circumstances to be re-elected. While inter-party differences remained, the post-war years were marked by broad policy consensus between the major parties (Kavanagh and Morris 1989). Under the conditions of the long post-war economic boom, the consensus seemed to have 'solved' Britain's economic and social problems.

But the onset of recession in the early 1970s brought the consensus to an end. Keynesian policies no longer seemed to allow governments to control the economy effectively. Unemployment and inflation both grew rapidly. By the end of the decade, monetarism was emerging as the new 'accepted wisdom', at least on the right of the political spectrum. Policy was geared to keeping inflation down, and to reducing state 'interference' in the running of market economies (Johnston 1993). The price was the abandonment of full employment as a policy goal: unemployment grew to levels unprecedented in the post-war period.

Labour in government in the 1970s had been reluctant converts to monetarism. But the Conservatives in opposition adopted it enthusiastically. Their programme (which some termed Thatcherism, after Margaret Thatcher) broke with the post-war settlement. Thatcherism contained two strands: 'the free economy and the strong state' (Gamble 1988). The 'free economy' reflected a belief in the greater efficiency and desirability of markets than of government intervention in the economy. Over the next eighteen years Conservative governments tried to cut state expenditure (not always successfully), and privatised state-owned industries. The 'strong state' referred to the maintenance of law and order, and to increasing political centralisation in Whitehall during the 1980s. Thatcherism transformed Britain irrevocably. Nor were the changes restricted to the balance between state and private sector. By the late 1990s, a new political consensus was emerging.

In the early 1980s, however, a new consensus seemed remote. The Conservatives in government moved right and Labour in opposition moved left (Seyd 1987). A group of prominent Labour right-wingers left the party in 1981 to form the Social Democratic Party (Crewe and King 1995). With the Liberals, as the Alliance, for a brief period in 1982 they seemed set to eclipse both Labour and Conservatives. By the 1983 election, support for the Conservatives had recovered. But Labour only just saw off the Alliance challenge.

The 1983 election result was Labour's worst performance in terms of vote share since 1918. A radical rethink was required. The party moved to the right (Shaw 1994). In part, Labour's policy review was a recognition that voters saw the party as dangerously extreme in 1983: by moderating its position, the party hoped to move closer to the 'centre ground', winning votes along the way. In part, too, however (and especially after 1987), the policy review was a recognition that the changes wrought by the Conservatives were in many cases irreversible. Rather like the Conservatives after 1945, Labour had to adapt to the new circumstances in order to stand a realistic chance of governing again. But it was not until the 1992–7 parliament that the party was perceived to have changed sufficiently for it to win an election: under its new leader, Tony Blair, the party abandoned Clause 4 of its 1918 constitution and made clear it was no longer a 'tax and spend' party. Helped by Conservative disarray over the economy and Europe, Labour won a landslide victory in 1997 (Pattie *et al.* 1997).

By the 1997 election, there was little to choose between the major parties in terms of most major policies: both were committed to free markets, to relatively limited state involvement, to low taxation, and so on. A new political consensus had emerged. They diverged primarily on constitutional matters: Labour was committed to devolution for Scotland and Wales (as were the Liberal Democrats), while the Conservatives were strongly opposed.

The movement of Britain's political elite from old (social democratic) to new (post-Thatcherite) consensus is clear. But public support for the post-war consensus remained strong (Crewe 1988; Pattie and Johnston 1996). In 1997, 92 per cent of the population felt that government should spend more to eradicate poverty, and 61 per cent felt that income and wealth should be redistributed (British Election Study 1997). Support for welfare state institutions such as the National Health Service (NHS) remained almost universal.

But support for the welfare state was mediated by private concerns over prosperity (Sanders 1996). Many vote in line with their views on the state of the economy, and of their own personal financial situation: if they feel prosperous, they are more likely to support the government, even if that means supporting a party whose policies on the welfare state they feel uneasy about. But the corollary also holds that if they feel the economy is doing badly, they will hold the government responsible and vote against it. This creates electoral pressures which make it difficult to reverse social inequalities, and which may, in fact, exacerbate them (Galbraith 1992).

The analysis of public opinion also reveals a (dis)united political geography. Focusing on the crucial years between 1979 and 1997, support for some aspects of the welfare state was widespread (Table 16.1). Large majorities in all regions thought that government should spend money to alleviate poverty. Although there were some differences across regions (with people living in the South outside London being slightly less convinced than people living in London and the North), these were not large, and there is clear evidence in all regions of growing public support for government action between the 1980s and 1990s.

But there were substantial cross-regional and cross-temporal variations in public support for some other aspects of the post-war settlement. Unsurprisingly in a climate of privatisation, support for the reverse process, nationalisation, was very low in all areas. But it grew in all regions after 1987 (in part a response to growing public disquiet over some of the privatised utilities). Furthermore, it was consistently higher in the North than in the South (excluding London). Similarly, support for the redistribution of income from rich to poor was backed by a majority in the North, while in the South it was supported by a minority (and a declining one between 1979 and 1992). Finally, despite government rhetoric focusing on 'scroungers' living off welfare benefits, there is little evidence that voters felt provision of benefits had 'gone too far' after 1979. In part, this reflects persistently high levels of unemployment in the 1980s especially. But even here, there is some evidence of a

TABLE 16.1 Support for the post-war consensus, by region, 1979–97

	1979	*1983*	*1987*	*1992*	*1997*
Support for nationalisation (% wanting more)					
Scotland and Wales	25.2	21.0	23.7	24.9	36.7
North	16.0	22.0	19.6	28.6	30.2
Midlands & East Anglia	15.3	15.9	14.2	20.4	26.5
South West	13.1	12.2	11.4	20.2	22.3
South East	13.2	12.9	12.5	16.8	20.5
London	20.8	18.2	20.4	30.0	31.4
Government should get rid of poverty (% agree)					
Scotland and Wales	87.7	89.3	91.4	93.3	95.1
North	83.8	85.8	89.2	93.3	93.6
Midlands & East Anglia	82.8	86.3	87.3	91.8	91.9
South West	80.4	82.0	87.3	90.6	88.5
South East	83.2	83.8	86.2	93.7	88.6
London	82.2	85.7	85.0	92.7	91.6
Income and wealth should be redistributed (% agree)					
Scotland and Wales	68.0	59.4	54.8	56.7	70.9
North	58.6	53.1	48.8	54.8	64.1
Midlands & East Anglia	55.5	46.5	40.8	46.1	62.1
South West	46.7	44.3	34.2	44.5	55.5
South East	47.4	38.2	36.7	37.5	49.7
London	51.6	39.9	43.4	47.6	61.8
Welfare benefits have gone too far (% gone too far)					
Scotland and Wales	40.5	15.9	21.7	15.8	
North	48.4	16.9	23.8	18.3	
Midlands & East Anglia	51.2	20.3	26.6	16.3	
South West	51.0	27.0	34.8	13.0	
South East	56.0	21.4	23.6	17.1	
London	52.0	21.5	22.6	16.3	

Source: British Election Studies, 1979–97.

(narrow) North–South divide in attitudes during the 1980s, which closed somewhat in the 1990s (unfortunately, the question was not repeated in 1997).

This suggests that the new elite political consensus has only shallow roots in the population, and that support for it has been higher in the South than in the North. The ideological divides of the period of 'high Thatcherism' (1979–87) were reflected in regional divides in public opinion (see also Pattie and Johnston 1990). During this period, Britain became more 'disunited'. But since 1987, the North–South divide in political attitudes has narrowed (Curtice 1996). National disunity has decreased, at least as far as public opinion is concerned.

TABLE 16.2 Perceptions of prosperity by region, 1992–7

	1992		*1997*	
	Got stronger (%)	*Got weaker (%)*	*Got stronger (%)*	*Got weaker (%)*
Scotland and Wales	10.6	55.9	27.8	28.7
North	11.6	55.3	33.6	25.3
Midlands & East Anglia	14.2	56.4	39.2	23.0
South West	6.2	59.8	40.7	23.5
South East	15.6	51.3	47.6	18.2
London	13.8	58.3	51.1	16.9

Source: British Election Studies.

During the 1980s and 1990s there were also regional divides in how people evaluated the state of the economy, and these divides reflected real differences (Table 16.2; Hudson and Williams 1995). Reflecting the recession conditions of the 1992 election, a majority of voters then thought the national economy had got worse. But people in the South East (outside London) were still slightly more likely to feel things had got better over the preceding year than were voters living in the North, despite the region's economic downturn. By 1997, with the economy in recovery, especially in the South, the regional differential in perceptions of economic performance had widened again to produce a pronounced North–South split: only 28 per cent of Scottish and Welsh voters thought the economy was stronger than in the previous year (though this was in itself an improvement on the situation five years earlier), compared to 48 per cent of voters in the South East. Compared to 1992, the 1997 figures represent a return to the status quo of the 1980s: residents in the South East are optimistic about the state of the economy relative to residents further north.

The geographies of public opinion and of public perceptions of economic well-being combined to influence the geography of the vote. Once again, the country became more disunited during the 1980s, and then more united again in the 1990s. One way of showing this is to compare the ratio of Conservative to Labour votes in the regions across a series of elections. Where more people support the Conservatives than Labour, the ratio is greater than 1; where more support Labour than Conservative, it is less than 1 (Figure 16.1). A North–South divide in voting patterns is a long-established feature of British elections, reflecting the underlying geography of social class. But in the 1980s that divide grew very rapidly. Labour's vote fell throughout the South, while the Conservatives did relatively poorly in the North (and Labour held on there). By 1987, it seemed that Britain was 'a nation dividing' along grounds of political allegiance (Johnston *et al.* 1988). Recession-hit northern voters rejected the Conservatives while affluent southern voters rewarded it. But the process was not ineluctable. As Labour moved back into the centre ground in the 1990s and as renewed recession hit the previously affluent voters of the South, so the electoral divide narrowed again at both the 1992 and 1997 elections. However, it is worth noting that the regional electoral divide in 1997, while not as wide as in 1987, was still wider than it had been in the 1950s and 1960s, the years of the post-war consensus. Britain remains divided electorally, although the trends are towards unity rather than disunity.

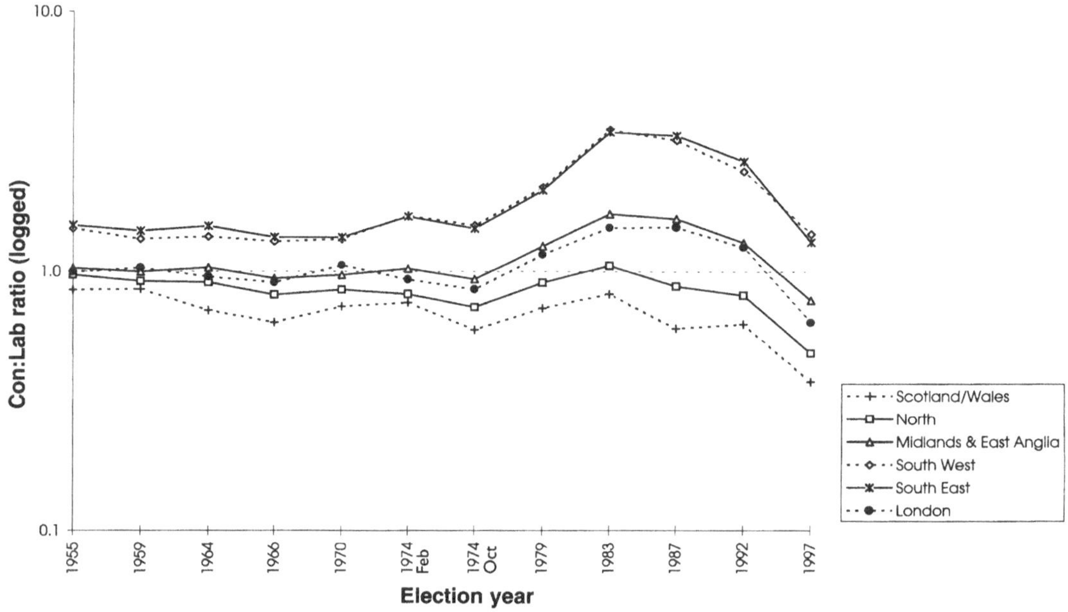

FIGURE 16.1 The changing regional geography of the vote, 1955–97, Conservative/Labour vote ratios

Nationalist challenges

Ironically, while the 1992 and 1997 elections reveal a closing of the political divide between the regions, and the major parties share more common ground than at any time since the late 1960s, other forces are at work which may weaken the bonds between different parts of the United Kingdom. Crucially, these forces operate in the main remaining area of substantial difference between the major parties, the constitution, and in particular, devolution of power to the regions. But renegotiating the relationship between Westminster and the regions means walking a fine line on issues of national sovereignty.

Northern Ireland

Ireland was the exception in the development of a British sense of national identity. Under British rule it was in some ways more a colony than part of an economically advanced nation-state. When a sense of Britishness was being forged in the eighteenth century, the Catholicism of much of the Irish population was at odds with British Protestantism. A history of repression and neglect fostered a series of nationalist risings, and, in the latter half of the nineteenth century, in political movements for greater Irish autonomy (Foster 1988). The 'Irish question' was an important factor in the politics of the late nineteenth and early twentieth centuries. In arguments which were echoed almost a century later in discussions over Scottish and Welsh devolution, some claimed home rule would fatally weaken the

Union, and others that it would save it. By 1914, it seemed the Liberal government was finally about to legislate for home rule, but the outbreak of the First World War put an end to the plan. Some Irish nationalists resorted to more militant, military, measures. Britain was forced to concede Irish independence.

The home rule debate was complicated, however, by divisions within Ireland itself. A substantial Protestant minority, especially in the north-east around Belfast, was strongly opposed to home rule, fearing it would be the first step to outright independence. To accommodate this minority, Ireland was partitioned, and the six counties of the north-east, where it was in a majority, remained in the UK. Ironically, given Unionist opposition to home rule for the whole island, Northern Ireland was given its own parliament, Stormont, and substantial devolution of power within the UK. Stormont was dominated by Unionist politicians. The Catholic minority in the north was discriminated against in employment, in political representation, and in access to social goods, especially housing (O'Leary and McGarry 1996). The 1922 settlement contained the seeds of future conflict.

Stormont's violent handling of the Catholic civil rights movement in the late 1960s brought renewed civil disorder. In response to a rising tide of inter-communal violence (mostly directed against Catholics, and sometimes involving the Northern Irish police), the British army was deployed on the streets of Northern Ireland in 1969. In 1972, Stormont was suspended and direct rule from Westminster was instituted. With one abortive exception, the 1973 Sunningdale Agreement (O'Leary and McGarry 1996: 198ff.), no serious attempt was made to devolve power back to the province for a quarter of a century.

Politics in Northern Ireland has been polarised to a greater degree than in any other part of the United Kingdom. In part, this was reflected in the climate of sectarian violence, with paramilitary organisations responsible for terrorist acts both in the province and beyond. But it also found a reflection in party politics. Whereas elsewhere in the UK the major parties developed along class lines, in Northern Ireland they reflected communal ties. Unionist parties draw their support almost exclusively from the Protestant community, while nationalist parties draw support from Catholics. There is little or no cross-communal party support.

But divisions exist within as well as between the two main camps (Evans and Duffy 1997; Graham 1998). The once solid bloc of Unionism fractured in the late 1960s and early 1970s over the extent to which nationalist demands should be accommodated. The Official Unionist Party represents the old, traditional unionism, while Ian Paisley's Democratic Unionist Party (DUP) draws support from a more hard-line, working-class constituency. The Social Democratic Labour Party (SDLP), which emerged as the major voice of nationalism in the 1970s, has faced growing competition for Catholic votes from Sinn Fein (often seen as the political wing of the IRA) since the latter party decided to abandon its opposition to participation in elections in the 1980s. One consequence has been to make compromise difficult, as more moderate politicians have had to take account of the likely impact on their support of concessions. Equally striking has been the almost total absence of the 'mainland' parties from the province's politics. At the 1992 and 1997 elections, some Conservative candidates stood for Ulster seats, but they did so without the full backing of their party.

Nor has there been consensus over desired constitutional arrangements (McGarry and O'Leary 1995). In the nationalist camp, the SDLP advocates a parliamentary road to Irish reunification. While campaigning on behalf of the nationalist community in Westminster, and advocating an increased 'Irish dimension' to politics in the province, the SDLP has seen unification as a long-term goal. By contrast, Sinn Fein (SF) advocated until recently a dual strategy of military and electoral struggle – the armalite and the ballot box. SF MPs, when elected, have refused to take up their seats (though the party has played a more active role in local politics). Unionists too have been divided (Graham 1998). One model has been to win the re-establishment of a Stormont parliament – in other words, devolution within the UK, with no (or limited) concessions to nationalists. Another – minority – view has been that Northern Ireland should become independent of both the UK and Ireland. Yet others have advocated 'normalising' relationships with Britain. Under this model, Northern Ireland would be no different to the other UK regions, the major UK parties would compete there, and (as a side-effect) the threat of Irish unification would be removed at a stroke.

But Northern Ireland's problems are not due simply to internal factors: the British and Irish government are also implicated. Solutions are elusive, but most recent attempts have involved both governments. Beginning with the 1985 Anglo-Irish agreement, the Irish and British governments have increasingly co-operated with each other. A growing recognition in the 1990s that neither the British army nor the paramilitaries – both republican and unionist – could achieve an outright military victory also helped foster a climate within which dialogue could take place. From small beginnings in 1991, talks began between the various parties involved in Northern Ireland. Despite sometimes severe setbacks, all parties – including those linked to paramilitaries, once ceasefires had been announced – became involved. While not accepted by all parties to the negotiations (the DUP and some Official Unionists remained opposed), an agreement was reached on Good Friday 1997.

All sides made concessions. An elected Northern Ireland assembly was to be established, with a power-sharing executive of twelve ministers, guaranteeing both communities a say in the government of the province. The Irish government agreed to hold a referendum on changing clauses 2 and 3 of the Irish constitution, which laid claim to Northern Ireland and have proved a significant sticking point for unionists. At the same time, in a concession to nationalists, a Ministerial Council would be established, consisting of ministers in both the Belfast and Dublin governments, to foster joint policy-making on areas of common interest, as would a British/Irish council, involving politicians from all the parliaments in the islands – Dublin, London, Belfast, Edinburgh and Cardiff. Finally, the Good Friday Agreement was to be put to referendum in both parts of Ireland. The vote, when it came, was overwhelming: on a huge 81 per cent turnout, 71 per cent of voters in Northern Ireland agreed to the terms of the deal (on a lower, 56 per cent turnout, it was also endorsed by 94 per cent of voters in Ireland).

The success of the Agreement is not guaranteed. Elections for the new Northern Ireland Assembly, conducted in June 1998, revealed large splits within the unionist camp, for instance. But the arrangements for power-sharing and for the involvement of both British and Irish governments represent a considerable constitutional shake up, with concessions for both sides. Furthermore, although disputes remain over the decommissioning of weapons, the main paramilitary groups seem set on maintaining their ceasefires. The Good Friday Agreement provides one of the best prospects for peace in a generation.

Scotland

Unlike the other countries of the United Kingdom outside England, Scotland had a long history as an independent and consolidated state, from the thirteenth- and fourteenth-century wars of independence to the 1707 Act of Union. Even after Union, many of the institutions of an independent Scottish state remained: Scottish law differs from English law; the Presbyterian Church of Scotland is the 'state church', not the Episcopalian Church of England; the education system, from school to university, is different; there is a distinctive and flourishing national press, and so on. And since 1885, Scottish affairs have been represented in government by a separate, powerful department of state, the Scottish Office (Mitchell 1996).

A shared sense of Scottish nationhood has never been far below the surface of Scottish politics. The 1970s saw the start of the most recent upsurge in demands for Scottish home rule, with the election of Scottish Nationalist Party MPs. While the SNP never achieved majority support (even at its October 1974 peak, it won only 30 per cent of the Scottish vote, and took eleven out of seventy-one seats: Figure 16.2), it did play a considerable part in putting the issue of Scottish self-government back onto the political agenda, forcing the minority 1974 Labour government to introduce legislation for Scottish and Welsh elected assemblies. A referendum was held on the issue in 1979. Although a small majority of those voting favoured devolution, the margin fell below a threshold of 40 per cent of the electorate stipulated in legislation. The devolution proposals fell.

The incoming Conservative government opposed devolution. If anything, it further centralised power. During the following eighteen years, Scottish self-government was off the national (UK) political agenda. But the issue did not go away.

Ironically, it was the very electoral success of the Conservatives throughout the 1980s and early 1990s which helped rekindle demand for a Scottish parliament. The beliefs and policies espoused by the Conservatives under Mrs Thatcher were particularly unpopular in Scotland (see Pattie and Johnston 1990; Bennie *et al.* 1997), and the Scottish economy did

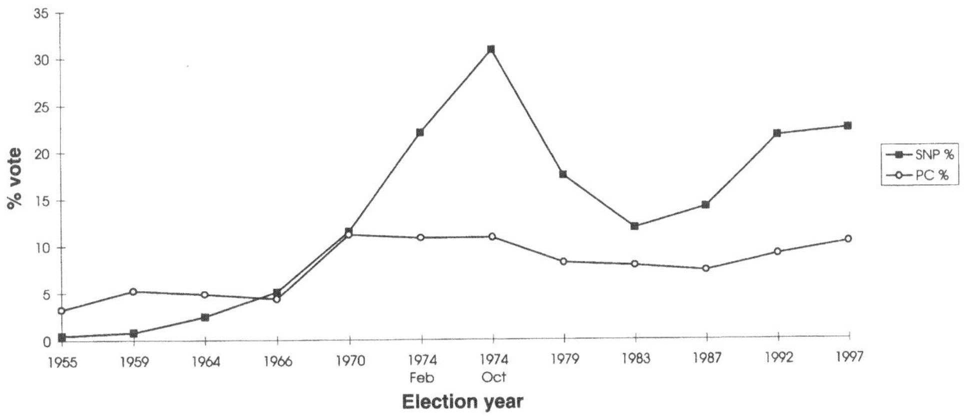

FIGURE 16.2 Vote share for Scottish and Welsh nationalists, 1955–97

SNP = Scottish Nationalist Party; PC = Plaid Cymru

not perform well enough, relative to the South of England, to compensate. Furthermore, the Conservatives gained a reputation in Scotland (how deserved is open to debate) as being at best uninterested in, and at worst actively antagonistic to, Scottish interests.

The net effect was to increase Scottish resentment of the Conservative government and to foster demands for a separate Scottish parliament. Prime Minister John Major believed that his anti-devolution appeal for the preservation of the Union, during the 1992 General Election campaign, was instrumental both in snatching an overall victory for the Conservatives and in effecting a slight improvement in the party's fortunes in Scotland. This led him to repeat his warnings about the possible break-up of the UK during the 1997 election campaign. In fact there is little evidence that Major's appeals for the preservation of the constitutional status quo made much impact on the Scottish electorate. The much-vaunted recovery of the Conservatives' Scottish vote in 1992 was in fact very slight.

None the less, successive Conservative Secretaries of State for Scotland made much of their ability to win concessions from government. Nor were they averse to playing the 'tartan card'. For instance, Michael Forsyth, the incumbent Secretary of State before the 1997 election, wore a kilt at the premier of the film *Braveheart* (which romanticised the exploits of William Wallace, a hero of Scotland's wars of independence in the late thirteen century and the early fourteenth; *Braveheart* was also used as a rallying device by the SNP), and was responsible for the well-publicised and highly symbolic return of the Stone of Destiny to Scotland.

A renewed commitment to devolution from both Labour and the Liberal Democrats was one manifestation of a revived home rule campaign in the late 1980s and early 1990s. The Scottish Constitutional Convention was another. Launched in 1989, the Convention involved most of the groups campaigning for a Scottish parliament, including the Labour and Liberal Democrat parties in Scotland, the Scottish TUC, and the Scottish churches (although the SNP, worried about its pro-independence principles being diluted, declined to participate). It provided a cross-party forum within which proponents of a parliament could meet, agree common goals, and conduct common campaigns.

Labour's 1997 landslide election victory put the issue back at the forefront of the Westminster agenda. Subject to ratification in the referendum, Scotland was to be offered a relatively strong parliament, with powers over many aspects of daily life. Furthermore, proposals were laid to give the parliament highly symbolic (if in practice rather limited) powers to raise some of its own tax revenue. As a consequence, the Scottish referendum asked two questions, therefore: whether voters supported the idea of a Scottish parliament; and whether the parliament should be given tax-varying powers. Attempts by the SNP to draft a question giving voters three options – the status quo, a Scottish parliament within the UK, and outright independence – came to nothing, however. The future of the Union was not in question in the referendum.

Pro- and anti-devolution campaign groups were formed (see Mitchell *et al.* 1998). Formally, these were non-partisan organisations drawing together supporters and opponents of home rule from across the political spectrum, but in fact the composition of the groups largely reflected the previous positions of the main parties on the issue. Labour, the Liberal Democrats and the SNP came together under the aegis of Scotland Forward to campaign for a 'Yes' vote on both referendum questions. In itself, this was a remarkable achievement. Not only had the SNP refused to join the Constitutional Convention, but relations between Labour and the SNP are poor, as the latter vies to replace the former as Scotland's 'majority' party. But the parties agreed to co-operate in the 1997 referendum campaign, avoiding potentially damaging splits. The main 'No' campaign group, meanwhile, was Think Twice. Although 'No' campaigners were anxious to dispel any notion that they were all Conservatives and that Think Twice was merely a Conservative 'front' organisation, that, in fact, is fairly close to the truth.

The referendum was held on 11 September 1997. The result on the first question was never seriously in doubt. On a respectable 60 per cent turnout, a substantial majority (74 per cent) of Scottish voters supported the establishment of the first separate Scottish parliament since 1707, a much more substantial majority than that achieved in 1979. And whereas in 1979 majorities in some parts (notably the Borders and the Northern Isles) of Scotland had actually opposed devolution, in 1997 majorities in favour were returned from all parts of the country (Figure 16.3).

There were doubts over the second question, on tax-varying powers. Prior to the 1997 election, the Conservative Secretary of State for Scotland had mounted an effective

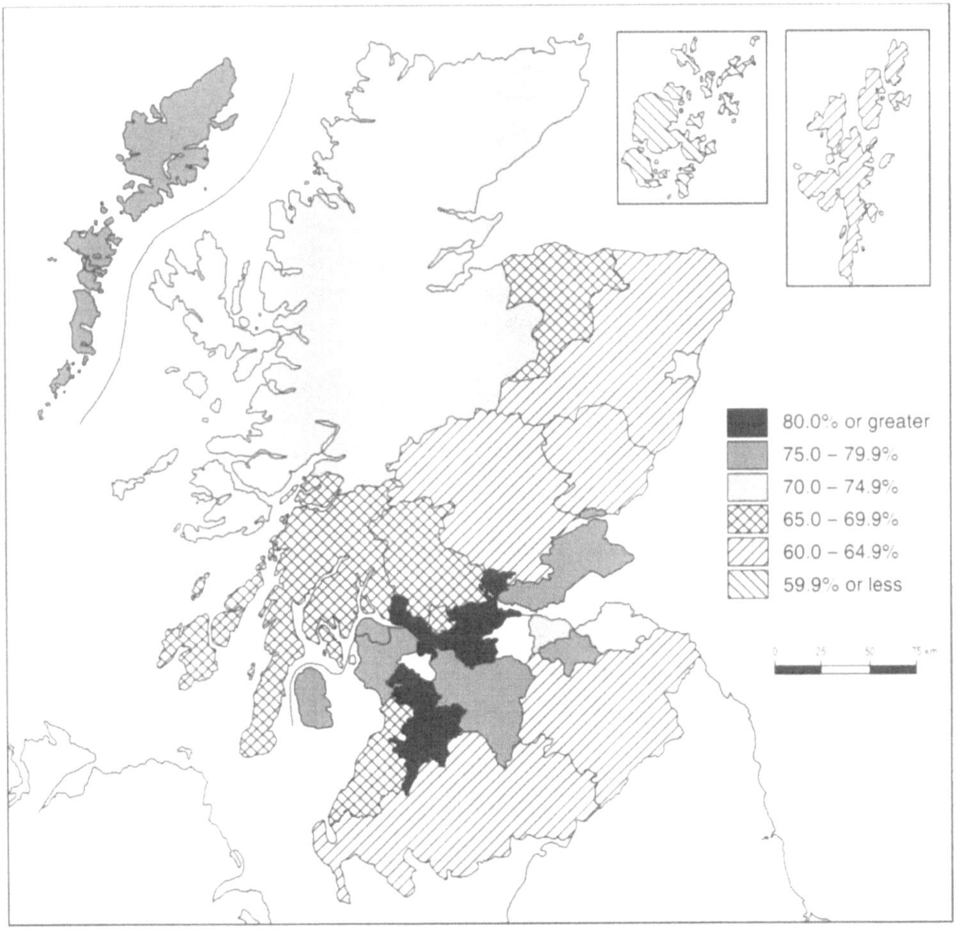

FIGURE 16.3 Percentage voting 'Yes' to a Scottish Parliament

campaign against what he dubbed the 'tartan tax', arguing a Scottish parliament would mean a higher tax burden for business in Scotland compared to the rest of the UK. The tax issue remained the Achilles' heel of Labour's proposals. Polls prior to the referendum suggested that public support for tax-varying powers was ebbing away. But in the event, the second referendum question produced majority backing for tax-varying powers too (albeit by a narrower though still substantial margin, 63 per cent).

The referendum gave strong public endorsement for the establishment of a powerful Scottish parliament. Opponents of the measure were forced to come to terms with the changed political environment. With widespread public support north of the border for a parliament, and with elections due in 1999 before the next UK General Election (which, in any case, Labour seemed likely to win, given its unprecedented landslide majority in 1997), there was little chance of retrospectively abolishing the new body once it had been set up. The Conservatives accepted the result of the referendum, and the inevitability of a new constitutional settlement for Scotland. One of the ironies of the legislation for the new assembly was that, by adopting an additional member proportional representation system for electing MSPs (Members of the Scottish Parliament), it almost certainly provides a route back into the Scottish political mainstream for the Conservatives, who will be guaranteed some MSPs at a time when they have no Scottish Westminster MPs.

The long-term consequences of a Scottish parliament are not yet clear. Opponents of the measure have seen devolution as a step to independence. They argue that conflicts between the Edinburgh and London parliaments are inevitable and will increase pressure for a final dissolution of the Union. Most advocates, on the other hand, have argued that by recognising the desire of the Scots to have a government responsive to Scottish needs, devolution will prevent the further development of resentments which might have otherwise led to demands for independence. (However, some advocates, in the SNP especially, do see devolution as a stepping stone to independence.)

In the year after the referendum, support for the SNP grew in opinion polls (McCrone 1998). Shortly after the referendum, around 48 per cent of voters said they would vote Labour for the Edinburgh parliament, with about 30 per cent supporting the SNP. But by early 1998, the two parties were neck and neck, and by May 1998, support for the SNP had outstripped that for Labour: about 45 per cent said they would vote SNP, while 35 per cent said they would vote Labour. These trends led some to suggest that the tide was turning towards independence.

But that judgement was premature. Over the same post-referendum period, Labour remained Scotland's most popular party for the UK parliament (McCrone 1998). Although SNP support climbed (from just above 20 per cent to around 30 per cent), the party still lagged far behind Labour as a popular choice for Westminster. The great majority of Scots still support pro-Union parties when it comes to UK elections. Furthermore, the rise of support for the SNP after the devolution referendum needs to be put into the context of Labour's difficulties in Scotland, with allegations of corruption and misconduct against some prominent Scottish Labour local authorities and MPs. The SNP's rise in the Scottish opinion polls may reflect a 'conventional' swing from government to opposition during the parliamentary cycle (in Scotland, of course, the SNP is, in effect, the main opposition alternative to Labour).

Support for outright independence remains a minority concern (Table 16.3). Before the referendum, it ran at just over a quarter of Scottish voters. After the referendum support

TABLE 16.3 Constitutional preferences in Scotland and Wales, 1997

	Scotland (%)	*Wales* (%)
Total independence	8.3	1.9
Independence within European Union	19.5	8.6
Parliament in UK, with tax-varying powers	44.1	36.0
Parliament in UK, with no tax powers	9.9	18.0
No elected body	18.1	35.6

Source: British Election Studies.

TABLE 16.4 Scottish and Welsh national identity

	Scotland		*Wales*
	1992 (%)	*1997* (%)	*1997* (%)
Scottish (Welsh) not British	19.3	23.2	12.3
More Scottish (Welsh) than British	40.4	38.4	28.8
Equally Scottish (Welsh) and British	33.0	26.9	25.2
More British than Scottish (Welsh)	3.4	4.4	8.7
British, not Scottish (Welsh)	2.7	3.6	18.0
None of these	1.2	3.4	7.1

Source: British Election Studies.

was slightly higher, at just below a third of voters (a very few polls have suggested much greater support for independence, but these seem to be 'rogue' results). But a Scottish parliament within the UK remained the most popular option, steady at just under 50 per cent of the Scottish electorate (McCrone 1998). Furthermore, Scots' sense of national identity, while confirming the importance of 'Scottishness', hardly suggests a strong desire for separation from the rest of the UK (Table 16.4). True, the proportion feeling 'Scottish only' seems to have risen slightly in the years between 1992 and 1997 – from 19 per cent to 23 per cent. And the vast majority of those living in Scotland feel Scottish to some degree: only between 4 per cent (1992) and 7 per cent (1997) of adults living in Scotland did not. But equally a large majority of Scots continue to identify with Britain as a whole: while most feel more Scottish than British, around 70 per cent felt a sense of dual nationality – Scottish and British. As yet, there remains a large degree of attachment to the UK.

Wales

Unlike Scottish nationalism, Welsh nationalism does not draw on a history of independence. While there were several Welsh-speaking 'statelets' and fiefdoms before conquest by the

English, there was never a consolidated Welsh state. Furthermore, Wales lacks the 'quasi-state' institutions of a separate legal and educational system: English law covers Wales too. Up until the late twentieth century, Wales was governed from Westminster in much the same way as any English region. It was not until 1964 that a separate government department was established for Wales (Jones 1997). Compared to the Scottish Office, however, the Welsh Office remains relatively small and weak.

Instead, Welsh nationalism draws strongly on cultural traditions, and in particular on the preservation of the Welsh language. Once spoken throughout the Principality, Welsh went into decline after the Industrial Revolution, and is now spoken by only around 20 per cent of the Welsh population. The language is now confined to the north-west and west coasts of Wales, and to some middle-class enclaves in the major cities of the south, especially Cardiff (Giggs and Pattie 1992). Welsh society is split between Welsh- and English-speakers, and between the Welsh-born and a large (relative to Scotland and Northern Ireland) 'incomer' population from other parts of the UK, primarily from England. Furthermore, and again in contrast to Scotland, the major roads and railways within Wales do not provide a well-integrated 'national' transport network. Most major routes run east–west, meaning that North Wales is better connected to the English North West, and South Wales is better connected to the English Midlands than either part of Wales is to the other.

Welsh national identity and nationalism are therefore rather more complex and mediated than their Scottish counterparts. Like the Scots, a sizeable majority of people living in Wales have a sense of dual nationality, Welsh and British (Table 16.4). But respondents in Wales were less likely to say they felt 'only Welsh', or even 'more Welsh than British' than were Scots asked about their Scottishness. Furthermore, a far higher proportion of those living in Wales than in Scotland reported feeling only a British national identity, or being neither Welsh nor British.

The complexities of Welsh national identity are reflected in the development of political nationalism in Wales. Plaid Cymru (PC), the Welsh nationalist party, has its roots in the Welsh language movement. Like the SNP in Scotland, it was not until the 1960s and 1970s that it moved from the fringes into the electoral mainstream. But unlike the SNP, Plaid Cymru has always found it hard to break out of a geographically concentrated base in Welsh-speaking north-west Wales. The party has generally done very badly in the more populous, English-speaking south of Wales. Even at its peak in 1970, PC won only 11 per cent of the vote in Wales.

But because its vote is concentrated in the north-west of the Principality, PC has been able to return two or more MPs at each election since February 1974. As a result, it too was important to the minority 1974–9 Labour government. Furthermore, the Welsh language issue had become increasingly politicised over the preceding decades. As a consequence, plans for devolution to Wales were set in train at the same time as the plans for Scotland. As in Scotland, they were subject to ratification through a referendum. But whereas a narrow majority of Scots voted for devolution in 1979, a decisive majority (80 per cent) of those living in Wales rejected it. Furthermore, devolution was opposed even by some Welsh Labour MPs. The 1979 referendum seemed to have ended prospects for Welsh devolution.

As in Scotland, the experience of Conservative government between 1979 and 1997 seems to have been a catalyst for change, however. The Welsh repeatedly rejected the Conservatives (and the party's share of the vote fell there, till in 1997 there were no

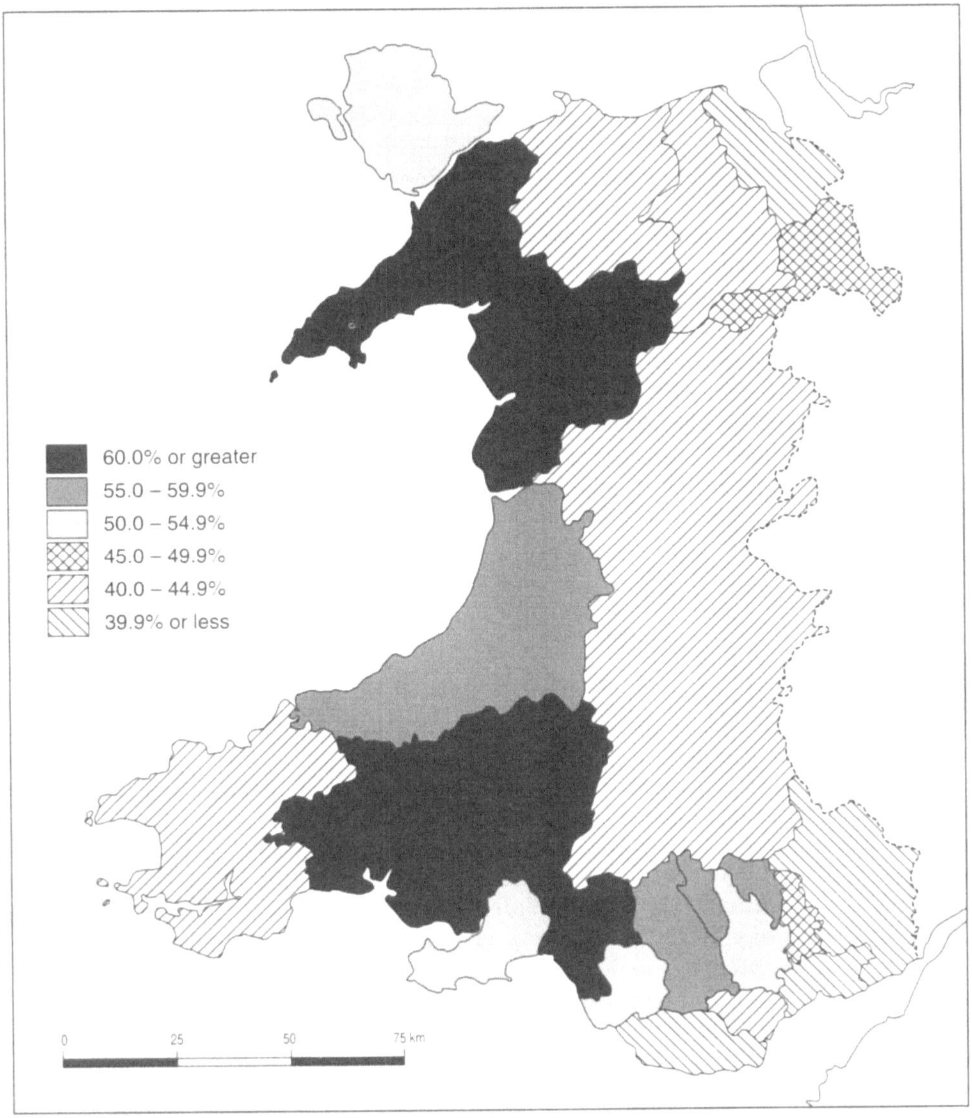

FIGURE 16.4 Percentage voting 'Yes' to a Welsh Assembly

Conservative MPs from Welsh seats), but found itself governed by a party it had not voted for. By 1997, support for Welsh home rule had strengthened (Table 16.3). Even so, it lagged behind support in Scotland, both in quantity and strength, and a still-sizeable group (by no means restricted to the non-Welsh living in Wales) was adamantly opposed. Perhaps tellingly, there was no Welsh equivalent of the Scottish Constitutional Convention. The broad consensus in favour of devolution which developed among Scottish opinion formers (including the Scottish press) in the 1980s and 1990s was not so well developed in Wales.

Welsh devolution remained the official policy of Labour, the Liberal Democrats (and the Alliance before them), and, of course, PC. But – and again the comparison with Scotland is telling – New Labour's 1997 proposals for a Welsh assembly were for a weak debating chamber, with no primary legislative or tax-raising powers. Furthermore, while Labour in Scotland managed to silence its anti-devolution wing in 1997 (a major change from 1979), many in the Welsh Labour Party remained vociferously opposed.

The new Labour government put its proposals to a referendum (McAllister 1998), but the timing of the Welsh referendum proved a bone of contention. Rather than hold both the Scottish and Welsh votes on the same day, the government decided that Welsh electors would vote a week after the Scots. The intention seems to have been to use the result in Scotland (where a substantial 'yes' vote was widely and confidently expected) to create a groundswell in favour of change in Wales, where the result was much more on a knife edge. Welsh voters went to the polls on 18 September 1997.

The referendum results could not have been more dramatic. There was no equivalent of the Scottish landslide. Neither camp was at all confident of victory. Not until the last local authority area had declared its result was the overall outcome clear – a 'yes' vote, but by the narrowest of margins (50.2 per cent voting in favour of an elected Welsh Assembly; only 5,900 votes separated the 'yes' and 'no' results).

Furthermore, the geography of the results emphasised the divisions within Wales (Figure 16.4). Along the Welsh Marches, in populous South Wales, and in Pembrokeshire, the result was a consistent majority against an Assembly. In North and West Wales (and in some of the former mining valleys of South Wales), however, the result was a 'yes' majority. Welsh-speaking and Anglophone Wales seemed at odds once more.

And the English regions . . . ?

Arrangements for the government of the 'Celtic fringes' have proved controversial over recent years, therefore, and the measures introduced by the Labour government elected in 1997 represent a major constitutional shake-up in the affairs of a once highly centralised state. But there has been very little equivalent discussion of arrangements for the English regions. In some respects, this is surprising. There are, as we have seen, substantial differences in political attitudes across the English regions, and also large variations in local conditions, both social and economic. It is not obvious that residents of the North East share the same hopes and expectations as those living in the South East, for instance. But with no clearly established structure of regional government there is no intermediary between local authorities and Westminster (Mawson 1997).

Furthermore, it is not clear that sufficient public support exists for the devolution of power to even the most distinctive of English regions (Elcock 1997). To a large extent, the English 'regions' lack coherence. The now familiar standard planning regions, for instance, have their origins in administrative convenience rather than in historically and socially united communities. The South West, for instance, stretches from the western end of the affluent M4 corridor and the south coast resorts, to economically depressed Cornwall. With no generally agreed views on what constitutes a 'region', it becomes hard to mobilise active public support for regional government.

But this does not mean there is no regional devolution whatever in England, nor that pressure for more does not exist – and will not increase in the future. Some members of the local political elite in the North East have become involved in the Campaign for a North of England Assembly, for instance. In part, the impetus comes from devolution plans for Scotland. There are concerns in the North East at the prospect of a powerful Scottish government with economic investment powers drawing much-needed investment, while the North East, lacking similar regional government structures, would be unable to respond. In the late 1970s, North East Labour MPs played a prominent part in the opposition to their own government's proposals for Scottish government. In the late 1990s, overt opposition was almost non-existent. But, for some at least, the strategy had changed: if Scotland was to get a parliament, one should be established in the North East too.

Pressure for a degree of devolution to the English regions comes in part, then, from concerns about the knock-on effects of Scottish and Welsh devolution, and a desire not to be left out. A related issue concerns the so-called 'West Lothian question', first raised by Tam Dalyell, Labour MP for West Lothian and long-time opponent of devolution. A potential inequity would exist if devolution was applied to some parts of the UK and not to others, especially if no change was made to the numbers of MPs each region returned to Westminster. For instance, establishing a Scottish parliament with no other changes would mean that Westminster MPs for Scottish seats (and hence Scottish voters) would have a say in legislation affecting England, but English MPs and voters would have no reciprocal say on Scottish affairs. As Labour's electoral prospects began to improve in the early 1990s, and the party seemed to have its first realistic chance of forming a government since the very early 1980s, it was forced to confront the West Lothian Question. One solution was to return to the policy espoused by the Liberals in the early years of the century of 'home rule all round'. In a 1991 consultation paper, the party pointed to 'the democratic deficit at the regional level in England' and proposed a rolling programme of regional devolution in England (Mawson 1997: 184). The timetable for change remained unclear, however, and no firm commitments were made, in contrast to the plans for Scotland and Wales.

The Conservatives, meanwhile, under both Margaret Thatcher and her successor John Major, were opposed to the establishment of elected regional assemblies in England as well as in Scotland and Wales. Assemblies, they argued, would introduce an unnecessary and largely unwanted extra layer of bureaucracy. However, this did not preclude a degree of administrative devolution. In 1994, the government established ten regional offices (Government Offices for the Regions, or GROs), which tried to create Whitehall outposts in the regions, allowing local businesses and public bodies to have access to an integrated local service. In addition, local authorities in most parts of the country were coming together in relatively informal 'regional organisations' to plan and campaign for common goals – for instance to provide co-ordination of land-use planning at the regional level (e.g. the South West Planning Conference) and the East Midlands Forum (which gave itself a wider remit). Four of the nine regional associations also established offices in Brussels, hoping to gain access to EU investment (Mawson 1997: 191).

In opposition again after the 1992 election, Labour continued to plan for elected English regional assemblies, which would take over powers from the GROs. But again, no precise timetable was set out, and the need for public support in each region was reiterated, suggesting that referendums would be carried out, as in Scotland and Wales. But by the 1997 election, devolution to the English regions had slipped down the party's list of priorities. The party now placed its hopes in Regional Development Agencies (RDAs), bodies which would be charged with the task of co-ordinating economic development strategies. Proposals for the establishment of RDAs were announced at the end of 1997. RDAs still fall short of a commitment to elected regional bodies: members are to be government appointments. But they represent a 'half-way house' option which does not require a referendum mandate. Labour thus avoids the risk of seeing its proposals rejected by local people.

London government has been the major exception to Labour's otherwise tentative approach to devolution within England. Since the abolition of the Greater London Council in 1986, the capital has had no city-wide elected government. Labour supported the re-establishment of a London authority. Its 1997 election programme contained radical proposals for a directly elected London mayor. Unlike the largely ceremonial role played by mayors in other British local authorities, the London mayor was to be a powerful executive figure drawing a direct mandate from London voters. Although technically a local government reform, the plans for London had, in some ways, more in common with plans for Scotland and Wales, not least in that London's population is larger than Scotland's.

Once again, the government used the referendum device: the vote was held on 4 May 1998. But, unlike the Scottish and (to a lesser extent) Welsh referendums, turnout was poor (34 per cent). Although a majority of voters (72 per cent) supported the measure, it hardly represented an overwhelming endorsement of the government's plans: only a quarter of London residents actively voted for an elected mayor. This is perhaps indicative of a more widespread lack of interest on the part of English voters. Certainly, consultations on the reform of English local government a few years earlier, although badly handled, produced even less public interest and involvement, suggesting that most are indifferent to the arrangements for the government of their area (Johnston and Pattie 1996).

Demands for regional level government in England do not seem deep seated, therefore. Rather, in so far as they exist, they operate at the level of the political elites in particular areas. And, even then, the elites are not fully united. Of course, as devolution to Scotland, Wales and Northern Ireland comes into operation and matures, there may be a growth in the demand for similar measures in England. But in the absence of clearly articulated regional loyalties to match the senses of national identity found in the Celtic fringe, proponents of elected English regional government face an uphill task. Unelected RDAs may well represent the limits, for the time being, of the devolution of governance in England.

Conclusion

As it enters the new millennium, the UK faces new challenges. Although its political geography is less polarised than in the recent past, pressure for substantial devolution of power to some of the member nations has grown, resulting in new parliaments in Scotland, Wales and Northern Ireland. In all three cases, relative economic disadvantage has helped fuel demands for change. But while an important catalyst, it is not in itself sufficient to fuel demands for self-government. After all, many of the English regions have weaker economies than the South East but have not spawned widespread demands for devolution. A distinctive national identity (in addition to 'British') is also necessary – and even then may not be enough, as the Welsh referendum shows.

In the short term, there will be significant adjustments as the new bodies interact with each other, and establish their relationships with the older power centre of Westminster. There is a learning curve to be climbed. It is as yet unclear, for instance, how the new arrangements will cope with (say) a Conservative government in Westminster and a Labour-dominated government in Edinburgh. In the longer term, will the UK survive in its current form? Advocates of devolution argue that it will preserve the Union, while opponents have argued that it will lead to break-up. Devolution, of course, is now a fact. But, at the moment, neither the Scots nor the Welsh seem on the brink of abandoning their British identity. It is worth remembering, of course, that a degree of regional autonomy is common in many European states, without creating secessionist pressures It is unusual only in the context of British politics (and then only if we ignore the many 'exceptions' – Stormont from 1922 until 1972; the Isle of Man; the Channel Islands, and so on). The greatest innovation, and the greatest challenge, is in the arrangements for Northern Ireland. Power-sharing is fraught with potential difficulties and it remains to be seen whether good will is sufficient to make the new arrangements work there. But the referendum results suggest overwhelming public support for the peace process. That is a good base to start from.

References

Agnew, J. (1987) *Place and Politics*, London: Allen and Unwin.

Anderson, B. (1983) *Imagined Communities*, London: Verso.

Bennie, L., Brand, J. and Mitchell, J. (1997) *How Scotland Votes*, Manchester: Manchester University Press.

Colley, L. (1992) *Britons: Forging the Nation 1707–1837*, New Haven, Conn.: Yale University Press.

Crewe, I. (1988) 'Has the electorate become Thatcherite?', in R. Skidelsky (ed.) *Thatcherism*, London: Chatto and Windus.

Crewe, I. and King, A. (1995) *The SDP: The Birth, Life and Death of the Social Democratic Party*, Oxford: Oxford University Press.

Curtice, J. (1996) 'One nation again?', in R. Jowell, J., Curtice, A., Park, L., Brook and K. Thomson (eds) *British Social Attitudes: the 13th Report*, Aldershot: Dartmouth.

Elcock, H. (1997) 'The north of England and the Europe of the Regions, or, when is a region not a region?', in M. Keating and J. Loughlin (eds) *The Political Economy of Regionalism*, London: Frank Cass.

Evans, G. and Duffy, M. (1997) 'Beyond the sectarian divide: the social bases and political consequences of nationalist and unionist party competition in Northern Ireland', *British Journal of Political Science* 27: 47–81.

Foster, R.F. (1988) *Modern Ireland 1600–1972*, Harmondsworth: Penguin.

Galbraith, J.K. (1992) *The Culture of Contentment*, London: Sinclair-Stevenson.

Gamble, A. (1988) *The Free Economy and the Strong State: The Politics of Thatcherism*, London: Macmillan.

Giggs, J.A. and Pattie, C.J. (1992) 'Croeso i Gymru. Welcome to Wales. But welcome to whose Wales?', *Area* 24: 268–82.

Graham, B. (1998) 'Contested images of place among Protestants in Northern Ireland', *Political Geography* 17: 129–44.

Hudson, R. and Williams, A. (1995) *Divided Britain* (2nd edn), Chichester: John Wiley.

Johnston, R.J. (1993) 'The rise and decline of the corporate-welfare state: a comparative analysis in global context', in P. Taylor (ed.) *Political Geography of the Twentieth Century: A Global Analysis*, London: Belhaven Press.

Johnston, R.J. and Pattie, C.J. (1996) 'Local government in local governance: the 1994–1995 restructuring of local government in England', *International Journal of Urban and Regional Research* 20: 671–96.

Johnston, R.J., Pattie, C.J. and Allsopp, J.G. (1988) *A Nation Dividing? The Electoral Map of Great Britain 1979–1987*, London: Longman.

Jones, B. (1997) 'Wales: a developing political economy', in M. Keating and J. Loughlin (eds) *The Political Economy of Regionalism*, London: Frank Cass.

Kavanagh, D. and Morris, P. (1989) *Consensus Politics from Atlee to Thatcher*, Oxford: Basil Blackwell.

McAllister, L. (1998) 'The Welsh devolution referendum: definitely, maybe?', *Parliamentary Affairs* 51: 149–65.

McCrone, D. (1998) 'Opinion polls in Scotland July 1997–June 1998', *Scottish Affairs* 21: 143–50.

McGarry, J. and O'Leary, B. (1995) *Explaining Northern Ireland*, Oxford: Blackwell.

Mawson, J. (1997) 'The English regional debate: towards regional government or governance?', in J. Bradbury and J. Mawson (eds) *British Regionalism and Devolution: The Challenges of State Reform and European Integration*, London: Jessica Kingsley.

Mitchell, J. (1996) *Strategies for Self-Government: The Campaign for a Scottish Parliament*, Edinburgh: Polygon.

Mitchell, J. Denver, D. Pattie, C. and Bochel, H. (1998) 'The devolution referendum in Scotland', *Parliamentary Affairs* 51: 166–81.

O'Leary, B. and McGarry, J. (1996) *The Politics of Antagonism: Understanding Northern Ireland* (2nd edn), London: Athlone Press.

Pattie, C.J. and Johnston, R.J. (1990) 'Thatcherism – one nation or two? An exploration of British political attitudes in the 1980s', *Environment and Planning C: Government and Policy* 8: 269–82.

Pattie, C.J. and Johnston, R.J. (1996) 'The Conservative Party and the electorate', in S. Ludlam and M. Smith (eds) *Contemporary British Conservatism*, London: Macmillan.

Pattie, C.J., Johnston, R.J., Dorling, D., Rossiter, D., Tunstall, H. and MacAllister, I. (1997) 'New Labour, New Geography? The electoral geography of the 1997 British General Election', *Area* 29: 253–9.

Sanders, D. (1996) 'Economic performance, management competence and the outcome of the next General Election', *Political Studies* 44: 203–31.

Seyd, P. (1987) *The Rise and Decline of the Labour Left*, London: Macmillan.

Shaw, E. (1994) *The Labour Party Since 1979: Crisis and Transformation*, London: Routledge.

Taylor, P. (1991) 'The changing political geography', in R.J. Johnston and V. Gardiner (eds) *The Changing Geography of the United Kingdom* (2nd edn), London: Routledge.

Further reading

For a recent overview of the changing nature of British politics up to and including the election of the Blair government and devolution to Scotland and Wales, see I. Budge, I. Crewe, D. McKay and K. Newton (1998) *The New British Politics* (London: Longman) especially chapters 3, 6 and 15. A good overview of the social and economic factors underlying regional political trends can be found in R. Hudson and A. Williams (1995) *Divided Britain*, 2nd edn (Chichester: John Wiley).

Two chapters in P. Dunleavy, A. Gamble, I. Holliday and G. Peele (eds) (1997) *Developments in British Politics 5* (Basingstoke: Macmillan) give good overviews of recent developments in British regional politics: P. Dunleavy, chapter 7, 'The Constitution', and I. Holliday, chapter 11, 'Territorial Politics'. J. Mitchell (1996) *Strategies for Self-Goverment: The Campaign for a Scottish Parliament* (Edinburgh: Polygon), contains a good history of twentieth century campaigns for a Scottish parliament. B. O'Leary and J. McGarry (1996) *The Politics of Antagonism: Understanding Northern Ireland*, 2nd edn (London: Athlone Press), meanwhile, provides one of the most thoughtful analyses of the development of the 'Northern Ireland' problem. A review of the theoretical literature on nationalism can be found in chapter 5 of P.J. Taylor (1993) *Political Geography: World Economy, Nation State and Locality* (London: Longman).

Chapter 17

Human occupance and the physical environment

David Jones

Introduction

The term 'physical environment' has traditionally been taken to cover the physical conditions and influences under which any individual or thing exists: air, water and land at the most basic level, to which most authors would also add flora and fauna. In the British Isles, the interrelationship between human societies and their physical surroundings has evolved in complex ways through time, but never more fundamentally than in recent decades when the term 'environment' has come into everyday usage and taken on new and varied meanings. Traditional constructions of the physical environment as a vast stage for human activity, providing both valued attributes (resources) and threats (natural hazards) in an ambivalent way, have come to be replaced by new constructions that increasingly focus attention on interdependence, change, threat, loss, uncertainty and concerns for the future. It is important, therefore, to begin this review with a brief overview of these changes before going on to examine some dimensions of the present interrelationships between human society and the physical environment in the UK.

A historic overview

Human occupance of the United Kingdom land area has not been achieved without cost to society. Early, pre-industrial populations were forced to overcome often severe constraints imposed by difficult terrain conditions (steep slopes, windswept uplands, swampy flood-prone lowlands), dense forest, unpredictable weather, wild animals and disease (including malaria which survived in Kent until 1918). Although it can be argued that the constraints imposed by the physical environment on human activity have been relaxed over time, through the application of science and technology, extreme events (natural hazards) continue to impose costs on society in terms of death, injury and distress, but more significantly as economic losses resulting from physical damage, loss of production and the general disruption to 'normal life'. Indeed, it must be recognised that increased technological ability does not necessarily make society less prone to hazard losses and in certain circumstances can result in growing vulnerability and escalating loss potentials. This is best illustrated by society's ever-increasing dependence on electricity, where any events that result in widespread power-failure, be they high winds (South East England in October 1987) or ice-storms (northern Scotland, January 1998), cause massive disruption and huge costs.

But this has not been a one-way process. Ten thousand years of human occupancy have resulted in profound alterations to the UK environment as have been widely detailed in the literature. These changes were slow and modest at first but gathered in pace and severity as the population grew, organisation improved and technology evolved. Developments in agriculture, industrialisation, urbanisation and the evolution of transport networks have all contributed, directly and indirectly, to modifications of landforms, atmospheric composition and behaviour, water movements, vegetation cover and animal populations, ranging in scale from the subtle to the extreme. Thus, the changes wrought in the last few decades, conspicuous and dramatic though they may appear, must nevertheless be viewed in their true perspective as merely representing the most recent phase in a long history of alteration.

The best-documented human impacts have concerned the effects of increasing human domination on the floral and faunal components of the ecosystem. This process of transformation, sometimes referred to as the 'diminution of nature', is most obvious in the widely developed urban environments or 'townscapes' where humans increasingly live, work, travel and recreate in controlled artificial environments set within radically transformed physical environments. Indeed, to a growing proportion of urban dwellers the concepts of physical environment, natural environment and nature are increasingly associated with the rural countryside. But it must be recognised that virtually none of the rural landscape can be described as 'natural', except for limited areas of 'wildscape' surviving in the remoter highlands and islands. The contemporary countryside is largely the product of culture and bears the imprint of a wide range of human activities; the varied agricultural landscapes that make up 'farmscape' have evolved through drainage of marsh, clearance of forests and the variable impact of the Enclosure Act. Change continues today, most particularly in the expansion of housing, industry and commercial activity and the removal of copses, hedges (until recently destroyed at a rate of 6,400 km per year: see Chapter 20) and heathland to provide larger, more efficient cultivation units. The ecology has changed as habitats have been altered. Species diversification carried out through the purposeful introduction of exotic trees, shrubs, plants, birds and animals (e.g. rabbits in the twelfth century), accidental releases (coypu, mink, parakeets, wild boar) and through the development of domesticated strains has, in part, been counteracted over the last few decades by the spread of expansive monoculture, pollution and the widespread application of chemical fertilisers and pesticides. Although some species have adapted well to these ecosystem changes and prospered (e.g. the seagull, house sparrow, starling, pigeon, collared dove, rabbit, fox, nettle) or have been actively encouraged (e.g. conifers in plantations), others have suffered serious decline or extinction (various raptors, Dartford warbler, smoothsnake, otter and numerous plants, butterflies and moths). Such examples of indigenous species decline, which are frequently highlighted in calls for nature conservation measures and controlled reintroductions (e.g. capercaille, red kite) to maintain biodiversity (see Chapter 20), thus merely represent some of the more obvious repercussions of human interaction with the environment.

Anthropogenic environmental change similarly affects all aspects of the abiotic physical environment: land, air and water. In fact, changes to any one usually result in response in the other systems. For example, the progressive changes in land use from original forest to the contemporary agricultural and urban environments have resulted in significant changes in surface conditions (roughness, water balance, thermal character) and thereby altered the micro-climate and run-off (drainage) characteristics, the most significant responses being increases in soil erosion, water pollution (suspended solids), river sedimentation and flooding. Evidence for anthropogenic induced soil erosion associated with early forest clearance, a process that would have been assisted by the arrival of the plough in circa 5000 BP, exists in the widespread occurrence of shallow, immature or truncated soils in upland areas, in the fact that most lowland floodplains are underlain by thick sequences of fine-grained alluvium laid-down during the last 9,000 years, and in the sedimentary sequences preserved in lakes which often reveal rapid increases in deposition at various times between 5,000 and 1,000 years ago.

Similarly, surface changes have collectively altered the thermal, hydrological and dynamic properties of the overlying air, and the addition of pollutants considerably changed its composition. The most obvious repercussion of these changes was that growing towns and cities created distinctive 'urban climates' with their characteristic 'heat islands', 'dust domes', and generally poor visibility conditions with frequent smoky fogs or smogs. Such features are of considerable antiquity (the term 'smog', i.e. smoke and fog, was coined in London in 1905). London is known to have grown large enough to modify the local climate by the mid-thirteenth century, largely due to use of sea-coal, first in forges and then in lime kilns. John Evelyn remarked in 1661: 'The weary traveller, at many miles distance, sooner smells than sees the city to which he repairs', an observation later supported by Gilbert White who noted the 'dingy smoky' appearance of the air in dry weather as far downwind as Selborne, Hampshire. The characters in the novels of Charles Dickens were frequently depicted groping around in the dense yellow fogs – or 'London particulars' as they were known – that plagued the capital, and Byron referred to a 'huge dun cupola' over the city. In fact Victorian London came to be affectionately known as 'The Smoke'. The Industrial Revolution led to particularly marked alterations, with widespread and intense air pollution in the industrialised coalfield areas, such as the Black Country, the Potteries and South Yorkshire. What has changed during the twentieth century is that the urban areas have expanded dramatically and altered markedly in form, while at the same time pollution emissions have changed in composition (see Chapter 18) with a decline in the traditional constituents (smoke and SO_2) and their replacement by motor vehicle exhaust gases.

Although the Clean Air Acts of 1956 and 1968 are often given the credit for the dramatically improved cleanliness of urban atmospheres and the associated improvements in visibility and recorded sunshine which, together with building cleaning programmes, urban renewal and pedestrianisation, have helped to make inner-city environments much more pleasant, in reality it was socio-economic and cultural changes post-1945 that were the dominant influences: the adoption of central heating, the growth in dependence on electricity (otherwise known as 'the electrification of society'), changes in the pattern of electricity generation, nuclear energy, the increased use of oil, the exploitation of North Sea gas, the modernisation and contraction of the railways and the progressive collapse of the heavy engineering industry. It was these factors that combined progressively to reduce the demand for coal (see Chapters 3 and 18, and p. 370 in this chapter), while at the same time causing an increasing proportion of coal burning to take place in large, efficient, high-temperature furnaces producing limited amounts of smoke which are discharged through high chimneys (tall stacks) and dispersed by the wind. As a consequence, the traditional cold weather, unhealthy smoke-fogs (12,000 Londoners are thought to have died prematurely due to the famous 'Great Smog' of 5–9 December 1952, 4,000 during the smog and 8,000 subsequently) that plagued cities during winter months, and occasionally became persistent and disruptive (i.e. the London smogs of 1873, 1880, 1882, 1891, 1892, 1905, 1948, 1952, 1956 and 1962) have been eradicated, only to be replaced by a new, summer, day-time, photochemical version first recorded in 1975, which is produced by the conversion of motor exhaust gases into a range of substances including PAN and ozone (see Chapter 18). Thus, human activities have continued to modify the atmosphere with the result that urban climates, in particular, continue to evolve.

While such indirect or accidental changes may be of great significance, the impact of direct or purposeful changes to the earth's surface is visually much more impressive. Human-made landforms of widely differing form and antiquity are ubiquitous. Created for an enormous variety of purposes, erosional forms range from the innumerable small pits that pock the landscape (37,000 were once recorded in Norfolk alone) produced through the removal of a few cubic metres of material and now often preserved as small ponds and copses, to enormous 'long-life' extraction sites where removal is measured in tens of millions of tonnes (see Chapter 4). Particularly impressive are the Norfolk Broads, a collection of twenty-five freshwater lakes created by the removal of 25.5 million cubic metres of material in peat diggings prior to AD 1300. Ancient depositional features range from the thousands of small prehistoric burial mounds (tumuli) and earthworks, to Silbury Hill, Wiltshire, at 40 metres the tallest prehistoric mound in Europe. Equally impressive are the ridge and furrow landscapes of the English Midlands which are thought to represent the survival of the medieval strip-field pattern. A survey of North Buckinghamshire in the late 1950s revealed that this patterning still covered 343 square kilometres or 28 per cent of the surface, each square kilometre of disturbance representing the movement of 62,000 cubic metres of earth. However, such legacies of past sculpturing pale into insignificance when compared with the products of the Industrial Revolution. The expansion of the railway network in the mid- and late nineteenth century led to prodigious anthropogeomorphic activity, including the creation of earthworks of such magnitude that Ruskin was moved to speak of 'your railway mounds, vaster than the walls of Babylon'. Even these were to be eclipsed in scale by the thousands of colliery spoil heaps that grew to dominate the surface of the coalfields. Spoil production expanded dramatically from the middle nineteenth century as deeper and thinner coal seams were exploited and mechanical extraction techniques developed, and the employment of mechanical tipping led to the creation of hundreds of huge conical heaps, many more than 50 metres high, that characterised coalfield landscapes until remodelled in the 1970s and 1980s. Now only the gleaming white sand mounds of the china clay industry – the so-called 'White Alps of Cornwall' – survive as significant examples of a once widespread landform, the towering conical spoil heap (see Chapter 4).

Human impact has not been confined to sculpturing the existing land surface. Coastal marshes and fens, as well as the lower reaches of most river valleys, were progressively reclaimed from the sea by the construction of dykes and levees, so that hundreds of square kilometres of land could be brought under cultivation. The Somerset Levels and Romney Marsh are just two examples of a widespread phenomenon which began 800 years ago and reached its most dramatic scale in the Fens, where 153 square kilometres had been reclaimed by 1241 and a further 500 added since 1640. Further modifications were achieved through the excavation of harbour basins and the dredging of channels, much of the material being used locally to extend the land area.

It is against this background of long-term and extensive environmental modification that the significance of recent and contemporary changes have to be assessed. Population growth, technological development and the changing patterns of urbanisation and economic activity have ensured that human occupance continues to exert evermore varied modifying influences on the physical environment, as has often been catalogued by geographers. But one important recent difference has been the growing appreciation of the magnitude of changes, which has arisen from increased interest in, and monitoring of, the physical environment. This has stemmed from the dramatic expansion in environmental awareness which has its roots in the 1930s, although the popular movement has only really flourished since the late 1960s. The intensity, complexity and spatial extent of human impacts are now known to have progressively increased with time, as is well illustrated by the example of atmosphere pollution which has changed from local (e.g. smoke) via the transnational to the regional scale (e.g. acidification), then to the sub-hemispherical level (e.g. stratospheric ozone depletion) and finally to the global scale (e.g. global warming). This knowledge has, in turn, fuelled concerns that the unchecked continuation of processes of transformation could result in unforeseen consequences severely detrimental to humankind, possibly even catastrophic as portrayed in popularist doomsday scenarios, thereby resulting in growing calls for controls on environmental change, the maintenance of biodiversity and the adoption of sustainable development. Suddenly the interrelationship between human society and the physical environment in Britain is no longer a parochial matter of academic interest only but has become a small element of global debates on the future of the ecosphere as testified by the Rio Earth Summit (1992) and the Kyoto Conference on Climate Change (1997).

The recognition of global environmental change has resulted in other re-evaluations of the interrelationship between human societies and the physical environment, most especially in the perceived ability of the physical environment to act as a benign framework for human activity and a limitless depository of waste. The progressive change in relationship stimulated by the Industrial Revolution has been referred to as 'The Great Climacteric' and has seen traditional, relatively simple, local environmental problems replaced by new, 'complex', multi-faceted, spatially extensive, 'first order' problems (e.g. global warming) which are sometimes referred to as 'problematics', 'concatenations' or 'syndromes'. This change from easily identifiable localised hazards to complex, uncertain international problems has also been termed the 'risk transition' and features a new class of hazards: diffuse, cumulative, slowly developing and insidious processes and changes with enormous potential for harm, known as 'elusive hazards'.

As the growing scale of human alteration to the physical environment has become recognised, so have the terms 'natural environment' and 'nature' come to be progressively replaced by 'environment'. While 'nature' was seen to be natural (i.e. non-artificial), reliable and ambivalent, the word 'environment' has increasingly come to mean human-modified, questionable and potentially problematic. The evolution of social constructions of the physical environment has resulted in the present situation where environment is increasingly seen to be risk-laden, containing unseen or invisible or elusive threats that can only be identified, determined and delimited by scientists. At the close of the millennium, therefore, the physical environment has achieved a greater than ever importance, in stark contrast to many predictions earlier in the century that envisaged human activities as becoming progressively divorced from the physical surroundings as technological ability came to dominate over the natural realm.

Many of the issues raised in this introductory overview are considered at greater length elsewhere in this book, so the remainder of this chapter will concentrate on human-induced alterations to the physical environment and the growing importance of environmental hazards, commencing with the latter.

Environmental hazards and risk

Hazard is an attribute that is definable as the propensity to cause harm, loss or adverse consequences. It is a human construct attributable to objects, substances, activities, processes or circumstances that result in harms, losses or costs to humans and what they value. Hazards are, therefore, defined by humans and not nature. They are cultural constructions and as human societies evolve and knowledge increases so too do the criteria for determining hazard and the range of phenomena that receive the label. As a consequence, hazard must be envisaged as a dynamic concept, in many regards similar to the concept of resource: indeed, Zimmerman's famous statement about resources can be adapted to hazard to yield 'hazards are not, they become'.

Traditionally, three broad groups of hazard have been recognised, 'social', 'technological' and 'natural', with much of the geographical literature focusing on the so-called 'natural hazards' which can be defined as 'those naturally occurring elements of the physical environment harmful to humans, human activity and the things that humans value'. Two broad categories are normally distinguished, geophysical and biological, which are capable of considerable further subdivision (e.g. atmospheric, geomorphological, floral, faunal, etc.). Attention on 'natural hazards' has been reinforced during the 1990s due to the United Nations proclaimed International Decade for Natural Disaster Reduction (IDNDR), which has sought to reduce the costs to global society arising from geophysical events (the so-called 'natural tax') by the international pooling of knowledge and expertise on how to mitigate adverse impacts.

However, the validity of the fundamental threefold division of hazards has also come to be questioned as research has shown that many hazard events are surprisingly complex in the sense that they involve combinations of 'social', 'technological' and 'natural' elements. The basic 'four phase' model of hazard – Incubation–Trigger–Primary Hazard–Consequences – reveals that while the trigger event and primary hazard may be easily categorised into one of the three main groups (although not necessarily the same group), the incubation process and the resulting consequences (impacts, further hazards, benefits) often involve complex combinations. This is well illustrated by the nuclear fall-out from the Chernobyl nuclear accident (28 April 1986) which was the result of technological failure due to poor operating practices (social), with the adverse impacts on the sheep-rearing industry in Britain arising because the radioactive material had been transported by the wind and deposited by rain so as to produce a spatially variable pattern of contamination (Figure 17.1). How to classify this event is clearly problematic and the question has to be asked as to whether the end results would have been significantly different had the same magnitude of

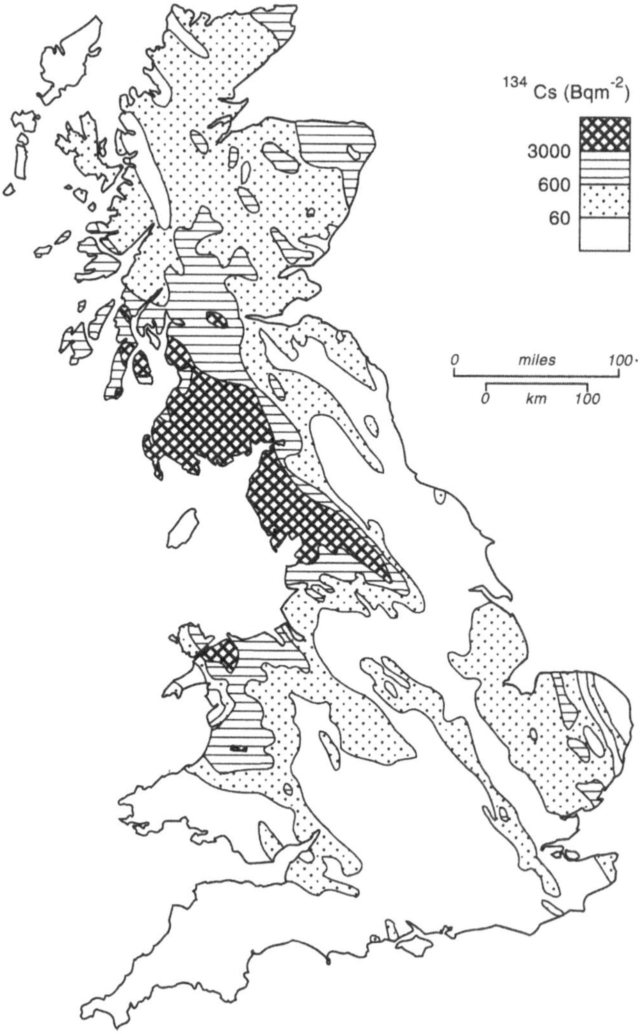

FIGURE 17.1 Generalised map of fall-out of Chernobyl-derived ^{134}Cs over Great Britain, April–May 1986

nuclear accident been achieved by a terrorist's bomb, poor maintenance, a design fault, corrosion or an earthquake? The answer is 'no', revealing that different triggers can result in similar primary hazards and adverse consequences (convergence), thereby indicating that while the classification of hazards on the basis of the trigger event may be crucial to hazard management, it is of less importance in risk management (see pp. 343–50).

The blurring of the distinctiveness between the three fundamental groups of hazard has for long been recognised. Human modification of the physical environment has produced changes in the magnitude, frequency and spatial extent of certain groups of extreme events (floods, landslides, avalanches, etc.) resulting in the identification of quasi-natural hazards where naturally occurring phenomena have been triggered or in other ways exacerbated by human activity. Pollution episodes involving naturally occurring substances are in the same category: for example, all rainfall is slightly acid and 'acid rain' is merely where the acidity of rainfall has been increased as a consequence of human activity. Where more exotic chemicals are introduced into the environment due to war, terrorism, accident, technological processes, waste disposal or as chemically synthesised pesticides, the resulting hazards clearly have social, technological and even natural triggers (e.g. the presence of concentrations of toxic chemicals due to carelessness, accident or landslide, flood, etc.), revealing a more complex interrelationship between the three groups. As a consequence, an increasing proportion of hazards are now interpreted as hybrid in origin (Figure 17.2), including those traditionally called quasi-natural hazards. Further, the term 'natural hazard' has become disputed as more and more geophysical phenomena are seen to reflect the increasingly profound influence of human activities on physical environmental systems, and has come to be replaced by 'environmental hazards' which can be defined as 'the threat potential posed to humans and what they value by events and circumstances originating in, or transmitted by, the physical environment, including the built environment'. Environmental hazards, therefore, represent a very broad collection of phenomena, including truly 'natural hazards' (earthquakes, tornadoes, radon), 'quasi-natural hazards' (landslides, floods, acidification, erosion) and a wide variety of hybrid hazards operating within the physical environment such as pesticides, atmospheric pollution, photochemical smogs, radioactive fall-out, oil-spills and other forms of water pollution, contaminated land, pests and disease.

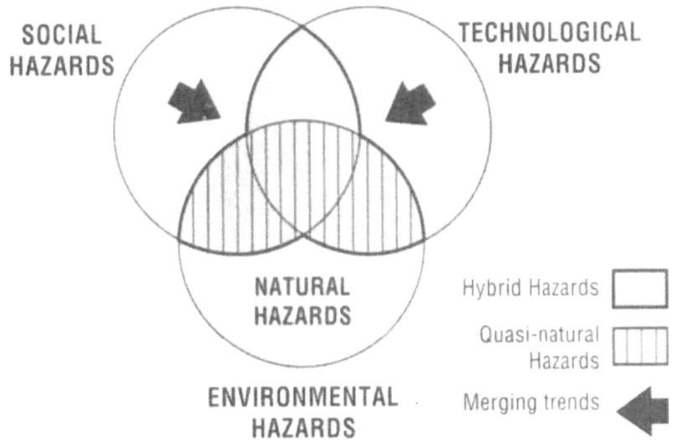

FIGURE 17.2 The hazard spectrum
Source: After Jones *et al.* (1993a).

Clearly, the multiplicity and diversity of environmental hazards render detailed consideration of their growing significance in Britain beyond the scope of a single chapter, so this contribution will concentrate on the 'natural' and 'quasi-natural' groupings of geophysical hazards, otherwise known as geohazards, with other aspects addressed in Chapters 18, 19 and 20. However, before embarking on this review it is necessary to briefly consider the topic of environmental risk, a subject that has rapidly gained in prominence over the past decade.

Risk has emerged as one of the most widely used and fashionable concepts of the 1990s, so it is not surprising that the terms 'hazard' and 'risk' are frequently confused in the literature. Both are cultural constructions, but whereas hazard is concerned with the actual and potential causes of loss or harm, risk is concerned with measuring exposure to the chance of loss or, to put it another way, the likelihood of differing levels of loss resulting from hazards. Exposure is often, incorrectly, assumed to be simply a function of the magnitude-frequency characteristics of hazard (i.e. event probability) but, in reality, it is mainly determined by vulnerability (V_u), or the propensity of individuals, groups, objects, systems and organisations to suffer harm, loss or detriment. Vulnerability is therefore the fundamental link between hazard and risk. It is a complicated concept that focuses on the value humans place on objects, activities, surroundings and even life, and the proportion of total value (V) potentially adversely affected by a hazard of specified magnitude (H_s). Thus total risk (R) for any context (individual, group, community, organisation, society, area) can be envisaged as the sum total of all specific risks (R_s), each of which may be expressed as $R_s = H_s*V*V_u$ where H_s is the probability of specified hazard of particular magnitude, or greater, occurring in a given period of time.

Risk is ubiquitous, for no human activity is risk-free. It is also a complex and abstract concept, for risk exists for events which have yet to happen or may never happen (e.g. a meteor impact destroying London), and for hazards that may never exist (e.g. a genetically engineered organism capable of extinguishing human society). As a consequence, there are many and varied interpretations of risk, and radically differing perspectives as to its nature and component parts. Of greatest relevance to this discussion is environmental risk which is, simply stated, the risk arising from environmental hazards and can be defined as 'an amalgam of the probability and scale of exposure to loss arising from hazards originating in, or transmitted by, the physical or built environment'. Clearly the accurate establishment of environmental risk for any context is an exceedingly long and complex process beset with four fundamental problems: lack of information on hazards, scientific uncertainties, the inability to foresee all possible outcomes, and the difficulty in placing agreed values on many outcomes. Thus it must be recognised that current pressures to increase the production of Environmental Risk Assessments (ERAs) for hazardous operations (e.g. chemical plants) and as an addition to the Environmental Impact Assessment (EIA) process, are not seeking complete appraisals but merely partial assessments relating to the probabilities of specific outcomes (e.g. the annual probability of death from a radiation leak in a specific nuclear power station) as determined by a process known as Quantitative Risk Assessment (QRA). However, the process of calculating environmental risks is still in its relative infancy and great advances can be anticipated in the early decades of the new millennium.

The significance of geohazards

The United Kingdom is not an area considered prone to geohazard impacts. There are no active volcanoes, significant earthquakes are infrequent (the last major one was the Colchester earthquake of 1884) and small by global standards, tsunami are even rarer, true hurricanes have yet to reach these islands despite the rhetoric generated by the October 1987 storm, and there has been no major meteorite impact in historic times. Thus few examples exist of dramatic, rapid onset, high energy, catastrophic events, thereby leading to notions that the physical environment is passive and of little relevance to the nation's economic performance. Indeed, risk assessments indicate that loss of life/injury is very much more likely to occur through industrial, transportational or domestic activity than due to so-called 'natural' events. Nevertheless, and despite technological development, geohazards continue to cause significant and mounting costs to society through destruction, damage, delay and loss of production. The vast majority of cost-inducing events are either directly or indirectly due to extremes of the weather: high winds, dense fogs, intense cold, blizzards, drought, landslides and river and coastal flooding. Often the effects are extensive rather than intensive, with the costs of impact absorbed by large numbers of individuals and organisations, but most especially electricity supply organisations, local authorities and the 'natural perils' sector of the insurance industry.

The repeated occurrence of geohazard impacts has resulted in management decisions ranging from no-action (loss-bearing) due to poor perception of risk, financial constraints or lack of technical knowledge, through a wide variety of adjustments which seek to limit the impact of similar events in the future. Such adjustments can be broadly subdivided into two groups: structural (engineering) solutions which seek to constrain/control the hazard or to protect the threatened population and infrastructure (dams, levees, flood walls, building codes, etc.), and non-structural (planning) solutions which seek to reduce hazard impact potential by means of spatial, temporal or financial adjustments (land-use zonation, forecasting, warning systems, insurance, disaster relief, etc.) (Figure 17.3). Alternatively, a distinction may be drawn between attempts to constrain or control hazard (hazard management), and measures focused on protecting society (vulnerability management), with both contributing to risk management. Irrespective of terminology, it must be recognised that the costs of geohazards are not restricted to the actual losses attributable to hazard impacts but must also include the costs incurred in determining threats, making adjustments and providing protection.

Chosen responses to geohazards are variable, with choice of adjustments (Figure 17.3) largely dependent on perceptions of risk, itself a complex function of hazard (size and character), frequency of occurrence and scale of expected losses. Studies have shown that risk assessments by individuals and groups result in risks being sub-divided into two broad groups, 'acceptable' or 'tolerable' risks (road accidents, lung cancer from cigarettes, etc.) and

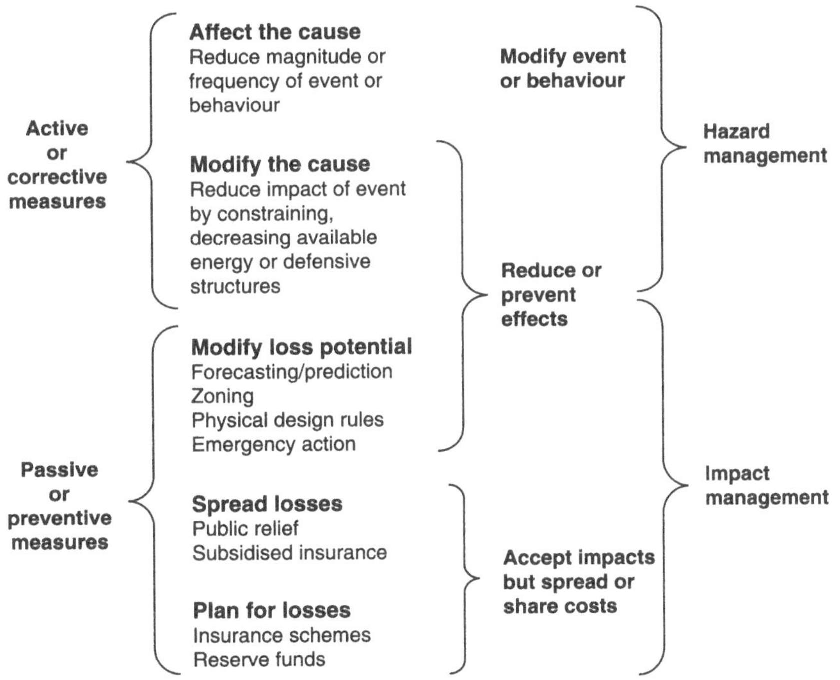

FIGURE 17.3 Adjustment choices

'unacceptable' risks (e.g. radiation from nuclear power stations), a perceptual division that is made on the basis of freedom of choice (i.e. freely chosen risks are much more acceptable than imposed risks) and the balance between perceived risks and benefits, irrespective of the statistical probabilities of harm. The ongoing concerns over the reduction in strength of the stratospheric ozone shield through the release of fluorocarbons is an example of this phenomenon, for the postulated increases in ultra-violet ray induced skin cancer are of a similar order to those resulting from two weeks' sunbathing on the Mediterranean coast, a risk which has yet to arouse the passions of a pressure group. Geohazard impacts generally tend to be placed in the 'unacceptable' category by the majority of UK citizens, because of the prevailing view that the organisational and technological ability of an advanced society should be sufficient to minimise (control) the impact of dynamic events in the physical environmental systems. This assessment illustrates the widespread failure to comprehend that hazard losses are as much the result of society's vulnerability as nature's violence.

Over the last seventy years the UK has suffered a surprising number of costly and often visually impressive geohazard impacts: severe river flooding, particularly in the English West Country (1952, 1968), the Scottish Moray area (1956, 1970), and Wales (1987), as well as along the Rivers Eden, Severn, Thames and Trent; a number of severe droughts which used to be ranked in terms of agricultural impact (i.e. 1976, 1934, 1944, 1938, 1974, 1964 and 1943) but more recently (1989, 1990, 1991, 1995, 1996) have been marked by claims on insurance companies for subsidence damage to housing (Figure 17.4) and water-supply shortages; blizzards and long snow-lie in the winters of 1939/40, 1941/2, 1946/7, 1950/1 (upland areas), 1954/5, 1962/3; 1969/70, 1978/9, 1985/6/7; wind damage, especially in 1950 (eastern England), 1965 (Yorkshire), 1968 (Scotland, twenty killed), 1987 (southern England, nineteen killed), 1990 (southern England), 1997 (Wales and Scotland) and 1998 (South West England), and frequent coastal storm damage. However, there have only been a handful of events of sufficient magnitude to qualify as natural disasters. The

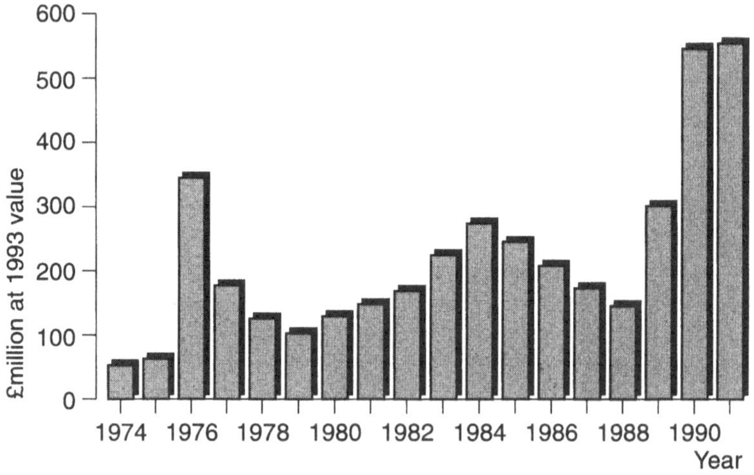

FIGURE 17.4 The cost of insurance claims in respect of building 'subsidence' damage in the UK, 1974–91
Source: After Doornkamp (1995).

eight most conspicuous are the Exmoor storm and resulting Lynmouth flood of 15 August 1952 (thirty-four killed, total cost £9 million at currently prevailing prices, including damage to ninety houses and 130 cars); the London smogs of 5–9 December 1952, 3–5 December 1957 and 3–7 December 1962, which caused the premature deaths of 12,000, 1,000 and 750 Londoners respectively; the East Coast floods of the night of 31 January/ 1 February 1953, which inundated 850 square kilometres, killing 307 people and causing estimated losses of £40 million, including 24,000 houses damaged; the Aberfan landslide disaster of 21 October 1966, when a spoil heap perched 150 metres above the South Wales colliery village suffered slope failure, the resultant enormous flowslide consuming several houses and a school, claiming 144 lives; the drought of 1975–6 which resulted in major losses to agriculture, 584,000 outdoor fires – some of which caused widespread destruction to forest and heathland – and some £100 million damage to over 20,000 houses and buildings due to subsidence, often because of extraction of moisture by tree roots; and most recently, the intense storm of 16 October 1987 when high winds gusting to over 160 kph blew down power-lines, telephone lines and 15 million trees, thereby virtually paralysing South East England for 24 hours and causing costs of about £1,000 million at 1987 prices.

These events may appear insignificant when compared with the losses inflicted by geohazards elsewhere on the globe (e.g. the 1988 Armenian earthquake) or the costs sustained during wartime, or even with the UK road casualty figures (over 250,000 killed during the last fifty years). Nevertheless, the scale of local devastation and/or the broader economic ramifications are more than comparable with those resulting from the most severe anthropogenic disasters such as the major mining accidents that punctuated the first half of the century, the explosions at Flixborough (1974) and Platform 'Piper Alpha' (1988), the Zeebrugge ferry tragedy (1988), the *Marchioness* riverboat disaster (1989), the Clapham rail crash (1988) and the Lockerbie air disaster (1989), the 1981 inner-city riots, the Hillsborough Stadium tragedy (1988) or the Docklands and Manchester bombs of 1996 and the Omagh bomb of 1998. Thus, the shock of these 'natural disasters', reinforced by the general view that geohazards pose 'unacceptable' risks which should be minimised by a technologically advanced society, stimulated investigations which resulted in management decisions of considerable environmental significance. The 1952 smog led to the establishment of the Beaver Committee whose Report (1954) identified the importance of smoke in generating toxic and persistent urban fogs and directly resulted in the Clean Air Acts of 1956 and 1968. The East Coast Floods of 1953 led to the Waverley Committee Report (1954) and investment in a wide variety of coastal protection measures, including the Thames Barrier Project. The Aberfan disaster (1966) provided a major stimulus for the lowering and reshaping of colliery spoil heaps, thereby leading to great decreases in visual intrusiveness and spontaneous combustion. Increasing awareness of the overall importance of flooding as one of the main cost-inducing hazards has resulted in the widespread implementation of flood alleviation projects, albeit on a somewhat *ad hoc* and piecemeal basis.

Finally, the growing recognition of the significance of the climatic factor has resulted in improvements in the accuracy of weather forecasts and attempts to achieve better ways of communicating prognostications. The latter have involved developments in the media (especially the graphics used in TV forecasts), the establishment of local forecasting offices, and the preparation of forecasts for specific purposes, including such hazards as adverse road conditions, severe weather, flooding and tidal surges. However, despite significant advances, forecasting accuracy is still bedevilled by difficulties stemming from the complexity of atmospheric processes and dynamics, and the fact that hazardous weather impacts can be produced by phenomena of vastly differing scale and duration, ranging from a 2,500-kilometre wide depression with frontal systems, to a small short-lived storm. Thus, the attainment of the necessary precision with respect to scale, intensity, timing and location are goals that are extremely difficult to achieve, and even the switch from analogue forecasting techniques (using past patterns of weather activity as a guide to the future development of synoptic situations) to quantitative computer-based numerical methods, combined with the increased employment of satellite imagery and radar, have only recently improved the detail of short-range (24-hour) and medium-range (1–7 day) forecasts. Despite the Meteorological Office's continuing optimism that the acquisition of new and evermore powerful computers, together with new methods of analysis, will result in dramatic improvements in forecasting accuracy, the failure to issue adequate warnings of severe events, such as the October 1987 storm, has revealed that many scientific, technical and organisational problems remain to be resolved. As a consequence, 'weather-sensitive' industries such as transport, farming, and building and construction, continue to suffer heavy costs through production loss, delay and the inefficient utilisation of resources, while damage, destruction, injury and loss of life will still be produced by the unexpected or unanticipated occurrence of infrequent high magnitude events.

Despite technological developments in investigating, monitoring and limiting the impact of geohazards, there is no evidence that hazard costs have declined over the past century. In fact, the reverse is probably true, although, as with other advanced nations, the situation is a complex one requiring information on the temporal changes in the levels of hazard (i.e. magnitude, frequency and distribution) and the exposure of society to harm (i.e. variations in the scale of losses likely to result from hazard impact as determined by wealth and vulnerability). Unfortunately, few reliable data are available to inform conclusions. Evidence for changing patterns of hazard occurrence over time is mainly restricted to atmospheric parameters (e.g. wind speed, temperature extremes, rainfall and drought) and flooding of certain rivers and coastal areas, supplemented by detailed studies of certain phenomena such as debris slides and coastal landslides (Figure 17.5). In many cases these do reveal increases in event occurrence through time which can be explained by natural and quasi-natural changes, most especially the growing influence of human activity in altering environmental conditions so as to actually increase the frequency and intensity of certain hazardous events (e.g. landsliding, river floods, coastal inundations because of sea-level rise due to global warming). However, this simple cause–effect relationship must be treated with caution as there are also examples where threat is diminishing due to human actions (e.g. reduced flooding in heavily regulated river systems, urban smogs). More important still, it must be recognised that while contemporary records may be relatively accurate, detailed and complete, the quality of past (historical) information is usually markedly inferior and becomes increasingly patchy and vague with increasing antiquity. This applies equally in the cases of field evidence and historical archived data. Thus there is an inevitable in-built bias to the present which results in an apparent incease in the frequency, and possibly magnitude, of recorded events with time, thereby sustaining the view that geohazards are increasing in significance and severity with time.

These features are clearly displayed in two recent surveys of hazard events. Lamb (1991) investigated the occurrence of storms in north-west Europe over the period 1570–1989 (Figure 17.5c) and revealed an apparent increase in storminess to the present

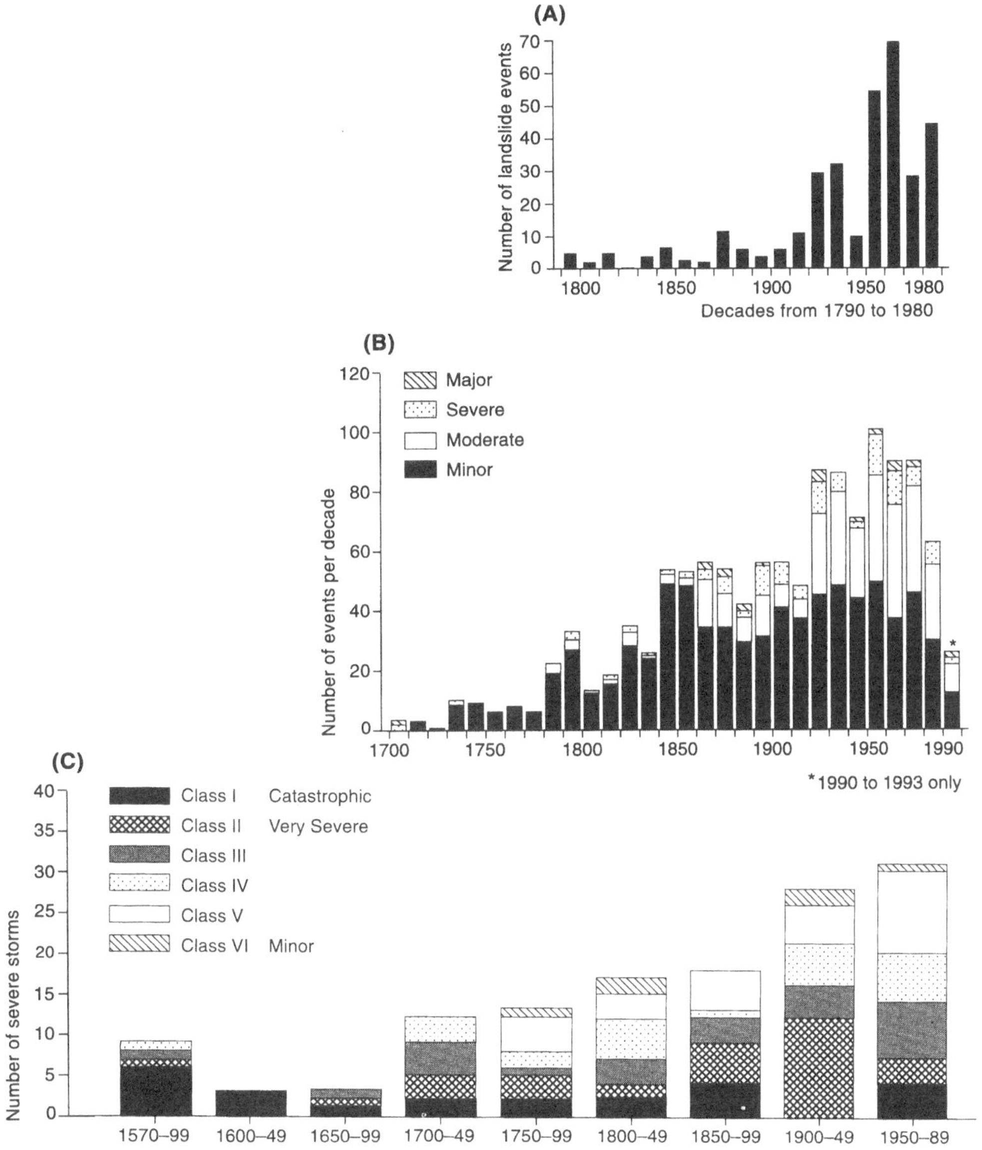

FIGURE 17.5 Changing patterns of recorded hazard events over time: (A) the frequency of coastal landsliding on the Isle of Wight, 1790–1980 (adapted from Brunsden 1995); (B) the frequency of erosion, deposition and flooding events, 1700–1993 (after DoE 1995); (C) storms of different severity class, 1570–1989 (after Lamb 1991)

with three particularly stormy periods: prior to 1650, 1880–1900 and since 1950. Recorded storms were divided into six categories on the basis of maximum wind speed, spatial extent and duration, and the resulting categorisation reveals that the numbers of recorded intermediate (classes III and IV) and smaller storms (classes V and VI) diminish rapidly with increasing time before present (Figure 17.5c). Thus it is difficult to draw firm conclusions regarding variations in the total number of storms over time, while the evidence for severe storms (classes I and II) suggests fluctuating occurrence rather than a consistent trend.

A second survey of literature sources on erosion, deposition and flooding events for the period 1700–1993 (DoE 1995) revealed a similar pattern (Figure 17.5b) with generally increasing recorded occurrences to the 1950s since when they have been approximately constant. While it is recognised that such reports only represent a sample of actual events and that no great credence should be placed on the apparent increase in number of events from 1700 to 1950, it is of interest to note that reports of 'major events' do not appear until 1860 (apart from one report in the first decade of the eighteenth century), since when they have been recorded in the majority of decades, averaging one per decade this century. Reports of 'severe events' show a similar pattern but appear earlier (in the 1790s apart from earlier reports in the 1700s and 1730s) and are recorded more frequently in recent decades, averaging seven per decade this century. Thus there appears to exist evidence in support of increasing scale of impacts through time.

Data on the costs of impacts are few and largely restricted to recent decades, when the increased involvement of the insurance industry in providing cover for natural perils has resulted in improved hazard-loss accounting. Even so, loss assessments remain vague and generalised, the true levels of impact remain difficult to estimate because of their complexity, and many adverse consequences are intangibles which cannot be easily assessed or valued in monetary terms (anxiety, ill-health, loss of irreplaceable 'valued' objects, reductions in amenity value, etc.). Thus, there is little evidence for the true costs of hazard impacts, let alone information on changing levels of impact through time.

Nevertheless, the results of the DoE study and analogues from elsewhere suggest some general trends. A reduction in loss of life due to increased awareness, better forecasting and warning procedures, well-developed emergency procedures and advances in medical care, has been offset by rising economic losses, for four main reasons. First, population growth and redistribution have caused pressure on land and resulted in the siting of an ever-increasing range of hazard-sensitive activities in hazard-prone zones (e.g. floodplains – see next section), despite the post-war growth of planning controls, thereby increasing the likelihood and scale of potential impacts. Second, the growing complexity and inter-dependence of commercial and industrial activity has meant that damage and disruption at one location can have massive repercussions elsewhere. Third, technological development sometimes leads to increased vulnerability, especially in the transport sector where the significance of climatic hazards has increased with motorway construction, railway electrification and air travel. And fourth, the evidence that human activity has increased the magnitude and frequency of certain hazardous events (see earlier). The main question that remains to be answered is whether or not the total cost of geohazards is rising faster or slower than the rate of wealth creation. If it is the former, then geohazards are really a growing problem.

River flooding

Despite the efforts of a succession of drainage and river management authorities and agencies, flooding continues to be a feature of both the summer and winter months. Prolonged heavy precipitation and/or rapid snowmelt are the main causes of high winter discharges, often resulting in widespread flooding. Classic examples include the catastrophic snowmelt floods of March 1947 which seriously affected much of Wales, the South West, the Midlands and the Thames Valley, and resulted in prolonged inundation of the Fenlands; the West Country and Welsh floods of 1960 and 1987; the widespread flooding of 1968; the inundation of Chichester in January 1994; and the severe South Midlands floods of 9–12 April 1998. Responses to this hazard have ranged across a variety of structural and non-structural adjustments, including the deepening, widening, straightening and regrading of channels (e.g. the Avon through Bath); channel maintenance; the construction of levees and floodwalls along low-lying tidal reaches; the employment of floodplain zonation so that the areas most liable to repeated inundation have land-uses with low-loss potentials (e.g. recreation), with more vulnerable land-uses (e.g. housing) placed on higher ground (i.e. 'set back'); the construction of large flood-relief channels to protect those urban areas particularly susceptible to flooding (e.g. Exeter, Spalding, Walthamstow); the designation of floodable areas (washlands), sometimes in connection with movable barriers, so as to protect vulnerable urban areas (e.g. Tonbridge), and the building of reservoirs. While no dams have been constructed purely for flood control purposes, there are several multi-purpose reservoirs with both water supply and floodwater storage functions. The impact of these reservoirs on flooding depends on the available storage capacity at the time of flood generation, and the volume of the flood wave. Research has shown that the scale of reduction of flood discharges increases with increasing reservoir size but diminishes with increasing size of floodwave. Thus, dams may reduce the volume of frequent low-magnitude floods by up to 70 per cent but have little or no effect on infrequent high-magnitude events. The application of combinations of adjustments to rivers particularly prone to flooding, such as the Dee, Exe, Severn, Trent and the systems draining the lowlands around the Wash and Humber, has resulted in considerable modifications to both their physical and discharge characteristics. Indeed, all main rivers are regulated to varying degrees and the overwhelming proportion of the drainage network has been modified due to 8,500 kilometres of capital works and major improvement schemes, with a further 35,500 kilometres benefiting from periodic dredging and vegetation control (Figure 17.6; Brookes and Gregory 1988).

Nevertheless, flooding continues to have a significant economic impact despite investment in flood protection schemes, although the true costs are unknown because they are spread throughout the economy, and published estimates are considered conservative. The reasons for this significance are complex. First, structural measures have been unevenly applied across the country. Second, the cost-effectiveness of protection measures diminishes with increasing size of flood. Consequently, structures are rarely designed to protect against discharges with recurrence intervals greater than a hundred years (i.e. flows likely to be equalled or exceeded *on average* once in a hundred years). Thus, there will always be floods whose magnitude exceeds the capacity of the provided protection measures. Third, a significant proportion of floodplain inhabitants and planners still have poor perception of flood hazard, with many of the former considering that the provision of any structural

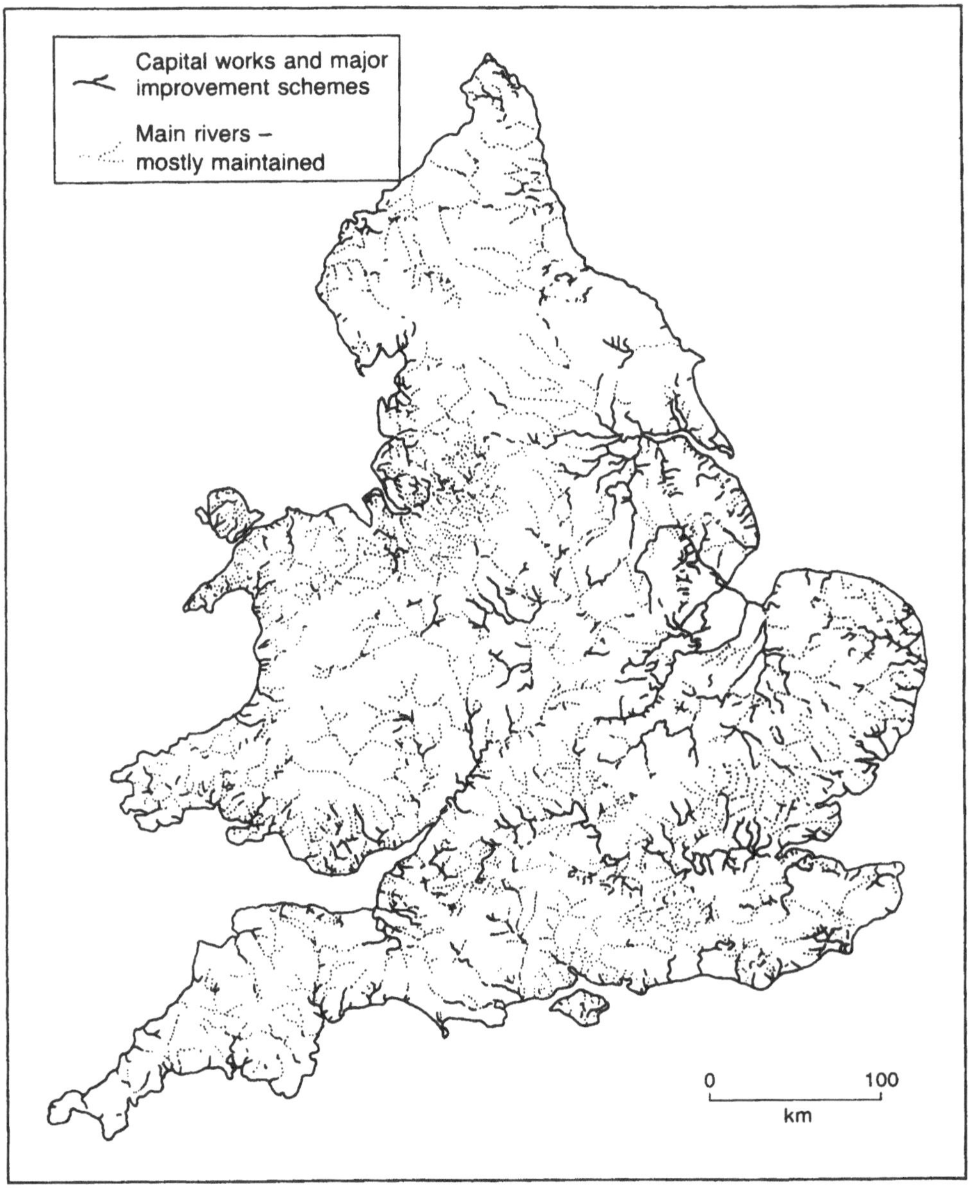

FIGURE 17.6 River channelisation in England and Wales
Source: After Brookes *et al.* (1983).

protection measures guarantees safety (the levee syndrome). Last, and probably most important, pressures on space have resulted in the continued development of floodplains due to an estimated 90,000 planning applications a year, many of which are approved despite the flood threat, so that zonation policies have only been poorly applied. Prime examples of the

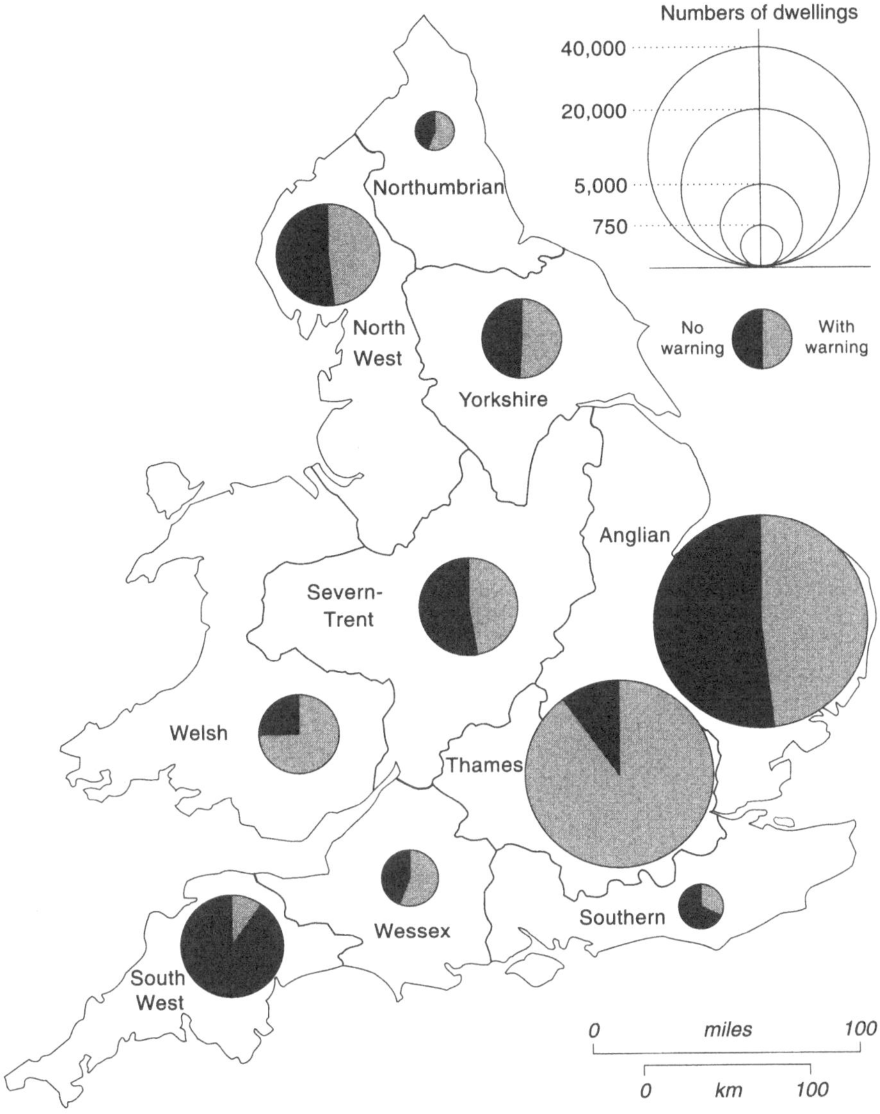

FIGURE 17.7 The number of dwellings in England and Wales in the mid-1980s within zones affected by 1:50 to 1:100 year flood levels
Source: Redrawn from Penning-Rowsell and Handmer (1988).

consequences are the Trent at Nottingham, where there was severe community disruption in 1990, and the Thames near Maidenhead which has proved so floodprone as to require an £80 million defence scheme, equivalent to £14,000 per house protected. All of these factors have tended to counterbalance the benefits stemming from improved monitoring, better forecasts and warning systems, and the increasing use of predictive computer-based models. Indeed, up to 450,000 people are thought to live in areas liable to flooding (Figure 17.7) with a recurrence interval of 50–100 years, many without the benefit of a flood-warning service. Loss-bearing and dependence on relief funds are still common responses to floods, although the relatively recent increased use of flood insurance cover has improved the situation. Uncertainty regarding the magnitude–frequency characteristics of geophysical events has always caused insurance companies to be reluctant to provide cover for natural perils – especially flooding, where loss claims can reach catastrophic levels. Thus, while storm damage cover has been available since 1929, it was only after the widespread 1960 floods that combined pressure from the then Ministry of Housing and Local Government and the building societies overcame the reluctance of the insurance companies and resulted in the availability of flood insurance cover. However, uptake remains patchy and many people still suffer great losses from unexpected events, such as in Chichester in 1994.

The second main type of river flood, the flash flood, can result from dam failure but is more characteristic of small, steep catchments where rapid run-off is caused by intense precipitation, usually during the summer months. In fact, summer storms often result in over 100 mm of rainfall in a few hours over a localised area (Figure 17.8) and account for the surprising fact that August has the worst record for catastrophic floods, closely followed by July. The magnitude of the resultant flooding is dependent on a number of variables relating to precipitation (area, duration, intensity, total), antecedent moisture conditions, and drainage basin characteristics (geology, land use, drainage network shape, topography, channel slope). Thus, the most dramatic rainstorm in recent history, the Hampstead storm of 14 August 1975 (Figure 17.8) which deposited an estimated 2,000 million gallons (8 million tonnes) of water (up to 170 mm of rain) over an extremely localised area of North London in under three hours, only caused local problems of surface ponding and the flooding of basements, mainly because much of the water was quickly transported away by stormwater sewers. By contrast, similar storms on small impervious clay-floored catchments can result in severe, albeit short-lived, flooding. However, the most spectacular and destructive floods occur when heavy rains fall on upland regions with short, steep catchments. The most notable examples include the Moray (NE Scotland) floods of 1956 and 1970, the Mendip floods of 1931 and 1968, and the Somerset and Devonshire floods of 1952 and 1968. The last mentioned includes the catastrophic Lynmouth floods of 15–16 August 1952, when up to 300 mm of rain fell on Exmoor in 24 hours (Figure 17.8), resulting in such feats of erosion by the diminutive River Lyn that fallen trees and boulders created temporary dams behind bridges, the breaching of which greatly intensified the devastating nature of the floods downstream. Such extreme floods are difficult to forecast because the storms that create them are usually the product of localised topographic or urban conditions. The short response times of the rivers (basin lag) means that forecasting is problematic and that warnings either cannot be given in time or fail to reach the entire population at risk, and emergency services are often hampered by inadequate preparation time. Planning for such events is also impracticable because their rarity (long recurrence intervals) means that there are few available comparable records to use as a guide, while the structural adjustments required to cope with potentially very high magnitude floods are prohibitively expensive and totally non-cost-effective. For example, the Lynmouth flood had a calculated recurrence interval of 50,000 years, due to the diminutive River Lyn having swollen to a size equivalent to the Thames in spate at Teddington and in so doing moved more than 100,000 tonnes of boulders, some weighing up to 7.5 tonnes. Such catastrophic events will continue to happen through the chance combination of factors and, as the past record of intense storms shows

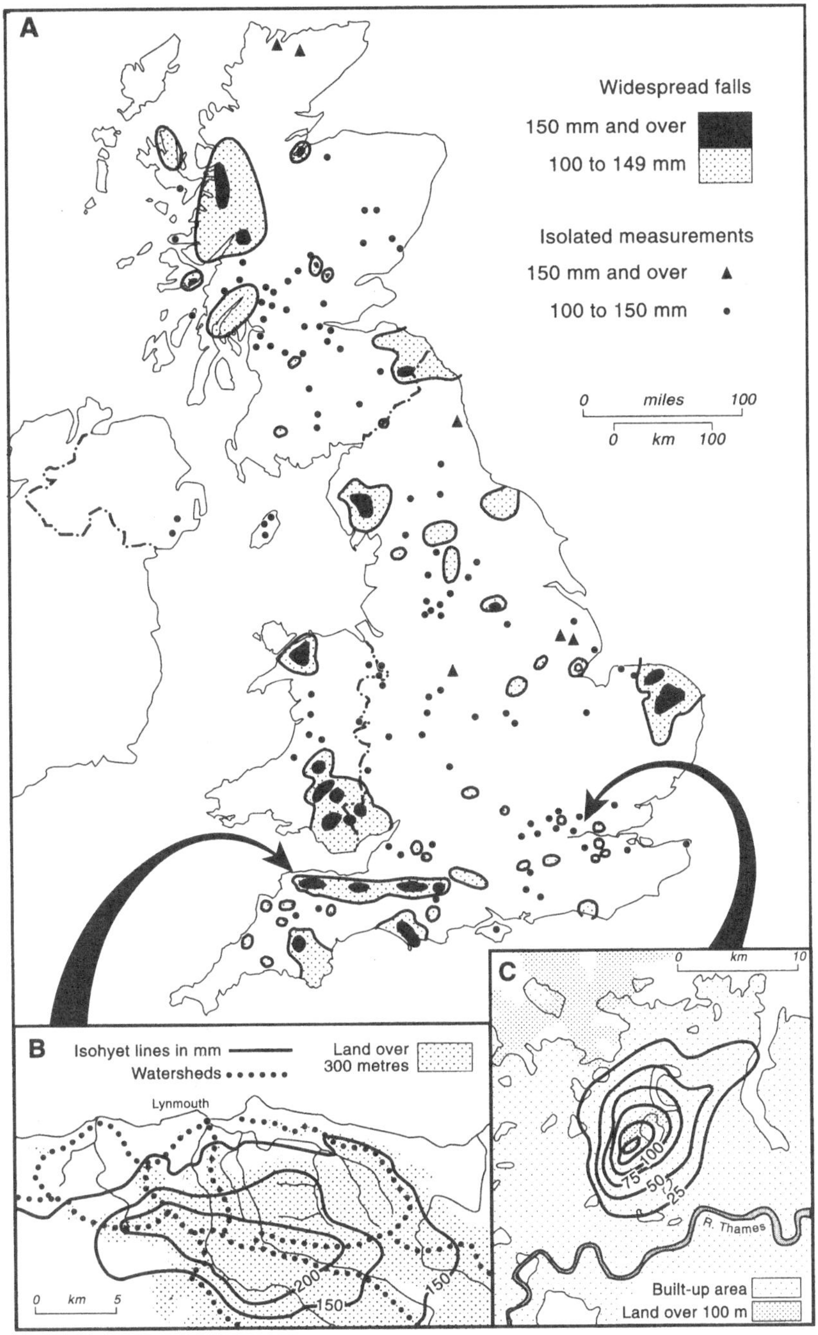

FIGURE 17.8 Rainstorm hazard in the UK: (A) distribution of daily rainfalls in excess of 100 mm, 1863–1960; (B) isohyets of Exmoor storm, 15 August 1952; (C) isohyets of Hampstead storm, 14 August 1975

(Figure 17.8), could occur virtually anywhere. Thus, the only feasible loss-reduction solution is to concentrate on saving lives and minimising damage to property by the employment of sophisticated monitoring techniques (e.g. radar, and interrogable rain-gauge and river-level systems) and the development of efficient warning and evacuation procedures.

While the majority of floods are generated by 'natural' mechanisms, human activity has been of considerable significance in changing the discharge characteristics of rivers and, under certain circumstances, has contributed to flood generation. Urbanisation has had a particularly great impact, for the sustained rapid rate of land conversion, which reached 26,000 ha per year in the early 1970s, results in a wide variety of hydrological consequences (Figure 17.9). Erosion during construction can cause suspended sediment concentrations to rise by factors of two to ten, and sometimes up to one hundred, the resultant sedimentation downstream increasing flood risk and raising channel maintenance costs. Once completed, the urban areas provide radically different surface conditions, for the buildings, roads and integrated stormwater and sewer systems cause reductions in interception and infiltration to soil and groundwaters, and result instead in the rapid run-off of an increased proportion of

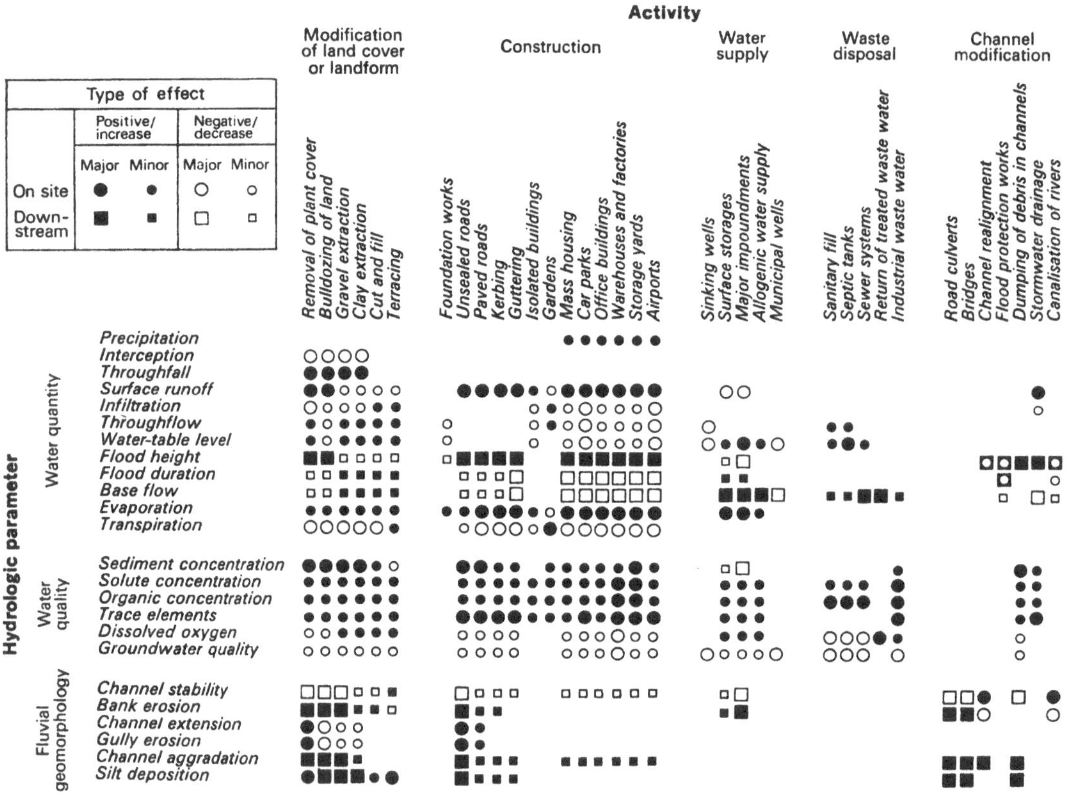

FIGURE 17.9 The potential range of changes to hydrological parameters produced by urbanisation
Source: After Douglas (1976).

precipitation. As a consequence, even partially urbanised catchments show increased annual water yields, higher peak flows and more rapid responses to precipitation events (reduction in basin lag-times). This increase in flashiness is particularly apparent for summer floods on small catchments, which may experience up to an elevenfold increase in peak discharge, thereby contributing to the widely reported enlargement of channels downstream of urban areas. However, the urban influence is only apparent on small and medium-sized floods, for there are few indications that discharges with recurrence intervals in excess of twenty years are affected, possibly because high flows result in a 'throttling' (holding back) of water in urban stormwater sewers.

Urbanisation is not merely associated with locally increased flooding, for water supply and effluent disposal can produce significant changes in river regime. Reductions in discharge result from reservoir construction, direct abstraction and the drying-up of springs due to the over-exploitation of underground aquifers, with the result that many chalkland streams have shrunk, dried-up or become only periodically active. In the case of the Chalk beneath London, aquifer water-table lowering amounted to up to 60 metres in the period 1875–1965. However, growing concern about undergroundwater resources has resulted in recent careful regulation of extraction. This, together with the employment of artificial recharge schemes (e.g. near Newbury) has halted the trend in many cases and is actually causing some water tables to rise again, most significantly beneath London where the rise has been so dramatic as to pose a real threat to basements and buried infrastructure, including the underground railways. Conversely, increased base-flow can result from the release of large volumes of water redistributed by piped water supply systems as inter-basin transfers (see Chapter 2). The scale of such allogenic water supplies is most apparent in the case of Birmingham where used-water, originally derived from the upper Wye in Wales and piped to the Midlands, is discharged into the Trent catchment. Many of the larger rivers appear to have a base-flow that is predominantly effluent, the River Tame being 90 per cent effluent in its 95 per cent duration flow.

Anthropogenic influences on flooding are not restricted to urbanisation, for recent agricultural developments have also had important hydrological consequences. The expansion of arable cultivation, the enlargement of fields by the removal of hedges and copses, the use of machinery which compacts soil thereby reducing infiltration, and the massive post-war growth in the use of tile-drains, have all contributed to the increasingly rapid transport of larger and larger proportions of precipitation from slopes into channel systems. Thus, the flashiness of most streams and small rivers in lowland Britain has been increased during the last half century.

Coastal flooding

A very different flood-hazard threat exists around the coastline of Britain, where adverse weather conditions combined with slowly changing land–sea relationships result in occasional breaches of defensive structures and the inundation of adjacent lowlands by saline water. The most important recent examples of such events are the Uphill (Somerset) floods of 13 December 1981, the Selsey (Sussex) floods of 4–5 January 1998 and the Towyn (Clwyd) inundation of 26 February 1990. The Towyn flood was the most significant of these events, inundating 2,800 homes and resulting in the evacuation of 5,000 people from the

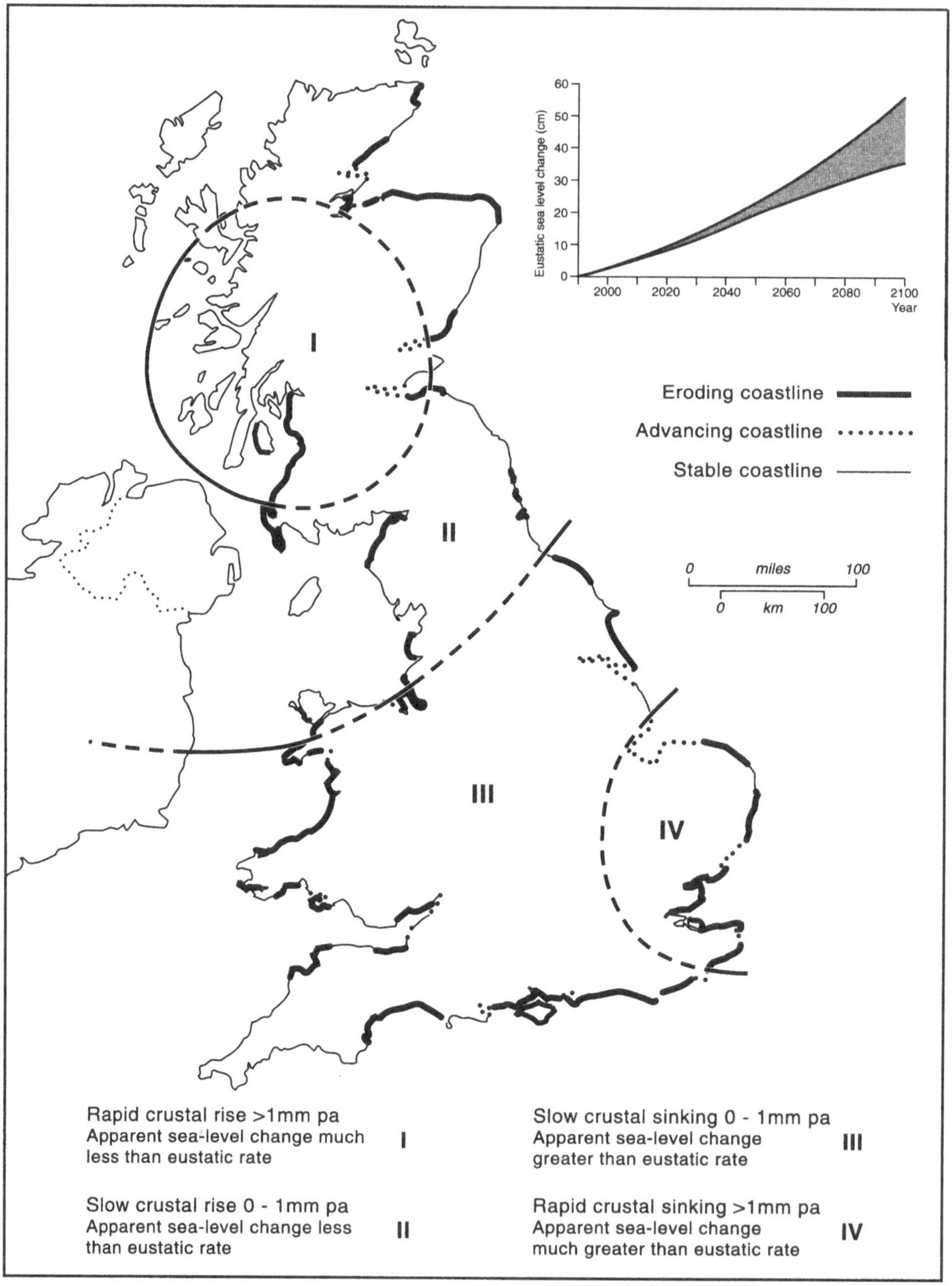

FIGURE 17.10 Generalised pattern of predicted sea-level change
Sources: Modified from Jones *et al.* (1993b) and Wigley and Raper (1992).

town itself. It also highlighted many of the features of coastal flooding: the high proportion of old people affected because of the tendency to 'retire to the coast', the low proportion of inhabitants with contents insurance, and the high levels of trauma which result in significant numbers of post-event deaths (fifty-four in the Towyn case).

It is along the coastline of eastern and southern England that the threat is greatest with documented records of local coastal flooding along the east coast dating back to 1236, including at least seven inundations that were of disastrous proportions. Here, local tectonic movements, including the progressive downwarping of the North Sea basin, together with the global (eustatic) rise in sea-level (currently *c.* 1.2 mm per annum) and changes in tidal configuration due to dredging, dumping, reclamation and coastal defences, have combined to produce significant rates of local sea-level rise. Analysis of tide-gauge records in the South East indicate that sea-level is currently rising at varying rates of 140–540 mm/100 years (Figure 17.10). The higher figures are from the Solent, Dover Straits and Lower Thames because of changing tidal characteristics. This situation of gradually rising high-water levels is exacerbated by the occurrence of storm surges generated by deep depressions tracking over the North Sea or along the Channel. Sea-surface levels can be raised over 2m through the combined effects of pressure reduction (305 mm rise for each 30 mbs reduction in atmospheric pressure) and the piling of water against the coast by strong on-shore winds. Coastal defences become severely stressed when these surges coincide with particularly high (Spring) tides, and can be over-topped and breached by high waves generated by the strong winds (e.g. Selsey, Sussex, January 1998). The risks are even greater in estuaries, for any reduction in wave-heights resulting from sheltering is more than compensated for by the raising of surge level due to funnelling, and the water surface may be dramatically raised still further if a surge coincides with a period of high river flow into the estuary. As extensive coastal tracts bordering the North Sea and the Channel stand at elevations little above mean sea-level, the potential for disastrous inundations is great. The economic consequences of such an event could be enormous and include the saline contamination of agricultural land and groundwaters, the swamping of residential areas, the paralysis of waterside industries, the dislocation of communications, and the severe disruption of activity in London.

The dramatic events of the night of 31 January/1 February 1953 graphically illustrated the problem and were to have particularly great repercussions in terms of risk perception and the implementation of remedial measures. Stormforce winds generated by a deep depression (minimum 968 mbs) drifting south-eastwards across the North Sea caused a surge up to 2.5 metres high to pass southwards down the east coast. Fortunately, the passage of the surge did not quite coincide with the progression of only a moderately high tide and the rivers were not in spate. Nevertheless, 850 square kilometres were inundated (Figure 17.11) along nearly 500 kilometres of coastline (the so-called Isle of Thanet reverted to a true island for a short time), 24,000 houses were swamped, cement works, factories and power stations on Lower Thameside were inundated and brought to a standstill, and 307 people were drowned. The subsequent committee of enquiry focused attention on the factors causing flooding and stressed that the situation would get progressively more serious due to regional subsidence. The initial reaction was to strengthen and raise the level of sea defence (levees, walls, etc.) from Yorkshire to Kent so as to allow them to withstand 1953 surge levels, a task that took fourteen years (1954–68) and cost £55 million. The new defences were completed none too soon, for although the initial recurrence interval calculated for the 1953 surge level was 200 years, the 1953 heights have subsequently been

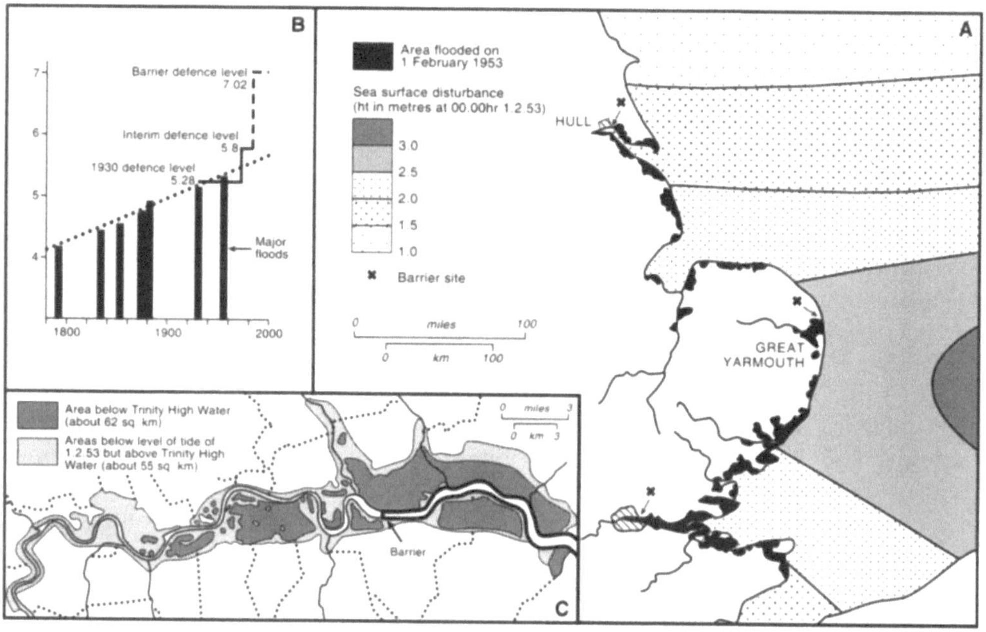

FIGURE 17.11 East coast flood hazards: (A) the 1953 flood; (B) rising surge levels and defences along the lower Thames; (C) the hazard zones in London

locally equalled or exceeded on three occasions: 29 September 1969, 3 January 1976 and 11/12 January 1978. The 1978 event involved a surge of up to 1.73 metres coinciding with the highest tide of the fortnightly cycle and produced levels which were locally well in excess of those recorded in 1953. In this instance, flooding was most severe around the Wash and in Lincolnshire (King's Lynn, Wisbech and Cleethorpes), with damage estimated at £200 million (but only one death). The relatively high damage costs indicate that although the defences were much stronger and restricted flooding, there had been only limited implementation of land-use zonation in association with the engineering works. Thus, residential, commercial and industrial activity continued to be located on the high-risk, low ground behind the defence, in the belief that such areas are safe because they are protected (the so-called 'levee syndrome'). On the other hand, the low death toll reflected the improved forecasting and warning systems that now operate and the better-disciplined evacuation procedures that stem from greater hazard awareness. Particularly significant in this context is the Storm Tide Warning Service established in 1954 by the Meteorological Office at Bracknell, which delivers phased warnings to relevant organisations; an 'alert' twelve hours before expected high-tide followed by 'confirmations' (or cancellation) at four hours and 'danger warnings' at 1–2 hours. As from September 1998 this operation will be known as the Storm Tide Forecasting Service to reflect the fact that it does not issue general warnings and to distinguish it from the Environment Agency which does issue warnings.

The protection of the coast represents a costly but technically less difficult problem than that of defending the land flanking the Lower Thames (Figure 17.11). No less than 61.4

square kilometres of London lies below Trinity High Water Mark (+3.48 metres OD) and 116 square kilometres below the 1953 flood level (+5.48 metres OD). Fortunately, the 1953 surge only affected 1,100 houses in east London because vast volumes of water had escaped from the confines of the Thames estuary through the numerous breaches along the Essex shore. A repeat of the 1953 surge, but with no such water escape, would be catastrophic for the capital city, with a currently estimated damage potential well in excess of £5,000 million. Unfortunately, the area at risk is growing steadily due to the combination of local and regional subsidence, progressively accelerating eustatic rise due to global warming (Figure 17.10) and changing tidal configuration. Thus, although the medium tide level at Tower Pier is rising at 433 ± 82 mm/100 years, only slightly more than the outer estuary figure of 344 ± 70 mm/100 years recorded at Southend, the high-tide level at Tower Pier is rising more rapidly (744 ± 116 mm/100 years) due to the mean tidal amplitude growing at 344 ± 70 mm/100 years.

The management response has been to construct a flood barrier in the Long Reach at Silvertown, near Woolwich, and to raise flood defences downstream of this structure. Defences upstream through Central London to Teddington Weir had already been raised to +5.28 metres in 1930–5, following the disastrous Thames flood of 1928 (+5.15 metres at London Bridge, fourteen drowned). These were further raised to +5.80 metres between April 1971 and December 1972 as an interim protection measure (removed in 1986) and a movable barrier constructed at Bow Creek to avoid raising banks along the highly industrialised Lee Valley. However, the main protection is derived from the movable drum barrier across the Thames designed to give protection against the 1,000-year flood (+7.2 metres OD). The choice of the Long Reach site was governed by the need for firm foundations (Chalk) and a straight section of river to ease navigation problems – a need that has subsequently declined with the closure of the Port of London. Work was begun in 1974 and completed in November 1983, over two years late because of technical problems and labour disputes. The cost had similarly escalated dramatically due to inflation and spiralling labour costs, the 1960 estimate of £10 million rising to £242 million in 1978 and £500 million in 1983. The total cost of works, including the raising of downstream defences to Southend and the construction of a barrier across Dartford Creek, came to about £700 million at 1983 prices.

Continuation of present physical trends, plus the accelerating rise in eustatic sea-level anticipated through CO_2 induced climatic warming (Figure 17.10), makes it inevitable that even higher defences and more ambitious schemes will be required in the near future. The Thames Barrier is currently raised about twice a year for defence purposes, but by 2030 the figure will have risen to about ten times per year and consideration is being given to establishing a new, higher line of defence to cope with a rise in mean sea-level of at least 0.5 metres by 2100 ('Project 2100'). Elsewhere, sea-defences and river levees will have to be raised and more movable barriers constructed across inlets, similar to that completed across the River Hull at Hull. The number of people threatened by sea-flooding, currently estimated at over 200,000, will have undoubtedly increased greatly thereby raising alarm in the insurance industry and highlighting the perceived need for additional defence measures. The key question is whether engineered measures designed to defend every metre of vulnerable coast will be cost-effective or desirable in the long term, thereby raising the possibility of an alternative strategy involving a phased retreat from the most threatened sections of coastline.

Human occupance and climate

The interactions between human society and climate are complex and continuously evolving. As already mentioned, despite technological development, weather phenomena remain a major source of cost to the UK economy. Heavy precipitation, atypical prolonged snowfall (south-east England in 1985/6 and 1986/7), unusually intense cold (–28°C, Kent, February 1985) and drought (1976) have all extracted significant levels of 'natural tax' in recent years. Rainfall and freezing conditions are probably the most widespread and pervasive problems, causing repeated heavy costs to the agricultural, transportation and construction sectors. Locally intense weather features also remain problematic, with thunderstorms causing scattered problems due to lightning strikes, hailstone damage, violent downbursts of wind and heavy rain (Figure 17.8). Even tornadoes occur, with an average of thirty-two reports a year, mainly in southern and central England (Figure 17.12), with the highest frequency and severest examples (e.g. the Selsey Tornado of 7 January 1998) occurring during the period September–January when the temperature contrasts between polar and tropical air-masses along frontal zones are at their greatest. But the most problematic hazard remains the wind.

The British Isles are amongst the windiest areas on the globe and gale damage is estimated to affect an average of 230,000 buildings a year, resulting in costs of £15–20 million (at 1987 prices). High-magnitude/low-frequency windstorms (tempests) achieve much greater impacts. The January 1953 gale blew down 5 per cent of the total standing value of coniferous timber in the UK (Figure 17.13); the January 1968 gale damaged 340,000 houses in the Glasgow area, and the gale of 2 January 1976 caused widespread damage throughout England, Wales and Ireland. More recently, the storm of 16 October 1987 had winds gusting up to 166 kph which devastated much of south-east England and represented the most severe event to affect that area since the 'greatest storm on record' (26 November 1703) which killed about 8,000 people. This time there were a mere nineteen deaths, although the destruction, damage and disruption was enormous as 15 million trees in full leaf, with root holds weakened by the saturated soil, were either broken or blown down wholesale by the violent gusts (Figure 17.13). It is fortunate that the most violent winds passed over the land during the night (2–4 a.m.) when most people were safely in their homes.

Since then there have been a number of powerful storms produced by deep depressions rushing eastwards over the British Isles, each of which has resulted in significant adverse impacts at regional levels because heavy snowfalls, freezing rain, flooding and wind damage have acted in varying combinations to disrupt electricity supplies, transportation and communications, to the detriment of social life and commercial activity. Significant examples include the 25 January 1990 'Windstorm Dora' which affected southern England, the *Braer* storm of January 1993 and the storms of December 1997 and January 1998 which particularly affected south-west England, Wales, Cumbria and south-west Scotland.

Despite the widespread appreciation that weather events are a significant cost factor to the UK economy, several fundamental interrelated questions remain to be answered, even in the most general terms. First, the true cost of weather events has yet to be ascertained. The main problems here concern defining the temporal and spatial limits for impact assessment and the complexity of the various intangibles; for example, what are the full costs resulting from a one-hour delay to a train due to fog, frozen points or freezing rain? Second, the extent

FIGURE 17.12 Distribution of severe tornadoes on record (until 1980) in England and Wales
Source: Adapted from Perry (1981).

to which costs associated with different specific weather conditions are rising or falling has yet to be evaluated. Third, whether evolving society is becoming increasingly or decreasingly vulnerable to weather-induced costs is little understood. Finally, the extent to which human activity has changed the weather is only poorly appreciated, despite growing concerns about the possible effects of global warming which could affect both patterns of weather (i.e. warmer winters) and the magnitude–frequency characteristics of extreme events (e.g. storm winds, tornadoes, thunderstorms, hail) (see Chapter 19). Some of these points are developed further in the following sections.

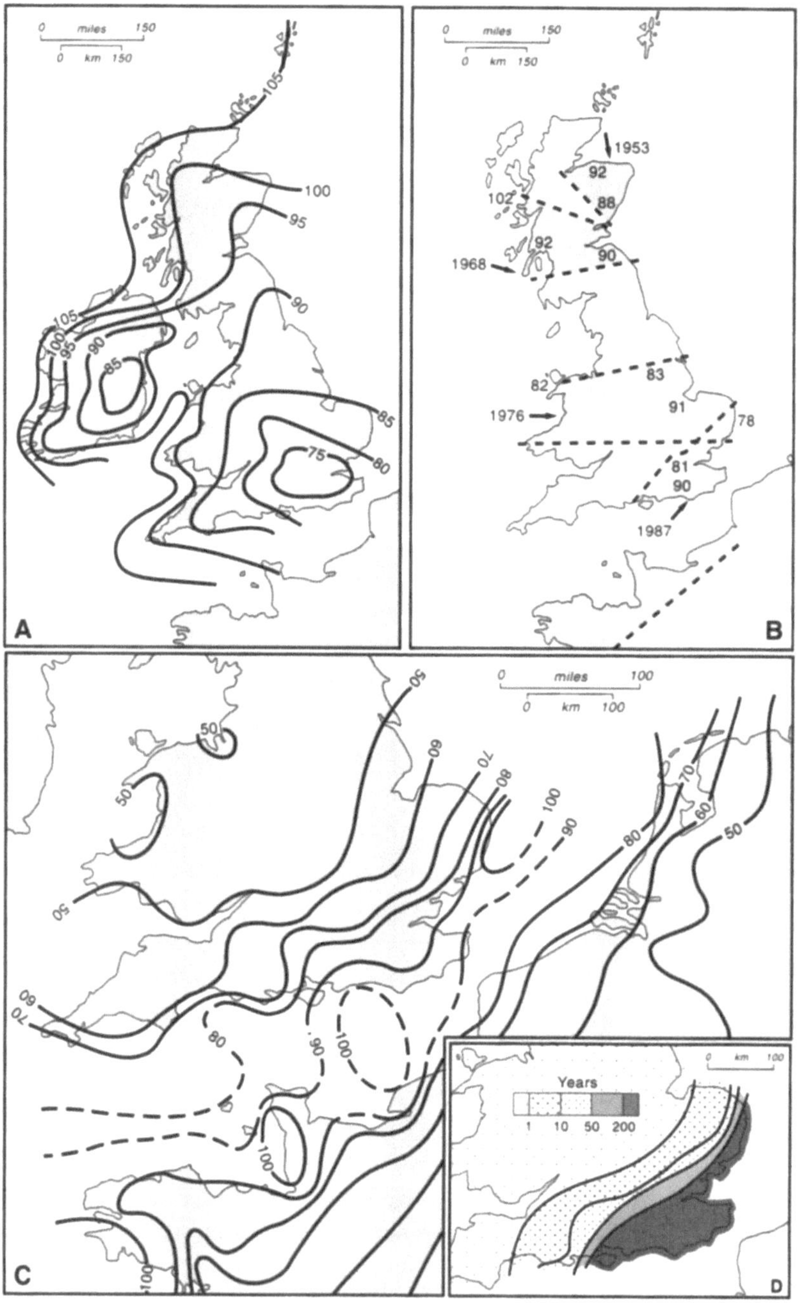

FIGURE 17.13 Wind hazard in the UK: (A) gust speed in knots with recurrence period of fifty years; (B) regions affected by catastrophic damage to forests, 1945–87; (C) highest reported gusts (knots) for storm of 16 October 1987; (D) approximate return period (recurrence interval) of highest gusts recorded on 16 October 1987

Local climate change

Centuries of economic activity and associated land-use changes have greatly altered the thermal, hydrological, dynamic and chemical properties of the overlying air, and consequently rainfall, temperature, visibility and windiness. Over most rural areas the increasing openness and better ground drainage has possibly resulted in slightly reduced mistiness and the accentuation of frost pockets, the latter due to greater radiation cooling at night and better air drainage. Similarly, the progressive removal of trees and hedgerows has increased near-ground wind speeds, thereby raising the potential for soil erosion (see p. 372). In general, however, recent changes in rural micro-climates have been relatively minor and difficult to detect.

By contrast, the urban-industrial environments continue to create marked and readily observable modifications to local climate, due to urban growth, changes to the physical nature of the urban fabric (high rise building, canyon streets, etc.) and variations in the volume and composition of pollutants. Although atmospheric pollutants are fully described in Chapter 18, their micro-climatological significance is such that they must also be briefly discussed here.

The nature and variation of urban climates was revealed first by studies in London in the 1950s and 1960s and then in other cities including Birmingham, Manchester, Nottingham and Leicester. The main features identified were (1) generally higher air temperatures than surrounding rural areas; (2) increased incidence of poor visibility; (3) decreased average wind speeds but increased gustiness; and (4) increased cloudiness and slight changes in precipitation characteristics. It is necessary first to examine the characteristics of urban climates as they existed in the 1950s and 1960s before looking at recent changes and assessing likely future trends.

Urban heat-islands

Of all the changes wrought by urban areas to their atmospheric environment, the most frequently studied have been the changes to air temperatures. The development of a shallow dome of slightly warmer air above cities results from (1) the greater capacity of concrete and brick structures to absorb, store and re-radiate energy, (2) the large amounts of heat produced by industrial, commercial and domestic activities, and by surface trans-port, and (3) the heat-absorbing characteristics of the overlying polluted atmosphere. The resultant 'heat-island' effect is variable both in time and space, for it is a function of urban size, topographic setting, synoptic conditions, urban fabric, industrial activity, atmospheric pollution load and prevailing macro-economic conditions. Large cities, especially those located in topographically sheltered positions, have the potential to show stronger thermal 'prints' than small towns, and studies have revealed that heat-island intensity is generally related to the logarithm of population size with a threshold population of about 10,000. However, land-use patterns are particularly significant, for building density appears more important in determining heat-island intensity than does city size. Similarly, extensive parks (e.g. London) reduce intensity, while the presence of large estuaries has a variable effect depending on season. Nevertheless, it is important to recognise that macro-climatological conditions determine whether a heat-island will form, and profoundly influence its shape and intensity.

Heat-islands develop best at night under calm anticyclonic conditions characterised by low-level temperature inversions and low near-surface wind speeds (<5–6 m/sec). Under these circumstances, rural–urban temperature differences of up to 10°C have been recorded for both London and Birmingham in the 1970s. Such heat-islands are fairly shallow (<180 metres), the pattern of isotherms being gently domed over the urban area with a marked gradient along the urban–rural boundary and show a tendency to form either during early summer (May–June) or during autumn and early winter (September–late January). While heat-islands may also exist by day, especially during summer months, they tend to be less intense because of higher wind speeds and greater turbulence. Increasing wind speeds eventually result in heat-island extinction, the threshold mean velocity depending on urban size, land-use patterns, building densities and recent thermal history. Investigations have revealed figures ranging from 4–7 m/s (14–25 kph) for Reading to 12m/s (43 kph) for London. Heat-islands are therefore transitory and repetitive rather than permanent features of the thermal map, and may be poorly developed or absent for long periods and even be replaced by short-lived negative anomaly patterns. Nevertheless, their fluctuating presence traditionally resulted in city-centres having higher average temperatures than both suburbs and rural areas, the inner city/rural contrast being 1.02°C for London (1931–60) (1.9°C for minimum temperatures). As a consequence, frosts and snowfalls were less frequent, snow-lie periods shorter and relative humidities less, although absolute humidities were sometimes greater.

Airflow

The flow of air over urban areas is affected by the greater roughness and frequently higher temperatures of the urban fabric. The highly differentiated skyline exerts a powerful frictional drag on airflow and turbulence is increased. High-rise blocks generate strong eddies and air is channelled and accelerated along canyon streets, so that wind speeds and wind directions are variable and fickle. Studies in London suggest that mean wind speeds in excess of 14.4 km/hr (4 m/s) are reduced, while below this figure velocities may actually be increased by turbulent exchange with faster-moving air at higher levels. These effects are most marked in central areas and diminish outwards, indicating that the scale of wind-speed modification is a function of city size and urban morphology. Small towns, may, therefore, exert little or no influence on airflow.

Gustiness is a particularly destructive aspect of airflow which becomes exacerbated in areas where surface roughness causes turbulent exchange. It is a characteristic and problematic feature of urban airflows where marked fluctuations in wind speed are a function of building form and layout, wind strength and variation with height, street widths and the relationships between wind direction and street orientation. Gustiness is usually measured by the gust ratio (i.e. the ratio of maximum gust speed in a particular period to the mean wind speed). Gust ratios vary from 1.5 at coastal sites to 1.5–2.0 over open farmland, 1.7–2.1 in well-wooded country and open suburbs, and 1.9–2.3 in the central parts of large cities (i.e. a 20 kt wind would be associated with gusts of 30, 30–40, 34–42 and 38–46 knots respectively; 1 kt equalling 1.151 mph or 1.852 kph). While this measure indicates a marked acceleration of maximum wind speeds in cities, the effect diminishes at higher wind speeds. The more generally experienced unpleasant conditions of sudden gusts of driving rain, blowing dust and flying litter, are due to the increased difference between maximum and minimum speeds in urban areas. This is measured by the gust factor and reaches a maximum in urban areas largely due to weaker lulls.

Visibility

Published studies of the incidence of reduced visibility have traditionally shown a strong correlation with urban and industrial areas, despite the reduced relative humidity due to the 'heat-island' effect. This was certainly true for the categories of Mist/Haze (visibility of 1,000–1,990 metres) and Moderate Fog (400–990 metres), although the patterns with respect to Fog (200–390 metres), Thick Fog (41–190 metres) and Dense Fog (<41 metres) were more complex. In the latter cases, records show greater frequency and duration in rural areas than in city-centres, but greatest of all in the suburbs. Thus, the overall lowering of relative humidities due to heat-island development, which averaged 5 per cent for London in the 1960s but with occasional urban–rural contrasts of 20–30 per cent, actually reduced the incidence of thick and persistent fogs in central areas. The increased fogginess of suburbs, on the other hand, was caused by slightly lower air temperatures, higher relative humidities, and the greater concentration of pollution aerosols (dust, smoke, etc.) from domestic, commercial and industrial sources. A significant proportion of these particulates formed hygroscopic condensation nuclei and thus aided in the formation of small airborne water droplets. As concentration of pollutants was greatest at times when conditions were most suitable for fog formation because (1) emissions were at a maximum in generally cool/cold conditions, (2) domestic emissions of smoke were cool and so did not disperse easily, and (3) dispersion was low because of slack winds and low-level temperature inversions, it is not surprising that fogs frequently developed. In addition, the dirtiness of suburban fogs made them more persistent because they absorbed a larger proportion of solar radiation than their rural equivalents, thereby assisting in the creation of smoky fogs, smogs or 'pea-soupers'.

Cloud cover and precipitation

Urban climate studies indicate that cloud and precipitation amounts are slightly greater over urban/industrial areas because of (1) heat-induced thermals, (2) mechanically induced turbulence due to buildings, (3) high absolute humidities and local concentrations of water vapour produced by industry, and (4) higher concentrations of condensation nuclei due to pollution. There is much qualitative evidence to support these assertions, such as the long lines of clouds ('streets') extending downwind of urban areas (particularly with south-west winds) and the cloud plumes created by power-station cooling towers, but few quantitative data, although satellite imagery and radar monitoring are beginning to provide valuable information on the distribution and intensity of rainfall. Evidence suggests that the cloud cover developed over large urban areas causes downwind suburbs to be significantly more cloudy than their upwind equivalents. Comparisons of observations at Kew and Heathrow, 9.7 kilometres apart, show Kew to have 5 per cent more cloud at 15.00 h, 6.2 per cent less clear days and 5.9 per cent more overcast days, with the greatest contrasts occurring in summer. However, the magnitude of the urban effect is difficult to quantify, for it cannot be reliably distinguished from orographic effects and the natural variability of cloud cover.

Cloud thicknesses are known to be increased both over and downwind of large urban areas, and increases in precipitation amounts and intensities, and thunderstorm frequency have all been recorded. Studies have shown that mechanical and thermal turbulence generated by built-up areas as small as Reading, can reinvigorate convective storms in unstable atmospheres. The intense Hampstead storm of 14 August 1975 (Figure 17.8) exemplifies what can happen, although the extent to which this was an urban-induced phenomenon remains unclear. Certainly thunderstorms have increased in frequency over London since the 1730s as the city has expanded, and there is evidence that thunderstorm rainfall is greater for London than elsewhere in the region. However, the local significance of topographic controls must not be overlooked, for there are local peaks elsewhere which can only be partly explained by urban influences. Even for London, detailed analysis suggests that urban stimulation of cumulonimbus clouds can only be detected in summer, and then for only six of twenty-seven possible synoptic situations.

Recent changes

Urban climates vary not only with size and character of urban areas but also with time. Thus, the continued expansion of towns and cities since 1950, despite the imposition of green-belt girdles and other planning controls, has significantly increased the extent and intensity of the urban influence. This, together with the dramatic morphological changes resulting from the construction of high-rise blocks and the increased insolation consequent on visibility improvements, has resulted in greater mechanical and thermal instability, so that cloud cover and precipitation levels, frequencies and intensities, have all been affected.

It is with respect to improvements in urban visibility that the greatest changes have occurred. Several factors have contributed: changing patterns of pollution emissions (especially the reduction in coal smoke), more efficient pollution dispersion due to increased turbulence, and processes of economic change. Particular significance has tended to be attached to the Clean Air Acts (1956, 1968) which have sought to restrict smoke emissions, particularly from domestic sources, in the belief that it was smoke that caused the heavy death tolls associated with smogs. Although subsequent research has revealed that cardio-respiratory problems were probably due to the combination of smoke with SO_2 and SO_3, the reduction in smoke emissions from 1.75 m/tonnes in 1960 to 0.375 m/tonnes in 1976, 0.28 m/tonnes in 1980 and under 0.20 m/tonnes since 1984 has dramatically improved urban atmospheres. This is especially true for London where well over 90 per cent of domestic properties are now covered by smoke control orders, and where the mean annual values for near-ground smoke have fallen from 190 mg/m^3 in 1955 to 55 mg/m^3 in 1967, 32mg/m^3 in 1976 and below 20 ug/m^3 since the early 1980s. However, some questions remain as to the actual scale of reduction because of problems with the siting of monitoring stations and the analysis techniques used, particularly as the traditional cheap 'smoke-shade test' tends to underestimate the growing contribution of motor-vehicle smoke compared with the more expensive gravimetric method.

The climatological significance of domestic smoke stems from (1) its production in large quantities (2.5–5.25 per cent of coal burnt by weight in open fires passes into the atmosphere as smoke) at times favourable to the production of poor visibility, and (2) its discharge at low temperature and low elevation where the dispersion efficiency is low. In consequence, domestic smoke aerosols were important in the generation of dense and persistent fogs. The fall in smoke production has greatly reduced the frequency and duration of haze and fog in urban areas and led to dramatic increases in recorded sunshine, especially during the winter months. Sunshine hours in Glasgow have increased by well over 60 per cent since 1960, and similar changes have been recorded for Edinburgh (Figure 17.14), Manchester, Birmingham and Leicester. Whereas central London received up to 30 per cent less winter sunshine than adjacent rural districts in the 1940s, these differences have now been eliminated (Figure 17.14). Indeed, there is growing evidence that the best winter visibility conditions now occur in the central parts of the larger cities.

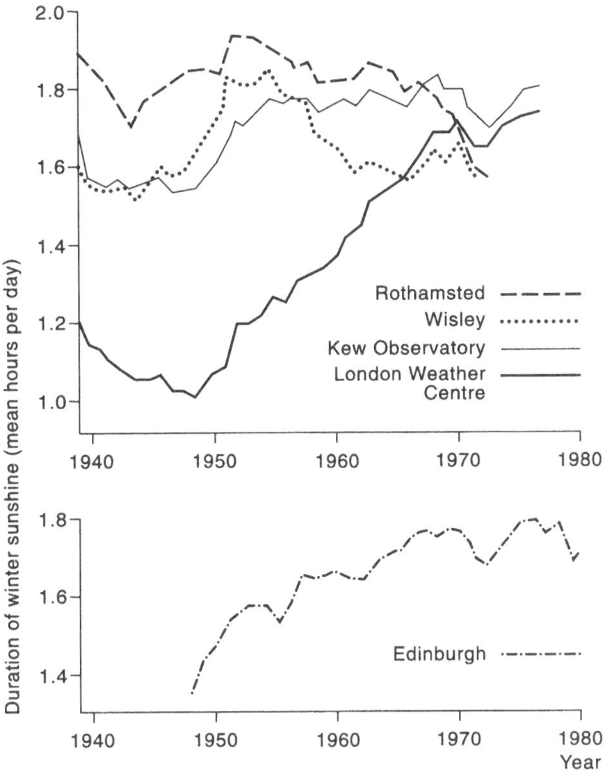

FIGURE 17.14 Increased winter sunshine hours in central London as compared with suburban (Kew), suburban fringe (Wisley) and rural (Rothamsted) locations; ten-year running means 1930–77 plotted on the last year (redrawn from Thornes 1978). The lower diagram shows the equivalent graph for central Edinburgh

While these dramatic changes have considerably improved the appearance and amenity value of our cities, it is important to examine the possible causes. The reduction in particulate concentrations in urban/industrial areas may be partly due to better dispersion, either because of increased turbulence as urban morphology has become more accentuated, or through the employment of the 'tall stack' policy whereby industrial emissions are carried to higher elevations for better dilution.

More importantly, the claimed significance of the Clean Air legislation has to be critically reassessed, especially as downward trends in smoke concentrations began well before 1956 and appear to be a national phenomenon irrespective of the extent to which smoke control orders have been implemented. Although the first Clean Air Act was passed by parliament in 1956, implementation of legislation through the creation of smokeless zones was the responsibility of local authorities. Delays were widespread for a number of reasons: shortages of smokeless fuel, civic lethargy, the impracticability of denying miners their 'perk' of free coal, and the costs involved in changing industrial processes have all been quoted as reasons for the subsequent slow spread of smoke control areas. This is well illustrated by the case of London – often claimed as the model example of the legislation's effectiveness. The City of London was covered by its own legislation from 1 October 1955, but the date of the first smoke Control Order in the wider London County Council (LCC) area was three years later on 1 October 1958, after which control spread slowly and sporadically, covering 35 per cent of the new Greater London Council (GLC) area by 1966, 60 per cent by 1970, 85 per cent by 1975 and 93 per cent by 1982. As a consequence, any significant improvements in air quality resulting from the legislation could not be anticipated until the late 1960s, and yet smoke levels had been falling since the 1920s, sunshine hours had been rising since the 1940s (Figure 17.14) and the last reported smog was in December 1962. That similar trends were recorded in other cities, irrespective of whether smokeless zones had been implemented or not, shows that the legislation was in reality working in unison with strong underlying processes of cultural and economic change. While it would be incorrect to suggest that the Clean Air Acts have had no effect, for they undoubtedly speeded change and have certainly blocked any reversion associated with the contemporary fashionability of open fires in gentrified areas, it is important to recognise that the underlying causes of change were the progressive diversification from a one-fuel to a four-fuel economy, and particularly the growth in use of electricity. Thus, in reality, the improvement in environmental conditions owes much to the so-called 'flight from coal' in the 1950s when shortages of coal due to the rundown state of the mining industry and labour disputes ('the coal gap') stimulated the search for alternative, dependable energy sources, a process subsequently enhanced by the exploitation of North Sea gas and oil supplies in 1967 and 1975 respectively. Once competition had been established, the desire for economies led to important changes in the industrial fuel-mix. For example, the dramatic reductions in smoke levels in the Potteries, and the attendant visibility improvements, are largely the product of the change from the traditional solid-fuel 'bottle ovens' to electric ovens. The growth in electricity generation, coupled with the development of the 'national grid' and 'supergrid' distribution networks, resulted in the closure of numerous small inefficient urban power stations (e.g. Battersea and Bankside in Central London) which had previously added generously to the urban air-pollution load, and their replacement by large, modern, efficient power stations often located in rural or semi-rural locations (see Chapter 3). As a consequence, over 70 per cent of present coal production is now burned in large, very efficient furnaces, many fitted with particulate screening devices so that smoke production is a minute fraction of that produced by domestic open fires. Other important planning and economic reasons for reduced urban smoke are the decline of port functions, railway modernisation and the resultant switch from steam locomotives to diesel and electric traction, urban renewal programmes, industrial re-location, decline of primary industry, and changes in industrial and domestic space heating.

While 'heat-islands' may have been the most dramatic and well-documented feature of urban climatology, there is uncertainty as to how they are evolving in response to economic factors and planning changes. This is because urban climates are composed of a kaleidoscope of site microclimates where temperatures can vary in response to very local conditions determined by a wide range of interacting influences. It seems likely that 'heat-island' intensities have reduced recently because of urban renewal programmes, improved insulation of buildings, changes in building materials, inner-city dereliction, increased turbulence and industrial decline and restructuring, although the maintenance of green belts has doubtless accentuated the bounding thermal gradients. Reductions in smoke pollution and haze have increased night-time infra-red radiation, so that most urban areas have become increasingly prone to frosts, snowfalls and longer snow-lie. Growth in the use of motor vehicles has resulted in increasing volumes of warm exhausts containing NO_2 which creates O_3 through photo-dissociation. As NO_2, O_3 and SO_2 (produced by diesel vehicles and through oil burning) all absorb some solar radiation, it is likely that heat-islands are being strengthened during summer months but weakened in the winter. Certainly, urban areas have become markedly warmer than surrounding country areas during the summer, both by day and by night.

Human occupancy and the land

The traditional focus of geographical enquiry into the relationship between humans and the earth's surface (land surface or ground) has been on morphological changes arising from human activity (see p. 339). Recently, however, emphasis has shifted towards assessing the significance of ground hazards, especially in the context of global warming. Both aspects are considered in the following section.

Morphological changes

Anthropogenic landform modification continues apace. Although the most conspicuous features remain the products of purposeful changes (*human-made landforms*) such as quarries, spoil heaps, motorway cuttings and embankments, there have also been widespread unintentional or 'accidental' changes consequent upon modifications to environmental conditions, either because of altered rates of operation of surface processes so as to yield *human-modified landforms*, or through the creation of *human-induced landforms* (i.e. naturally created features whose location in time and space is wholly human-dependent, such as deltas in reservoirs, shingle accumulations behind groynes, gullies on tips). These 'accidental' changes, although individually small-scale, are of considerable significance in terms of the total work achieved. As a consequence, it is now believed that population redistribution, economic growth and technological development have combined to make human activity the most potent instigator of contemporary geomorphological change.

Urbanisation has a particularly significant impact on surface form extending far beyond the physical limits of urban areas. Urban developments have been largely responsible for the dramatic growth in surface mineral extraction (see Chapter 4), while the increased production of rubbish (refuse) yields return flows to country areas to facilitate the reclamation of derelict land (see Chapter 4) thereby converting irregular holes into unnatural looking low hills. Similarly, river regime and channel form have been altered because of urban influences on discharge characteristics (see pp. 356–7) and increased erosion downstream of reservoirs (clear-water erosion).

To these must be added those changes produced by the increased run-off of water and sediment from agricultural land and the widespread channel modifications (enlargement, straightening and cleaning) undertaken by management bodies to reduce flooding which, in many cases, are superimposed on the effects of canalisation in the seventeenth and eighteenth centuries. Thus, it is true to say that no lowland river is in anything remotely like its natural state (see Figure 17.6).

Changes in agricultural practices have also accelerated soil losses. The trend to expansive monoculture, the abandonment of the practice of 'claying' light soils and the progressive switch to crops which leave much of the ground surface bare during the relatively dry early summer months (e.g. sugar beet), have resulted in significantly increased wind-induced soil erosion. Duststorms have been increasingly recorded since the 1920s in the Fenlands, East Yorkshire, the Brecklands and Lincolnshire, and plumes of blowing dust have recently become commonplace on the light Bunter (Sherwood Sandstone) soils of the Midlands. Water erosion has also increased on exposed clayey soils due to increased surface run-off, often exacerbated by the use of heavy machinery which destroys soil structure through compaction thereby reducing infiltration. As a consequence of both these processes, 7,700 square kilometres or 44 per cent of arable soils are now considered at risk from erosion by wind and water, with characteristic annual losses in the range of 1–5 tonnes/ha, and maximum rates of soil removal due to extreme events which were estimated at 17–18 tonnes/ha per annum in the early 1970s have been increased to 50–100+ tonnes/ha per annum. Such erosion seldom produces obvious features, except for the rills developed during severe storms. Gullies rarely develop on agricultural land in the UK, being mainly confined to spoil heaps, disturbed ground (construction sites, deforestation sites, etc.) and areas of great recreational activity (usually well-known viewpoints, sandy heathlands or elevated peatlands) where the intense use of paths results in vegetation destruction and soil compaction through trampling. However, such accelerated soil erosion is extremely significant in terms of channel and reservoir sedimentation (a sample of southern Pennine reservoirs showed losses of capacity of between 4 per cent and 75 per cent in around a century) and has profound implications for long-term soil resource conservation. The latter is clearest seen in the case of the Fens where agriculture and land drainage since 1848 have so increased the potential for wind erosion that only 25 per cent of the original area of peat now remains.

Coastal changes have also increased. Reclamation, often using urban refuse or dredged material, has resulted in useful increments of land in several areas, including Belfast, Cardiff, Liverpool, Portsmouth and Southend. Defence works have ranged from the stabilisation of eroding dune systems using wind-breaks and rapidly rooting psammophytic grasses (e.g. Braunton Burrows, Cublin Sands), via the artificial recharge of eroding beaches, to the erection of complex systems of walls, groynes and wave energy dissipation structures to protect eroding cliff lines (e.g. east of Brighton, Scarborough). While coastal engineering schemes are often impressive, they may produce unforeseen consequences, both seaward of structures (accelerated erosion) and elsewhere along the coast where reduction in beach volumes due to restricted littoral drift (under nourishment) can expose cliffs to wave attack and result in accelerated retreat. Thus, the inhibition of natural longshore shingle movement by the development of Folkestone Harbour over the period 1810–1905 is thought to have contributed to the dramatic landslides at Folkestone Warren in 1915 which moved the Folkestone to Dover railway line seawards by 50 metres. Similarly, the numerous groynes on the Sussex coast and the large breakwater at Newhaven are thought to be responsible for the rapid Chalk cliff retreat at both Seaford Head and the Seven Sisters (0.91 metres per annum), brought into focus by a huge fall at Beachy Head, January 1999. Further modifications include the regrading of potentially dangerous cliffs, the removal of unstable stacks (e.g. Thanet coast), the dumping of spoil (e.g. the small extension of land between Folkestone and Dover made from material produced in boring the Channel Tunnel) and the extraction of gravel for aggregate. Shoreline or beach extraction is carefully controlled since the disaster at Hallsands, Devon, where the removal of 660,000 tonnes of shingle in 1887 for construction work in Plymouth Dockyards resulted in beach lowering by an average of 4 metres, the abandonment of Hallsands Village and cliff retreat of up to 6 metres in the period 1907–57. Research into sediment transport rates and nourishment status is now a prerequisite for granting extraction licences in the coastal zone, and investigations in the English Channel have revealed that submarine extraction at depths greater than 18 metres below the low-water mark should have no impact on beaches. As a consequence, submarine extraction has expanded off many coasts (see Chapter 4), although tonnages are more than matched by the weight of sediment dredged from estuaries and harbours, much of which is dumped at sea (33.7–40.8 million tonnes per annum over the period 1987–91).

Ground hazards

An important development over the past three decades has been the growth in interest in ground hazards stimulated by the recognition that ground movements can be exceptionally costly to the construction, insurance and transportation industries. As a consequence, a range of reviews were undertaken by the then Department of the Environment into landsliding (1987), mining instability (1991), seismic risk (1993), natural underground cavities (1994) and foundation conditions (1994), with a view to consolidating knowledge and establishing the best technical and procedural approaches for minimising the adverse impacts arising from these hazards through improvements in impact assessments, site investigations and planning processes.

Limitations of space preclude more than minimal consideration of most of these topics. Seismic activity has long been recognised to affect Britain, although the magnitude of shocks is minor by global standards. However, significant damage-inducing shocks in 974, 1246 and 1382 (both in the Dover Straits), 1769 and 1816 (Inverness), 1884 ('The Great English Earthquake' centred on Colchester), 1896 (Hereford), 1932 (The Dogger Bank) and 1990 (Wrexham) reveal that seismic risk needs to be considered in the development of vulnerable activities (e.g. North Sea oil and gas extraction) or where catastrophe potential is great (e.g. the building of nuclear power stations, the siting of underground nuclear waste depositories). Studies have revealed the existence of fault zones susceptible to movement (e.g. the Great Glen shear zone) and areas where moderate seismic shocks occur most frequently (e.g. the Welsh block, southern Pennines and the margins of the Lancashire basin – see Figure 17.15).

Studies of mining risk have focused on areas where solid materials have been removed from beneath the ground surface, thereby leading to subsidence. The legacy of the underground extraction of coal, iron ore (e.g. Northamptonshire), magnesian limestone (e.g. Yorkshire) and salt (e.g. Cheshire) is to be found in sudden or progressive subsidence, which not only damages buildings and structures but also results in the creation of lakes (flashes), increased riverine flooding and landsliding. To these human-induced subsidence features must be added the natural underground solution of limestone, including the Chalk of southern and eastern England, which can result in a range of collapse structures with adverse consequences for buildings, roads and buried infrastructure. Further, the presence of surficial clays, particularly those containing clay minerals especially susceptible to hydration and dehydration with consequently exaggerated properties of expansion and contraction, is a hazard to property developers, the building industry and insurance companies because of the damage wrought to building foundations (see Figure 17.4).

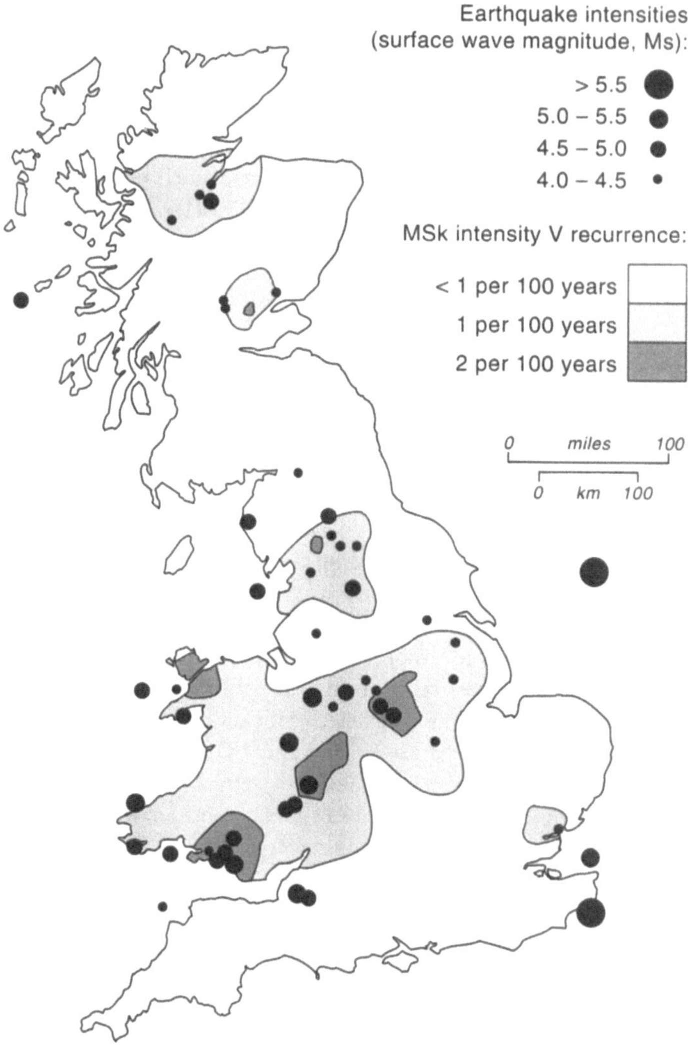

FIGURE 17.15 Historic earthquake magnitudes and intensities in and around Great Britain

Landsliding has for long been a recognised feature of cliffed coastlines developed in sedimentary rocks (e.g. west Dorset, southern Isle of Wight, Isle of Sheppey, Holderness), but is less associated with inland areas, apart from one or two conspicuous exceptions (e.g. Mam Tor in Derbyshire). However, the dramatic expansion in construction of dual carriageway bypasses and motorways begun in the late 1950s soon resulted in the unexpected discovery of numerous examples of 'ancient' landslides concealed by vegetation or surficial resculpturing, which were easily re-activated by excavation or loading. As similar costly problems were subsequently encountered by major housing developments and other forms of construction involving extensive earthworks, the results were rapid developments in preliminary ground investigation techniques designed to identify threat at an early stage of projects, including the adoption of modified forms of morphological and geomorphological survey, and stimulated research interest into the legacy of ancient landslides inherited from times when different environmental conditions were conducive to widespread inland slope failure.

The DoE review of knowledge on landsliding (1987) included a desk study census of recorded landslides which revealed the existence of a surprisingly wide variety of different types of slope instability, with locally marked concentrations of failures on 'soft rock' coasts, escarpment zones in lowland Britain (e.g. the Cotswolds) and dissected uplands (e.g. southern Pennines) (Figure 17.16). No fewer than 8,835 reported landslides were recorded (1,302 on the coast) of which 2,129 had occurred within the last one hundred years or were the sites of ongoing instability (Jones and Lee 1994). It has to be recognised that the census represents only a partial survey because it focused on knowledge of landslides rather than on the actual distribution of slope failures, thereby biasing the results to aspects and locations/areas that happen to have been subject to detailed investigation. Nevertheless, the resulting information on nature, types, movement characteristics, distributions, relationships between landslides and geological units (Figure 17.17), and evidence as to the relative importance of causal factors, has proved crucial to the development of a better understanding of this hitherto little appreciated hazard which occasionally makes the headlines in dramatic fashion, as in the Aberfan disaster of October 1966 or the visually dramatic progressive collapse of the Holbeck Hall Hotel, Scarborough, in June 1993 (Jones and Lee 1994). There are now much higher levels of vigilance regarding landsliding, despite the fact that the true extent and costs of the problem remain unknown, partly because landslide hazard is rarely catastrophic in the UK but dominated by huge numbers of small events whose impact is distributed widely within the economy, and partly because landsliding continues to be confused with subsidence. How the threat will change in the future, especially as a consequence of global warming, remains uncertain. Slope instability along 'soft rock' coastlines in southern Britain could well increase due to sea-level rise (Figure 17.10), but the effects on inland landsliding are more difficult to determine and will depend on (1) the ways in which global warming is translated into local changes of rainfall (intensity, quantity and distribution) and temperature, (2) changing patterns of ground resculpturing by humans, and (3) the extent to which careful ground management practices are adopted (Jones *et al.* 1993b). In this particular instance, it is humans that will largely determine the scale of events to come.

A further ground hazard that has risen to prominence in the past two decades is radon, a naturally occurring radioactive gas produced by the decay of uranium and thorium in rocks, especially igneous rocks. Radon-222 was discovered in 1900 but thought harmless until the 1960s when research revealed that the gas decays into minute solid particles which, if breathed in, significantly increase the risk of lung cancer. Recent studies have shown that radon represents about 50 per cent of the annual average individual dose of radiation from all sources in the UK and is thought to result in about 2,000 premature deaths from lung cancer each year, or about 5 per cent of lung cancer deaths, thereby making it the next most important cause of the disease after smoking.

The average annual whole body dose from natural radiation in the UK is estimated to be 2.2 millisieverts (mSv), above the International Commission for Radiological Protection's recommended limit of 1 mSv which has been accepted by the government (1995) and is

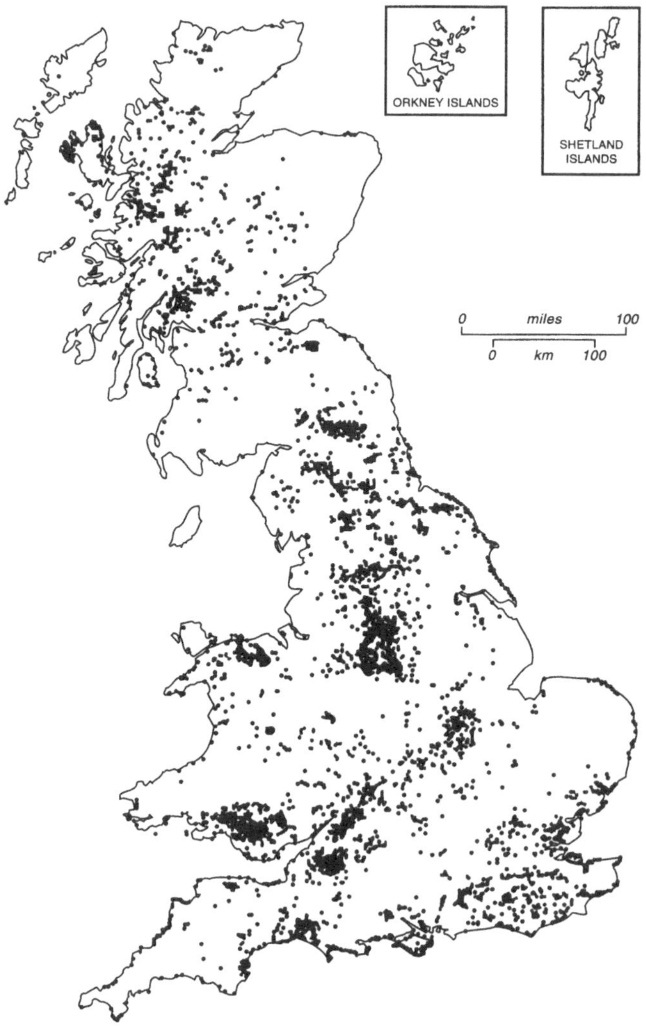

FIGURE 17.16 Distribution of the 8,835 reported landslides recorded in the DoE-commissioned National Landslide Database
Source: From Jones *et al.* (1993b).

likely to become law in the near future. Radon gas is easily dispersed in the air but becomes concentrated within houses, where it poses a direct threat to humans. A sample survey of houses conducted in 1988 revealed an average indoor concentration of 20 Bq/m^3 (where one becquerel (Bq) is equivalent to one atomic disintegration per second) indicative of an average annual dose of 1.3 mSv. However, in certain areas recorded levels ranged to in excess of 8,000 Bq/m^3 and the government recommended that an estimated 100,000 householders with levels above 200 Bq/m^3 should take remedial action to reduce concentrations (improved ventilation, sealing floors and windows, etc.). The distribution of radon-affected

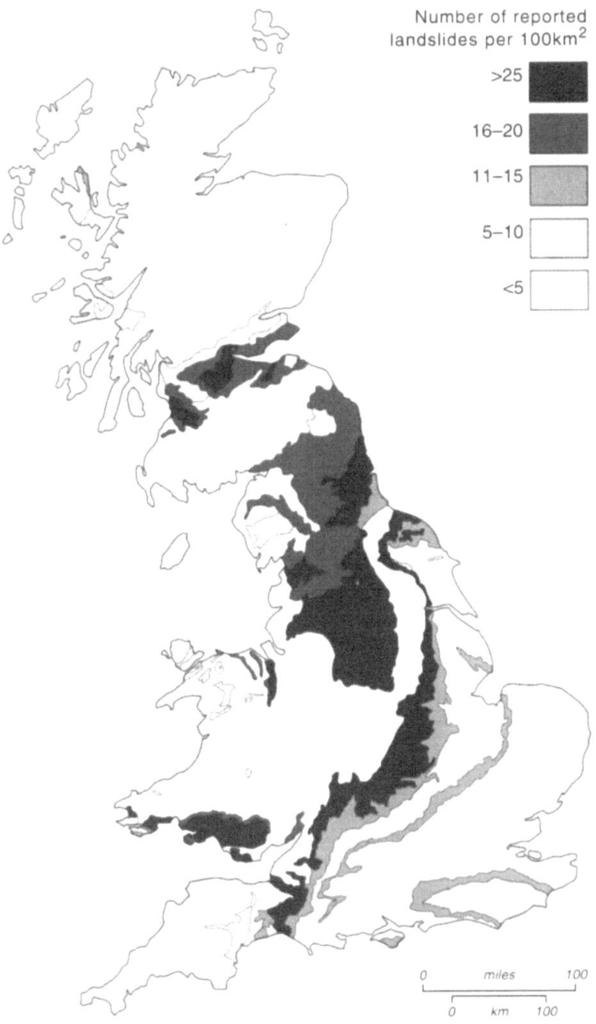

FIGURE 17.17 Distribution of 'landslide-prone' geological strata based on the number of reported landslides on each exposed outcrop

Source: After Jones *et al.* (1993b).

Note: It should be noted that the actual distribution of landslides on these outcrops will reflect topographic conditions, lithological variations and the extent to which the solid geology is obscured by surficial deposits.

areas (Figure 17.18) reveals eleven distinct 'hot spots'; two in Ulster, two in Scotland and seven in England affecting the counties of Cornwall, Devon, Somerset, Northamptonshire and Derbyshire, with especially widespread and acute problems in Devon and Cornwall where an estimated 60,000 dwellings have concentrations above the 200 Bq/m^3 Action Level. Radon levels in nearly 300,000 dwellings have been measured since 1987, most

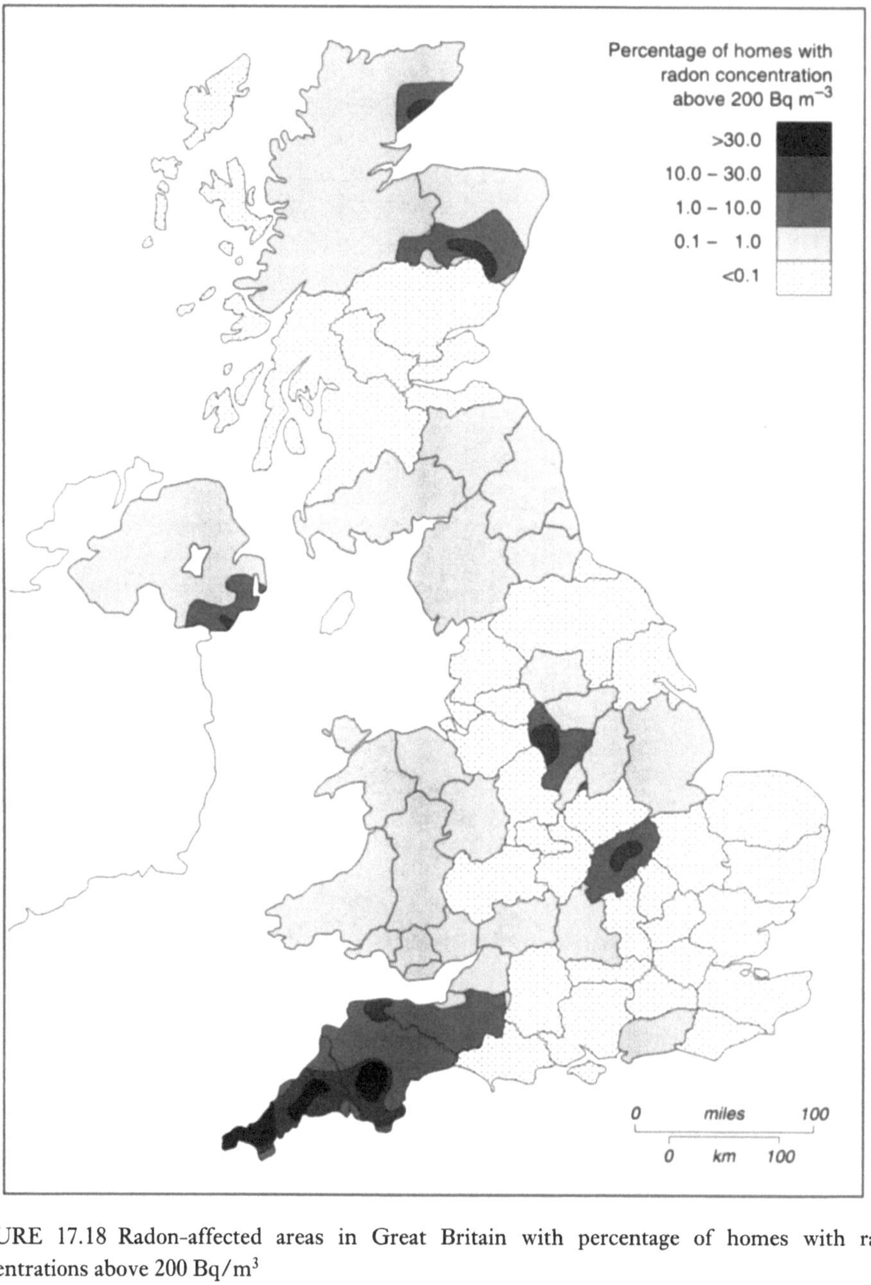

FIGURE 17.18 Radon-affected areas in Great Britain with percentage of homes with radon concentrations above 200 Bq/m^3

recently as part of a programme of free measurement of radon levels in the 200,000 dwellings most likely to have high concentrations, which was introduced in February 1996.

Radon is an unusual 'new' hazard worthy of further study. Growth in awareness of radon clearly illustrates the dynamic nature of hazard, while its undetectability using human sense organs (sensory amodality) shows how society is becoming increasingly dependent on scientists and instrumentation for information on environmental threats. These features, coupled with the carcinogenic (cancer-forming) effects of the gas and the known life-shortening consequences, should combine in widespread expressions of public concern and demands for action to protect threatened communities. The fact that none have arisen reflects human perceptions of risk, blame and responsibility, combined with the realisation that this elusive hazard is characteristic of areas and not an externality problem arising from market failure.

Conclusion

As has been shown, the two-way interaction between human society and the physical environment has become increasingly complex over the last half-century. Several reasons have been advanced: population growth and redistribution, technological development, increased mobility, and economic growth have led to major changes in the urban, industrial, transportation and agricultural sectors, many of which have had important consequences in terms of local climate, geomorphology and hydrology. In addition, there has been increasing appreciation of the scale of the better-known environmental hazard impacts (especially flooding and wind), partly due to better accounting procedures revealing the vulnerability created by the growing organisational complexity and interdependence of commercial activity, and partly because of the visual impact achieved by post-war developments in television reporting. This, in combination with the growth in environmentalism, has led to greater awareness as to the possible ramifications of human–environment interactions and increased sensitivity (both apparent and real) to a growing range of recognised environmental hazards, thereby resulting in some concomitant development of appraisal and monitoring programmes, including the increased use of environmental impact and risk analysis as part of development projects.

The power of society to create change in the physical environment continues to increase, both in scale and variety, as do the threats posed by environmental hazards. While there continue to be advances in understanding regarding the nature of physical environmental systems and their responses to changes, there remains a considerable need for further research in order to provide an adequate information base, so that the potential for adverse changes can be recognised at an early stage, and the necessary enforceable environmental legislation formulated and enacted when and where required. This aspect has attained particular significance in the last decade as the potential ramifications of Global Environmental Change have become apparent. Local problems have become merged with global concerns and the emphasis of geographical enquiries has begun to switch from explaining present patterns and characteristics to predicting future changes and consequences.

Relatively suddenly the state of the physical environment and the processes of change have become important in the context of sustainable development, thereby achieving a relevance in terms of politics and economics that would have been unthinkable thirty years ago. However, the human transformation of the physical environment is merely one component of the changing relationship between society and environment, the other significant development being changing attitudes towards the physical environment as testified by the growing emphasis devoted to environmental risk.

References

Brookes, A. and Gregory, K.J. (1988) 'Channelization, river engineering and geomorphology', in J.M. Hooke (ed.) *Geomorphology in Environmental Planning*, Chichester: Wiley and Sons.
Brookes, A., Gregory, K.J. and Dawson, F.H. (1983) 'An assessment of river channelization in England and Wales', *The Science of the Total Environment* 27: 97–111.
Brunsden, D. (1995) 'Learning to live with landslides: some British examples', in T. Horlick-Jones, A. Amendola and R. Casale (eds) *Natural Risk and Civil Protection*, London: E. & F.N. Spon.
Department of the Environment (1987) *Review of Research on Landsliding*, Geomorphological Services Limited.
Department of the Environment (1991) *Review of Mining Instability*, Arup Geotechnics.
Department of the Environment (1993) *Preliminary Assessment of Seismic Risk*, Ove Arup and Partners.
Department of the Environment (1994a) *Review of Natural Underground Cavities*, Applied Geology (Central) Limited.
Department of the Environment (1994b) *Review of Foundation Conditions*, Wimpey Environmental Limited.
Department of the Environment (1995) *The Occurrence and Significance of Erosion, Deposition and Flooding in Great Britain*, London: HMSO.
Doornkamp, J.C. (1995) 'Perception and reality in the provision of insurance against natural perils in the UK', *Transactions of the Institute of British Geographers* NS 20(1): 68–80.
Douglas, I. (1976) 'Urban hydrology', *Geographical Journal* 142(1): 65–72.
Jones, D.K.C. and Lee, E.M. (1994) *Landsliding in Great Britain*, London: HMSO.
Jones, D.K.C. *et al.* (1993a) 'Environmental hazards: the challenge of change', Collection of papers in *Geography* 339(78): 161–98.
Jones, D.K.C. *et al.* (1993b) 'Earth surface resources management in a warmer Britain', Collection of papers in the *Geographical Journal* 159(2): 124–208.
Lamb, H.H. (1991) *Historic Storms of the North Sea, British Isles and Northwest Europe*, Cambridge: Cambridge University Press.
Penning-Rowsell, E.C. and Handmer, J.W. (1988) 'Flood hazard management in Britain: a changing scene', *Geographical Journal* 154(2): 209–20.
Perry, A.H. (1981) *Environmental Hazards in the British Isles*, London: Allen and Unwin.
Thornes, J.E. (1978) 'London's changing meteorology', in H. Clout (ed.) *Changing London*, Slough: University Tutorial Press.
Wigley, T.M.L. and Raper, S.C.B. (1992) 'Implications for climate and sea-level of revised IPCC emissions scenarios', *Nature* 357: 293–300.

Further reading

This chapter has covered many and diverse topics which cannot be pursued without reference to a range of published sources. For general reviews of human impacts on the physical environment see A.S. Goudie (1993) *The Human Impact on the Natural Environment*, 4th edition (Oxford: Blackwell) and J.G. Simmons (1996) *Changing the Face of the Earth*, 2nd edition (Oxford: Blackwell), and for specific details regarding the landscape of the UK reference must be made to the three classic texts by: W.G. Hoskins (1955/70) *The Making of the English Landscape* (London: Hodder and Stoughton/

Penguin), L.D. Stamp (1964) *Man and the Land*, 2nd edition (London: Collins) and O. Rackham (1986) *The History of the Countryside* (London, J.M. Dent).

The nature of urban climates is examined in J.T. Chandler and S. Gregory (1976) *The Climate of the British Isles* (London: Longman, 307–29), while urban influences on rainfall are further discussed by B.W. Atkinson (1981) 'Precipitation', in K.J. Gregory and D.E. Walling (eds) *Man and Environmental Processes* (London: Butterworths, 22–37). For a historical review of urban atmospheric pollution see P. Brimblecombe (1987) *The Big Smoke* (London: Methuen), and readers interested in the implementation of clean air policies are recommended to A. Ashby and M. Anderson (1981) *The Politics of Clean Air* (Oxford: Oxford University Press).

The best general introduction to the subject of environmental hazards is K. Smith (1996) *Environmental Hazards*, 2nd edition (London: Routledge), but the only text focused on the UK remains A.H. Perry (1981) *Environmental Hazards in the British Isles* (London: Allen and Unwin). Details on river flooding are available in K. Smith and G.A. Tobin (1979) *Flooding and the Flood Hazard in the United Kingdom* (London: Longman) and E.C. Penning-Rowsell and J.W. Handmer (1988) 'Flood hazard management in Britain: a changing scene', *Geographical Journal* 154: 209–20. Flooding along the east coast and the nature of adopted defence measures are discussed in J.A. Steers (1953) 'The east coast floods, January 31 to February 1 1953', *Geographical Journal* 119: 280–95 and in R.W. Horner (1979) 'The Thames Barrier project', *Geographical Journal* 145: 242–53. For discussions of the October 1987 storm see special issues of *Weather* (March 1988) and *The Meterological Magazine* (April 1988). Full information on ground hazards is to be found in the reference list, with the exception of radon where the following are recommended: L.A. Owen (1993) 'Radon – a new environmental hazard', *Geography* 339(78): 194–8; T.K. Ball, D.G. Cameron, T.B. Colman and P.D. Roberts (1991) 'Behaviour of radon in the geological environment; a review', *Quarterly Journal of Engineering Geology* 24: 169–82, and J. Peto (1990) 'Radon and the risk of cancer', *Nature* 345: 389–90.

The best general scientific introduction to risk is Royal Society (1992) *Risk: Analysis, Perception and Management*. J. Adams (1995) *Risk* (London: UCL Press) is a readable partial review, while the best sociological discussion is U. Beck (1992) *Risk Society* (London: Sage). The subject of environmental risk is examined in S. Gerrard (1995) 'Environmental risk management', in T. O'Riordan (ed.) *Environmental Science for Environmental Management* (Harlow: Longman 296–316) and in Department of the Environment (1995) *A Guide to Risk Assessment and Risk Management for Environmental Protection* (London: HMSO).

Chapter 18

Atmospheric, terrestrial and aquatic pollution

Jonathan Horner

> Great progress has been made over recent years in improving air quality in the UK. However, air quality varies between regions and localities. In some areas it fails to achieve national and international guidelines, particularly in certain towns and cities.
>
> (Environment Agency 1997: 44)

> Since 1990, water quality has improved in 25.8% of the total length of rivers across England and Wales. The poorest classes have reduced from 14.3% of the total to 9.8%.
>
> (Environment Agency 1997: 20)

> When a pollutant is attacked at the point of origin – in the production process that generates it – the pollutant can be eliminated; once it is produced, it is too late. This is the simple but powerful lesson of the two decades of intense but largely futile effort to improve the quality of the environment.
>
> (Commoner 1990: 38)

Introduction

Environmental pollution is an important aspect of the changing geography of the United Kingdom and one which has seen considerable temporal and spatial change in recent years. Its importance can be demonstrated by summarising some of the effects which pollutants can have on the environment. One of the more obvious effects is on human health. Accidental and routine discharges of pollutants can cause fatalities and human health problems which may reduce life expectancy. For example, cases of asthma reported by doctors, which appear to be linked with air pollution, have been increasing in the UK in recent years. Animals and plants are also affected by pollution; for example, various estimates have suggested that crop damage caused by air pollution is costing UK farmers well in excess of £100 million annually. Buildings and materials are also damaged by pollutants, for example damage to UK materials caused by ozone(O_3) has been estimated to be in the range from £100–£345 million (PORG 1997). The quotations at the beginning of this chapter suggest that UK pollution is still a serious problem in the 1990s but that improvements have been made in recent years. The objective of this chapter is to examine this suggestion in the context of the changing geography of the UK.

It is important to clarify the meaning of the term 'environmental pollution', for which various definitions exist. One of the most widely recognised is that

> pollution is the introduction by man into the environment of substances or energy liable to cause hazards to human health, harm to living resources and ecological systems, damage to structures or amenity, or interference with legitimate uses of the environment.
>
> (Holdgate 1979: 17)

In the 1990s it would be more politically correct to omit 'by man'. This would have the added advantage of encompassing so-called natural pollutants which are emitted by natural sources. For example, radon gas, which is covered in more detail in Chapter 17, is naturally emitted from rocks and soil. It occurs in potentially dangerous concentrations in some areas, notably parts of south-west England and the south Midlands. Structural modifications to buildings have been made to facilitate dispersion of this carcinogenic gas. Naturally elevated concentrations of some toxic heavy metals in soils occur in some areas reflecting underlying mineralisation, for example lead in parts of Derbyshire and Yorkshire. Concentrations have often been further elevated by mining. In coastal areas, sea salt deposition can cause crop damage. It is important to be aware of such natural pollutants so that damaging effects can be avoided or minimised. However, unlike anthropogenic emissions, which will be the main focus of this chapter, it is not usually possible to reduce emissions of these pollutants and so they are usually permanent features of UK geography rather than changing ones.

There have been many reported examples of pollution incidents of different types, and occurring in different parts of the United Kingdom, during the 1990s. Some major ones are referred to in Figure 18.1, as are some examples of introduced pollution control measures. Together these emphasise the impacts and costs of pollution and demonstrate the wide geographical spread of pollution incidents and problems. Despite improvements in pollution control, occasional accidental discharges are inevitable. It is important that risk assessments of such potentially polluting activities are conducted. This enables controls to be employed, or emergency action plans to be in place, to eliminate or reduce risks to acceptable levels, taking into account financial, social and technological factors. Likewise it is vital that pollutants which are more routinely emitted are reduced to the lowest possible levels. All too often in the past, concentrations of pollutants considered to be safe have been revised downwards as more knowledge about their effects has been gained. A classic example of this is radiation, for which exposure limits set by the International Commission on Radiological Protection have been reduced on several occasions.

Legislation is particularly important for reducing emissions, and hence environmental levels, of pollutants. The next section of this chapter deals with recent changes in pollution control legislation as it has affected the UK. Coverage extends to both European and

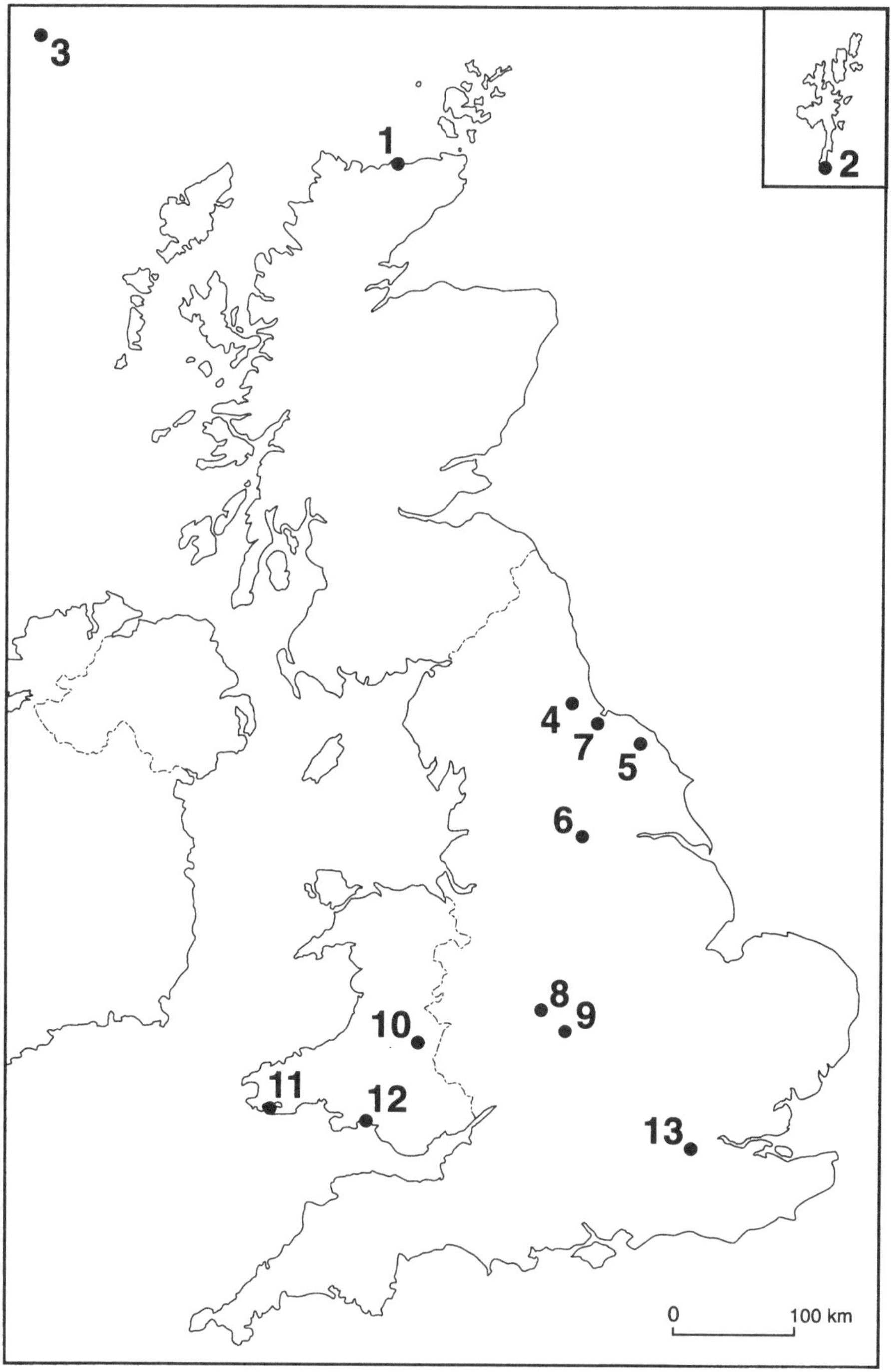

FIGURE 18.1 Examples of UK pollution incidents and control measures in the 1990s
(See key opposite)

1 1998: Announced that the UK's fast breeder reactor at Douneray will be run down, shortly after controversial acceptance of waste for reprocessing from Georgia.
2 1993: Tanker *Braer* ran aground at Garth's Ness spilling 80,000 tones of crude oil which caused *c*. £35 million damage to local salmon farms.
3 1995: Deep-water disposal of Shell's Brent Spar oil loading buoy in the North Feni Ridge abandoned after a European consumer boycott.
4 1996: *c*. 1,500, drums of liquid toxic waste illegally dumped in Consett – *c*. 120 tonnes of contaminated soil removed.
5 1996: *c*. £200 million spent at Dupont, Wilton to reduce emissions from the UK's biggest 'greenhouse gas' emitter.
6 1996: Sulphur trioxide released from a chemical plant in Staveley caused breathing difficulties in local workers.
7 1996: Road tanker crash on A19 spilt nitrobenzene, polluting a drain feeding Stainsby Beck – *c*. 2,000 tonnes of polluted soil removed costing over half a million pounds.
8 1995: 'Brumcan', a professionally run volunteer-based recycling operation, established to recycle a range of materials in Birmingham.
9 1996: A record high ozone concentration at nearly double the national health limit was recorded in Leamington Spa.
10 1995: Ferric sulphate escaped from a water treatment works, polluting the River Elan near Rhyader and killing *c*. 35,000 fish.
11 1996: *Sea Empress* ran aground in Milford Haven, discharging over 100,000 tonnes of crude oil and threatening the local ecology and tourist industry.
12 1998: £62 million sewage works for Swansea Bay opened.
13 1996: Fire at a paint manufacturer in South London polluted a tributary of the River Wandle – fish stocks saved by prompt action by firemen.

international laws because they are increasingly influencing national legislation. The remaining sections of the chapter will consider how and why changes have occurred in the recent distribution and concentrations of atmospheric, terrestrial and aquatic pollutants in the United Kingdom. Atmospheric pollutants often return to the earth's surface, sometimes transformed into secondary pollutants. They may be deposited in areas remote from the source of primary pollutant emissions, leading to terrestrial and aquatic pollution. For these reasons, atmospheric pollutants often have more serious and wider-reaching consequences than do terrestrial and aquatic pollutants and so they will be examined first and in more depth. Furthermore, aspects of terrestrial and aquatic pollution have also been considered in Chapters 2 and 17.

Pollution control legislation

Pollution, particularly of the atmosphere, does not recognise political or geographical barriers. Pollutants emitted from one country often have impacts in other countries. Local and national emissions of air pollutants, and to a lesser extent aquatic pollutants, may have both regional and global effects. For example, it is well known that atmospheric pollutants emitted from UK power stations have contributed to acid deposition damage in

Scandinavian countries. For this reason, pollution monitoring and control need to be on an international rather than on a national scale. This is also sensible in view of the fact that pollutants tend to be emitted from large numbers of globally dispersed sources. In the 1990s there is therefore increasing emphasis on regional and global aspects of pollution and a growing influence of international regulations on controlling pollution in the United Kingdom. The changing geography of UK pollution will therefore be examined in the context of international developments in pollution control before recent national progress is considered.

As the serious and wide-ranging nature of environmental pollution problems has become more evident, so has the need for greater international co-operation to develop effective pollution control policies. International agreements have recently been forged to control a number of important pollutants. For example, at the June 1992 United Nations Conference on Environment and Development ('The Earth Summit' held in Rio de Janeiro) delegates from many countries signed the Framework Convention on Climate Change, and over 160 countries have now signed. The ultimate objective of the Convention is to stabilise 'greenhouse gas' concentrations at levels that will prevent anthropogenic influence on the climate system. The 1986 Montreal Protocol, subsequently amended at meetings of the countries involved (the ninth meeting taking place in 1997) to bring in more stringent controls, agreed international action to reduce emissions of ozone (O_3) depleting chemicals. The United Kingdom is committed under a United Nations Economic Commission for Europe (UNECE) Protocol to reduce emissions of volatile organic compounds (VOCs) by 30 per cent between 1988 and 1999 with the aim of reducing tropospheric ozone concentrations (Great Britain 1996). In 1994 the government signed an international protocol committing the UK to reducing sulphur dioxide (SO_2) emissions from all sources by 80 per cent by 2010 compared with 1980 levels (Great Britain 1996). These major air pollutants all present serious environmental threats and will be considered in more detail in the next section.

In the 1990s, the European Union (EU) has been increasingly influencing pollution control policies in member states. In July 1987, the Single European Act came into force and called for unified environmental standards and pollution prevention measures. EU Directives have, for example, set air-quality standards for sulphur dioxide (SO_2), smoke, nitrogen dioxide (NO_2), lead and ozone. They have also required licensing systems for waste management and disposal operations, and have set water-quality standards for bathing and drinking water. These Directives are binding on member states, although each country is free to devise its own methods to achieve the standards – usually by a specified date. The European Environment Agency (EEA) began operating in October 1994 and acts to integrate environmental protection and improvement information in Europe. In August 1995, the EEA published the most detailed and comprehensive review available of the state of the environment in Europe – *Europe's Environment – The Dobris Assessment* (Stanners and Bourdeau 1995). The report highlighted twelve 'prominent European environmental problems', which are listed in Table 18.1. Examination of these shows that they are all either specifically pollution problems or include pollutants as contributing factors to the problem. These prominent environmental problems will be used as a focus for discussion and for selecting atmospheric, terrestrial and aquatic pollutants for consideration in the following sections of this chapter.

During the 1990s, there have been some significant changes in the organisational structure for controlling pollution in the United Kingdom. The 1990 Environmental

TABLE 18.1 Twelve 'prominent European environmental problems' identified by the Dobris Assessment

1	Tropospheric ozone and other photochemical oxidants
2	Acidification
3	Climate change
4	Stratospheric ozone depletion
5	Major accidents
6	Waste reduction and management
7	Chemical risks
8	The management of freshwater
9	Coastal zone threats and management
10	Urban stress
11	Loss of biodiversity
12	Forest degradation

Source: Stanners and Bourdeau (1995).
Note: The Dobris Assessment was produced in response to the First European Environment Minister Conference held at Dobris Castle near Prague in June 1991 and reinforces concerns raised at the conference. It was organised by Josef Varousek, First Minister of the Environment of Czechoslovakia (today split into the Czech and Slovak Republics), and attended by Ministers or their deputies from almost all European countries. The report covers the state of the environment in nearly fifty countries, presenting data (including data on pollution), and identifies a number of serious environmental threats. It confirms poor environmental quality in many parts of Europe, often caused by the emission of pollutants. The twelve problems listed above were identified as the twelve most pressing environmental problems currently facing European countries.

Protection Act introduced Integrated Pollution Control (IPC), a system which requires a single authorisation to be obtained for polluting processes which encompasses atmospheric, terrestrial and aquatic pollution. Authorisation is based on the Best Available Technology Not Entailing Excessive Cost (BATNEEC) being used to control emissions. On 1 April 1996, the Environment Agency for England and Wales took up its statutory duties in this respect, bringing together functions previously conducted by Her Majesty's Inspectorate of Pollution (for air pollution), the National Rivers Authority (for water pollution) and the sections in local and central government dealing with waste regulation and contaminated land (for terrestrial pollution and waste disposal). Legal structures in Scotland and Northern Ireland differ from those in England and Wales, and the Scottish and Northern Ireland Offices have their own Environment Departments.

In 1990, the UK government published its first White Paper on the environment (Great Britain 1990), which has been followed by annual reviews. The major UK political parties have paid more attention to environmental issues in recent years and the apparent commitment by consecutive governments to improving environmental performance and to reducing environmental pollution is possibly a major cause of the relative demise of the Green Party. Having received 15 per cent of votes cast in the European parliament elections in June 1989, the Green Party saw its share of the vote cut to less than 2 per cent in the May 1997 General Election. However, whilst the Labour Party's 1997 manifesto included many environmental proposals, environmental groups have expressed concern that the Queen's Speech for the opening of parliament referred to few of the proposals.

Increasingly, companies are paying more attention to environmental issues and making attempts to reduce emissions of pollutants. As Porritt (1991) pointed out:

> in contrast to the kind of green public relations and promotional gimmicks which we saw so much of in the mid 1980s, businesses are now genuinely working to clean up their act.
>
> (Porritt 1991: 193)

In the UK, the Advisory Committee on Business and the Environment was jointly established by the Department of Trade and Industry and the Department of the Environment in 1991. It provides advice on environmental policy issues and encourages dialogue between government and business on environmental matters. Leading companies have adopted environmental policies which aim to improve environmental performance, for example by reducing emissions of pollutants. The policies are implemented by establishing environmental management systems which can be subjected to environmental auditing. This has enabled companies to adopt a more proactive approach to pollution control compared with the more reactive approach generally adopted in the 1980s. Companies are increasingly recognising that risks and liabilities associated with pollution can translate into a tarnished public image, legal costs and loss of competitive advantage. In 1995, the first registrations of UK companies under BS 7750 (a UK-developed environmental management system) and the EU Eco-Management and Audit Scheme took place. In 1997, thirty UK companies in the *Financial Times* Top 100 companies published annual environmental reports separate from their annual accounts, compared with twenty companies in 1993. In 1997, fourteen of the companies set quantifiable targets, whilst none did so in 1993 (KPMG 1997).

Whilst it would be generally true to state that increasing concerns for environmental quality, and improvements in pollution control, referred to by the Environment Agency in this chapter's opening quotes, have continued in the United Kingdom in the 1990s, there are certainly exceptions to this. UK government figures show that during this decade complaints about noise pollution have increased, that the amount of derelict land (which is often polluted) in the South East has increased, and that there have been some falls in compliance with EC Bathing Water Directives in the Northumbrian, Southern and South West Water Regions (DETR 1997). The Environment Agency (1997) showed that there was a steady rise in reported water pollution incidents in England and Wales during the first half of this decade. However, it should be borne in mind that increases in reported incidents or complaints of pollution could arise from increased awareness and tendency to complain about it. The following sections examine recent changes in UK atmospheric, terrestrial and aquatic pollution, respectively.

Atmospheric pollution

UK improvements in industrial pollution control and reductions in emissions from domestic sources, to a great extent initiated by the Clean Air Acts of 1956 and 1968, have continued during the 1990s as further legislation has been introduced. However, an increasing number of motor vehicles on UK roads, reported in Chapter 6, has led to this source of emissions becoming increasingly significant in contributing to the changing patterns of UK air pollution. More effort to control motor vehicle emissions, or to reduce the UK's reliance on motor cars, is clearly needed if motor vehicles are not to be an increasingly significant pollution source in the next millennium. Pollutants, notably smoke and SO_2, previously monitored to measure overall air quality, have declined in concentrations, particularly in urban areas. However, they have been replaced by increasing concentrations of some other air pollutants, notably by O_3 and VOCs, sometimes in rural areas. This will now be explained, starting with air pollutants of more local concern and then moving on to consider pollutants which are more of an international problem.

Smoke and SO_2 have been the traditional indicators of air quality in the UK. These pollutants were largely responsible for infamous urban smogs, most notably the so-called Great London Smog of 1952 which was estimated to have caused 4,000 premature deaths during a five-day period in December 1952. They also cause damage to vegetation, reducing crop yields, and to buildings, corroding metal and soiling brickwork. A general decline in emissions of these pollutants has continued during the first part of the 1990s (Figure 18.2). Between 1980 and 1995, emissions of SO_2, about three-quarters of which come from large combustion plants (LCPs), declined by 52 per cent. Over the same period black smoke

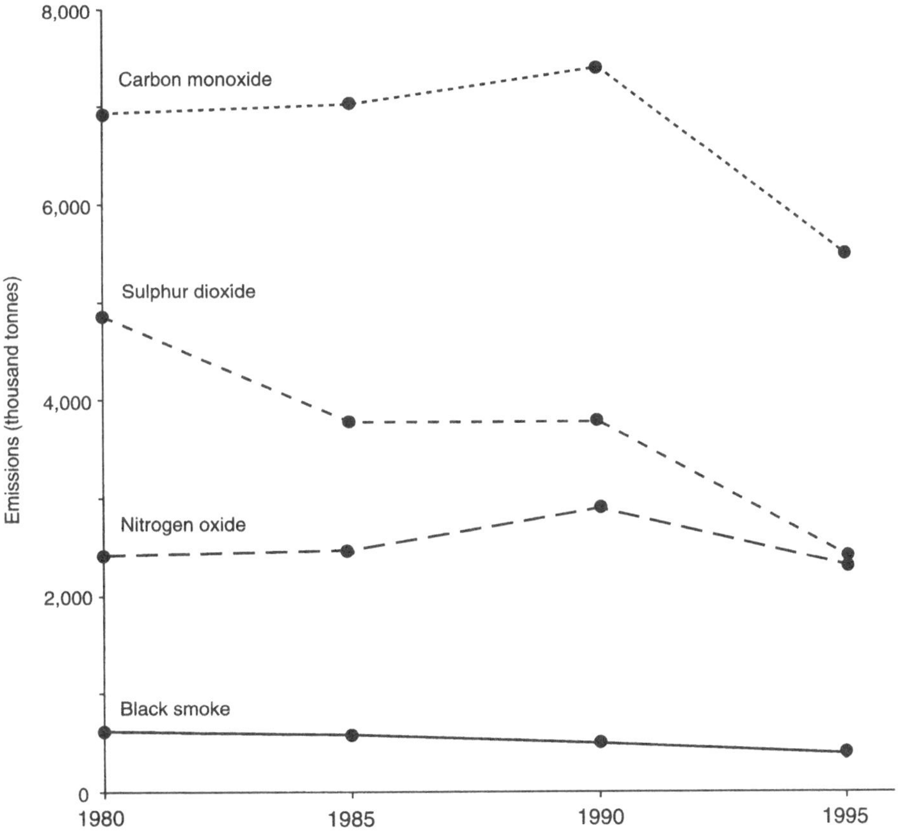

FIGURE 18.2 Total UK emissions of selected local air pollutants, for the period 1980–95
Source: DETR (1997).

emissions, of which diesel engine vehicles have become the major emission source, declined by 40 per cent. Black smoke emissions from domestic sources decreased by 79 per cent during this period, reflecting increasing use of gas rather than coal for heating and further introductions of smoke control areas. The EC LCP Directive required the UK, taking 1980 as the baseline, to have reduced emissions of SO_2 from LCPs by 20 per cent by the end of 1993, by 40 per cent by the end of 1998 and by 60 per cent by the end of 2003. The latter target is already close to being attained, with recent reductions having been achieved by commissioning new gas-burning power stations. However, with natural gas supplies in relatively short supply it might be difficult to maintain this reduction as the next millennium progresses.

With declining UK concentrations of smoke and SO_2, increasing attention has been paid to other air pollutants. The recent Dobris Assessment, referred to in the previous section of this chapter, identified four problems in its list of twelve 'prominent European environmental problems' which are specifically caused by atmospheric pollutants. These are 'tropospheric O_3 and other photochemical oxidants', 'acidification', 'climate change' and 'stratospheric O_3 depletion'; each will be considered in turn. Other problems identified, notably 'urban stress' caused partly by motor vehicle emissions, 'major accidents', 'forest degradation' and 'chemical risks' also have atmospheric pollutants as major contributing causes or as being the cause of resulting effects. There have been increasing episodes of poor urban air quality, characterised by elevated concentrations of nitrogen oxides (NO_x), CO and VOCs, and of tropospheric O_3 formation, both of which are largely caused by motor vehicle emissions. These pollutants are now more appropriate than black smoke and SO_2 as indicators of air quality. Concentrations of NO_x and CO shown in Figure 18.2, each of which is emitted in significant quantities by motor vehicles, have declined to a much lesser extent between 1980 and 1995 compared with black smoke and SO_2. During this period total NO_x emissions, for which road transport is now responsible for about half of total emissions, decreased by just 5 per cent. Oxides of nitrogen irritate the respiratory tract (for example triggering asthma attacks), cause damage to vegetation (reducing crop yields) and contribute to photochemical smog and acid rain formation. The EC LCP Directive required the UK to reduce NO_x emissions from LCPs, taking 1980 as the baseline, by 15 per cent by 1993 and by 30 per cent by 1998. The latter target had already been achieved by 1995, with a 45 per cent reduction. However, the decrease in emissions from LCPs during this period was almost matched by an increase in emissions from other sources, notably road transport.

Carbon monoxide emissions, for which road transport now accounts for about three-quarters of total emissions, decreased by 21 per cent, and 'VOC' emissions, which almost exactly matched NO_x emissions, declined by just 3 per cent between 1980 and 1995. Carbon monoxide emissions are only a health threat in poorly ventilated confined situations, as reviewed by Horner (1998). This invisible and odourless gas preferentially binds with haemoglobin in the red blood corpuscles, reducing the amount of oxygen which can be transported. Road transport accounts for about 30 per cent of total VOC emissions, the remainder coming from solvent use and industrial processes. VOCs contribute to photochemical smog formation and some are known carcinogens. The decline in CO and VOC emissions has been partly due to increasing numbers of diesel cars on the road which emit less of these pollutants than do petrol-engined vehicles. Increasing use of catalytic converters on petrol-engined vehicles has also reduced emissions. However, diesel-powered vehicles have been linked with emissions of carcinogenic organic compounds, so increasing use of diesel-engined vehicles in the UK should be carefully considered.

Emissions of lead from petrol-engined vehicles have declined substantially with the reduction in lead content of petrol and the increasing use of unleaded petrol. Unleaded petrol, which first appeared in the UK in 1987, accounted for 69 per cent of total UK petrol consumption by the end of 1996 and total lead emissions from motor cars dropped from 7,500 tonnes in 1980 to just 1,000 tonnes in 1996 (DETR 1997). Of greater concern in the 1990s is the need to consider hazards from lead in old water pipes and old paint work. For example, Horner (1996) reported lead concentrations in paint from surfaces in an old school to be well in excess of the 5,000 ppm safety level recommended by the World Health Organisation. Lead is a proven neurotoxin and has been shown to reduce IQ levels in children.

Whilst smogs caused by smoke and SO_2 are largely a thing of the past in the UK, they have been replaced by photochemical smogs. The formation of such smogs, largely from reactions between vehicle-emitted NO_x and VOCs, has been of increasing concern in the 1990s. Hospital admissions in London have increased after surges in pollution levels. Photochemical smogs are characterised by increased concentrations of tropospheric O_3 during the summer, the effects of which are summarised in Table 18.2. No other compound has atmospheric levels so close to toxic levels as commonly as O_3. Ozone generating reactions require sunlight and take some time to occur. This is one reason why highest O_3 concentrations are often found in rural areas several hundred kilometres downwind from emission sources. Concentrations of O_3 exceed thresholds for effects on vegetation and human health throughout the UK. The largest and most frequent exceedences occur in southern England, especially in rural areas of the South East (PORG 1997). Figure 18.3 shows the extent to which UK O_3 concentrations exceeded recommended safety levels for the years 1990–5. According to the Dobris Assessment, O_3 concentrations in the northern hemisphere are expected to continue rising at a rate of between 0.5 and 1 per cent a year.

TABLE 18.2 Effects of tropospheric ozone and other photochemical pollutants

Target	*Effects*
Materials	All organic molecules, including natural polymers such as cotton, cellulose, leather and rubber, and synthetic polymers, such as dyes, elastomers (used to make tyres), nylon, paints and plastics, are degraded
Plants	Visible injury to foliage Early symptom is shiny water-soaked regions Chlorotic flecking or stippling of leaves commonly develops and these may turn red or bronze Growth may be reduced, hence crop yields decreased Predisposition to disease and insect attacks Synergistic interactions with other pollutants, notably NO_x, SO_x and acid rain
Humans	Irritation of the eyes, nose, throat and respiratory tract Premature lung ageing caused by damage to bronchiolar and alveolar walls Possible triggering of asthmatic attacks Exercise exacerbates the effects (hence, at-risk groups such as children should be moved indoors during high pollution episodes)

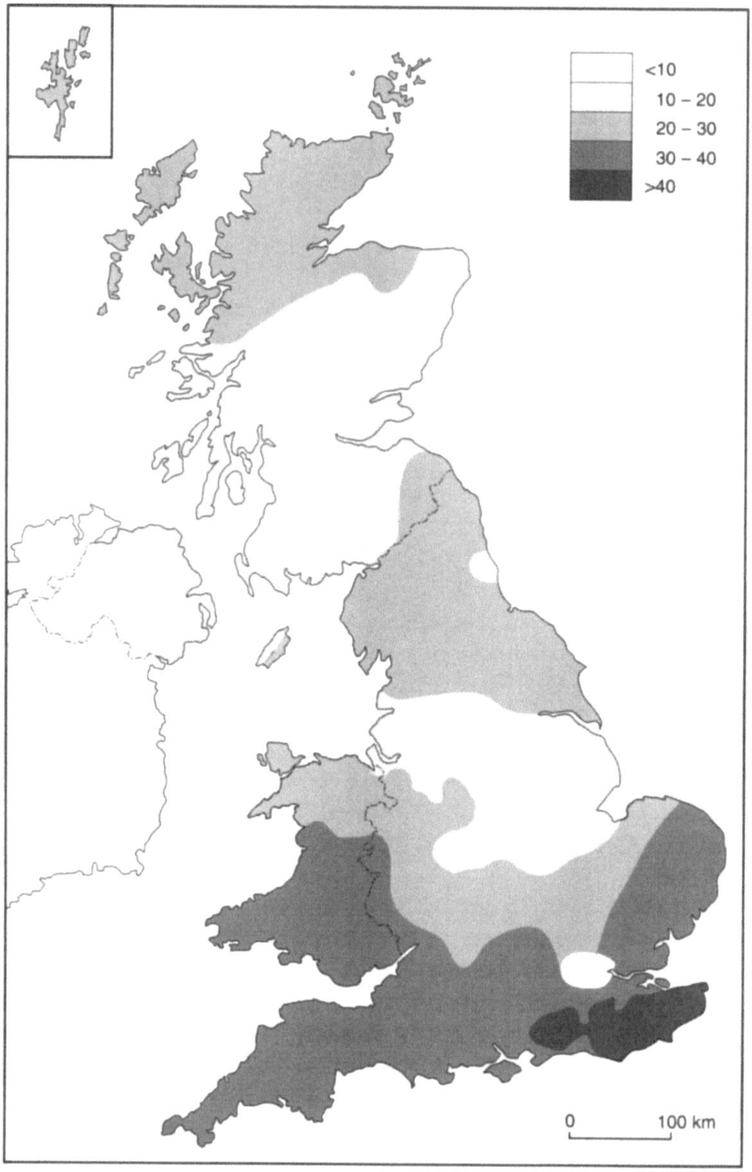

FIGURE 18.3 UK tropospheric ozone levels, showing the number of days with eight-hour periods of ozone exceeding fifty parts per billion, for the period 1990–5
Source: DETR (1997).

The role of NO_x in forming 'acid rain' and the problems which this phenomenon causes were widely reported in the 1980s. Sulphur oxides (SO_x) and NO_x, mainly from the burning of fossil fuels, react in the atmosphere to form sulphuric and nitric acids. Some of the major effects of acid deposition are summarised in Table 18.3. Widespread damage to

TABLE 18.3 Effects of acid deposition

Target	*Effects*
Materials	Corrosion of steel and other metal structures Erosion of limestone buildings Fading and damage to fabrics and paintwork
Plants	No direct damage to above-ground parts from 'acid rain', except in extreme events and where synergistic interactions occur with other pollutants SO_x and NO_x inhibit growth, significantly reducing crop yields, and contribute to tree damage, acting synergistically with other pollutants Growth reduced in poorly buffered soils owing to nutrient leaching and metal toxicity Decreased resistance to droughts and frosts Increased sensitivity to pests and diseases
Rivers and lakes	Increased toxicity of aluminium, causing fish deaths Decrease in sensitive birds, amphibians and invertebrates Increased *Sphagnum* moss growth
Humans	No known direct effects of 'acid rain' NO_x and SO_x irritate the lungs and impair functioning Toxic heavy metals, notably aluminium, in drinking waters (especially well waters)

European forests has been caused, including in the UK, and serious damage to lakes in Scandinavia, Scotland and north-west England has occurred, with major conservation implications (see Chapter 20). It is important to distinguish between dry deposition of the primary gaseous pollutants, which tend to have more local effects, and wet deposition of the acids ('acid rain'), which may be transported for several thousand miles before being deposited. The concept of 'critical loads' has been used increasingly in the UK in terms of predicting and controlling damage from acid deposition. A critical load is the estimated load of deposition for a specified location, below which no harmful effects are predicted. By comparing with actual deposition, critical load exceedences can be predicted for UK soils and fresh waters. For soils, 32 per cent, and for fresh waters 17 per cent, of the area of the UK exceeded the estimated critical load for total acidity in the period 1989–92 (DoE 1996). The most vulnerable areas are the uplands of north and west Britain. Acidification of soils and freshwaters can create nutrient deficiency and metal toxicity problems, hence limiting the range of sensitive fauna and flora. Based on the apparent frequency of UK reporting, concern about acid deposition would appear to have diminished in the 1990s. This partly reflects the fact that wealthier countries like the UK, where damage was largely reported in the 1980s, have begun to introduce emission control measures. In the UK several power stations, including the largest coal-burning station at Drax in South Yorkshire, have introduced flue gas desulphurisation equipment to reduce emissions of SO_x. However, whilst this reduces atmospheric pollution problems, it creates new pollution problems of a largely terrestrial nature. There is a need to quarry limestone (calcium carbonate) (see Chapter 4) which is needed for the cleaning process, and then the waste product, gypsum (calcium sulphate), needs to be disposed of. This partly explains why relatively clean gas-fired power stations, which were referred to earlier in the section, have been a favoured UK electricity generation method in recent years. Additionally, low NO_x emitting burners have been fitted in many power stations and NO_x emissions from cars have been reduced by catalytic converters. A major worry must be that the problems caused by acid deposition are now being repeated in the industrialising nations of Asia, Eastern Europe and Latin America where there are lesser controls. Concerns about acid rain may well resurface in the UK media as the next millennium approaches.

Last to be considered in this section are the truly global air-pollution-related problems – climate change and stratospheric ozone depletion. It is becoming increasingly clear that emissions of certain pollutants can now affect processes on a planetary scale. Increasing concentrations of 'greenhouse gases' from human activities are expected to cause a significant change in the earth's climate. This may have important consequences for the landscape of the United Kingdom (Table 18.4), considered in Chapter 19, and for water resources, considered in Chapter 2. The main gases involved are carbon dioxide (CO_2), methane (CH_4), nitrous oxide (NO), chlorofluorocarbons (CFCs) and ozone (O_3). These gases absorb outgoing infra-red radiation which has been reradiated from the earth's surface. Observed global temperature increases over recent years are consistent with estimated increases caused by increased greenhouse gas concentrations. Although a molecule of CO_2 is less potent than molecules of other greenhouse gases, the quantity of emissions is relatively so large that CO_2 is the major contributor to global warming. Under its Climate Change Programme, the UK is committed to reducing CO_2 and CH_4 emissions to 1990 levels by the year 2000. Total UK emissions of CO_2 fell by 7 per cent, and of CH_4 by 15 per cent, between 1990 and 1995, largely reflecting the decreasing use of coal in power stations, and the

TABLE 18.4 Potential effects of global climate change in the UK

Target	*Effects*
Low-lying land	A sea-level rise of an estimated 22 cm by the year 2050 (Houghton *et al.* 1990) could inundate some areas if sea defences are not strengthened Increased flooding in coastal and estuarine areas Erosion of beaches, cliffs and dunes Salinisation of soils and waters
Other land	Changes in land-use patterns caused by rainfall changes Changes in erosion and sedimentation patterns Inland flooding owing to river gradient changes Diminution of aquifers and drought conditions
Plants	Changes in productivity affecting competition, hence leading to biodiversity changes Changes in crop plant ranges (for example new areas would be available for maize if temperatures rise) Changes in diseases and pest incidences
Animals	Plant changes would have 'knock-on' effects for animals
Humans	Changes in disease incidences

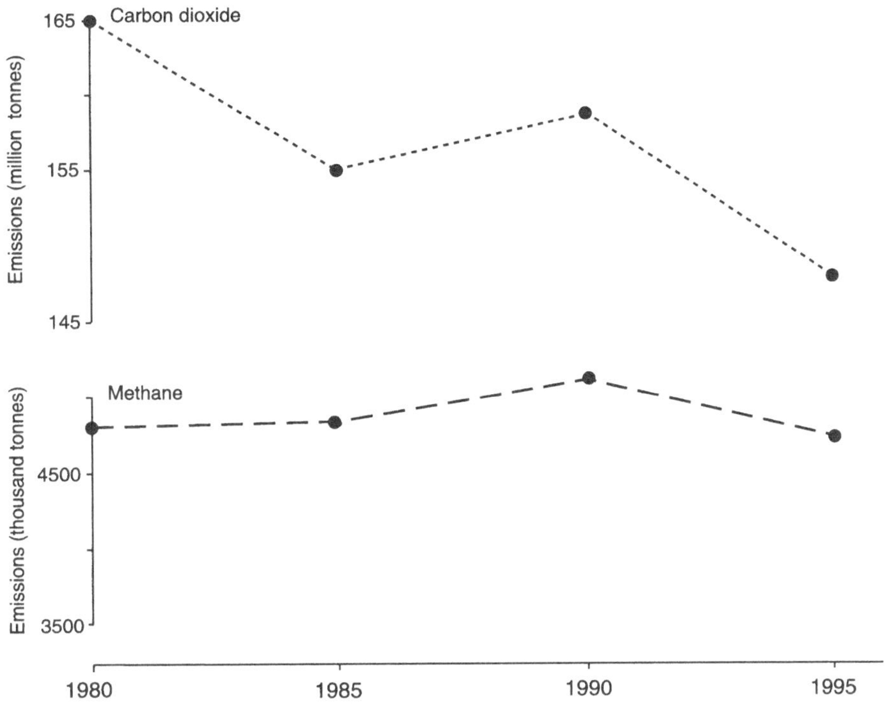

FIGURE 18.4 Total UK emissions of carbon dioxide and methane, for the period 1980–95
Source: DETR (1997).

decrease in deep mining of coal, respectively (Figure 18.4). However, international action is needed if global atmospheric concentrations of greenhouse gases are to be reduced.

Stratospheric O_3 absorbs harmful ultraviolet radiation, preventing it from reaching the earth's surface. Depletion of this 'ozone layer' allows more ultraviolet light to reach the earth's surface, causing the effects summarised in Table 18.5. Increases in atmospheric chlorine and bromine concentrations have been shown to be the major cause of O_3 depletion. These are derived from CFCs, which are used mainly as aerosol propellants, refrigerator coolants and foam-blowing agents, and from bromofluorocarbons (halons), which are used in fire extinguishers. Measurements over Lerwick in the Shetland Islands and Camborne in Cornwall have shown a general trend for decreasing O_3 levels since the early 1980s (DoE 1996). Although O_3 levels above the UK have not diminished sufficiently to identify significant effects on human health or crop yields to date, calculations have indicated that a UK child's lifetime risk of developing non-melanoma skin cancer under current depletion rates is 10–15 per cent higher than the risk under an intact O_3 layer (Diffey 1992). Ecological damage has been identified closer to the Poles where more serious O_3 depletion has occurred. Concerns in the UK have led the government to set up an Ultraviolet B (UVB) Measurements and Impacts Review Group and to initiate a £1.5 million research programme to consider UVB's potential impact on skin cancer (Great Britain 1996). International action is

TABLE 18.5 Effects of increased exposure to ultraviolet light

Target	*Effects*
Materials	Degradation of wood and plastics leading to discolouration and reduced strength
Plants	Increased cuticular reflectivity Reduced leaf area Reduced crop yields Early flowering
Marine phytoplankton	Adverse effects on nitrogen metabolism, photosynthesis, orientation and mobility could threaten survival or productivity, which would have 'knock-on' effects up food chains
Humans and animals	Skin cancer deaths would rise – a 10 per cent decrease in stratospheric O_3 could cause 300,000 extra non-melanoma and 4,500 extra melanoma skin cancers globally each year (UNEP 1991) Corneal damage and eye cataracts would increase The skin's immune system would be impaired, leading to increased incidence and severity of skin diseases

needed to reverse the trend for O_3 decrease and to respond to the global threat of O_3 depletion. The 1986 Montreal Protocol referred to in the previous section has successfully resulted in reductions in the production and consumption of CFCs. In 1992, the sixty-five signatory countries agreed more stringent measures to stop global emissions of CFCs, carbon tetrachloride and methyl chloroform before 1996. Production of halons was phased out in 1994. However, because of the long lifetimes of all these chemicals, consequent O_3 depletion will continue well into the next millennium. Of particular concern will be ensuring that the commitments are complied with, controlling any black market in the chemicals and providing aid for poorer countries to help them discontinue use of the chemicals. In the UK, the supply of CFCs fell by 97 per cent between 1986 and 1995. However, the supply of HCFCs, which are being used as transitional replacements for CFCs but which do have some O_3 depleting potential of their own, increased by 285 per cent (DETR 1997).

Terrestrial pollution

Perhaps the most serious terrestrial pollution problems faced by the UK and many other countries relate to waste disposal. The Dobris Assessment specifically identified 'waste production and management' as one of the twelve 'prominent European environmental problems', so this topic will be used as the focus for discussion in this section. Existing waste processing and disposal capacity is unlikely to be adequate for much longer. Many areas of the UK, notably south-east England and other large urban areas, have already run out of land for controlled tipping and have to transport their waste considerable distances for disposal. For example, much of London's waste is disposed of in either the Bedfordshire

TABLE 18.6 Environmental impacts of landfill waste disposal

- Noise from machinery and transport
- Odours
- Visual intrusion
- Windblown litter and dust
- Insect pests and vermin
- Leaching of toxins into groundwater and watercourses
- Leaching of nutrients into groundwater and watercourses
- Air pollution from burning refuse
- Methane gas from biodegradation of organic refuse
- Subsidence

brickworks or in reclaiming land from the sea in Essex. Landfill creates a number of environmental problems, particularly on poorly managed sites, as listed in Table 18.6.

About 400 million tonnes of waste are generated annually in the UK. Much of this waste is relatively innocuous, including waste from mining, quarrying and agriculture, which mainly remains on site. Demolition and construction waste is mainly inert and much waste from industry is recycled (Figure 18.5). Of major environmental concern is Special Wastes, defined in the Special Wastes Regulations 1996, which consist of substances dangerous to life (such as toxic chemicals). Each household in England and Wales has been estimated to generate 21 kilograms of waste per week. Only about 7 per cent of UK household waste is currently recycled, by far the majority (about 83 per cent) still being landfilled (DETR 1997). However, a growing amount of such waste is being recycled, and the UK government has set a target of increasing recycling to 40 per cent by the year 2005.

As pointed out by the Dobris Assessment:

> there are few environmental problems in Europe that cannot be traced back to some sort of excessive loading by chemicals.

Estimates suggest that the current annual world production of chemicals is about 400 million tonnes, with 10,000 different chemicals being commercially produced (Lonngren 1992). These may be routinely released into air or water, or onto land, or may be discharged in accidents. Exposure to chemicals may adversely affect ecosystems, and in humans there may be harmful effects on health. There has been increasing concern about health hazards posed by hazardous chemicals from waste disposal and other industrial activities. Attempts to identify potentially contaminated sites have been made in some European countries, but in the UK measures to require local authorities to make registers of potentially contaminated land were postponed in 1992. The most common toxic soil pollutants include heavy metals, organic chemicals, oils, tars, pesticides, radioactive materials and asbestos. The Environment Act 1995 made the Environment Agency responsible for producing a national report on contaminated land and for advising local authorities on the problems. The UK distribution of derelict land, which is often contaminated, and its reclamation is considered further in Chapter 17.

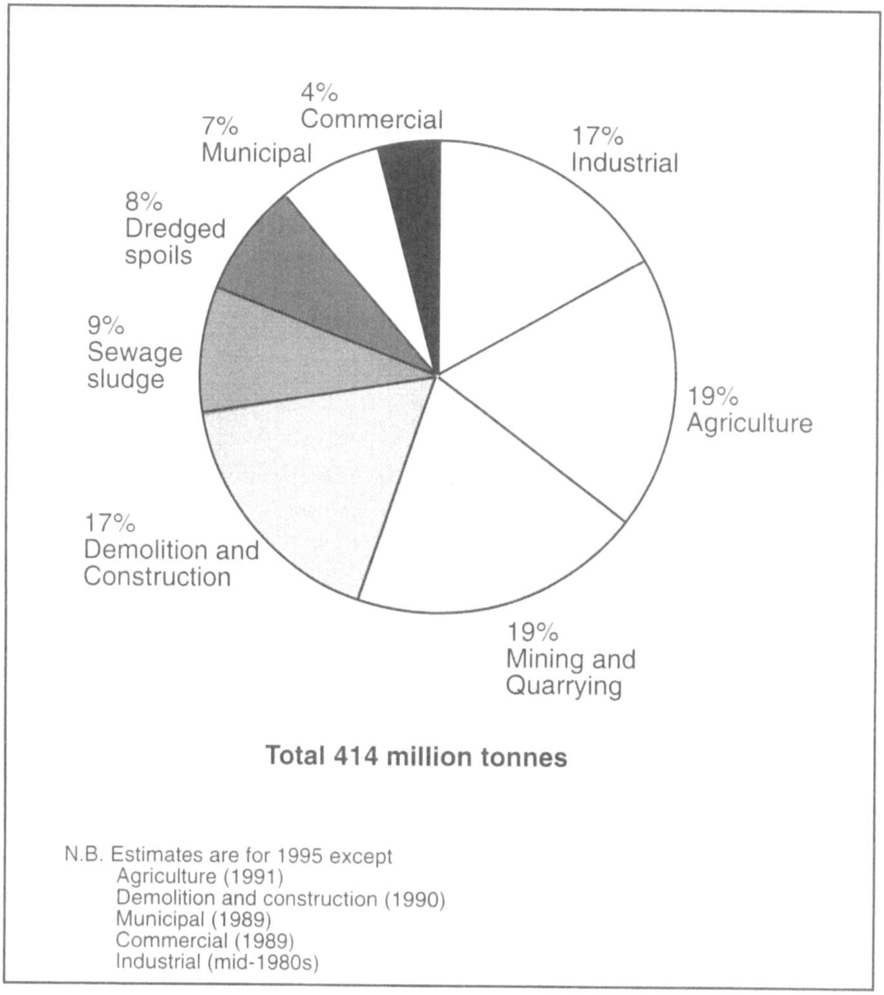

FIGURE 18.5 Estimated UK waste production by sector for 1995
Source: DETR (1997).

Old landfill sites in the UK are a potential 'time bomb'. Leaching of toxic chemicals, such as heavy metals from old batteries or organochlorine compounds from old pesticide containers, may contaminate drinking-water supplies and pose a health risk where housing has been developed on or near old disposal sites. Leachate may also contain high concentrations of nitrates and phosphates leading to eutrophication of surface waters in surrounding areas. There have been a number of UK incidents in which houses built on old landfill sites have been damaged or destroyed by explosions. The threat of explosions has led to large property price reductions and the blighting of some areas in terms of house sales. Methane and carbon dioxide emitted also contribute to the greenhouse effect considered in the previous section, and landfill emissions have been estimated to account for a quarter of total methane emissions (Hewett 1995).

About 9 per cent of waste generated in England and Wales in 1995/6 was incinerated (DETR 1997). Incineration creates the problem of disposing of ash, which includes contaminants such as heavy metals. If the process is not properly controlled toxic gases are emitted into the atmosphere, one notable group being dioxins which are produced if the temperature drops too low. Municipal waste incinerators were required by EC legislation to comply with stricter emission standards by 1 December 1996. Thirteen UK incinerators were unable to comply with the new standards and had to be closed, leaving just five up-graded plants in operation. Therefore, the incineration disposal option has recently decreased in significance in the UK. Proposals to construct new plants such as incinerators, which have major potential environmental impacts, invariably encounter considerable local opposition (the so called NIMBY – Not In My Back Yard – syndrome). This is true for any development proposals involving any form of waste handling, processing or disposal.

The UK generates about 2–2.5 million tonnes of Special Waste each year (DoE 1996). Of particular concern is the international trade in such wastes. This is worrying and unethical when the recipient is a poorer country, lacking adequate treatment or disposal facilities, which has been tempted by financial incentives. International co-operation is needed to achieve full control of such transfrontier waste movement. However, the UK has already acted, under the 1996 Management Plan for Export and Import of Waste, to ban the export or import of waste for disposal.

The disposal of radioactive waste (considered in more detail in Chapter 3) and fall-out of radioactive isotopes from nuclear installations have long been a cause for concern in the UK. Most accidents which have occurred at nuclear power plants have resulted from human errors, the most serious being the April 1986 Chernobyl accident in the Ukraine (see Chapter 17). Radioactive caesium-137 was deposited over wide areas of Europe and led to restrictions on livestock sales from some parts of the UK. For the average UK individual, less than 0.1 per cent of their radiation dose comes from routine operation of nuclear installations. Of more immediate concern is the hazard posed by naturally occurring radon gas referred to in the introductory section of this chapter and discussed in greater detail in Chapter 17.

Aquatic pollution

Two of the Dobris Assessment's 'prominent European environmental problems' specifically involve aquatic pollution, namely the 'management of freshwater' and 'coastal zone threats and management'. Others, notably 'major accidents', 'acidification', and 'chemical risks' can all affect aquatic ecosystems.

Major pollutants of UK freshwaters include sewage, leachates from fertilisers and manures, silage effluent, pesticides, industrial chemicals and mine drainage. Most UK rivers are classed as of good or fair quality (chemically, 91 per cent in England and Wales, 99 per cent in Scotland and 88 per cent in Northern Ireland; biologically, 93 per cent in England and Wales, 98 per cent in Scotland and 100 per cent in Northern Ireland in 1995). Differences between regions largely reflect differences in industrialisation (referred to in Chapter 7) and population distribution (referred to in Chapters 9 and 10). Quality is generally improving, mainly owing to improved sewerage and sewage treatment (see Chapter 2) and industrial pollution controls. There was a net upgrading in biological and chemical quality of rivers and canals in England and Wales of 26 per cent and 28 per cent respectively between 1990 and 1995. However, there were still significant stretches classified as poor to bad (DETR 1997). For example, 9 per cent of rivers and canals in England and Wales, and 12 per cent in Northern Ireland, were described as poor or bad for chemical quality in the period 1993–5. The number of reported water pollution incidents in England and Wales rose steadily between 1988 and 1995 (Figure 18.6). However, in 1996 the number of reported incidents fell for the first time in eight years. Drinking water quality appears to have improved in recent years, as discussed in Chapter 2.

The most important pollutants of marine waters include sewage in coastal bathing areas, oil, synthetic organic compounds such as polychlorinated biphenyls (PCBs) and pesticide residues, litter, heavy metals and radionuclides. Routine operation of ships and oil rigs, as well as accidents involving them, regularly create oil pollution incidents. The Irish Sea is renowned for its high levels of radioisotopes, routinely discharged from the Sellafield nuclear fuel reprocessing plant, and for various chemicals such as PCBs discharged from industries in north-west England which have been implicated in the deaths of seabirds. In the 1970s, Sellafield was easily the most significant site in Western Europe routinely discharging radioisotopes into the sea. However, stricter discharge authorisation limits have resulted in a 91 per cent reduction in discharges to water from UK nuclear plants between 1983 and 1993, despite an increase in nuclear power generation of 84 per cent (DoE 1996). Reductions have resulted from improved working practices and technology and increased emphasis on storing radioactive wastes rather than discharging them into the environment. However, there is concern that the UK does not have a fully developed strategy for disposing of radioactive wastes.

Since 1988 there has been an underlying trend for the quality of UK bathing waters to improve (Figure 18.6). Work is continuing to improve the remaining 10 per cent of sub-standard waters through a £10 billion investment programme centred on sewage treatment (Environment Agency 1997). Compliance varies on a geographical basis, ranging from 100 per cent compliance for Northern Ireland bathing waters to only 61 per cent for bathing waters in north-west England. This largely reflects the population densities of the regions involved, with more-populated areas producing more sewage and hence tending to have poorer-quality waters. For example, major sewage discharges into Swansea Bay have led to bathing-water quality test failures and nuisance complaints from the public. A new £82 million sewage treatment works looks set to improve the area from 1998 (Environment Agency 1997). British rivers contribute about 20 per cent of pollutants entering the North Sea. About 9.3 million tonnes of sewage sludge and 5.1 million tonnes of industrial waste from the UK were dumped at sea in 1990, but sea dumping is being phased out (Hewett 1995).

Oil pollution incidents, such as those caused by the *Braer* and *Sea Empress* referred to in the opening section of this chapter, are a regular occurrence around the UK coasts. For every major incident such as these there are many smaller incidents, some of which involve the deliberate discharge of oil into water. Coastwatch Europe Surveys, which began in 1987, have shown that coastal litter pollution is a major problem. Some forms of litter – for example sanitary waste, medical waste and glass – pose a direct hazard to human health. Other forms – for example fishing gear and plastic bags – float in the sea and may entangle and kill sea birds and marine mammals (Coastwatch 1994).

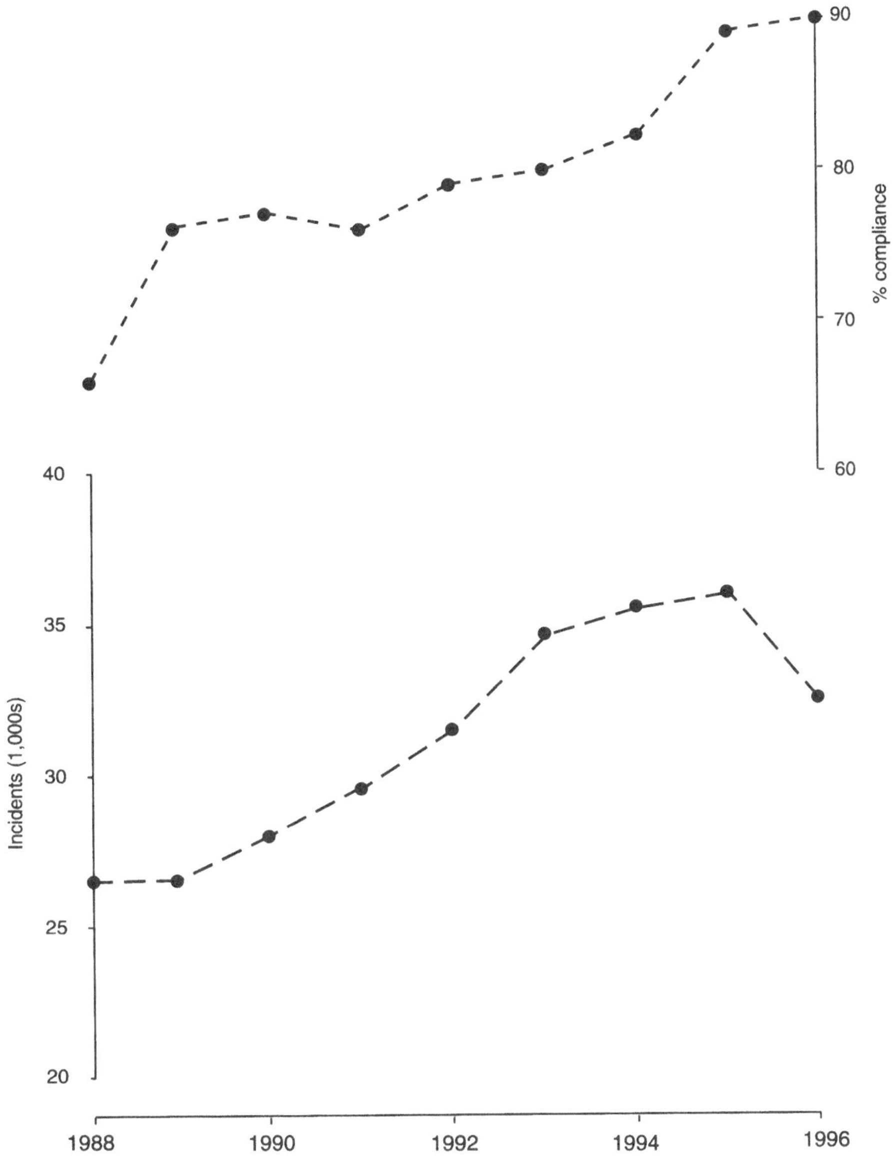

FIGURE 18.6 Reported water pollution incidents in England and Wales, and UK compliance with EC Bathing Water Directive 76/160/EEC, for the period 1988–96
Source: DETR (1997).

The future

The foregoing account supports the contention of the opening quotes that significant improvements in UK pollution control have been made in the 1990s. The account also demonstrates that improvements are not universal and that levels of some pollutants in certain areas have been increasing in recent years. In particular, it is becoming apparent that emissions from motor vehicles are playing an increasingly significant role in atmospheric pollution in the UK and that urgent controls are required. Improvements in control for individual vehicles have tended to be outweighed by the ever-increasing number of vehicles on our roads. There was a 58 per cent increase in motor vehicle traffic on UK roads between 1980 and 1995, most of which was attributable to motor cars (DETR 1997). The changing pattern of transport in the UK has been considered in more detail in Chapter 6. As has been demonstrated in the present chapter, motor vehicles are now the major culprits in terms of UK emissions of black smoke, nitrogen oxides and carbon monoxide. They are also a serious problem for many people living near roads as regards noise nuisance. Complaints received by Environmental Health Officers about road-related traffic noise (including roadworks and road construction) increased more than threefold between 1983/4 and 1994/5 (DETR 1997), reflecting the general trend for increasing noise complaints from all sources. Further action is needed to reduce increasing photochemical oxidant formation to which motor vehicles are major contributors. Efforts are needed to reduce our dependence on motor vehicles and to increase the efficiency of individual vehicles. The need for comprehensive transport planning emphasising public transport systems is paramount. Electric trams, which have been recently introduced in Sheffield and which will soon appear in Croydon, and navigable urban waterways may both have a part to play, being relatively minor pollution sources when compared with motor vehicles. Planning policies, considered in more detail in Chapters 10 and 11, should aim to reduce the need for long-distance journeys and promote cycling and walking as environmentally and socially acceptable forms of transport. For example, the trend for increasing numbers of 'out-of-town' shopping centres (e.g. Meadowhall, Sheffield – see Chapter 1), which generates additional car journeys and associated pollution, not to mention adding to the demise of local town and village centre shops, really needs to be reversed.

In August 1997, the Deputy Prime Minister, John Prescott, made a government commitment to developing an integrated transport policy. It is to be hoped that Local Agenda 21, which was initiated at the 1992 Earth Summit, will facilitate more sustainable use of resources which will help to reduce pollution. For example, energy conservation measures in offices, shops, industry and homes would help to reduce the trend for increasing fuel use reported in Chapter 3. Insulation of buildings and the use of low-energy lighting are two simple ways of reducing fuel use, and hence emissions of CO_2 and other pollutant gases. In 1997 at 'Earth Summit II', the Prime Minister, Tony Blair, reiterated the UK's commitment to reduce CO_2 emissions by 20 per cent, compared with 1990 levels, by 2010.

Increasing shortages of land, particularly in south-east England and other urban areas, for the landfilling of waste as well as for housing development, is clearly a major issue as regards terrestrial pollution, which needs to be tackled as the next millennium approaches. More incentives need to be provided to promote housing developments in derelict urban areas. Developers will need to pay particular attention to remediating any soil contamination which might exist in such areas and more assistance could be provided for clean-up operations. As regards waste disposal, the UK needs to develop and optimise programmes to encourage a reduction in waste generation, the reuse of packaging and the recycling of waste. Further action to minimise packaging and to provide more recycling facilities is needed. Two simple examples will now be provided to demonstrate the sort of things which could be done. First, doorstep milk deliveries, which use returnable bottles and typically involve transport by electric vehicles, could be encouraged by incentives for such deliveries or by a form of tax on 'non-doorstep' milk. This would reduce the huge number of disposable plastic and paper milk cartons generated, as well as reducing air pollution associated with car journeys made to purchase milk. Second, individual household composting of organic waste could be encouraged by distribution of free composters, as has been attempted by a few local authorities, combined with an education campaign on this subject. As well as reducing pollution from the transport and disposal of household waste, this would also reduce pollution generated by the manufacture, packaging and transport of horticultural fertilisers.

The EC Council Directive 67/548/EEC on labelling, packaging and classification of dangerous substances looks set to form the basis of legislation for controlling chemical use and reducing packaging in the UK and other EC nations. For products to qualify for an EU logo they must have a reduced environmental impact when compared with usual products and must comply with specific ecological criteria. These take into account environmental impacts at all stages of a product's life cycle – from raw materials, through manufacture, distribution, use and final disposal. The UK, along with four other EC member states, has started product life-cycle analysis of paper, detergents, paints, hair sprays, refrigerators and washing machines.

There are signs that the quality of many UK rivers and coastal waters is slowly improving after many years of a general trend for decline. Recent years have seen the widely publicised re-emergence of otters and salmon in the River Thames, from which both species were virtually eliminated, mainly by pollution. A lesser-known species, the smelt, is also making a comeback in the Thames. In the eighteenth century, the smelt formed the basis of a major industry in Chiswick from which up to one hundred boats fished.

Four guiding principles for the prevention and safe management of waste are listed in the Dobris Assessment and could be usefully applied more generally to any process in which pollutants are generated. First, the 'prevention principle' states that waste production should be avoided at source. Second, the 'polluter pays principle' states that the polluter should stand the costs of waste disposal and pollution. Third, the 'precautionary principle' suggests that disposal options should consider all potential environmental impacts and ensure that no serious environmental harm is done. Fourth, the 'proximity principle' recommends that waste should be managed as near as possible to the source. These four principles should be considered when dealing with any form of pollutant in the UK as we enter the next millennium. It is to be hoped that the new Environment Agency successfully fulfils its statutory responsibilities for protecting and improving the environment, which would be demonstrated by continually decreasing levels of atmospheric, terrestrial and aquatic pollution in the UK as the next millennium progresses.

Acknowledgements

Thanks are due to Adrian Redfern (DETR) for providing permission to use data from *The Environment in Your Pocket*, and to Mary Mackenzie (Roehampton Institute, London) for drawing the figures.

References

Coastwatch (1994) *Coastwatch Europe: International Results Summary of the 1993 Survey*, Dublin: Coastwatch Europe Network, International Coordination, Trinity College.
Commoner, B. (1990) *Making Peace with the Planet*, London: Gollancz.
DETR (Department of the Environment, Transport and the Regions) (1997) *The Environment in Your Pocket 1997*, London: HMSO.
Diffey, B.L. (1992) *Stratospheric Ozone and the Risk of Non-Melanoma Skin Cancer in a British Population*, London: Greenpeace.
DoE (Department of the Environment) (1996) *Indicators of Sustainable Development for the United Kingdom*, London: HMSO.
Environment Agency (1997) *Environment Agency Annual Report and Accounts 1996–97*, Bristol: Environment Agency.
Great Britain (1990) *This Common Inheritance: Britain's Environmental Strategy*, London: HMSO.
Great Britain (1996) *This Common Inheritance 1996, UK Annual Report*, London: HMSO.
Hewett, J. (ed.) (1995) *European Environmental Almanac*, London: Earthscan.
Holdgate, M.W. (1979) *A Perspective of Environmental Pollution*, Cambridge: Cambridge University Press.
Horner, J.M. (1996) 'Lead in paint and dust from a children's nursery', *Environmental Management and Health* 6 (1): 5–9.
Horner, J.M. (1998) 'Carbon monoxide: the invisible killer', *J. Roy. Soc. Health* 118 (3): 141–5.
Houghton, J.T., Jenkins, G.J. and Ephraums, J.J. (eds) (1990) *Climate Change: The Intergovernmental Panel on Climate Change Scientific Assessment*, Cambridge: Cambridge University Press.
KPMG (1997) *The KPMG Survey of Environmental Reporting 1997*, London: KPMG.
Lonngren, R. (1992) *International Approaches to Chemicals Control*, Stockholm: KEMI.
PORG (Photochemical Oxidants Review Group) (1997) *Ozone in the United Kingdom, 4th Report*, London: UK PORG.
Porritt, J. (1991) *Save the Earth*, London: Dorling Kindersley.
Stanners, D. and Bourdeau, P. (eds) (1995) *Europe's Environment – The Dobris Assessment*, Copenhagen: European Environment Agency.
UNEP (United Nations Environment Programme) (1991) *Environmental Effects Panel Reports: Environmental Effects of O_3 Depletion, 1991 Update of 1989 Report*, Nairobi: United Nations.

Further reading

Bridgman, H.A. (1990) *Global Air Pollution: Problems for the 1990s*, Chichester: Wiley.
Elsworth, S. (1984) *Acid Rain*, London: Pluto Press.
Harrison, R.M. (ed.) (1992) *Pollution: Causes, Effects and Control* (2nd edn), London: Royal Society of Chemistry.
McCormick, J. (1997) *Acid Earth: The Politics of Acid Pollution* (3rd edn), London: Earthscan.
Middleton, N. (1995) *The Global Casino: An Introduction to Environmental Issues*, London: Arnold.
Pickering, K.T. and Owen, L.A. (1997) *An Introduction to Global Environmental Issues* (2nd edn), London: Routledge.
Wellburn, A. (1994) *Air Pollution and Climate Change: The Biological Impact*, London: Longman.

Of the texts listed in the References section, the DETR's is an annual publication which constitutes a useful booklet of broad-ranging UK environmental data, some of which is relevant to other chapters in this book. Students, in particular, will be delighted to learn that it is available free of charge – so there should be no excuse for not including up-to-date UK environmental data in assignments! Copies can be requested from DoE Publications Dispatch Centre, London, SE99 6TT (telephone 0181 691 9191). Recent data should also be accessible via the Internet at http:/www.open.gov.uk/doe/epsim. Another useful publication of recent UK environmental information is the Environment Agency's *Annual Report* which can be obtained from Rio House, Waterside Drive, Aztec West, Almondsbury, Bristol, BS12 4UD (telephone 01454 624 400). They too have an informative web site at www.environment-agency.gov.uk. To 'counterbalance' these government-based reports readers might like to refer to the web sites of Greenpeace and Friends of the Earth, which contain useful information on pollution. These can be found at http:/www.greenpeace.org.uk and http:/www.foe.co.uk, respectively.

For readers requiring a more widely based account (both geographically and content-wise) of the state of the environment in Europe, the Dobris Assessment includes very detailed and comprehensive accounts of pollutants. It would be a useful reference source in any university library where courses in Geography and/or Environmental Studies are taught. The Further Reading list reflects the interdisciplinary nature of pollution studies. As most readers of this textbook will be from a geographical background, Middleton's *Global Casino* or Pickering and Owen's *An Introduction to Global Environmental Issues* is most likely to appeal. They cover a broad range of environmental issues and so are relevant to other chapters in this book. Those with more biological or chemical backgrounds might find Wellburn or Harrison, respectively, more to their liking, and these texts do provide more detailed coverage of pollution.

Chapter 19

Climate change

Mike Hulme

Introduction

> Climate: the condition of a region or country in relation to prevailing atmospheric phenomena such as temperature, humidity, etc., especially as these affect animal or vegetable life.
>
> (*OED* 1973)

The climate of a *place* is one of its most important physical attributes. The unique sequence of weather events experienced at a place over a long period of time determines to a considerable extent the nature of the surface landforms, the mix of vegetation and the potential of the place to act as a habitat for different species, including humans. The climate of a *country* may often be one of the fundamental unifying characteristics of its national identity. The shared experience by a community of the march of the seasons, of weather extremes or of the climatic resources of a region, creates a common set of climatic perceptions and instincts that are almost a language in themselves. These two roles of climate have been recognised throughout human history, and the cultural response of humans to different climates – in both space and time – has left imprints on both the natural and social environment that are among the most distinctive. These imprints range from physical artefacts such as irrigation infrastructures or flood defences, to intellectual constructs such as climatic determinism (Huntington 1915) and the defining of certain racial characteristics of different nations. Although this close relationship between climate and its induced human response has long been recognised, awareness of the inverse relationship – the response of climate to different human activities – is of more recent origin. While the human modification of *microclimates* – whether inadvertent through expanding cities or deliberate through planting trees – has a substantial history, the understanding that human actions can modify larger-scale synoptic circulation, or even *global* climate, has only penetrated widely through the intellectual world during the latter decades of the twentieth century.

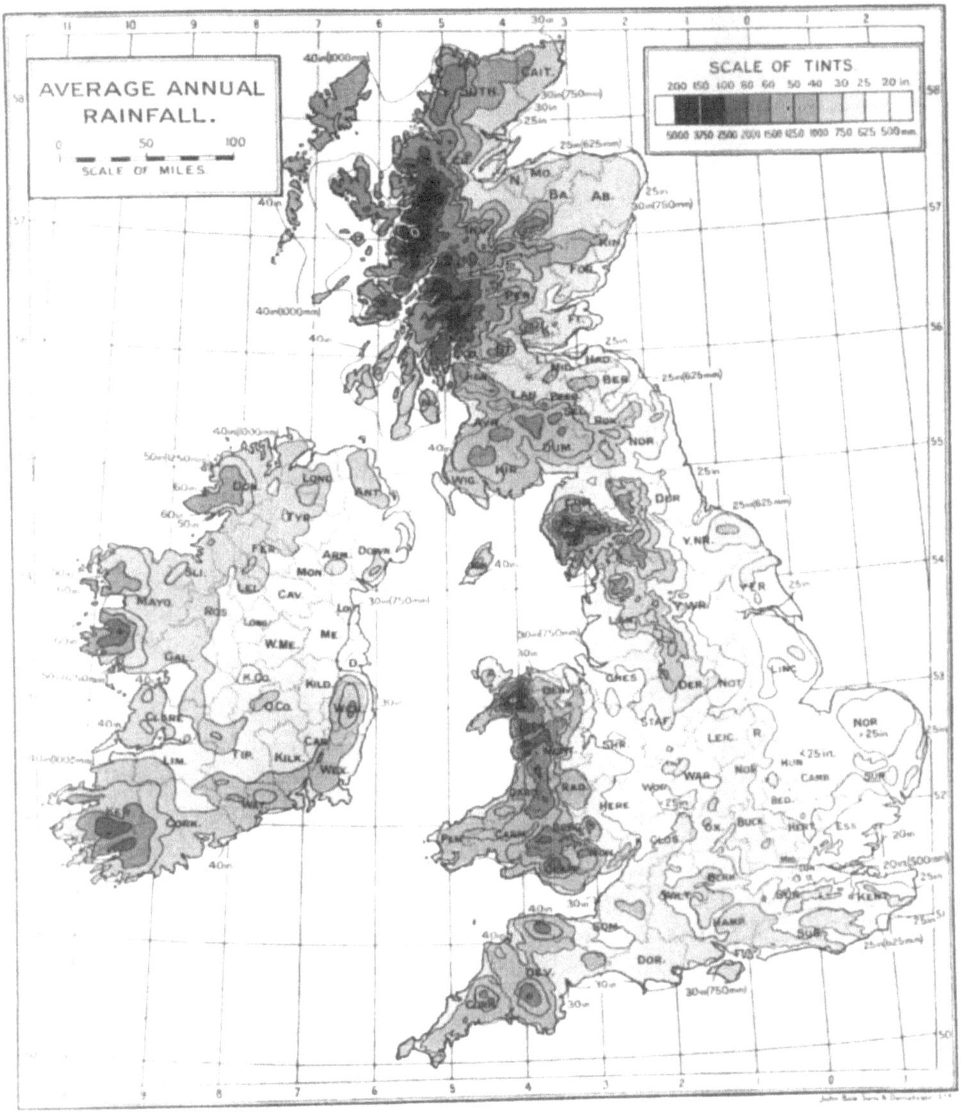

FIGURE 19.1 Average [*sic*] annual rainfall for the British Isles, based on 1880–1915 data, measured in inches

Source: Royal Meteorological Society (1926).

This changed perception of climate, and of its potential to be changed, is as true for the UK as it is for other nations. Thus for much of the present century, climate 'normals' for the UK were defined as thirty or thirty-five year averages (e.g. Royal Meteorological Society 1926; Figure 19.1) and these were commonly regarded as providing definitive descriptions of that most characteristic of British 'institutions', the climate. Similarly, the standard climate textbook of the 1970s, Tony Chandler and Stan Gregory's *The Climate of the British Isles* published in 1976, implicitly subscribed to this static view of climate through its title, even though the book devoted one concluding but rather brief chapter to climate change. Or taking the example of the present book, *The Changing Geography of the UK*, neither the first nor second editions, published respectively in 1983 and 1991, contained a chapter on climate change or even on climate, despite the book explicitly dealing with changes in the geography of the UK.

A number of intellectual developments and climatic events have led to this increasing awareness of the importance of climate variability, some specific to the UK and some related to the worldwide stage. The pioneering scholarly work of the late Hubert Lamb ranks highly among the former. For over half a century, Professor Lamb endeavoured to bring to the attention of a wider public his belief in the reality of climate change on human time-scales, an endeavour that culminated in the publication of Volumes I and II of his mammoth work *Climate: Present, Past and Future* in, respectively, 1972 and 1977 (Lamb 1972, 1977). He also founded, in 1972, the Climatic Research Unit at the University of East Anglia, a research centre that has developed into one of the world's leading centres for research into climate change. Among the latter set of reasons – notable climate events – the great drought of 1975–6 was a landmark climate anomaly in the UK, a drought of such significance that it prompted the publication of an atlas dedicated to the character and consequences of the drought (Doornkamp *et al.* 1980).

In the political sphere, the speech to the Royal Society delivered by the Prime Minister, Mrs Thatcher, in September 1988 marked a sea-change in the response of the UK government to the prospect of significant climate change. The core funding of the Hadley Centre for Climate Prediction and Research, opened in May 1990, was guaranteed and this centre now contains perhaps the world's leading climate modelling group. And at a global level, the series of increasingly warm years during the 1980s and 1990s (1981, 1983, 1988, 1990, 1995, 1997 and 1998 all set new annual records for global-average surface air temperature) provided the empirical impetus to the realisation that global climate was changing. This sequence of record warm years paralleled developments in the institutional framework of global climate science. In 1988, the World Meteorological Organisation (WMO) and the United Nations Environment Programme (UNEP) established the Intergovernmental Panel on Climate Change (IPCC), and four years later at the Earth Summit in Rio de Janeiro the Framework Convention on Climate Change (FCCC) was signed by over 150 nations under the auspices of the United Nations. The IPCC provided an emerging scientific consensus concerning the nature and causes of global climate change, while under the instruments of the FCCC a Protocol for limiting the growth of climate-warming greenhouse gases was first agreed at the Kyoto Climate Summit in December 1997 (Bolin, 1998).

As the century ends, therefore, not only is the physical reality of climate in the UK different from what it was just fifty or a hundred years ago – as this chapter will elucidate – but the intellectual construct of climate is also very different. Climate, within the UK and also globally, is now seen as continually changing. Past climate statistics, defined over thirty or so years, are no longer an adequate guide to the future. Private industries (such as those concerned with water and insurance) and public bodies (such as the Environment Agency and the Countryside Commission) commission studies and reports that quantify the likely range of climate change to be realised next century. These estimates are subsequently incorporated into their long-term strategic planning activities (see Arnell *et al.* 1997, as an example for the water industry). An important boundary condition of the geography of the UK is changing and can no longer be viewed as fixed. The components of the geography of the UK that are most sensitive to climate are therefore also subject to climate-induced change.

Before looking at some of these sectoral sensitivities (pp. 413–17) and summarising some of the key indicators of UK climate change (pp. 417–23), the chapter starts with a very brief survey of the historical change in UK climate over recent millennia and centuries (pp. 409–13). The fifth section (pp. 423–31) then looks at what the UK climate of the next century may be like and evaluates the significance of this prospect for the national UK economy and environment. The national and international climate policy initiatives relating to these changes are discussed on pp. 431–33, while the concluding section provides some reflections on future climate–society interaction in the UK. Throughout this chapter the terms *climate variability* and *climate change* are used with specific meanings. *Climate variability* is used to describe fluctuations in climate on all time-scales, but occurring due only to natural internal variability in the climate system. The term *climate change* is reserved for *climate variability* induced by some external forcing agent, whether natural (e.g. volcanic eruptions) or human (e.g. greenhouse gases) in origin.

A brief history of recent UK climate

The annual mean temperature of the UK over the most recent ten-year period (1989–98) has averaged about 8.9°C, very nearly 1°C warmer than the average UK temperature of the last 300 years. This has without doubt been the warmest decade in the period for which we have instrumental measurements, a decade in which the warmest single year – 1990, 1.45°C warmer than the 300-year average – also occurred. How this warmth compares with earlier decades or years in the UK, for example during the late medieval period of the thirteenth or fourteenth centuries or even earlier during the early Holocene about 8,000 years ago, is much harder to establish.

Climate variability in the Holocene

That the climate of the UK has varied substantially over these millennia is of course well established. One of the more enlightening methods for reconstructing temperatures in the UK over this period of time has been to examine the assemblages of particular beetle species, identified as dated remains in a number of late glacial and Holocene sites in the UK (Atkinson *et al.* 1987). Using the 'mutual climatic range' method the annual mean temperature of these sites has been deduced for more than fifty discrete time points within the period of transition from glacial to interglacial climates, a period from roughly 15,000 to 8,000 years BP. A further eight climatic 'snapshots' have also been reconstructed during the few thousand years before and after this glacial–interglacial transition. This reconstruction (Figure 19.2) provides quantitative evidence for major changes in UK climate as the Devensian ice-sheet retreated and the warmer climate of the Holocene became established. From this evidence, UK mean annual temperature was between –9° and –6°C during the last centuries of the Devensian period, rising to a temperature more typical of recent centuries

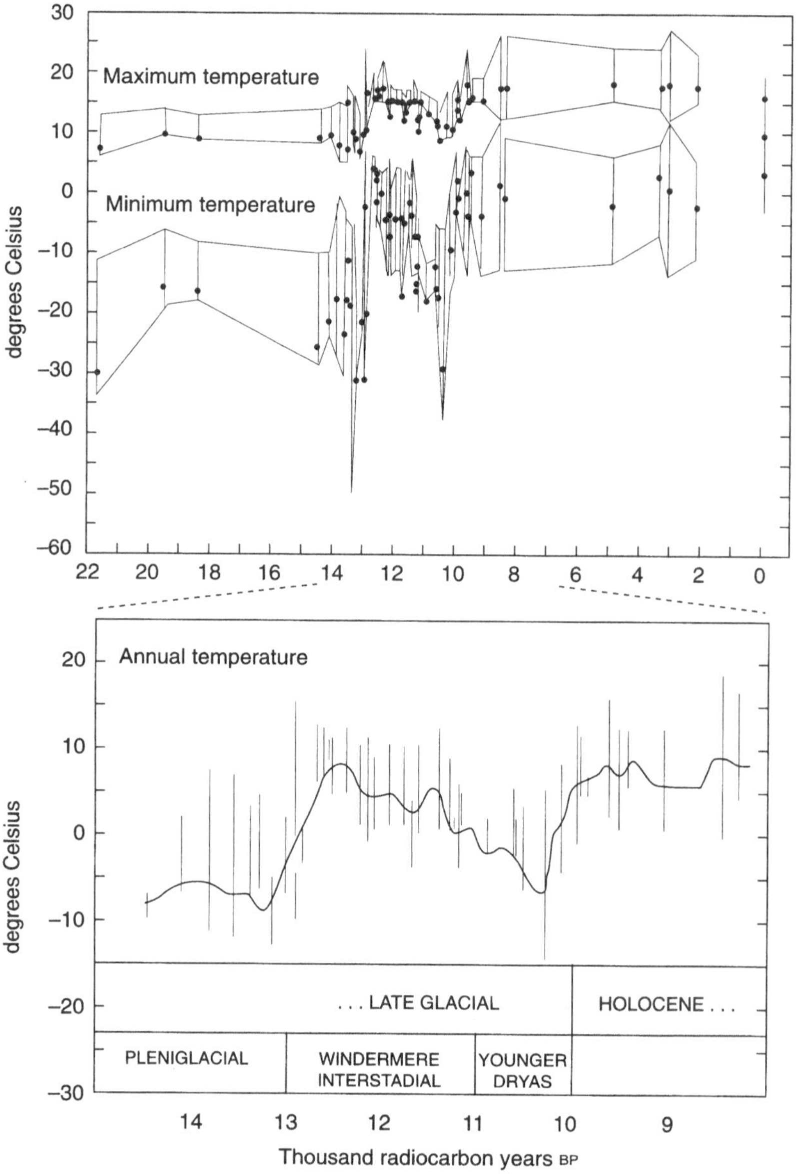

FIGURE 19.2 Temperature changes in the British Isles during the late glacial and Holocene periods estimated on the basis of beetle remains. Top panel: maximum and minimum temperature for 22,000 years BP to present; bottom panel: annual temperature for 15,000 to 8,000 years BP. The dots indicate the most probable values within the temperature ranges. The values plotted at 0 (i.e., 'today') in the top panel are the mean (dot) and maximum range (bar) of the Central England Temperature series over the period 1659–1995

Source: Briffa and Atkinson (1997).

(7° to 9°C) by the early Holocene at about 10,000 years BP. Since then, temperatures in the UK on at least century time-scales have probably not varied by more than about ±1°C, although we cannot say quantitatively what the spectrum of annual temperature variability has been on shorter, decadal or annual, time-scales. It is for this reason that we cannot be certain just how unprecedented the warmth of the 1989–98 decade in the UK has been.

It is even harder to reconstruct quantitatively the *precipitation* history of the UK over the Holocene. The changing distribution and abundance of certain vegetation species perhaps offer themselves as the most useful proxies for reconstructing moisture changes, although separating out the different effects of temperature and moisture change on species is not always easy. For example, a tree-ring-width chronology based on Irish bog oaks resolves the annual growth of this tree species in Ireland over the period from 7,500 to 2,000 BP. This index is a function of July/August growing conditions such that high growth is associated with warm/dry conditions and low growth with cool/wet conditions (Briffa and Atkinson 1997). Notably warm and dry conditions appear to have been more frequent between 7,000 and 6,400 and 5,900 and 5,450 BP, whereas longer cool/wet conditions are suggested between 6,400 and 5,900 BP, 4,200 and 3,900 BP and between 3,150 and 2,750 BP. These reconstruction efforts give support to the notion of meaningful climate variability having occurred in the UK on these multi-century time-scales, but without being able to quantify such variability fully or to resolve its interannual or interdecadal frequency spectrum.

Climate variability in the last millennium

Climate reconstruction using documentary sources from more recent history suffers from similar limitations. Various medieval annals, chronicles and account rolls of English manors reveal useful information about weather extremes and seasonal climate anomalies in the pre-instrumental period of the thirteenth and fourteenth centuries (Ogilvie and Farmer 1997). They are limited, however, in not enabling us to state with confidence how many warm years and decades there were during this so-called medieval warm epoch nor exactly how warm such years or decades were relative to today's conditions. For this precise quantitative information on early UK climate we have to rely on the instrumental record of climate. The longest homogeneous instrumental series in the world of monthly resolved mean temperature and precipitation originate from the UK. These commence in 1659 for mean temperature – the Central England Temperature (CET) record of the late Gordon Manley (1974) – and in 1757 and 1766 for precipitation – the Scotland precipitation series of Smith (1995) and the England and Wales Precipitation (EWP) series compiled by Tom Wigley and Phil Jones following earlier work by Nicholas and Glasspole (1931).

These three series are shown in Figure 19.3 and provide us with the longest quantitative context for evaluating the climate variability observed in the UK in recent decades. The average temperature of central England as defined by the CET is about 1.2°C warmer than for the UK as a whole (i.e., 10.1°C during 1989–98, compared to 8.9°C for the UK), but the interannual and interdecadal variations captured by the CET are representative of nearly the whole UK domain. The annual warming of UK climate over the period from 1659 has been about 0.7°C, with a stonger warming in winter (1.1°C) compared to summer (0.2°C; see Figure 19.7).

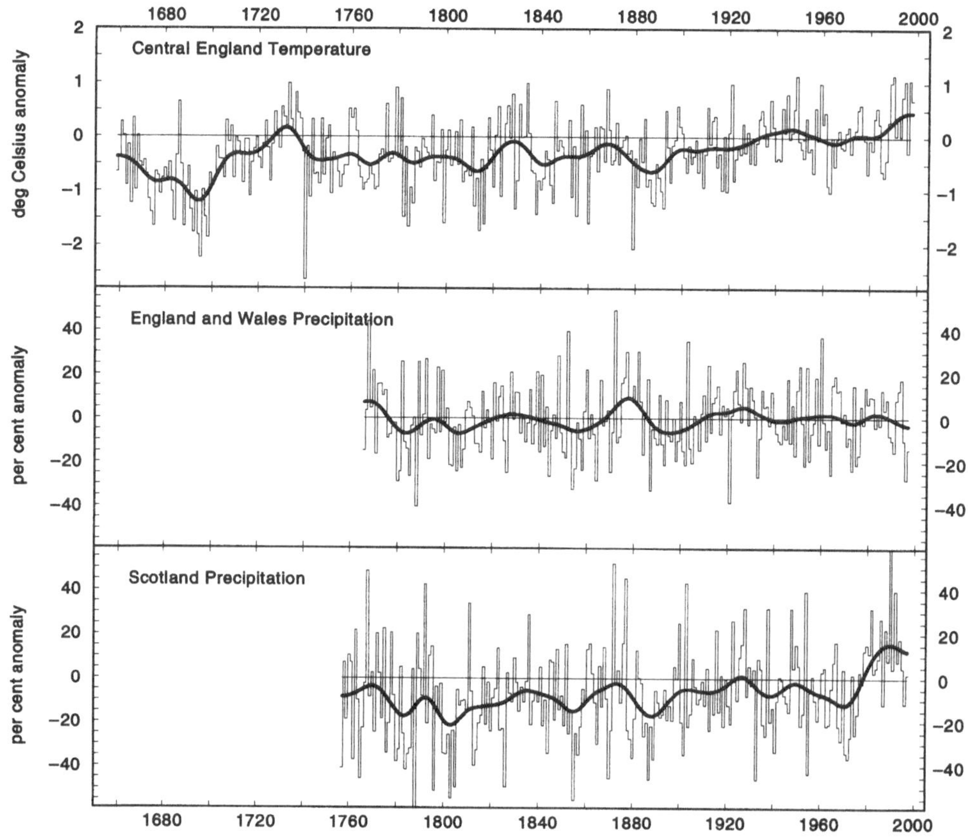

FIGURE 19.3 Annual anomalies for Central England Temperature (top), England and Wales Precipitation (middle) and Scotland Precipitation (bottom) with respect to 1961–90 mean (horizontal lines). The smooth curves emphasise variations on thirty-year time-scales. Last year is 1997

Analysis of the detrended CET series shows that the decadal variability of annual-mean temperature is about ±0.85°C and the thirty-year variability is about ±0.45°C (Table 19.1). We can calculate similar levels of 'natural' variability for precipitation, although in this case only the Scotland series displays any trend over time. The last twenty years in Scotland have been substantially (+20 per cent) wetter than any equivalent period in the last 240 years, with the wettest year on record occurring in 1990. Precipitation variability on a thirty-year time-scale is greater than ±10 per cent for both winter and summer, rising to over ±20 per cent for decadal-scale variability. These relatively large levels of 'natural' climate variability are of fundamental importance when considering the sensitivity of UK resources and institutions to climate change (see pp. 429–31).

This brief overview has provided evidence for the non-stationary character of UK climate over both shorter and longer time-scales. This climate variability has over the ages of millennia greatly shaped the landscape of our country and over the ages of centuries provided a constraint within which our economic, social and cultural heritage has developed.

TABLE 19.1 Levels of 'natural' climate variability in detrended seasonal and annual series for Central England Temperature (CET), England and Wales Precipitation (EWP) and Scotland Precipitation (SP) on ten-year and thirty-year time-scales

	Summer	*Winter*	*Annual*
CET 10-year (°C)	±0.85	±0.90	±0.85
CET 30-year (°C)	±0.40	±0.55	±0.45
EWP 10-year (per cent)	±23.5	±18.0	±10.5
EWP 30-year (per cent)	±11.5	±14.0	±4.0
SP 10-year (per cent)	±21.5	±29.0	±13.0
SP 30-year (per cent)	±13.5	±18.5	±6.0

In the next section we provide some examples of how different economic and environmental assets within the UK are sensitive to different magnitudes of climate change and variability.

The sensitivity of the UK to climate change and variability

Climatic determinism

It is a truism to state that environmental systems and human activities are influenced by climate. One only needs to think of the uneven abundance of surface water availability between the north-west and south-east of the UK, the siting of vineyards in the counties of eastern and southern England, or the summer timing of our outdoor national sporting events to recognise the role, sometimes almost subliminal, of climate in our national life. The study of climate–environment–society interactions has a long history and has, at times, led to rather exaggerated claims about the extent to which different climates can affect – or determine – human culture. Determinism is a reductionist philosophy that sees events and behaviour as controlled by a very limited set of physical factors. Ellsworth Huntington, the Yale geographer, has perhaps been the best-known proponent of such a role for climate. He argued in 1915 that, 'The climate of many countries seems to be one of the great reasons why idleness, dishonesty, immorality, stupidity, and weakness of will prevail' (Huntington 1915: 411). Although not always as strident or doctrinaire as Huntington, the importance of the climatic influence on our lives has been stressed by numerous thinkers, starting with the ancient Greeks and their supposedly uninhabitable, torrid and frigid 'climata'. The influence of climate has also been interpreted psychologically. In the middle of this century, for example, Manley (1952: 22) stated that, 'Appreciation of the British climate depends largely on temperament. That it has not been conducive to idleness has been reflected in the characteristics of the people' and, more recently, Beck (1993: 63) argued that 'the historical record is highly suggestive . . . that a mild climate in mid-latitudes helps to foster a tolerant society or that an extreme climate may predispose people towards intolerance'.

Without falling into the dangers of climatic determinism, it is important that we discover just how sensitive different environmental systems and human activities are to climate and hence to climate change and variability (Hulme and Barrow 1997). Or, put differently, to what climate regime are these functions best adapted and, if they are well adapted to the current regime, how vulnerable are they to differing magnitudes of climate change and variability? To take the three examples mentioned above – what would happen to UK water resources were there to be a shift in the patterns of precipitation across the UK? How would the British wine industry respond to a further 1°C warming in UK summer climate? And how would our summer sporting culture change should summers become drier and warmer? These are questions not of abstract intellectual curiousity but, given that we have demonstrated that UK climate varies on different time-scales quite naturally, are of major importance for national welfare. The questions become even more important when we consider the prospect of accelerated climate change, over and above natural climate variability, occurring as a consequence of human-induced changes in atmospheric composition. It is with these thoughts in mind that the UK government established in 1997 the UK Climate Impacts Programme (UKCIP) with a programme office at the University of Oxford. In the next part of this section we give some examples of climate sensitivity within the UK.

Sensitivity to climate variability and climatic extremes

The changing temperature of central England over past centuries has already been illustrated (Figure 19.3). In a study published in 1985, Martin Parry and Tim Carter showed how the risk of oat crop failure in the Lammermuir Hills of southern Scotland is sensitive to different climatic regimes within this period of record (Figure 19.4). The 1-in-10 and 1-in-50 risk of failure contours contract substantially during the mild period of 1931–80 compared to the colder conditions prevalent during the period 1661–1710. This later period was on average about 0.8°C warmer than the period in the seventeenth century. Such changes in agricultural risk have major implications for farm management practices and consequently for human settlement and the nature of the landscape.

Other relevant work has examined the consequences for the UK environment (Doornkamp *et al.* 1980; Cannell and Pitcairn 1993) and UK economy (Subak *et al.* 1999) of recent seasonal or annual climate anomalies. Thus Cannell and Pitcairn (1993) concluded that one of the main environmental impacts of the warm UK climate of 1989 and 1990 (a two-year mean temperature anomaly about 1.4°C above the 300-year average) was an increase in the abundance, activity and geographic spread of many insects. Some of these insects were pests – aphids or wasps – while others such as butterflies, moths and crickets may be said to have enhanced the natural environment. Increased honey production by bees during these two years was certainly welcomed by producers. Figure 19.5 shows the sensitivity of the activity and abundance of one moth species to the 1989 climate anomaly, in effect bringing forward its life cycle by between one and three weeks. Other environmental effects of this warm two-year period included the increase in nuisance blooms of blue-green algae on many lakes and reservoirs, lasting damage to mature broadleaved trees in the south-east of England, and the enhancement of fruiting and seeding of many warmth-loving herbaceous species.

A more comprehensive assessment of UK economic sensitivity to a climate anomaly was completed by Subak *et al.* (1999) following the record-breaking twelve-month period

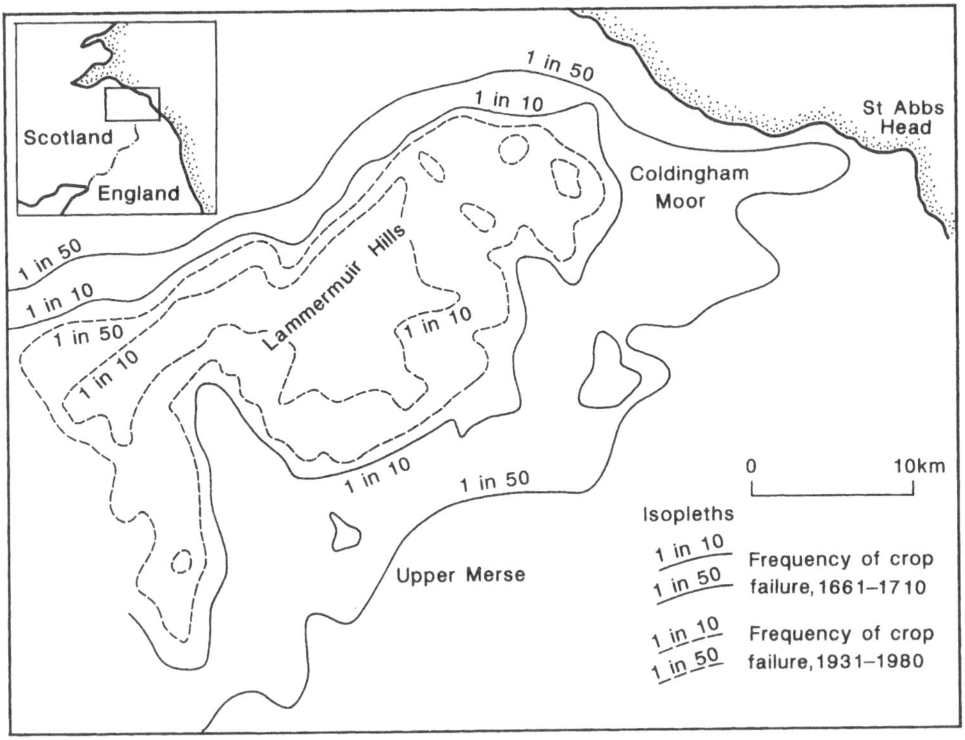

FIGURE 19.4 Locational shift of contours for risk of oat crop failure (1-in-10 years and 1-in-50 years) between cool (1661–1710) and warm (1931–80) periods in the Lammermuir Hills, south-east Scotland
Source: Parry and Carter (1985).

from November 1994 to October 1995. This period experienced a mean temperature anomaly of 1.9°C above the 300-year average, the warmest twelve-month period in the entire Central England Temperature series (Hulme 1997). The study was distinguished by its breadth of coverage, which included consideration of impacts on secondary industries as well as on tertiary and primary production sectors. These activities and sectors were agriculture, water supply, forestry, health, human behaviour, fires, energy, retailing and manufacturing, construction and buildings, transportation and tourism. Both negative and positive impacts of the warm weather were incurred within most sectors. Net positive impacts (to the general public) were found convincingly for energy and health, and clear negative impacts for agriculture, water supply and buildings insurance. It appeared that for most areas of human activity in the UK the greatest impacts were sustained from anomalous warm winter weather rather than from unusually hot summer weather.

Examples of water abstraction and domestic insurance

Two specific examples of sensitivity to such climate variability are shown in Figure 19.6 – water abstraction licences and subsidence-related domestic insurance claims. Water

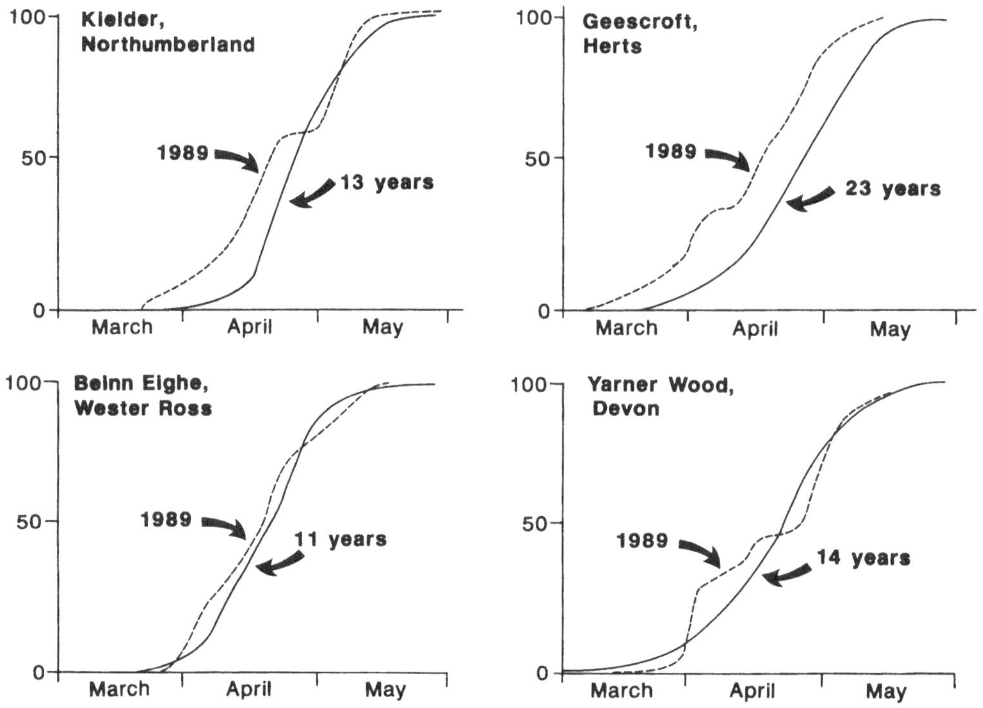

FIGURE 19.5 Total catch of the moth Hebrew character, *orthosia gothica* (L.) (Lep. Noctuidae) in light traps at four UK locations in 1989 compared to the mean for all previous years of trapping
Source: Cannell and Pitcairn (1993).

abstraction for agricultural applications in England and Wales increases dramatically during unusually hot and dry seasons or years. Abstraction rates more than trebled during the drought of 1989–92 and again during 1994–5 as farmers compensated for the lack of rainfall and high evaporation rates by spray irrigating their crops, particularly in the Anglian Water region. A similar pattern is evident for subsidence insurance claims. During the 1989–92 drought subsidence claims increased by a factor of six to about £600 million per year, and the dry weather of 1995 led to at least a trebling of claims. In contrast, the 1975–6 drought saw only a doubling of insurance claims for subsidence. This latter observation illustrates an important point about the sensitivity of different economic sectors to climate variability – the relative sensitivities of such sectors to climate variability themselves change over time. This may either occur as a function of changing economic or cultural circumstance (there were more home owners in 1991 than in 1976), but also as a result of specific adaptations to climate extremes (more homes were insured against subsidence in 1991 than in 1976). This consideration is most important when it comes to assessing the likely impacts of future climate change on either economic or environmental systems – past sensitivity is not necessarily a good guide to future sensitivity and a large array of adaptation strategies may lessen the impact of climate anomalies in the future (Subak *et al.* 1999).

We have demonstrated the inherent variability of UK climate on different time-scales and we have shown how different environmental assets and economic activities are sensitive

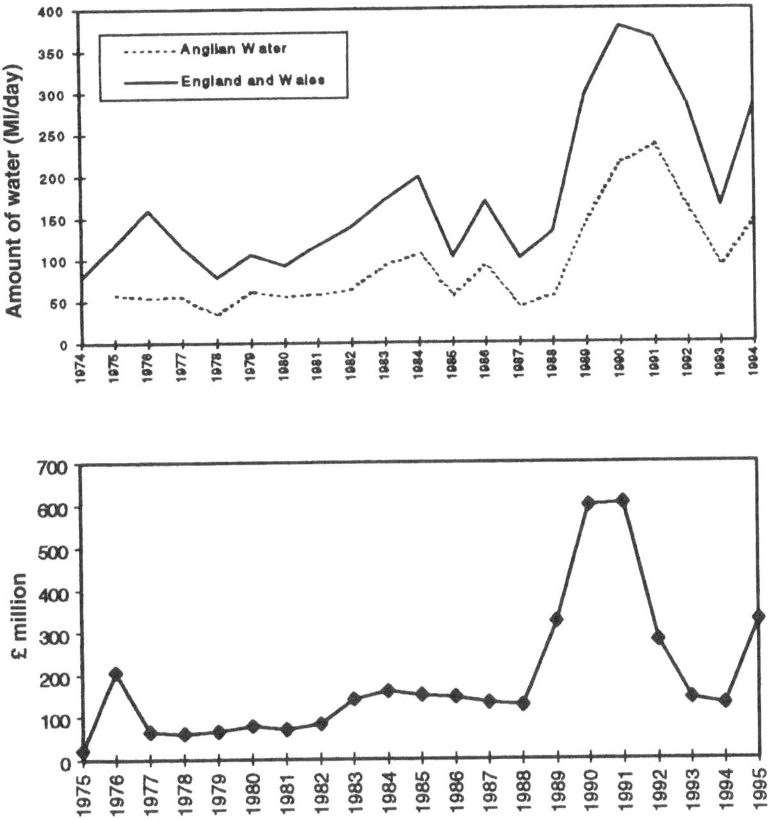

FIGURE 19.6 Top panel: water abstraction for agricultural use (spray irrigation in ml/day) for 1975–94 for England and Wales and for the Anglian Water region; bottom panel: subsidence-related domestic insurance claims at 1995 prices for 1975–95

Source: Subak *et al.* (1999).

Note: Many claims pertaining to the 1994–5 climate anomaly would appear in 1996 or 1997 data which are not shown.

to such climate variations. But just how rapidly is UK climate warming and what are the associated changes in non-temperature variables such as precipitation, gale frequencies and sea-level? The next section presents some indicators of climate variability over past centuries, while the fifth section (pp. 423–31) presents scenarios of UK climate change for the next century and a summary of their possible impacts.

Some key indicators of UK climate change and variability

The UK possesses some of the longest instrumental climate time-series in the world, the longest being the 340-year series of Central England Temperature referred to above. This presents a unique opportunity to examine climate variability in the UK on long time-scales based on observational data. It would be advantageous if we could treat these long time series as describing purely natural climate variability, thus enabling us better to identify what level of human-induced climate change may truly be significant. This would not be a correct interpretation, however, since – at least during the most recent century – human forcing of the climate system has been occurring through increasing atmospheric concentrations of greenhouse gases and sulphate aerosols. This has allowed scientists to detect a 'discernible human influence on the global climate system' (IPCC 1996: 4). What we observe therefore, and this is as true for the UK as it is for the world as a whole, is a mixture of natural climate variability and human-induced climate change, with the contribution of the latter increasing over time. It is nevertheless very instructive, before we progress to examine future climate change scenarios for the UK, to appreciate the level of climate variability that the UK has been subject to over recent generations.

Temperature indicators

The Central England Temperature series has already been referred to and an annual version of the series plotted in Figure 19.3. Here, we show the four seasonal components of this series smoothed with a 100-year filter to emphasise the century time-scale trends (Figure 19.7). The greater warming in winter compared to summer is clear, although most of this winter warming occurred before the present century commenced. Autumns over recent decades have averaged nearly 1°C above the full period mean, while the cluster of warm springs at the end of the series has meant that this season also has never been as mild as it has over the last decade. The unusual warmth of the decade 1989–98 in the UK is supported by the sequence of very warm months and seasons shown in Table 19.2.

The Central England Temperature series can also be used to examine changes in daily temperature extremes, although in this case data are only available since 1772. Figure 19.8 shows the annual frequencies of 'hot' and 'cold' days in Central England over these two centuries. There has been a marked reduction in the frequency of cold days since the eighteenth century, these frequencies falling from between fifteen and twenty per year to around ten per year over recent decades. There has not been a commensurate rise in the frequency of hot days, although as with annual temperature the last decade has seen the highest frequency of such days. For example, 1995 recorded twenty-six hot days compared to a long-term average of less than four, and the decade 1988–97 averaged 7.4 hot days per year, nearly double the 1961–90 average.

Other climatic indicators

The annual precipitation series for England and Wales and for Scotland are shown in Figure 19.3 and display little long-term trend. Here, we examine the seasonal distribution of precipitation with Figure 19.9 showing the proportion of annual England and Wales precipitation falling in winter (December–February). This index is a simple measure of the *continentality* of UK climate – the larger proportion of precipitation falling in winter the more continental the climate – and this seasonal distribution of precipitation has important

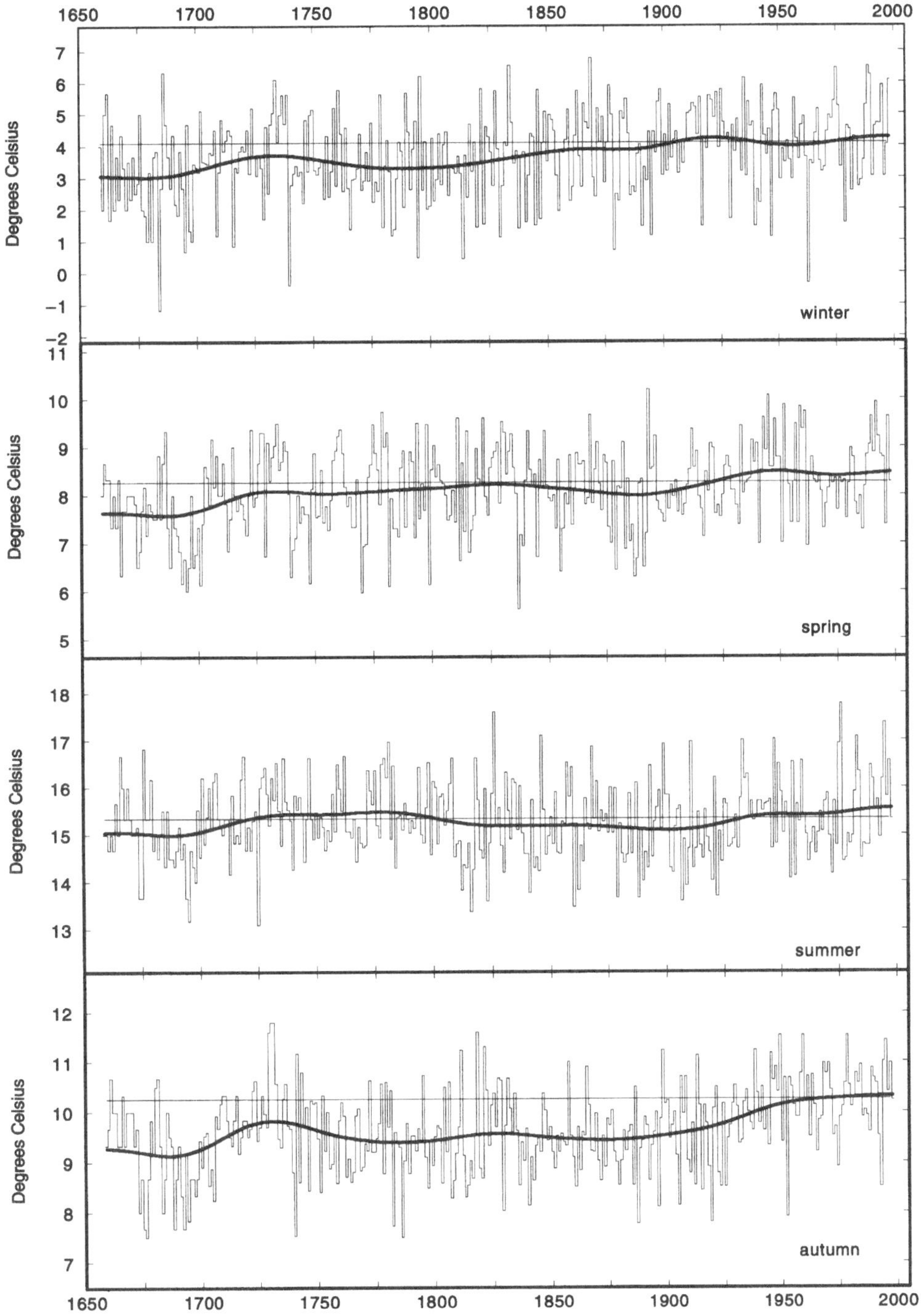

FIGURE 19.7 Seasonal Central England Temperatures for the period 1659 to 1997. Horizontal lines show 1961–90 means. Smoothed curves emphasise century time-scale variability

TABLE 19.2 Exceptionally warm months, seasons and years during the 1990s as indicated by the Central England Temperature series

	Largest anomaly in the 1990s		*Second largest anomaly in the 1990s*		*Third largest anomaly in the 1990s*	
January	1990	+2.7				
February	1990	+3.5	1998	+3.4		
March	1997	+2.7	1990	+2.6	1991	+2.2
April						
May	1992	+2.4				
June						
July	1995	+2.5				
August	1995	+3.4	1997	+3.1	1990	+2.2
September						
October	1995	+2.3	1990	+1.3		
November	1994	+3.6	1997	+2.0		
December						
Winter	1990	+2.2	1998	+2.0		
Spring	1992	+1.7	1997	+1.4	1990	+1.4
Summer	1995	+2.0				
Autumn	1995	+1.2				
Annual	1990	+1.15	1997	+1.06	1995	+1.05

Note: All the anomalies shown fall in the ten warmest respective months, seasons or years in the complete 340-year CET series. By chance, and assuming the monthly data are uncorrelated, one would expect only four months, seasons or years to appear in this table. In fact, twenty-five exceptionally warm anomalies have occurred during the 1990s. Over this period, only one month – June 1991 – has been exceptionally cold (using the inverse definition). Anomalies are shown in °C with respect to the 1961–90 mean. Data complete to July 1998.

implications for how water resources are managed. The proportion of precipitation over England and Wales falling in winter has increased over time, rising from about 23 per cent in the nineteenth century to a 1961–90 average of about 27 per cent. Two of the three years with the highest index values this century have occurred in the last decade – 1990 and 1995 – although there have also been recent years with quite low proportions of winter precipitation – 1992 and 1997. This shift in the seasonal distribution is a result both of increased winter precipitation and decreased summer precipitation (not shown) and is repeated in the Scotland series.

We also show (Figure 19.10) an index of gale activity over the UK, updated following the analysis of Hulme and Jones (1991). This series is shorter than those for temperature and precipitation and as with annual precipitation it shows no long-term trend. Gale activity is highly variable from year to year, however, with a minimum of two gales in 1985 and a maximum of twenty-nine gales in 1887. The middle decades of this century were rather less prone to severe gales than the early and later decades, whilst the most recent decade – 1988 to 1997 – has recorded the highest frequency of severe gales since the series began, 15.4 per year compared to an average of 12.3 (see also Chapter 17).

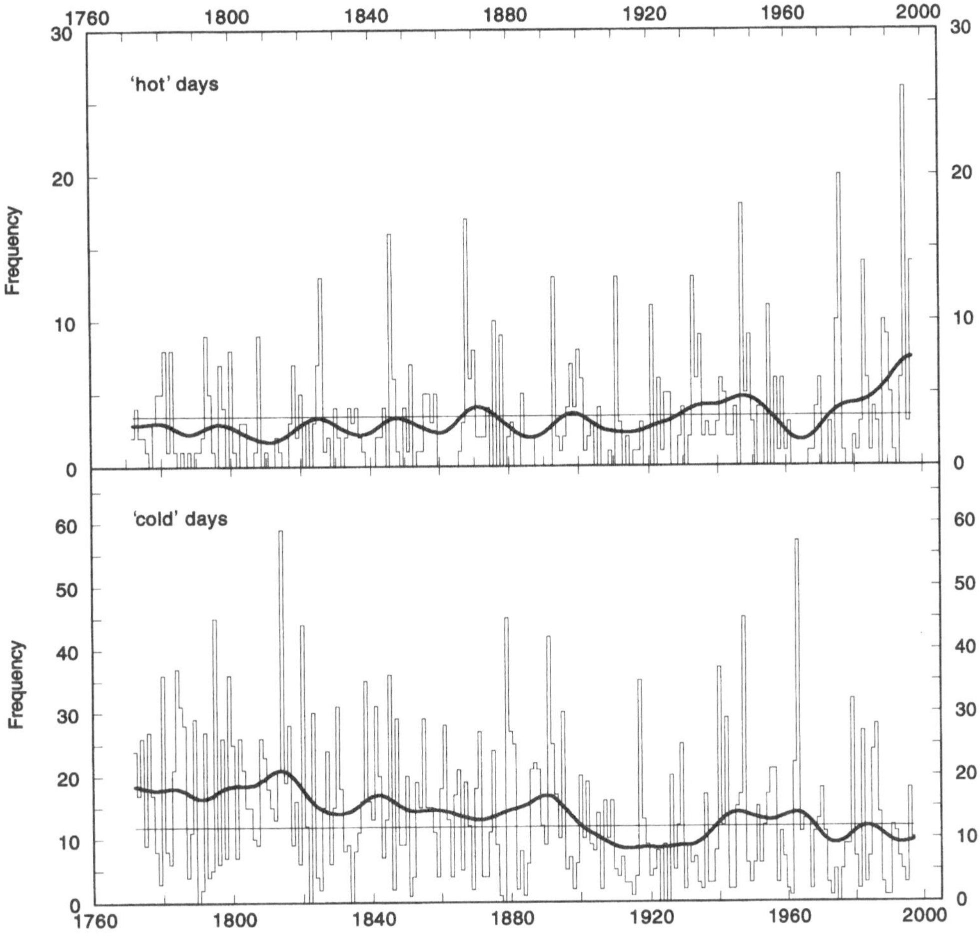

FIGURE 19.8 Frequency of 'hot' (Tmean >20°C) and 'cold' (Tmean <0°C) days extracted from the Central England Temperature record for the period 1772 to 1997. Horizontal lines are 1961–90 means (3.4 and 11.8 days per year respectively) and smooth curves emphasise thirty-year time-scale variability

There are many factors in the global climate system that have contributed to these observed changes, but one of the more important for the UK has been the behaviour of the North Atlantic Oscillation (or NAO). The NAO is a major disturbance of the atmospheric circulation and climate of the North Atlantic-European region, linked to a waxing and waning of the dominant middle-latitude westerly wind flow. The NAO Index we show here is based on the monthly mean sea-level pressure difference between various stations to the south (Azores and/or Iberian peninsula) and north (Iceland) of the middle-latitude westerly flow (Jones *et al.* 1997). It is therefore a measure of the strength of these zonal winds across the Atlantic. When the NAO Index is positive, the westerly flow across the North Atlantic and Western Europe is enhanced. During the winter half-year, the strengthened westerly

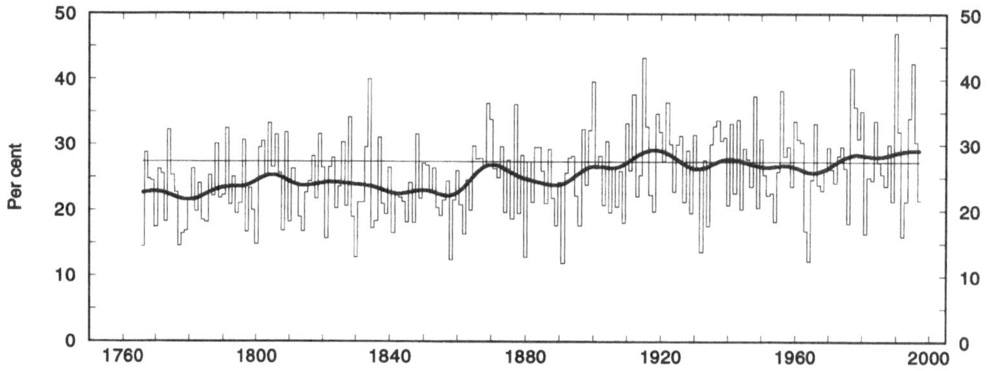

FIGURE 19.9 Percentage of precipitation over England and Wales falling in winter for the period 1766–1997. The horizontal line is the 1961–90 mean (27.4 per cent) and the smooth curve emphasises thirty-year time-scale variability

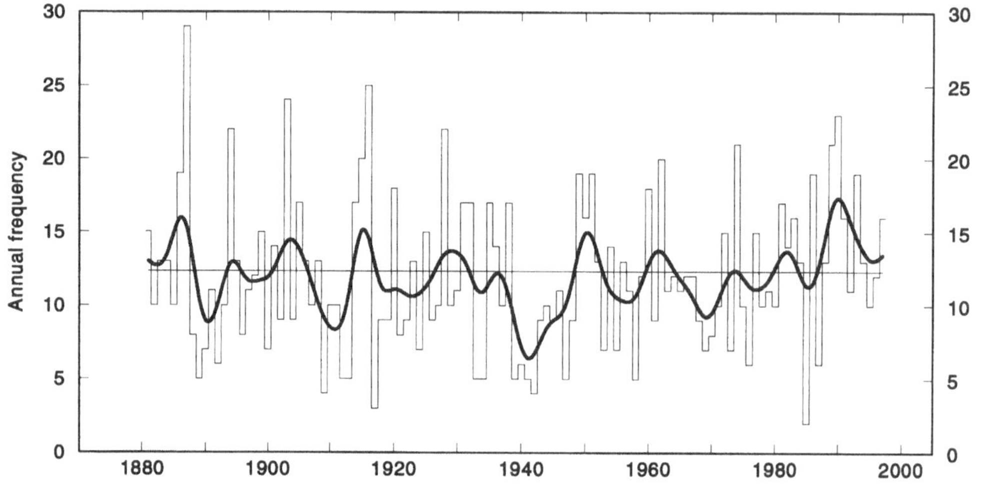

FIGURE 19.10 Annual frequency of 'severe gales' affecting the UK for the period 1881 to 1997. The horizontal line shows the 1961–90 mean frequency (12.3 per year) and the smooth curve emphasises decadal time-scale variability

winds bring warmer, maritime air over north-west Europe causing a rise in temperature and, usually, precipitation. When the Index is low or negative, the opposite occurs with temperatures falling over north-west Europe and rising over the north-west Atlantic. The net result is a 'see-saw' or oscillation in temperatures across the North Atlantic-European sector, as well as changes in other climate variables such as precipitation and sea-ice extent.

The NAO exerts a strong influence on year-to-year climate variability in the UK. For example, the correlation between NAO and winter temperature in the UK is about 0.67

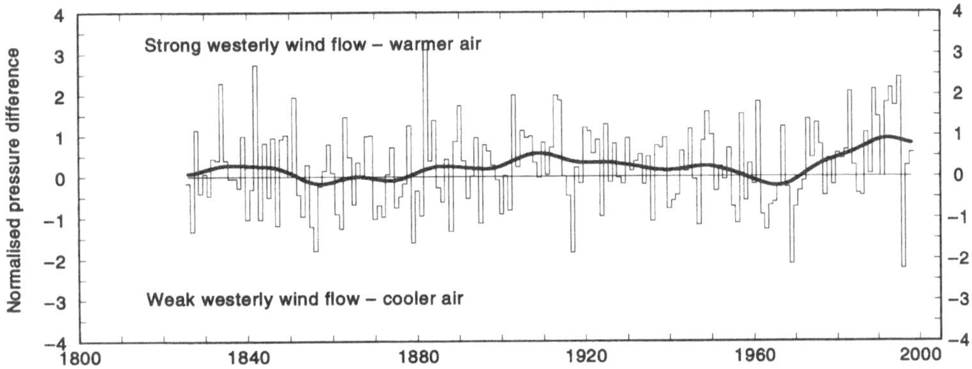

FIGURE 19.11 NAO Index for the extended-winter period (November to March) from 1825/6 to 1997/8 (dated by January). The Index is the normalised sea-level pressure difference between Gibraltar and south-west Iceland. Units are dimensionless. The smooth curve emphasises variations on time-scales of thirty years

(Jones and Hulme 1997). The winter (November to March) NAO Index is shown in Figure 19.11. The period from about 1970 recorded rising Index values, with the highest value being recorded in 1994/5. The change in NAO condition between the winters of 1994/5 and 1995/6 was quite remarkable – from the highest twentieth-century value to the lowest twentieth-century value in successive years. The very low Index value in winter 1995/6 was associated with a cold winter in the UK. The 1996/7 Index value was close to the long-term average and the UK winter was correspondingly milder. The NAO displays variations on a number of different time-scales, most of which may be unrelated to global warming. However, given its importance in determining winter weather over the UK, trends in UK climate cannot fully be understood without reference to the NAO (Osborn *et al.* 1999).

A final indicator of UK climate variability shown in this section is that of sea-level rise. Climate warming is anticipated to lead to a rise in global-mean sea-level, primarily because of thermal expansion of ocean water and land glacier melt. Figure 19.12 shows long-term series of tide-gauge data for five locations around the UK coastline. All of these series indicate a rise in unadjusted sea-level, ranging from 0.7 mm/yr at Aberdeen to 2.2 mm/yr at Sheerness (Woodworth *et al.* 1999). These raw estimates of sea-level change need adjusting, however, to allow for natural rates of coastline emergence and submergence resulting from long-term geological readjustments to the last glaciation. The adjusted net rates of rise resulting from changes in ocean volume range from 0.3 mm/yr at Newlyn to 1.8 mm/yr at North Shields, evidence of an expanding ocean at least around the shores of the UK (see Chapter 17).

Future changes in UK climate

Having established from proxy indicators and from the observed record that UK climate varies naturally on a variety of time-scales, we now wish to consider what the future may hold. To do so we have to consider the mechanisms which control UK climate, which

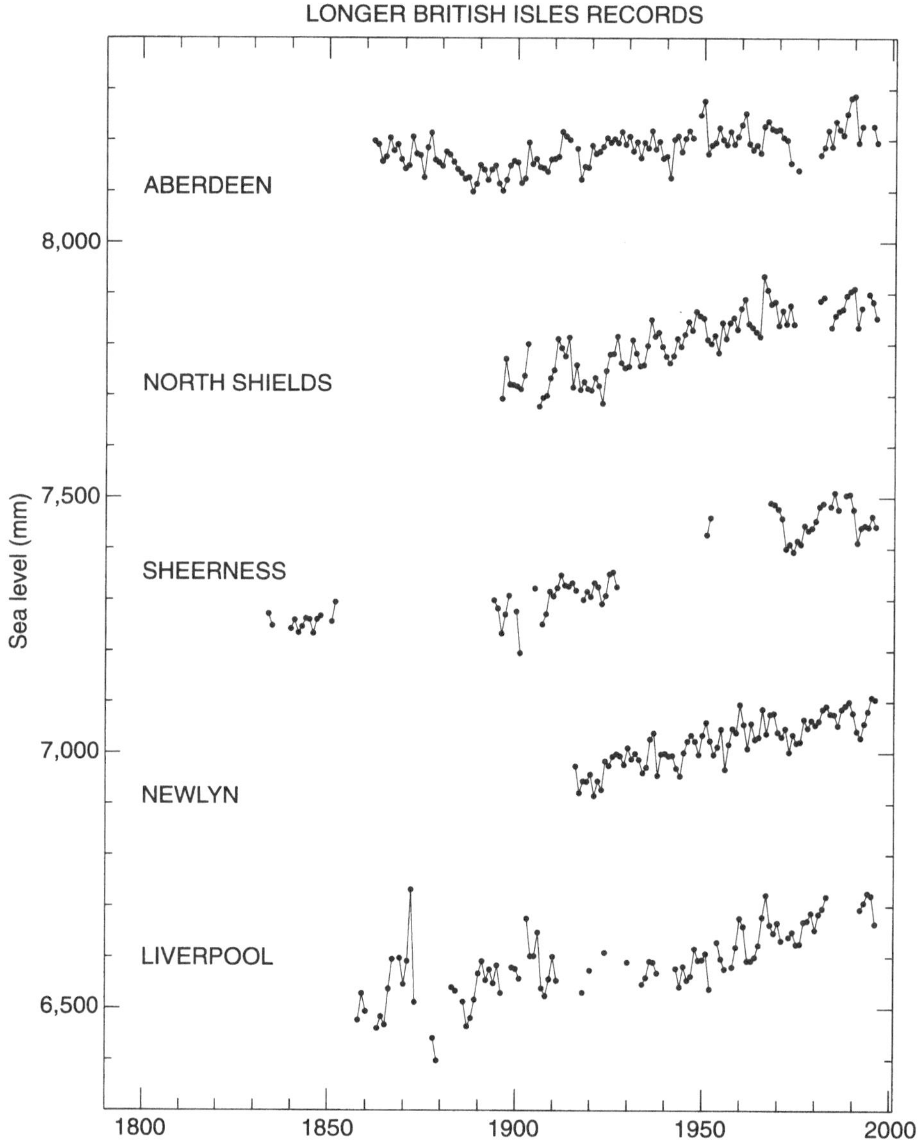

FIGURE 19.12 Relative changes in sea-level over the last 100–150 years as recorded by tide gauges at five UK locations. Last year of data is 1995 or 1996 and units are millimetres. Data are unadjusted for crustal movements
Source: Woodworth *et al.* (1999).

requires an understanding of two sets of factors. These are, first, the array of physical processes and interactions within the climate system that shape the variations in UK climate on these different time-scales and, second, the range of forcings that are external to the climate system but which have a profound effect on the way it functions. To achieve this understanding in the most comprehensive way possible we need to rely upon climate models. Climate models enable us, on a range of spatial and temporal scales, to investigate the variability of climate that results from internal feedbacks within the climate system. We can also use these models to examine the relative consequences for climate of imposing different magnitudes of external forcing on the system, whether this forcing be changes in solar activity, aerosols resulting from volcanic eruptions, or changes in the concentrations of greenhouse gases in the global atmosphere. In this section, we use results from global climate model experiments to present a range of future climate scenarios for the UK. In this approach we follow the science reported by the IPCC in their Second Assessment Report (IPCC 1996) and also used in the UK national climate change scenarios prepared for the UKCIP in 1998 (Hulme and Jenkins 1998).

Future climate without climate change

We consider first the possible UK climate of the period around the 2050s decade in the absence of any external forcing. That is, we assume that there are no further increases in greenhouse gas concentrations in the atmosphere and that there is no significant change in solar output nor any major volcanic eruptions between now and then – these are, of course, artificial assumptions, but it helps us to identify the range of possible climates that we would need to adapt to in the absence of climate change. For this exercise we rely upon the results of a 1,400-year simulation of unforced global climate by the Hadley Centre global climate model, HadCM2. By using this model simulation as a description of natural climate variability we can define the probability distribution of thirty-year mean climate states for the UK centred on the decade of the 2050s (Figure 19.13). While all of these seasonal distributions of mean temperature and precipitation are centred on zero (that is, there is an equally likely chance of mean temperature or precipitation increasing or decreasing by this period), it is the tails of the distributions that are most important for policy. Thus, even in the absence of any climate change, over south-east England there is a finite chance of, for example, (a) annual precipitation for the period 2040–70 averaging 8 per cent higher or lower than the 1961–90 average, or (b) mean annual temperature for this thirty-year period being 0.5°C warmer or colder than at present. Similarly, there is a 50 per cent chance that the annual climate of this part of the UK will be more than 2 per cent wetter or drier than now and a 50 per cent chance of it being more than 0.15°C warmer or colder. Furthermore, Figure 19.13 shows that the possible ranges for seasonal change – in particular summer precipitation – are considerably larger than these annual values, although over northern UK the ranges are slightly smaller than over southern UK.

We start our description of possible future UK climate with this analysis of unforced natural variability in climate on thirty-year time-scales because it is often assumed that the only thing we have to plan for is human-induced climate change (popularly referred to as 'global warming'). This is not the case, as we have shown. To take a more specific example, consider what level of summer rainfall variability the water industry has to be able to manage successfully. Without climate change affecting the UK and if, hypothetically, we were able to experience a large number of climate states for the 2050s, then *on average* summer rainfall during 2040–69 will be similar to what it has been during 1961–90. In reality, however, we only experience the 2050s once and Figure 19.13 shows that for south-east England there is

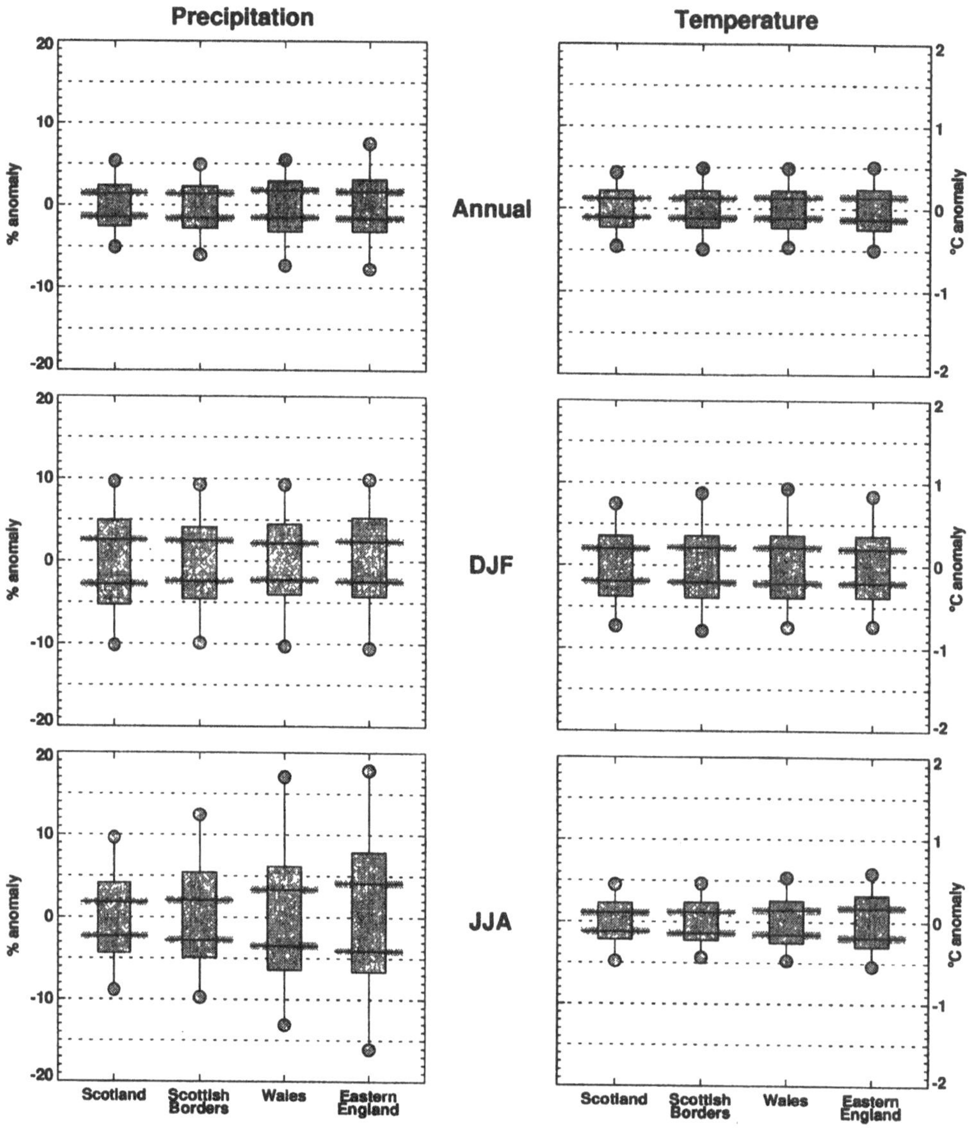

FIGURE 19.13 Natural variability of thirty-year means of annual and seasonal precipitation and temperature for different regions of the UK. Data are extracted from the 1,400-year control simulation of the HadCM2 climate model. Box plots show absolute range (dots), the ±10 per cent extremes (boxes), and the inter-quartile range (extended horizontal bars). Anomalies are with respect to the 1,400-year mean

a 10 per cent risk that over this period summer rainfall will be averaging 7 per cent less than the 1961–90 mean and a 10 per cent chance that it will be averaging 8 per cent more. Even before we start to factor in climate change, therefore, it is important that this level of thirty-year natural climate variability can be accommodated within our economy and institutions.

Furthermore, the level of natural variability to be expected on shorter time-scales, for example decades, is even larger.

Future climate with climate change

We know, however, that other external factors will cause global climate to change, quite apart from natural variability. The largest of these forcing factors over the next century is almost certainly going to be increasing atmospheric concentrations of greenhouse gases. Since we do not know at exactly what rate these gases will accumulate in the atmosphere and since we do not know exactly how sensitive the climate system is to such accumulation, it is usual (e.g. Hulme and Jenkins 1998) to calculate a range of climate change scenarios rather than present a single estimate. When we do this for the UK then by the 2050s mean annual temperature as a result of anthropogenic climate change may be between about 0.9°C and 2.3°C higher than the 1961–90 average, a warming rate of between 0.1°C/decade and 0.3°C/decade. This rate of UK climate warming should be compared with the annual warming rate as measured by the CET over the last two decades, which is about 0.15°C/decade. In other words, UK climate has been warming over the last thirty years at an average rate that we anticipate will be representative of the next one hundred years. We are already living under conditions of twenty-first century climate change.

We can show what this range of warming rates means for the climate of the UK in a different way. Table 19.3 shows the proportion of the UK land surface that falls into different mean annual temperature zones, first for 1961–90 (current climate) and then for the period centred on the 2050s (future climate), but under the four different climate change scenarios presented in the UKCIP98 scenarios (Hulme and Jenkins 1998). The small

TABLE 19.3 Percentage of the UK land surface falling in different mean annual temperature zones for 1961–90 climate and for the thirty-year period centred on the 2050s, assuming the four UKCIP98 climate change scenarios. The scenarios are termed 'low', 'medium-low', 'medium-high' and 'high'. The global warming shown is with respect to the 1961–90 average

		2050s			
	1961–90	*Low*	*Medium-low*	*Medium-high*	*High*
Global warming	0°C	0.9°C	1.5°C	2.1°C	2.4°C
UK temperature	8.3°C	9.3°C	9.6°C	10.1°C	10.4°C
Temperature zone					
<4°C	1	0	0	0	0
4–6° C	5	4	3	1	1
6–8° C	27	11	8	6	5
8–10°C	58	43	36	30	25
10–12°C	9	41	51	57	61
12–14°C	0	1	2	6	8
>14–16°C	0	0	0	0	1

proportion of land with mean annual temperature below 6°C contracts from 5 per cent to between 1 and 4 per cent, while at the other extreme the proportion of land with mean annual temperature above 12°C increases from zero under present climate to between 1 and 8 per cent by the 2050s. It is important to note the considerable range in these figures, a range that results from assuming different future emissions scenarios and different global climate sensitivities.

The changes in mean seasonal precipitation that accompany this climate warming are even more variable. Under all four scenarios annual precipitation increases by between 1 and 6 per cent, but there are marked seasonal and geographic differences. Thus winter precipitation increases by between 5 per cent ('low' scenario) to 15 per cent ('high' scenario), with slightly larger percentage increases in the South than in the North. Summer precipitation decreases in southern UK by between 2 and 18 per cent, but increases slightly over northern UK by up to 5 per cent. These changes in summer precipitation are not large in the context of the natural climate variations discussed above.

Given the above changes in mean temperature and precipitation in the UK, what associated changes in other climatic and non-climatic variables may occur, and what significance may they have for different sectors of the UK economy and environment? We continue by describing changes in atmospheric carbon dioxide concentrations, sea-levels around the UK coast and the frequency of storm events.

Changes in other climatic and environmental variables

Different emissions profiles of greenhouse gases over the next century will lead to different atmospheric concentrations of carbon dioxide. The emissions profiles assumed in the UKCIP98 climate change scenarios lead to estimates of carbon dioxide concentration of between 443 ppmv and 554 ppmv by the 2050s, an increase of between 33 per cent and 66 per cent over average 1961–90 concentration (Hulme and Jenkins 1998). The latter is close to a doubling of the pre-industrial carbon dioxide concentration of about 280 ppmv. These estimates do not allow for any deliberate policy-related reductions in carbon emissions, such as have been written into the Kyoto Protocol. However, even were this Protocol to be fully ratified and adhered to, reductions in atmospheric carbon dioxide concentrations by 2050 would be modest (Wigley 1998). This increase in concentration of a major atmospheric plant nutrient will lead to a general increase in photosynthetic rates for most UK plants and, due to closure of stomatal pores, a general increase in water-use efficiency. Combined with increased temperatures and nitrogen deposition from the atmosphere, these changes will generally be beneficial for plant growth (CCIRG 1996). The exceptions to this will be where plant growth is moisture-limited and where excessively high temperatures inhibit plant photosynthesis.

In the previous section we showed examples of rising sea-level around the UK coast as measured by tide-gauge data. Climate warming will contribute to accelerated rises in mean sea-level due primarily to thermal expansion of ocean water and to land glacier melt. Under the climate scenarios described here, mean sea-level is estimated to rise over average 1961–90 levels by between 13 cm and 74 cm by the 2050s (Table 19.4). As mentioned earlier, however, this mean rise due to climate change will be moderated by the background change in crustal movement. Thus coastline emergence in Scotland will reduce the net change by up

TABLE 19.4 Representative changes in sea-level (cm) around the UK coast by the 2050s due (a) to global climate change only ('Climate') and (b) to the combined effect of climate and natural land movements ('Net'). Changes are with respect to 1961–90 mean levels

	Low		*Medium-low*		*Medium-high*		*High*	
	Climate	*Net*	*Climate*	*Net*	*Climate*	*Net*	*Climate*	*Net*
West Scotland	13	2	20	9	28	17	74	63
East Scotland	13	8	20	15	28	23	74	69
Wales	13	18	20	25	28	33	74	79
English Channel	13	19	20	26	28	34	74	80
East Anglia	13	22	20	29	28	37	74	83

Source: Hulme and Jenkins (1998).

to about 11 cm, while coastline submergence in southern UK will increase the net change by up to about 9 cm (Table 19.4). The significance of such a change in mean sea-level will depend greatly on the accompanying changes in storminess and the local shoreline morphology. Figure 19.14 shows one example, for Harwich, of how the return period of a given tidal limit would shorten with a 41-cm rise in mean sea-level and assuming no change in storminess. A tide-level of 5.6 metres above datum with a return period currently of a hundred years, becomes a 1-in-10 year tide-level by the 2050s (see also Chapter 17).

Having mentioned possible changes in storminess, what would this scenario imply for storm frequencies across the UK? Although modest increases in mean annual wind speed occur (Hulme and Jenkins 1998), the changes in gale frequencies are very much less clear. Table 19.5 shows the changes in the frequencies of three different categories of gales – gales, severe gales and very severe gales – for three different periods over the next century. For the winter season, there is a suggestion that overall gale frequencies may decline in the future, although very severe winter gales may increase. Note that the changing sign of the changes in severe gale frequencies between the three different periods indicates that a clear anthropogenic signal in severe gale frequencies is not easily detected from the noise of natural variability. For the summer season gale frequencies are much lower (less than 2 per year) and consequently any changes in gale frequencies are less significant. Nevertheless, by the 2080s this scenario suggests a modest (approximately 10 per cent) increase in the frequency of all categories of summer gales.

Environmental impacts of climate change

The climate change scenarios described above would have a variety of impacts on different environmental assets and economic activities in the UK. We summarise a few of these based on the UK government's Climate Change Impact Review Group report published in 1996 (CCIRG 1996) and summarised by UKCIP (1998). These impacts are far from definitive and in most cases the studies upon which they are based have not considered the interactions that would occur across or within sectors; nor have they fully evaluated the adaptive responses that may be triggered. Nevertheless, they give us a first-order impression of what climate change may mean for selected aspects of the UK environment.

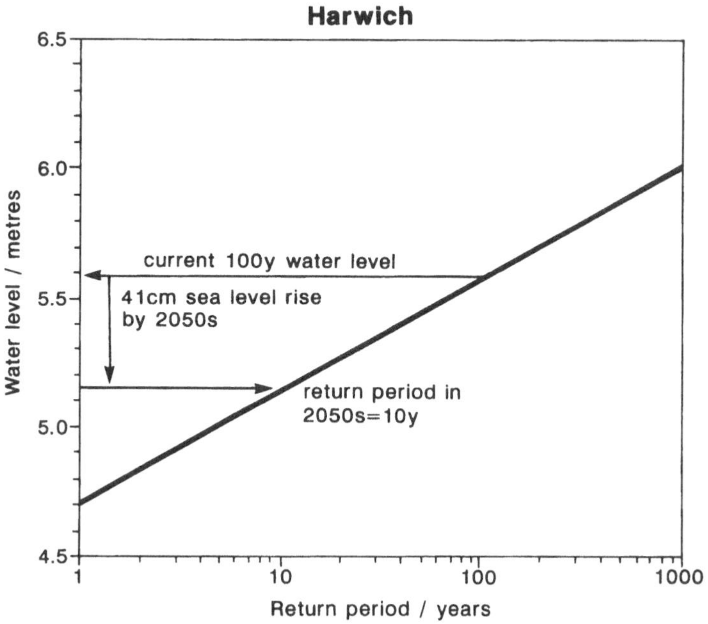

FIGURE 19.14 Estimated change in tide-level return period for Harwich in East Anglia assuming a 41-cm rise in mean sea-level by the 2050s. The example highlighted shows that the 100-year return tide-level of +5.6 metres becomes the ten-year tide-level by the 2050s. No allowance is made for changes in storminess
Source: Hulme and Jenkins (1998).

TABLE 19.5 Percentage changes from the 1961–90 reference period in seasonal gale frequencies over the British Isles for 2020s, 2050s and 2080s for the UKCIP98 Medium-high scenario. The 1961–90 frequencies are calculated from climate model outputs and not from observations

	1961–90 (frequency)	*2020s (% change)*	*2050s (% change)*	*2080s (% change)*
Winter gales	10.9	–1	–9	–5
Winter severe gales	8.5	–1	–10	–5
Winter very severe gales	1.4	+8	–10	+11
Summer gales	1.8	+3	0	+14
Summer severe gales	1.1	0	–2	+15
Summer very severe gales	0.1	+25	–16	+9

Source: Hulme and Jenkins (1998).

The climate warming would be likely to result in shifts in natural habitats for wildlife species (including diseases and pests) northwards by between 40 and 70 km/decade. The actual ability of a species to migrate will vary greatly, however, between, say, an insect and a tree. With an increase in mean precipitation over northern UK also being accompanied by an increase in the *intensity* of precipitation, river flooding frequencies in north-west England,

Scotland and Northern Ireland are likely to increase. In south-east England, however, reductions in summer precipitation would increase drought frequencies and/or drought intensities. More intense thunderstorms would cause flooding, but would fail to benefit either water supplies or soil moisture. Coupled with increased evaporation such a scenario would create a landscape of drying rivers, desiccated soils and forest fires in summer, creating problems for, among others, farmers, forest owners, pleasure boat operators, water companies, anglers and fire fighters (UKCIP 1998).

Natural habitats at risk from climate change include the heaths of southern England (which are vulnerable to fires), wetlands (such as the Scottish peat bogs and Sussex water meadows) and rare mountain habitats. CCIRG (1996: xii) warned that, 'a substantial number of the 506 species listed in the Red Data Book as being endangered, vulnerable or rare, could be lost as a result of climate change'. Substantial areas of mudflats and salt marshes would be at risk from inundation from rising sea-levels. Salt intrusion could be an even bigger problem than rising water levels themselves. A substantial proportion of the UK's Grade 1 agricultural land could become saline since nearly half of such land is below the 5-metre contour.

Another area that CCIRG (1996) reported on was health. Climate warming in the UK as described here would likely lead to a reduction in winter death rates, but an increase in heat-related mortality, ranging from asthma (exacerbated by rising urban air pollution levels) to heat-stroke and drowning. The risk of food poisoning from E-coli, salmonella and toxic chemicals in shellfish could increase. Hay fever would occur earlier in the year and could be more intense (UKCIP 1998).

Climate policy in the UK

There are two fundamental responses to the realisation that human-induced climate change may have potentially adverse consequences for the environment and society. These are *mitigation* and *adaptation*. Mitigation seeks to reduce the magnitude of the climate change by reducing emissions of climate-warming greenhouse gases or by strengthening the biospheric sinks of these gases. Adaptation seeks to intervene in environmental or economic systems at a management level to reduce the vulnerability of such systems to climate change. Both responses are important and both are being pursued as part of UK government policy.

Climate mitigation policy

UK national emissions of carbon dioxide in 1990 were about 158 million tonnes of carbon per year, less than 3 per cent of the global total. With about 1 per cent of the world's population, however, per capita carbon emissions in the UK were about three times the global average. About 95 per cent of these emissions come from the burning of fossil fuels for energy use. As part of the Kyoto Protocol, the UK has agreed to reduce emissions by 8 per cent from 1990 levels by the year 2012, a target which requires annual carbon emissions to fall to 145.4 million tonnes of carbon. During the 1990s, carbon emissions have already fallen by some 10 million tonnes, largely as a result of fuel switching from coal to gas and also as a result of lower energy consumption per unit of GDP. On the other hand, emissions from the transport sector have been steadily rising and it will be necessary to monitor future emissions levels carefully in the years to come to ensure that the Kyoto target is met by the UK.

Despite these successes in reducing UK greenhouse gas emissions, the overall success of a climate mitigation policy depends entirely on the collective actions of the world's major emitting nations. No climate benefits are gained by reducing emissions in one country if these are more than compensated for by increased emissions elsewhere. It is for this reason that the Kyoto Protocol agreed collective action by the world's industrialised nations (so-called Annex-1 countries) to achieve a net 5.2 per cent reduction in greenhouse gas emissions by the year 2012. The climate gains from such a global mitigation policy are likely to remain small, however, if expressed purely in climate terms. Figure 19.15 shows the effect on global temperature and sea-level of a literal and complete adherence to the Kyoto Protocol. Temperature change by 2100 is only about 5 per cent lower than it would otherwise be and the reduction in the rate of sea-level rise is even smaller. Such outcomes would do little to alter the climate changes and resulting impacts described in the previous section.

Substantial mitigation of human-induced climate change will therefore depend on two further criteria. First, the agreement by Annex-1 nations for more far-reaching emissions reductions beyond 2012 and, second, the inclusion of non-Annex-1 nations – the world's developing countries – within the Protocol. This latter condition is only likely to be met subsequent to Annex-1 nations achieving the initial Kyoto target and demonstrating a commitment to reducing their very high per capita carbon emissions to a level closer to the global average. Neither of these two criteria will be achieved easily.

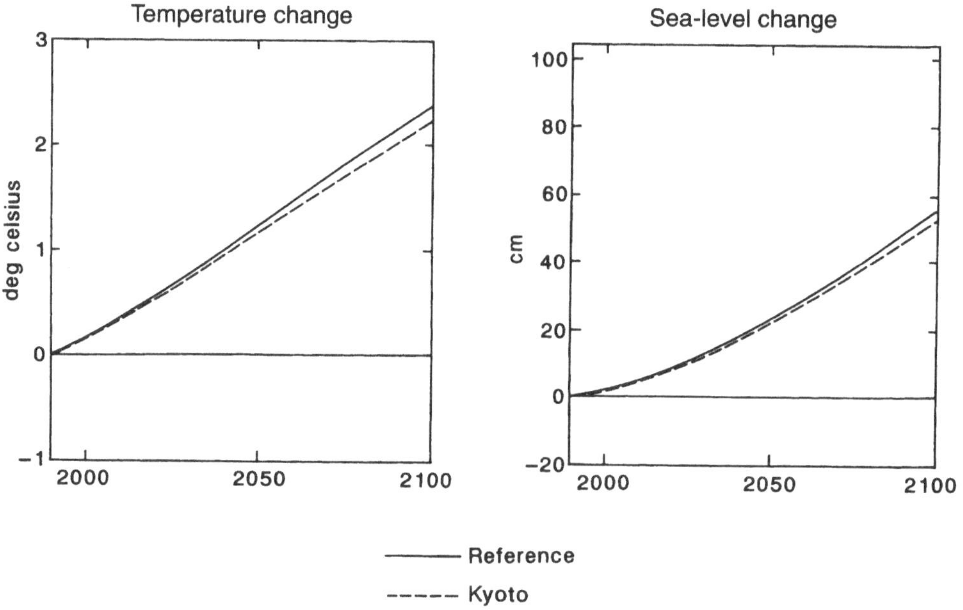

FIGURE 19.15 Simulated changes in global-mean temperature (left) and sea-level (right) from 1990 to 2100 assuming a reference emissions scenario (top curve) and a literal and full adherence to the Kyoto Protocol (bottom curve). As a consequence of the Protocol, temperature change by 2100 is reduced by about 0.15°C and sea-level rise by about 3 cm

Climate adaptation policy

For these reasons, a second and perhaps more important strand to an integrated national climate response policy is the need for adaptation. Despite the best efforts at either UK national or global scales to reduce greenhouse gas emissions, our historical emissions of these gases combined with the inertia of the climate system means that we are 'committed' to a substantial amount of climate change. This 'committed' climate change may not be greatly different from the changes described in the previous section. It is essential therefore that we also develop practical policies to adapt to the consequences of the climate change that will occur.

Adaptation to climate change is a more achievable policy response than climate mitigation. This is because it can be implemented by individual countries, or even within individual sectors within countries, and does not depend on international agreement. Furthermore, many adaptation options can be implemented using existing decision frameworks. Many decisions about investment levels, infrastructure design criteria and resource management are risk-assessment decisions. What level of risk are we prepared to live with and what level of investment do we need in order to achieve that level of risk? There are many examples of where climate variability already affects such decisions and where adaptation to climate change therefore simply requires an appropriate adjustment of the probabilities of certain weather or climate extremes occurring. Water resource management and coastal protection are two examples of where such adaptation is already occurring. The UK water industry has been issued with climate change adjustment factors to use in calculating future run-off in different UK catchments (Arnell *et al.* 1997). These factors allow for scenarios of temperature, evaporation and precipitation change caused by global warming and should ensure a wider safety margin in the future provision of adequate and timely water for domestic and industrial consumption. In the other example, the planning of new coastal protection infrastructure has been required to incorporate changes in the design criteria for certain projects, in this case an additional rise in mean sea-level at the rate of 3 mm/year.

Conclusion

The co-evolution of climate and society raises a number of paradoxes concerning their relationship. On the one hand, the development of a technologically advanced society has overridden some of the traditional controls exerted by climate on human activities. Thus in the UK we are no longer subject to climate-induced food shortages, or even fluctuations in food supply, and the growing contribution of the financial sector to the UK national economy is less affected by weather variations and climate changes than were the traditional manufacturing industries of the early century. At a trivial level, because of undersoil heating we can continue to play football on outdoor pitches in the middle of a freezing winter. Yet at the same time we have a growing sense of our vulnerability to climate extremes, whether this vulnerability is related to water shortages and hosepipe bans or the effects of severe windstorms or coastal flooding. This paradox is further compounded when we consider that it is the very advancement of our carbon-intensive technological society (supposedly making us more immune to the influences of weather) that in turn is causing changes to our climate regime (supposedly fuelling *angst* about our ability to cope with a warming climate).

The question that this paradox raises is not whether global (or UK) climate will change in the future, nor even by how much it will change. The question rather is whether on the relatively slow time-scales that climate will change – that is, slow relative to time-scales of political and technological change – we can adapt to such change by incorporating flexibility and resilience into our infrastructure and regulatory planning environments. Such a challenge will impinge on many sectors of the regulated UK economy and environment – water, land use, transport, energy, construction, insurance, coastal management, wildlife and agriculture. Organisations, charged with investment decisions, environmental management and the development of future planning strategies, need to include climate in their decision-frameworks as a key variable rather than as an assumed constant. If they do, then the twenty-first century need not hold the prospect of major climate-induced dislocations in UK economic activity or environmental systems.

This conclusion may not be true, however, for other world regions with a less regulated natural and economic environment and with less investment capital available to be directed into such adaptation objectives. As a signatory to the United Nations Framework Convention on Climate Change therefore, the UK will need to continue to articulate and implement climate mitigation policies both at home and abroad. This can and should be justified on the criteria both of international social justice and of the economic and environmental benefits to be gained from the use of more sustainable energy technology.

Ultimately, climate, and therefore climate change, should be seen as a resource to manage and to benefit from, and as essentially neutral. The notion of 'good' and 'bad' climates is a hard one to defend in any absolute sense. Temporal changes in climate, seen in this way, present societies with challenges to cope with and opportunities to exploit. These challenges and opportunities presented by *temporal* changes in climate are very much those that every colonising community throughout history has realised are posed by *geographical* differences in climate. This view of climate, whether implicit or explicit, has been true of, for example, the Mongols in Europe, the Vikings in Greenland, or the Europeans in Africa. For example, one may view the nineteenth-century history of the interaction between climate and society as one that concerned the ability of the European colonising powers to exploit geographical differences in climate in the Tropics – rubber in Malaysia, cocoa in West Africa or bananas in the Caribbean – and to manage the regional climate impacts that such exploitation might bring with it. A twenty-first century history of climate–society interaction may well be about the ability of different communities, nations or regions to exploit and manage the forthcoming temporal changes in climate brought about by human pollution of the atmosphere.

References

Arnell, N.W., Reynard, N., King, R., Prudhomme, C. and Branson, J. (1997) *Effects of Climate Change on River Flows and Groundwater Recharge: Guidelines for Resource Assessment*, UKWIR/EA Report No. 97/CL/04/1, London.

Atkinson, T.C., Briffa, K.R. and Coope, G.R. (1987) 'Seasonal temperatures in Britain during the past 22,000 years, reconstructed using beetle remains', *Nature* 325: 587–92.

Beck, R.A. (1993) 'Climate, liberalism and intolerance', *Weather* 48: 63–4.

Bolin, B. (1998) 'The Kyoto negotiations on climate change: a science perspective', *Science* 279: 330–1.

Briffa, K.R. and Atkinson, T.C. (1997) 'Reconstructing late-glacial and Holocene climates', in M. Hulme and E.M. Barrow (eds) *Climates of the British Isles: Present, Past and Future*, London: Routledge.

Cannell, M.G.R. and Pitcairn, C.E.R. (eds) (1993) *Impacts of the Mild Winters and Hot Summers in the UK in 1988–1990*, London: UK Department of the Environment, HMSO.

CCIRG (1996) *Potential Impacts and Adaptations of Climate Change in the United Kingdom*, London: Department of the Environment, HMSO.

Chandler, T.J. and Gregory, S. (eds) (1976) *The Climate of the British Isles*, London and New York: Longman.

Doornkamp, J.C., Gregory, K.J. and Burns, A.S. (eds) (1980) *Atlas of the Drought in Britain, 1975–76*, London: Inst. Brit. Geographers.

Hulme, M. (1997) 'The climate in the UK from November 1994 to October 1995', *Weather* 52: 242–57.

Hulme, M. and Barrow, E.M. (eds) (1997) *Climates of the British Isles: Present, Past and Future*, London: Routledge.

Hulme, M. and Jenkins, G.J. (1998) *Climate Change Scenarios for the United Kingdom*, UKCIP Technical Note No.1, Norwich: Climatic Research Unit.

Hulme, M. and Jones, P.D. (1991) 'Temperatures and windiness over the UK during the winters of 1988/89 and 1989/90 compared to previous years', *Weather* 46: 126–35.

Huntington, E. (1915) *Civilization and Climate*, Hamden, Conn.: Shoe String Press.

IPCC (1996) *Climate Change 1995: the Science of Climate Change* (J.T. Houghton, L.G. Meiro Filho, B.A. Callendar, N. Harris, A. Kattenburg and K. Maskell, eds), Cambridge: Cambridge University Press.

Jones, P.D. and Hulme, M. (1997) 'The changing temperature of "Central England", in M. Hulme and E.M. Barrow (eds) *Climates of the British Isles: Present, Past and Future*, London: Routledge.

Jones, P.D., Jonsson, T. and Wheeler, D. (1997) 'Extension of the North Atlantic Oscillation using early instrumental pressure observations from Gibralter and south-west Iceland', *Int. J. Climatol.* 17: 1433–50.

Lamb, H.H. (1972) *Climate: Past, Present and Future. Vol. I: Fundamentals and Climate Now*, London: Methuen.

Lamb, H.H. (1977) *Climate: Past, Present and Future. Vol. II: Climatic History and the Future*, London: Methuen.

Manley, G. (1952) *Climate and the British Scene*, London: Collins.

Manley, G. (1974) 'Central England Temperatures: monthly means 1659 to 1973', *Quart. J. Roy. Meteor. Soc.* 100: 389–405.

Nicholas, F.J. and Glasspole, J. (1931) *General monthly rainfall over England and Wales, 1727 to 1931*, British Rainfall (1931), UK Met. Office.

Ogilvie, A. and Farmer, G. (1997) 'Documenting the medieval climate', in M. Hulme and E.M. Barrow (eds) *Climates of the British Isles: Present, Past and Future*, London: Routledge.

Osborn, T.J., Briffa, K.R., Tett, S.F.B., Jones, P.D. and Trigo, R.M. (1999) 'Evaluation of the North Atlantic Oscillation as simulated by a climate model', *Climate Dynamics* (in press).

Parry, M.L. and Carter, T.R. (1985) 'The effect of climatic variations on agricultural risk', *Climatic Change* 7: 95–110.

Royal Meteorological Society (1926) *Rainfall Atlas of the British Isles*, London: Royal Met. Society.

Smith, K. (1995) 'Precipitation over Scotland, 1757–1992: some aspects of temporal variability', *Int. J. Climatol.* 15: 543–56.

Subak, S., Palutikof, J.P., Agnew, M.D., Watson, S.J., Bentham, C.G., Cannell, M.G.R., Crowards, T., Hulme, M., McNally, S., Sparks, T.H., Thornes, J.E., Turner, R.K., Waughray, D. and Woods, J.C. (1999) 'The impact of the anomalous weather of 1995 on the UK economy', *Climatic Change* (in press).

UKCIP (1998) *Climate Change Impacts in the UK* London: Department of the Environment, Transport and the Regions, HMSO.

Wigley, T.M.L. (1998) 'The Kyoto Protocol: CO_2, CH_4, temperature and sea-level implications', *Geophys. Res. Letts.* 25: 2285–8.

Woodworth, P.L., Tsimplis, M.N., Flather, R.A. and Shennan, I. (1999) 'A review of the trends observed in British Isles mean sea-level data measured by tide gauges', *Geophysical Journal International* 136: 651–70.

Further reading

Cannell, M.G.R. and Pitcairn, C.E.R. (eds) (1993) *Impacts of the Mild Winters and Hot Summers in the UK in 1988–1990*, London: UK Department of the Environment, HMSO.

CCIRG (1996) *Potential Impacts and Adaptations of Climate Change in the United Kingdom*, London: Department of the Environment, HMSO.

Hulme, M. and Barrow, E.M. (eds) (1997) *Climates of the British Isles: Present, Past and Future*, London: Routledge.

Hulme, M. and Jenkins, G.J. (1998) *Climate Change Scenarios for the United Kingdom*, UKCIP Technical Note No.1, Norwich: Climatic Research Unit.

IPCC (1996) *Climate Change 1995: the Science of Climate Change* (J.T. Houghton, L.G. Meiro Filho, B.A. Callendar, N. Harris, A. Kattenburg and K. Maskell, eds), Cambridge: Cambridge University Press.

Palutikof, J.P., Subak, S. and Agnew, M.D. (eds) (1997) *Economic Impacts of the Hot Summer and Unusually Warm Year of 1995*, Norwich: UEA/DoE, Climatic Research Unit.

Wheeler, D. and Mayes, J.C. (eds) (1997) *Regional Climates of the British Isles*, London: Routledge.

Chapter 20

Conservation and preservation

Chris Park

Introduction

The United Kingdom has long prided itself for its wildlife and landscape, much of which remains relatively natural, given the long history of land management and development within these crowded islands. Very few places are more than about 10 kilometres from countryside, and even in the most heavily built-up places there is abundant wildlife. The United Kingdom also has a long and distinguished history of taking care of its wildlife and landscapes, not simply in the twentieth-century planning and conservation legislation, but also through enlightened land-use decision-making, public interest and involvement, and national pride in natural heritage. The background to, and history of, such initiatives are clearly described in the first two editions of this book (Curtis and Walker 1982; Curtis 1991).

Wildlife and landscape resources in the United Kingdom, as elsewhere, are subject to many pressures, including agriculture, forestry, mineral working, transport and development. Changes in land use almost inevitably affect wildlife and landscape, generally for the worse, because they alter land cover and this in turn affects the visual appearance and ecological integrity of the area in question. Yet, despite being one of the most densely populated countries in the world, and after a long history of land-use changes and developments, the United Kingdom remains essentially rural (see Chapter 11). More than 85 per cent of the land remains free from major development and is still used for agriculture or forestry. But there is relentless pressure to convert open land to other uses – particularly housing – in response to changes in the size, structure and spatial distribution of the population (see Chapter 9), changing lifestyles (see Chapters 12 and 13) and rising living standards and expectations (see Chapters 13 and 14).

The focus in this chapter is on the radical changes which have occurred during the 1990s in the way wildlife and landscape in the UK are understood, interpreted, valued and managed. Many of these, as we shall see, are products of both national and international trends that collectively reflect a move towards more sustainable development. In earlier editions of this book, the equivalent chapter was titled 'conservation and protection'. 'Preservation' is preferred here because it more appropriately describes the dynamic, proactive nature of what happens.

Changing context

Attitudes to and the practice of wildlife conservation and landscape preservation in the United Kingdom changed radically during the 1990s (Adams 1996). It would be difficult to appreciate the causes, consequences and implications of these changes fully without reflecting on the wider context within which environmental decision-making is set.

Two themes are particularly relevant here – the emergence of sustainable development as a core objective in government policy, and restructuring of the institutional framework for environmental management. In both cases government decision-making during the 1990s had to reconcile tensions between national and international interests, in the light of subsidiarity within the European Union and in the wake of the 1992 Rio Earth Summit.

What emerges is nothing less than a major sea-change in national interest in and commitment to preserving its environmental heritage. The pace of change has quickened since Tony Blair's New Labour government came to power in May 1997, but the foundations had already been set in place, and the main directions established, during the dying days of John Major's Conservative government.

Sustainable development

Without doubt the most significant environmental event during the 1990s was the United Nations Conference on Environment and Development (UNCED) – widely referred to as the Rio Earth Summit – which took place in Rio de Janeiro, Brazil, in June 1992. The conference was the culmination of more than a decade of preparatory work, and it established the tone, pace and direction of the international environmental agenda for the foreseeable future (Park 1997). It built upon the success of the 1972 United Nations Conference on the Human Environment, held in Stockholm, and the publication in 1984 of *World Conservation Strategy* (jointly by the United Nations Environment Program, WorldWide Fund for Nature and the International Union for the Conservation of Nature). The 1986 report *Our Common Future*, produced by the World Commission on Environment and Development (widely referred to simply as the Brundtland Report), was a major turning point. Brundtland defined sustainable development as 'development that meets the needs of the present without compromising the ability of future generations to meet their own needs'.

By the late 1980s the axis of the environmental debate had moved away from nature conservation *per se*, towards the sustainable development of all natural resources (Owens 1993). But it needed the focus and momentum provided by Rio to turn ideals into achievable objectives. Participation in the Earth Summit helped to focus the UK government's attention on environmental issues. This commitment became tangible when the government signed The Convention on Biological Diversity (designed to preserve plant and animal life), adopted the Rio Declaration (a general statement of intent which recognises the need for sustainable use of natural resources and environmental protection), and agreed to implement Agenda 21 (the action programme).

The biodiversity convention calls on countries to identify endangered species and conserve the places where they live, and required each country to prepare a National Action Plan or Biodiversity Strategy – which should be realistic, measurable, achievable and realistic – by December 1993. Agenda 21 – an agenda for the twenty-first century – was designed to turn theory into practice by creating a national and international 'blueprint for action' on environment and development.

The United Kingdom government took seriously and acted quickly upon the challenges of Rio. Within less than two years it published *Sustainable Development: The UK Strategy* (UK Government 1994a). The strategy is based on maintaining a buoyant economy, making optimal use of non-renewable resources, sustainable use of renewable resources, and minimum damage to the carrying capacity of the environment, human health and biodiversity. It built upon the 1990 Environment White Paper *This Common Inheritance, Britain's Environmental Strategy*, which defined sustainable development as

> living on the earth's income rather than eroding its capital. It means keeping the consumption of renewable natural resources within the limits of their replenishment. It means handing down to successive generations not only man-made [*sic*] wealth, but also natural wealth, such as clean and adequate water supplies, good arable land, a wealth of wildlife, and ample forests.
>
> (UK Government 1990: 3)

Adoption of the principles of sustainable development has informed many aspects of government decision-making since the 1992 Rio Earth Summit, not just in terms of approaches to environmental legislation but also in terms of the government's own housekeeping and resource use.

Institutional change

The 1990s have seen sweeping changes in the way that the UK government approaches the challenges of conserving wildlife and preserving landscape and environment (Goldsmith and Warren 1993). These have been driven partly by a need to conform to international expectations and commitments (particularly post-Rio). But they also reflect a desire to take seriously the challenges of sustainable development, and recognition of the need to be good role models of enlightened environmental management (Sheail 1995).

The government is well aware of the importance of practising what they preach by trying to adopt proper procedures and approaches in their own activities. One illustration of this latter point is the radical change in attitude embodied in the 'Greening government' initiative introduced in the early 1990s. In the 1990 White Paper *This Common Inheritance* the government committed all departments to draw up strategies for environmental 'good housekeeping' by the end of 1992. Improvements in environmental performance were sought in all areas of government activity, including green procurement, energy efficiency, conservation, waste management, transport, land and estate management, environmental management systems, and compliance with EMAS (the EC Eco-Management and Audit scheme).

Beyond its own backyard, the government has taken a proactive approach to the challenges of sustainable development by restructuring the institutional machinery responsible for environmental management. This is seen in the creation of the Environment Agency, English Nature, and the new Department of Environment, Transport and the Regions. Driving each change has been a desire to create agencies and structures which are reactive, flexible and multi-sector, and which embody both the principles and practice of partnership with stakeholders.

Typical of this new approach was the creation of the Environment Agency under the Environment Act 1995 (Jewell and Steele 1996). The Act defines the aim of the agency as 'to engage in activities that protect or enhance the environment, within the context of achieving sustainable development'. The agency took over the functions of its predecessors – the National Rivers Authority, Her Majesty's Inspectorate of Pollution, Waste Regulatory Authorities and some parts of the Department of the Environment. The prime objective of creating a multi-functional 'super-agency' was to harmonise better and more effectively manage the UK environment. The end product is one of the most powerful environmental regulators in the world, that provides a comprehensive approach to the protection and management of the environment by combining the regulation of land, air and water.

Although the Environment Agency is not directly responsible for conservation of wildlife and preservation of landscape, it is required under the Environment Act 1995 to contribute to the conservation of nature, landscape and archaeological heritage in fulfilling all its functions. 'Contribute' in this sense means to have regard to conservation, and in some cases to further and promote it, and to take into account the beauty and amenity of the countryside when carrying out all of its functional responsibilities. Two other responsibilities of the Agency which also impact upon wildlife and landscape are to provide independent and authoritative views on environmental issues, and to liaise with international counterparts and governments to help develop consistent environmental policies.

The creation of English Nature in 1991 out of the former Nature Conservancy Council under the Environmental Protection Act 1990 runs oddly counter to this more harmonised, multi-function approach. The Countryside Commission was created under the Countryside Act 1968 to 'have regard to conserving the natural beauty and amenity of the countryside' although, as Curtis (1991) points out, its powers are mainly advisory rather than executive, and it provides advice to government and carries out research on landscape matters. Whilst the Countryside Commission remains the government's statutory adviser on countryside matters, English Nature is now the statutory service responsible for looking after England's wildlife, biodiversity and natural features. It is the government's main agent for nature conservation in England. It advises government, promotes nature conservation, selects, establishes and manages National Nature Reserves and – at a more local level – identifies and notifies Sites of Special Scientific Interest. It also provides advice and information about nature conservation, supports and conducts research, and works through and enables others engaged in conservation activities. The new organisation was intended to be more proactive, progressive and customer-oriented than its predecessor (Box 1994).

Treatment of nature conservation in England is curiously different from that in Scotland and Wales, because the Countryside Council for Wales and Scottish Natural Heritage perform all of the English Nature functions but also have responsibility for the wider countryside which is discharged in England by the Countryside Commission. Harmonisation and collaboration within and between countries is essential for both practical and political reasons, and English Nature works closely with the Joint Nature Conservation Committee, Scottish Natural Heritage, Environment and Heritage Service (Northern

Ireland) and the Countryside Council for Wales. This encourages a consistent approach to nature conservation throughout Great Britain, and ensures a co-ordinated approach towards meeting international obligations.

Over-arching these institutional changes, the new Labour government also recognised the need for a broader-based, better-integrated and more forward-looking administrative structure within which to embed its sustainable development aspirations. John Prescott, Deputy Prime Minister, wasted no time in merging the Departments of Environment and Transport in June 1997 to create a major new government department called the Department of the Environment, Transport and the Regions (DETR). His objective was to create a single super-agency Cabinet-level department, with integrated policies (including housing, construction, regeneration, countryside and wildlife, environmental protection, local government, planning, transport and health and safety). The Environment Agency is now part of the DETR.

It remains to be seen whether the DETR helps or hinders the conservation of wildlife and preservation of landscape, and how instrumental the creation of the super-agency might be in realising the goal of sustainable development. Viewed from soon after the event it holds great promise, and is a tangible sign of the new government's commitment to sustainable environmental management.

Inventory

One hallmark of the 1990s has been the better availability of information about the scale, character and distribution of environmental resources within the United Kingdom, and wildlife and landscape are no exception. Government, agencies and the general public now have access to more comprehensive, more reliable and more up-to-date environmental information than ever before thanks to better data collection and publication strategies. Increasing use is being made of the World Wide Web as a way of making official statistics and information widely available (often in downloadable format).

This information explosion is made possible by, and promotes, investment in environmental monitoring and analysis. For both wildlife and landscape, the need for reliable information bases is widely recognised. This includes baseline surveys to establish what exists at present, where it exists, and its condition. It also includes monitoring of changes through time, including repeated ground surveys and use of remote sensing imagery. Catalysts include the need to establish benchmarks against which to assess future changes, and to evaluate compliance with national and international standards and agreements.

The most useful single initiative designed to generate information about the state of the environment was the 1990 Countryside Survey (Barr *et al.* 1993) undertaken on behalf of the Department of the Environment. The survey was carried out by the Institute of Terrestrial Ecology and Institute of Freshwater Ecology (research institutes of the Natural Environment Research Council), based on field surveys in 508 one-kilometre squares. It followed similar surveys carried out in 1978 and 1984, and was designed to record the present character and status of rural land cover, and changes through time. As such it provides an important information baseline for animal and plant diversity and richness in different habitat types throughout Great Britain, and it can be used with earlier surveys to show pace and pattern of change in rural land cover. It is intended to update the survey at the end of the millennium.

In addition to the formal government surveys, a great deal of mapping and monitoring of many species of wildlife – particularly rare plants, breeding birds and mammals (including seals) – is undertaken by a wide range of organisations in the UK. Such surveys usually rely heavily on the help of large teams of amateur and professional naturalists.

Government sources

During the 1990s the government started to publish a number of summary reports on the state of the UK environment (both in paper and electronically). This reflects the government's commitment to more open governance and more accountable decision-making, to partnership with users and stakeholders and to international initiatives and agreements.

One of the earliest such reports was *The UK Environment in Facts and Figures*, which was first published in 1992 by the Department of the Environment. It provides a useful source of up-to-date statistics and information, intended for the non-specialist reader. In seventeen chapters it deals with key environmental resources (air, land, water and wildlife) and the pressures they face, as well as public attitudes towards the environment, and government expenditure on it. The Environment Agency also produces a *State of the Environment* report (in paper and on-line formats), covering England and Wales. This is designed to offer 'a snapshot look at the pressures on the environment and how the quality of the environment has changed over the last twenty-five or so years'.

Without doubt the most comprehensive government source is the *Digest of Environmental Statistics*, published annually by the Government Statistical Service (GSS). It brings together data on many aspects of environmental protection in the UK, including statistics on key trends over time, on geographical patterns and variations, and on performance in relation to policy targets and commitments (national and international). Annual editions of the *Digest* contain chapters on air quality, climate change, water quality and use, coasts and seas, radioactivity, noise, waste and recycling, land and wildlife. Amongst many other things, the *Digest* includes useful summary statistics on land cover (Table 20.1), preservation of landscape and conservation of wildlife (Table 20.2).

Indicators of sustainable development

In March 1996 the government published a preliminary set of *Indicators of Sustainable Development*. These indicators are intended to provide simple quantitative measures which are easily understood, in order to help people (including those in government, industry, and non-governmental organisations, as well as the general public) to better understand whether or not development within the UK is becoming more sustainable. They should also help the government to judge how well it is meeting its objectives as set out in the Sustainable Development Strategy, and its international obligations of reporting to the UN Commission on Sustainable Development.

The UK response is part of a wider international movement. Other institutions

TABLE 20.1 Information about land cover published in *Digest of Environmental Statistics* (1996)

Theme	*Table*
Agricultural land use	Agricultural land use: 1984–1994 Land by agricultural and other uses: 1994 Agricultural land use and crop areas: 1982–1994
Forest cover	Forest cover: 1980/1–1994/5 Forest cover, new planning and restocking: 1980/1–1994/5, Great Britain Forest health surveys, crown density: 1987–1995, Great Britain
Changes in rural land cover	Changes in rural land cover: 1984–1990 Environmental accounts for managed grass and tilled land cover: 1984–1990, Great Britain Environmental accounts for semi-natural land cover: 1984–1990, Great Britain Environmental accounts for woodland: 1984–1990, Great Britain Environmental accounts for built-up land in rural areas: 1984–1990, Great Britain
Residential land	Residential development on reused sites: 1987–1993, England Previous use of land changing to residential use: 1985–1993, England Previous use of land changing to urban uses: 1985–1991, England Previous use of land changing to urban uses in 1991: by county, England
Green belts	Green belts by region: 1979, 1989, 1993, England
Derelict land	Derelict land by region: 1988 and 1993 Derelict land by type of dereliction and by standard region at 1 April 1988 and 1993, England Derelict land by urban/rural location and standard region at 1 April 1993, England
Hedgerows	Hedgerow stock: 1984–1993 Hedgerow stock and changes: 1984, 1990 and 1993, England and Wales Hedgerow length by adjacent land-use type and hedgerow change by change type and adjacent land-use type: 1990 and 1993, England and Wales

engaged in trying to develop meaningful indicators of sustainable development include the Organisation for Economic Co-operation and Development (OECD 1994), the World Bank (1995), Eurostat (the Statistical Office of the European Communities) and the European Environment Agency. As yet there is no international consensus on how environmental indicators should be constructed and what they should cover, so at present each country and agency involved is pursuing its own agendas and objectives. This might optimise user-friendliness on home ground, but it makes international comparisons difficult if not impossible.

TABLE 20.2 Information about preservation and conservation published in *Digest of Environmental Statistics* (1996)

Theme	*Table*
Preservation	
Designated areas	Area of Sites of Special Scientific Interest: 1984–1995 Area of Sites of Special Scientific Interest: 1982–1995, Great Britain Designated areas, by region, at December 1995 Protected areas at 31 March 1995
Damage to Sites of Special Scientific Interest	Damage to Sites of Special Scientific Interest: 1990/1–1994/5 Damage to Sites of Special Scientific Interest (SSSIs): 1990/1–1994/5, Great Britain
Conservation	
Native species at risk	Native species at risk: 1995 Native species at risk: 1995, Great Britain Protected native species: 1981–1994, Great Britain
British mammals	The most numerous British mammals Estimated populations, geographic distribution and changes for mammals over last thirty years, by taxonomic Order/Family, Great Britain Estimated changes in population or range of native and non-native mammals over last thirty years, Great Britain Estimated populations, geographic distribution and changes for mammals over the last thirty years by species, Great Britain Status of grey and common seals: 1986–1994, Great Britain National otter surveys: 1977–79, 1984–86 and 1991, Great Britain
Native reptiles and amphibians	Populations of native reptiles and amphibians: 1995 Estimated populations and geographic distribution of native amphibians and reptiles: 1995
British birds	The most common British breeding birds: 1994 Changes in populations of British birds over the last twenty years Population and changes in breeding numbers for the most common British breeding birds: 1969–1994, Great Britain Changes in geographical distribution and population of bird species by habitat type between 1968–1972 and 1988–1991, Great Britain
Plant diversity	Plant diversity in Great Britain: 1978 and 1990 Changes in plant diversity within habitat types and linear features: 1978–1990, Great Britain Gross change in plant diversity in habitat types and linear features: 1978–1990, Great Britain Nationally Rare, Threatened or Scarce plants by habitat: 1970–1992, Great Britain Proportion of species of vascular plants which have decreased or increased by at least 10 per cent: 1952–60 and 1987–88, Great Britain

TABLE 20.3 Summary of environmental indicators: land cover and landscape

Indicator	*Theme*
s1	Rural land cover
s2	Designated and protected areas
s3	Damage to designated and protected areas
s4	Agricultural productivity
s5	Nitrogen usage
s6	Pesticide usage
s7	Length of landscape linear features
s8	Environmentally managed land

Difficult choices are needed to make sure that the chosen environmental indicators cover relevant themes, are workable and can be interpreted unambiguously. The UK response has been to propose a set of around 120 indicators clustered within twenty-one families of sustainable development issues. In alphabetical order the issues are – acid deposition, air, climate change, economy, energy, fish resources, forestry, freshwater quality, land cover and landscape, land use, leisure and tourism, marine, minerals extraction, overseas trade, ozone layer depletion, radioactivity, soil, transport use, waste, water resources, wildlife and habitats.

Two families of indicators from this list are most relevant to the present chapter – land cover and landscape (Table 20.3), and wildlife and habitats (Table 20.4). Choice of what indicators to use was influenced by various factors, including ease of measurement and interpretation, and relevance to government objectives. Key objectives of the UK government's approach to wildlife and habitats are 'to conserve as far as reasonably possible the wide variety of wildlife species and habitats in the UK, and to ensure that commercially exploited species are managed in a sustainable way'. Key objectives for land cover and landscape are 'to protect the countryside for its landscape and habitats of environmental value while maintaining the efficient supply of good quality food and other products'.

TABLE 20.4 Summary of environmental indicators: wildlife and habitats

Indicator	*Theme*
r1	Native species at risk
r2	Breeding birds
r3	Plant diversity in semi-improved grassland
r4	Area of chalk grassland
r5	Plant diversity in hedgerows
r6	Habitat fragmentation
r7	Lakes and ponds
r8	Plant diversity in stream-sides
r9	Mammal populations
r10	Dragonfly distributions
r11	Butterfly distributions

Data on these two families of indicators, contained in the *Indicators of Sustainable Development* report, have been used in constructing the following sections of this chapter.

Preservation and conservation

To treat landscape preservation and wildlife conservation as separate themes would be to deny their interrelatedness and co-dependency. It would also run counter to both the spirit and purpose of recent government initiatives – including the institutional changes described earlier – designed to preserve the environmental estate of the United Kingdom. A more sensible and realistic approach is to identify some core themes that are relevant to both landscape preservation and wildlife conservation, whilst recognising that some themes are more appropriate to one than the other.

Competition and conflict

A critical tension underlies and challenges all debates about how to manage the natural resources of the United Kingdom in a sustainable manner. It stems from the obvious fact that the land resources are required to meet a range of different objectives at the same time, not all of which are compatible. Hence 'a key sustainable development issue is to balance the protection of the countryside's landscape and habitats of value for wildlife with the maintenance of an efficient supply of good quality food and other products' (UK government 1994a).

Landscape and habitats are often created or altered incidentally, as the end result of other resource-use decisions. This makes it very difficult to ensure that the nation's environmental capital is not progressively eroded, without careful monitoring and control. That, in turn, creates many challenges for the agencies responsible for environmental protection.

Strategies

The traditional approach, certainly in post-war Britain, has been to use a combination of tools in the quest for an optimum land-use strategy, which would derive the largest possible benefits from its environmental estate consistent with preserving the 'natural capital' on which it is based. This toolkit includes making special efforts to prevent loss of specific landscapes features such as hedges and ponds, using planning legislation to control unsuitable development in natural and semi-natural areas, and designating and protecting particular landscapes and habitats which have regional, national or international importance. It also includes working in partnership with landowners and managers, environmental organisations and groups, and the general public in order to inform, advise and empower them to play their part in looking after the nation's environmental assets.

Changes in land cover

Results from the 1990 Countryside Survey show some interesting patterns of change in land cover, many of which have direct or indirect consequences for landscape and habitats. Land cover in Great Britain remains dominated by arable land, improved grassland and heath/moorland. But there were net reductions in area under each cover type between 1978 and 1990 – the area of arable land fell by 4 per cent, and area under improved grassland fell by 2 per cent – mainly because of urban development and new woodland. Crop patterns changed within the arable area between 1978 and 1990, with large increases in the area used to grow wheat and oilseed rape, a large increase in non-cropped arable land set aside from productive use, and a corresponding decrease in the area used to grow barley. These changes have in turn promoted changes in patterns and levels of use of fertilisers and pesticides, which also benefit wildlife.

Most of the changes in land cover between 1978 and 1990 were concentrated in lowland landscapes. In the lowlands arable was lost mostly to woodland, urban and semi-natural land cover types which were partly offset by the conversion of some improved grasslands to arable. Improved grassland was lost to arable, woodland, urban and semi-natural cover types. In upland landscapes the area of 'other semi-natural land' cover types fell by 3 per cent, whereas in lowland landscapes it rose by 10 per cent (largely because of increases in non-agriculturally improved grass, unmanaged tall grassland, and felled woodland).

Other cover changes between 1984 and 1990 include a major decline in well-managed grassland and corresponding expansion of weedy grassland (offering more habitat diversity for wildlife), and further decrease in the small surviving area of heath and moorland in lowland landscapes (mainly replaced by woodland).

Farmers are traditionally seen as custodians of the landscape, and not surprisingly their decisions can have significant impacts on landscape and habitats. Many of the adverse changes in UK landscapes in recent decades are the direct result of the post-war drive for increased food production that promoted more intensive farming. This is true particularly in lowland landscapes, where there have been major losses of features such as hedgerows and walls and semi-natural habitats such as ancient meadows, heaths and wetlands. But there are a growing number of environmentally friendly farmers and environmentally friendly farming practices playing significant roles in maintaining valued landscapes and habitats. Increased environmental awareness, coupled with reform of the Common Agriculture Policy in 1992, has reduced some of the incentives to intensify food production and this is encouraging the improved management of rural land (Bignal and McCracken 1996). Results from the 1990s Countryside Survey confirm the importance of urban development and new woodland as drivers of land cover change. Changing agricultural landscapes are discussed further in Chapter 5.

Arable set-aside

Nature conservation has benefited a great deal from European Union measures designed to reduce agricultural output, particularly of crops (Baldock 1994). The EC Set Aside Scheme was first introduced in 1988, and the 1992 reform of the Common Agriculture Policy introduced further measures and incentives. Since the early 1990s the UK government has introduced a range of measures intended to encourage the environmental management of set-aside land, which have proved beneficial to both agricultural land management and rural landscapes and habitats. One such measure was the voluntary set-aside scheme that was introduced in 1988 and replaced in 1992 by the Arable Area Payments Scheme (AAPS). Under the scheme participating farmers must set aside at least 15 per cent of their arable area each year, on a rotation basis. Most arable farmers have joined the scheme, which gives them financial incentives not to plough up permanent pasture for arable crops. But the AAPS scheme is proactive, because it includes compulsory rules designed to maximise the environmental benefits of set-aside land. Farmers are also offered free advice on how best to manage their set-aside land to benefit wildlife.

In 1995, there were some 6,400 square kilometres of arable land set aside in Great Britain, much of it concentrated in eastern England. The net effect has been to reduce the total area under crops significantly, helping to produce the desired decrease in food output and increase in habitats more attractive to wildlife.

Habitat fragmentation

Wholesale changes in land cover, particularly those involving complete replacement of one cover type with another (as happens with urban development), obviously cause major changes in the visual appearance and ecological value of an area. But problems are also created by fragmentation rather than wholesale removal, because smaller habitats suffer from proportionately larger boundary effects (so there is less homogeneous habitat within the unit) and dispersed habitats reduce seed sources for plants and make it difficult for animals to move around freely.

Unfortunately there is little nationwide data on the pace and pattern of habitat fragmentation through time. One of the few examples for which data exist (environmental indicator r6 in Table 20.4) is the fragmentation of calcareous grassland on chalk in Dorset. Between 1966 and 1983 this chalk grassland habitat was increasingly fragmented into more and smaller units, as the margins of surviving areas were progressively converted to other cover types. Fragmentation appears to have continued into the 1990s, putting at risk the survival of the existing remnants and further decreasing their ecological usefulness.

Grassland changes

Grassland of various types covers extensive areas of Britain, and in many areas the grassland is being converted or altered in ways which are harmful to wildlife. Many types of grassland within the broad habitat group 'semi-natural grassland', for example, are of particular importance to nature conservation yet they are scattered throughout the country in relatively small areas, and have been subject to a range of pressures.

A good example is calcareous grassland that grows on soils derived from chalk and limestone. According to the 1990 Countryside Survey it now covers less than 500 square kilometres (0.2 per cent of the total land area of Britain). Historical records show that the area of chalk grassland in Dorset shrank from about 1,170 square kilometres in 1793 to about

130 in 1815 (as corn growing replaced sheep grazing), and by the early 1990s it was down to about 30. The pattern has been repeated in most parts of Britain, so that by the late 1990s few large areas of chalk grassland remained other than Salisbury Plain and Porton Down in Wiltshire. Much (60 per cent) of the chalk grassland lost between 1966 and 1984 was ploughed, and most of the rest (32 per cent) was invaded by scrub after grazing ceased.

Attempts are being made to preserve surviving remnants of such ecologically important habitats under the two major farmland conservation schemes (Environmentally Sensitive Areas and Countryside Stewardship, see p. 454). About a third of the remaining calcareous grassland on chalk is located within Sites of Special Scientific Interest, where it is protected by appropriate management.

Another type of grassland that has been subjected to change and disturbance is semi-improved grassland, which owes its origin and character to many generations of farming use mainly as pasture. Semi-natural grassland is part of this semi-improved grassland category. In 1990 lowland semi-improved grassland covered an estimated 13 per cent of the total area of Great Britain, and it supported a wide variety of plant and animal species. Environmental indicator r3 (Table 20.4) describes changes in plant diversity in semi-improved grassland.

Results from the Countryside Surveys show a statistically significant reduction between 1978 and 1990 in plant diversity in semi-improved grasslands, reflecting more intensive management. This applies both in 'arable' landscapes (where the mean number of species per plot fell from 19.4 to 17.4) and 'pastural' landscapes (where it fell from 21.5 to 16.6). Declines were fastest among plants associated with unimproved meadows, including some rare grassland species. Traditional land management (which does not use fertilisers and herbicides) seems most appropriate for these cover types, and seems the most promising way of preserving the diversity of wildlife for which it provides a habitat.

Preservation of specific features

Many landscape features, particularly in the lowlands, have been radically altered during the twentieth century either deliberately or through neglect. Hedgerows, stone walls and ponds and lakes are the most obvious. They are attractive landscape features that add variety to the appearance of an area, and have helped create the traditional patchwork-quilt mosaic of many parts of the countryside. But their importance extends beyond the visual, because each offers important habitats for wildlife. Linear features such as hedges and stone walls also provide important environmental services as wind-breaks and shelter belts which act as barriers against soil erosion. Many hedges and stone walls (and to a lesser extent farm ponds) are also of historical importance. Streams and streamsides also offer important habitats to wildlife and they can be important components of the visual landscape.

Hedgerows and stone walls

One hallmark of twentieth-century landscape change in the United Kingdom has been radical change if not wholesale removal of hedges in the countryside. Modern farm machinery is used most efficiently on large fields, so there has been relentless pressure to uproot traditional hedge field boundaries to create larger working units. Between 1984 and 1990 hedgerow lengths in Great Britain decreased by an estimated 150,000 kilometres (25,000 per year). Nearly 87 per cent of the losses (130,000 kilometres in total, 22,000 per year) were in England and Wales and the rest (20,000 kilometres in total, 3,000 per year) were in Scotland. Between 1990 and 1993, the net rate of loss of hedgerows in England and Wales slowed to around 18,000 kilometres per year. Stone walls, another feature in the traditional farming landscape in many areas, have also been cleared to increase field sizes and improve farm efficiency. Between 1984 and 1990 the length of stone walls in Great Britain decreased by 21,000 kilometres from an estimated 214,000 to 193,000 (a rate of around 3,000 per year).

In the past, deliberate hedge clearance has been the major problem, but many surviving hedges are now suffering from neglect through lack of maintenance and conversion to other types of field boundary (particularly fences). Regular hedge management is costly and time consuming, and it contributes little if anything to the farm economy in the short term. In a climate of financial constraints within agriculture, hedge management can be suspended while more pressing problems are tackled. Natural regrowth within the neglected hedge can quickly turn it into a line of bushes or trees. Between 1984 and 1990 about a third (36 per cent) of the hedge loss in Britain was due to uprooting, about a quarter (23 per cent) due to management neglect, and the rest (41 per cent) due to conversion.

Some of the wholesale loss and decline of traditional hedgerows is now being offset by the restoration of degraded hedges and planting of new ones. Creation and proper management of hedgerows and stone walls is being promoted under the Countryside Stewardship Scheme and regulated under Section 97 of the Environment Act 1996. Results from a 1993 survey for England and Wales indicate that more hedges are being planted than uprooted – the rate of new planting exceeded the rate of removal by around 800 kilometres per year between 1990 and 1993. New hedges take many years to become fully established, but in the medium to long term they will increase the diversity of habitats for wildlife within the countryside.

But it is not just the quantity of linear features such as hedges and stone walls that matters to conservation, their quality is also vitally important. Field boundaries provide important habitats for birds, mammals and plants that are often rare locally, so they serve as reservoirs of biodiversity. They also provide seed banks of locally native plants, which under suitable conditions might allow natural regeneration of species-rich habitats in the future. Environmental indicator r5 (Table 20.4) records the changes in mean species numbers in linear plots recorded as hedgerows in 1978, based on 202 hedge plots sampled in 1978 and reliably relocated and recorded in the 1990 Countryside Survey. The 188 hedges that remained in 1990 supported high species diversity, although the number of species for hedge plots in 'pastural' landscapes fell significantly between 1978 and 1990. This was true particularly of plants associated with meadow and chalk grasslands, reflecting a move towards more intensively managed vegetation. There was no significant change in the species diversity of hedges in lowland 'arable' landscapes, although the survey shows a change towards plants more characteristic of arable fields.

Ponds and lakes

Lakes and ponds (water bodies generally smaller than about 2,000 square metres in area) are characteristic features of the landscape in many part of the United Kingdom, even though many of the smaller ones were artificially created. They provide important wildlife habitats, offering opportunities for aquatic species in what are essentially terrestrial landscapes. Many species of plants and animals live in the water or at the water's edge, or use them during their life cycle. In the past ponds were common on farms because they provided water for livestock, but many have been drained or filled in because they serve no practical purpose to modern farming.

Environmental indicator r7 (Table 20.4) provides estimates of the numbers of lakes and ponds in Great Britain between 1945 and 1990. Although the two surveys are not directly comparable, the results suggest a 30 per cent decrease in the number of static inland water bodies from about 470,000 in 1945 to about 330,000 in 1990. A survey by the Institute of Terrestrial Ecology estimates that between 4 and 9 per cent of water bodies disappeared between 1984 and 1990 (partly following the 1990 drought). The same survey suggests a 6 per cent fall in the number of ponds, which accounts for more than 90 per cent of the water bodies recorded in 1990.

As with hedgerows, this trend of decline has been ongoing since at least the mid-1940s, reflecting changing farming practices and more intensive land-uses. But the decline might be slowed down or even halted if a number of initiatives taken since the late 1980s prove to be successful. The two most important of these are changes in the Common Agriculture Policy (away from continued intensification of production) and the introduction of more environmental land management schemes (see p. 454).

Streams and stream-sides

There are more than 250,000 kilometres of rivers and streams in England and Wales, and both the waterways and their banks provide important habitats for wildlife and add variety to landscape. Whilst some rivers have been studied intensively, until recently nationwide evidence about river habitats has been patchy. The Environment Agency has established a River Habitat Survey (RHS) to provide a national inventory, based on 500 randomly selected lengths of river throughout England and Wales. Initial results indicate that only about 9 per cent of the river lengths are unmodified.

Water quality is regularly measured at a large number of sites by the Environment Agency to determine the scale and pattern of water pollution across the country (see Chapter 18.2). Biological indicators are another useful way of evaluating environmental quality, and the presence or absence of otters offers a good indication of freshwater quality. A survey of the distribution of otters in Great Britain was made in 1993, and it shows that otters are returning to many river systems (including the Severn, Avon and Teme) as water quality improves. The Environment Agency is encouraging this natural process of ecological recovery by taking steps to protect existing populations and facilitating the recolonisation of new stretches of river.

Like hedgerows, lakes and ponds, stream banks provide habitats for many plant and animal species and they offer important reservoirs of biodiversity. Stream banks are often affected by river management schemes (such as straightening and dredging), so the quality of bank habitats is not necessarily a reflection of water quality within the river.

The 1990 Countryside Survey provides useful information on changes in plant diversity in stream-sides (environmental indicator r8 in Table 20.4) between 1978 and 1990,

TABLE 20.5 Changes in plant diversity in stream-sides, Great Britain, 1978–90

	Mean species number per plot	
	1978	*1990*
Arable	16.1	14.6
Pastural	18.1	15.0
Marginal upland	20.7	19.5
True upland	23.9	20.7

based on 322 sample plots in four landscape types. Stream-side plant diversity (Table 20.5) is generally higher in upland landscapes, and it declined for all four landscape types between 1978 and 1990. The decline was statistically significant in 'pastural' and 'true upland' landscapes. Species typical of wet meadows and moist woodlands appear to have declined more than most other stream-side species. Some of the decline might be caused by the 1990 drought, but that would not explain why losses also occurred in upland streams (which were not affected by the drought) and why most of the species that disappeared were long-lived perennials.

Non-statutory protected land

The adoption of more environmentally friendly practices by farmers and encouragement to preserve particular features such as hedges, ponds and lakes, and stream-sides are extremely important to wildlife conservation and landscape preservation in the United Kingdom. But they must be seen as part of a broad approach to the problem of sustainable development (Adams *et al.* 1994). Another key ingredient in this multi-faceted approach is to designate particular areas of land for specific protection. Within them natural resources can be more effectively managed, specifically to preserve and enhance habitats and landscape, and potentially damaging cover changes or development can be properly handled. Within the UK these protected areas fall into two groups – those which form the statutory system (set up and controlled under legislation), and non-statutory areas.

TABLE 20.6 Protected areas in the United Kingdom as at 31 March 1995

Status	*Number**	*Area ('000 ha)*
Statutory		
National Nature Reserves	333	199
Local Nature Reserves†	487	25
Sites of Special Scientific Interest (SSSIs)†	6,178	2,041
Areas of Special Scientific Interest (ASSIs)‡	72	75
Marine Nature Reserves	2	3
Special Protection Areas (SPAs)	104	327
'Ramsar' wetland sites	88	362
Environmentally Sensitive Areas (ESAs)	38	3,108
Non-statutory		
Biosphere Reserves	13	44
Biogenetic Reserves	18	8

Notes: * Some areas may be included in more than one category.
† Great Britain only.
‡ Northern Ireland only.

The non-statutory areas are usually owned by the agency which establishes them, they are carefully managed, and public access (sometimes under controlled conditions) is generally encouraged (Box *et al.* 1994). Typical of such areas are the estates and land owned and managed by the National Trust in England and Wales, which cover a total area of around 2,390 square kilometres. Many have historical connections, most attract large numbers of visitors, and the National Trust invests a great deal of resources in managing and restoring habitats (Hearn 1994). Another important group of non-statutory areas are the seventy-six nature reserves (total area 487.83 square kilometres) managed by the Royal Society for the Protection of Birds (RSPB), and the 1,870 smaller reserves (total area 360 square kilometres) owned or managed by the Royal Society for Nature Conservation. In addition there are also the 4,000 square kilometres or so of woodlands managed by Forest Enterprise (formerly the Forestry Commission) in England and Wales, of which 40 per cent (1,600 square kilometres) are Forest Parks. Public access is allowed (on foot) to all of the forest and woodland areas.

There are two other important categories of non-statutory protected areas in the UK (Table 20.6) – thirteen Biosphere Reserves (covering a total of 440 square kilometres) and eighteen Biogenetic reserves (covering a total of 80 square kilometres). These are part of an international programme of nature conservation and sustainable development. Each reserve is designed to protect unique areas and their wildlife, but they are also used for research, monitoring, training and demonstration of best conservation practice (Price 1996).

Protected areas and statutory protected land

Large areas of rural land in the United Kingdom are protected because of their special interest, their importance as landscape or their value as wildlife habitat. Most are protected under national or international legislation, hence they are described as statutory. As outlined below, a range of designations has been introduced which offer different levels and types of protection. Some designations are designed primarily to protect landscape, while others are designed primarily to protect habitats (Idle 1995). Inevitably both types of designation benefit both landscape and wildlife, although tensions are often created in setting objectives and priorities within each.

It is sometimes wrongly assumed that designated areas are given unlimited protection, but in a crowded country like the UK there is usually pressure to allow appropriate types of activity to take place within the designated areas. 'Appropriate' in this sense does not simply mean non-damaging or environmentally friendly, because in some cases – such as the National Parks – maintenance of the character of a landscape is conditional upon continued economic activity. Clearly much tighter controls on development are appropriate in some designations, such as nature reserves. Throughout the system of designated areas within the United Kingdom, however, difficult decisions are required which balance the need to protect environment against the need for optimal use of available resources. This tension lies at the very heart of sustainable development.

All sites designated for landscape preservation or wildlife conservation are defined and delimited on the basis of the best available appropriate scientific information. Periodic reviews are undertaken to evaluate whether designations, site boundaries and management strategies need to be revised in the light of new information (e.g. about habitat condition or species diversity), changing circumstances, and changing objectives (including the need to comply with international agreements). Consequently the inventory of designated sites is never static.

Environmentally managed land

One part of the government's proactive approach to sustainable development of the countryside has been the introduction of a range of environmental management schemes, some designed to encourage more extensive farming methods. The two most important of these are the Environmentally Sensitive Areas (ESA) scheme and the Countryside Stewardship scheme. Both are voluntary schemes that offer landowners financial incentives to manage their land in more environmentally friendly ways. The objective is to conserve and where possible recreate valued landscapes and wildlife habitats, to offset some of the more environmentally damaging aspects of modern intensive farming. Particular attention is paid within both schemes to promoting better management of landscape features such as hedgerows and traditional stone walls.

The Environmentally Sensitive Areas (ESA) scheme was introduced in 1987 with the objective of protecting particular areas of England whose environments are regarded as nationally important, and whose conservation depends on adopting, maintaining or extending particular farming practices (Perkins 1996). ESAs are designated on the recommendations of the relevant conservation bodies (such as English Nature). Under the scheme landowners (including farmers) receive annual payments under voluntary ten-year management agreements to implement particular agricultural practices such as the traditional management of hay meadows. The area under ESAs (Table 20.7) trebled between 1991 and 1994 from 1,140 to 3,460 square kilometres.

The ESA scheme is complemented in England by the Countryside Stewardship scheme, which was introduced in 1991 to promote the conservation of landscape types outside ESAs. Between 1991 and 1994 the area of land covered by the Countryside Stewardship scheme (Table 20.7) more than trebled from 280 to 910 square kilometres. In October 1995 the government announced its intention (in the Rural White Paper) to increase funding for the Countryside Stewardship scheme. This would allow the scope of the scheme to be broadened to include (amongst other things) grant support for management of traditional stone walls and banks, and for conserving the remaining unimproved areas of old meadow and pastures on neutral and acid soils throughout lowland England.

TABLE 20.7 Environmentally managed land in England, 1991–4

	Hectares			
	1991	*1992*	*1993*	*1994*
Environmentally Sensitive Areas (ESAs)	114,300	128,400	266,500	346,400
Countryside Stewardship	28,100	58,700	78,900	91,400
Habitat	0	0	0	3,700
Total	142,400	187,100	345,400	441,500

By 1994, some 4,400 square kilometres of land in England – around 4 per cent of the total agricultural land area – was covered by management agreements under the ESA and Countryside Stewardship Schemes (Table 20.7).

Landscape designations

The most important designations for preserving landscape are National Parks, Areas of Outstanding Natural Beauty (AONB) in England, Wales and Northern Ireland, and National Scenic Areas (NSA) in Scotland (Table 20.8) (Figure 20.1).

National Parks

There are ten National Parks in England and Wales, established under the 1949 National Parks and Access to the Countryside Act (Curtis 1991). Each serves both conservation and recreational functions, and each contains a mixture of land cover and includes settlements, extractive industries and farming. Nearly a quarter of the Northern Region lies within National Parks (Table 20.8), whereas the South East and East Anglia have none although the Broads of East Anglia have a separate status equivalent to that of a National Park, and the New Forest in Hampshire has special protection in law. The ten National Parks cover nearly

TABLE 20.8 Designated areas in the United Kingdom, by region, at December 1995

	*National Parks**		*Areas of Outstanding Natural Beauty*†*		*Defined Heritage Coasts*
	Area ('000 ha)	*Percentage of total area in region*	*Area ('000 ha)*	*Percentage of total area in region*	*Length (km)*
Northern	362	23	226	15	128
Yorkshire and Humberside	315	20	92	6	82
East Midlands	92	6	52	3	—
East Anglia	0	0	91	7	121
South East	0	0	662	24	72
South West	165	7	712	30	638
West Midlands	20	2	127	10	—
North West	10	1	78	11	—
England	963	7	2,039	16	1,041
Wales	410	20	83	4	496
England and Wales	1,373	9	2,122	15	1,539
Scotland	—	—	1,002	13	—
Northern Ireland	—	—	285	20	—

Notes: Some areas may be in more than one category, and areas are estimated.
* Figures shown may differ from those previously published due to redefinition or remeasurement of some National Parks and AONBs.
† National Scenic Areas in Scotland.

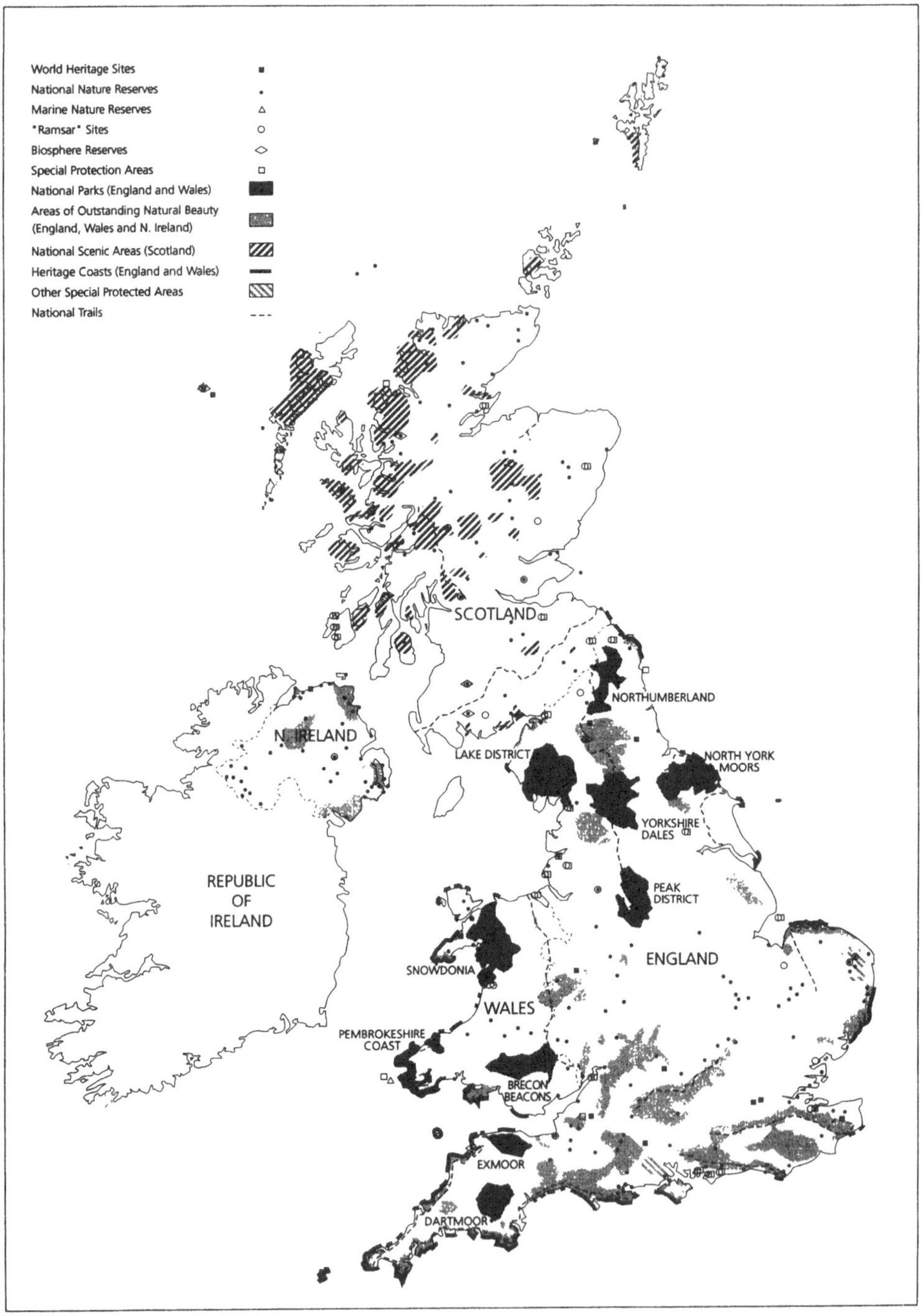

FIGURE 20.1 Protected areas in the United Kingdom as at 18 October 1993
Source: Countryside Commission.

14,000 square kilometres, roughly 7 per cent of the land area of England and 20 per cent of Wales.

Areas of Outstanding Natural Beauty

Areas of Outstanding Natural Beauty have also been designated under the 1949 legislation. These are smaller than National Parks, still protected by tight planning controls, and closer to the main centres of population and hence accessible to large numbers of visitors. The total area of AONBs in England, Wales and Northern Ireland rose from just over 19,000 square kilometres in 1984 to 22,740 in 1990 and 24,000 in 1995. By late 1995 the forty-one AONBs covered about 16 per cent of England, 20 per cent of Northern Ireland, and 4 per cent of Wales (Table 20.8). National Scenic Areas in Scotland, which covered just over 10,000 square kilometres (13 per cent of the land area) in 1995, serve the same purpose as the AONBs.

Heritage Coasts

In addition to the National Parks and AONBs there is a special landscape designation to protect attractive or important stretches of coastline. Heritage Coasts are defined following recommendations of English Nature or Welsh Nature, and they are managed by local authorities via the planning system. By late 1995 some forty-four Heritage Coasts had been defined, covering more than 1,500 kilometres (roughly a third) of the coastline of England and Wales.

Wildlife designations

The two most important designations for conserving wildlife are National Nature Reserves (NNR) and Sites of Special Scientific Interest (SSSI) in Great Britain, which are called Areas of Special Scientific Interest (ASSI) in Northern Ireland. These national designations form the basis for a network of nature conservation protection areas (Figure 20.1), which also includes sites protected in order to meet international obligations, such as Special Protection Areas for Birds (SPA) and Special Areas for Conservation (SAC) under the EC Birds and Habitats Directives.

Some areas belong to more than one category of designation; for example, some National Nature Reserves are also SSSIs, some SSSIs are also SPA sites, and many Ramsar sites (see p. 460) are also designated as SPA sites.

Sites of Special Scientific Interest

Sites of Special Scientific Interest are sites that contain important wildlife, geological or physiographic features. They are notified by English Nature, although most are privately owned or managed. About 40 per cent of the sites are owned or managed by public bodies

TABLE 20.9 SSSIs, Ramsar sites and Special Protection Areas in Great Britain, 1984–94

	SSSIs		*Ramsar sites*		*Special Protection Areas*	
	Area ('000 ha)	*Number*	*Area ('000 ha)*	*Number*	*Area ('000 ha)*	*Number*
1984	1,388	4,225	69	19	18	7
1985	1,434	4,497	69	19	18	7
1986	1,431	4,842	97	29	41	18
1987	1,518	4,734	99	32	43	22
1988	1,581	5,014	130	36	114	26
1989	1,646	5,207	174	41	127	33
1990	1,720	5,459	174	41	127	33
1991	1,786	5,697	179	45	135	40
1992	1,816	5,852	187	52	147	47
1993	1,948	5,999	228	59	192	69
1994	1,972	6,109	298	75	275	86
1995	2,041	6,178	308	86	327	104

such as Forestry Enterprises, the Ministry of Defence and The Crown Estate, or by the voluntary conservation movement. By mid-1995 there were more than 6,000 SSSIs covering more than 20,000 square kilometres in England and Wales (Table 20.6), up from 4,225 sites covering nearly 14,000 square kilometres in 1984 (Table 20.9). Areas of Special Scientific Interest (ASSIs) serve similar functions in Northern Ireland.

Special regulatory controls restrict the type and scale of development that can take place on SSSIs, and they require specific consultation between site owners and English Nature before approval is granted for developments that might alter the character and ecological value of the site. English Nature liaises with about 23,000 owners and occupiers of SSSI sites.

Despite the special protection offered to habitats and special features within SSSIs damage can and does occur from time to time. The countryside agencies – English Nature, Scottish Natural Heritage and the Countryside Council for Wales – regularly collect information on damage to sites such as SSSIs. Environmental indicator s3 (Table 20.3) summarises the extent of damage to designated and protected areas.

The records suggest that damage occurs on only a small proportion of the total number, and the extent of damage has decreased since the late 1980s (UK Government 1995). Most reported damage is short term, from which sites recover usually within three years given appropriate management. The most common cause of short-term damage is agricultural activities, but it can also be caused by factors such as pollution, unauthorised tipping and burning. In 1994–5 around 12.5 square kilometres – 0.05 per cent of the total SSSI area – was affected by long-term damage (which will take more than three years to recover). A small amount of damage is so serious that all or part of the SSSI is lost and gets denotified. Table 20.10 shows the breakdown of damage to SSSIs during 1996–7. English Nature concluded in 1997 that two-thirds of all SSSIs can be described as favourable or recovering, and a further quarter have stabilised from decline. The other 10 per cent were classed as unfavourable because of ongoing agricultural activities (such as overgrazing by

TABLE 20.10 Damage to Sites of Special Scientific Interest, 1 April 1996 to 31 March 1997

	Number
Sites with loss of whole (unit) feature – damage which may result in the denotification of the whole feature	2
Sites with partial loss of (unit) feature – damage which may result in the denotification of part of the feature	8
Sites with long-term recovery of features – a reduction in the special interest which will take more than three years for recovery	26
Sites with short-term recovery of features – a reduction in the special interest taking less than three years for recovery	16
Sites with unknown feature recovery period – damage where the likelihood of recovery cannot be assessed	13
Total number of sites	65

sheep on some upland grasslands), inadequate positive management (particularly on lowland heath, grasslands and wetlands), and individual damaging incidents (such as unauthorised ploughing or deep drainage).

But the damage records only show particular events, such as field drainage or road building. Evidence about progressive changes within SSSIs, such as those associated with long-term air pollution or unsuitable land management (e.g. overgrazing in the uplands) is much less readily available.

National Nature Reserves

National Nature Reserves (NNRs) are sites that have been designated for special protection because of their national importance as wildlife habitats. They are either owned or controlled by English Nature or held by approved bodies such as the Wildlife Trusts (Barkham 1994), and are carefully managed to conserve habitats and species. NNRs are declared by English Nature or its predecessors under Section 19 of the National Parks and Access to the Countryside Act 1949 or Section 35 of The Wildlife and Countryside Act 1981.

Whilst nature conservation is the primary objective in NNRs, many of them are managed to allow public access. Their wardens, conservation work and interpretative facilities are funded by English Nature. There were a total of 333 NNRs in March 1995, covering an area of nearly 2,000 square kilometres (Table 20.6). Most counties contain at least one NNR, and the reserve network includes representatives of most types of vegetation within the United Kingdom, including native woodlands, meadows, downlands, dunes and salt-marshes. NNRs protect many of the country's most scarce and threatened habitats, including chalk downs, lowland heaths and bogs.

Despite their title, National Nature Reserves are not necessarily owned by the nation. Neither are they necessarily permanent. Many NNRs survive by agreement between landowners and English Nature, and if these agreements break down then NNR designation and status can be removed. In a well-publicised case this is precisely what happened to Braunton Burrows NNR in September 1996. Braunton Burrows in Devon, widely regarded to be one of Britain's finest National Nature Reserves, is one of the country's three largest coastal sand dune systems and home to more than 440 plant species. It was listed as a UNESCO Biosphere Reserve. But because of a disagreement with the owner of the site, who refused to accept English Nature's recommendation that the dune be grazed to maintain the habitat, its status as a NNR was lost when English Nature was unable to renew its lease. It remains as an SSSI, but – without the landowner's agreement – grazing could only be restored if the site were bought under a Compulsory Purchase Order. Conservation groups, led by Friends of the Earth, called for a legal 'duty of care' for landowners that would require them to manage sites in accordance with conservation objectives.

Local Nature Reserves

Local Nature Reserves (LNRs) seek to conserve habitats and species that are important locally and regionally within the United Kingdom. As such, they represent 'First Division' conservation sites compared with the 'Premier League' National Nature Reserves. Local authorities, in consultation with English Nature, designate LNRs. All of the 487 LNRs, covering a total of nearly 250 square kilometres in 1995, were owned or controlled by local authorities. Some LNRs are also SSSIs.

Marine Nature Reserves

Marine Nature Reserve (MNR) designation is intended to offer particular protection to coastal sites that contain nationally important habitats and species. Two such sites (Table 20.6) have been declared by the Secretary of State for the Environment – the island of Lundy in the Bristol Channel, and the island of Skomer, off the south-west coast of Wales.

Ramsar sites and Special Protection Areas

There are two special categories of designations that reflect the UK government's commitment to international conservation initiatives. In 1973 it signed the so-called Ramsar Convention (more properly the Convention on Wetlands of International Importance), under which it is committed to designate 'Wetlands of International Importance' (Ramsar sites) and to manage the wetlands within its territory in sustainable ways. Between 1984 and 1995 the area covered by Ramsar sites increased from around 690 square kilometres to more than 3,000 (Table 20.9), and in March 1995 there were eighty-six Ramsar sites scattered throughout the United Kingdom.

The UK is also bound by the 1979 European Communities EC Directive on the Conservation of Wild Birds (79/409/EEC). This requires the government to designate Special Protection Areas (SPAs) to conserve the habitat of certain rare or vulnerable birds (listed under the directive) and regularly occurring migratory birds. Designated sites have to be protected from significant pollution, disturbance or deterioration. By March 1995 there were 104 SPAs in the UK, covering a total area of 3,270 square kilometres (Table 20.6). The previous ten years had seen continuous growth from seven sites covering 180 square kilometres (Table 20.9).

All sites listed as Ramsar sites and Special Protection Areas are also designated as SSSIs, and some sites qualify as both Ramsar and SPA sites. As with all SSSIs, English Nature has responsibility for identifying and designating these sites, and for consulting with owners, occupiers, local authorities and other interested parties.

Biodiversity

One of the buzzwords of nature conservation during the 1990s has been 'biodiversity', short for 'biological diversity'. It refers to the number and variety of different habitats and species of wildlife living within an area, and is a useful measure of quality of environment as well as ecological health. The UK government eagerly adopted both the rhetoric and reasoning of biodiversity, gave it a high profile in its sustainable development strategy, and is committed to successful implementation of the Rio Convention on Biodiversity (Raustiala and Victor 1996).

Natural capital

A key objective in the government's sustainable development strategy is to conserve, as far as possible, the biodiversity of the country's wildlife and habitats (Whitby and Adger 1996). This natural capital amounts to more than 30,000 native species of animals (excluding marine, microscopic and lesser known groups – such as mites and roundworms – of which there are many thousands), about 2,300 species of higher plants, about 1,000 species of liverworts and mosses, about 1,700 species of lichens, about 20,000 species of fungi, and about 15,000 species of algae.

The emergence of new species and decline of existing ones is an inherent part of the 'circle of life' as species evolve, adapt and become extinct. Natural environmental change promotes evolution and extinction, and the expansion and contraction of the range of individual species. As a result, the country's natural capital is never static. Climate change is likely to make the UK warmer and wetter in the twenty-first century, which might promote an increase in species diversity (Elmes and Free 1994). But since at least 1945 that capital stock has been depleted as many species and habitats have declined. Some decline is perhaps inevitable given that some native species are at the geographical limits of their natural range, or whilst the habitats that support some other species are small and scattered.

Change and decline

Biodiversity in the United Kingdom has been adversely affected in recent decades both directly and indirectly as a result of human activities – particularly urban expansion, transport developments, the intensification of agricultural production, and the growth of plantation forests. Many species of wildlife have become nationally extinct, or their populations have declined so much that their survival is threatened. It is estimated that around 170 species of plants and animals in the UK became extinct during the twentieth century, and that up to a quarter of native species of fish, invertebrates, plants, and mosses are threatened or nationally scarce.

TABLE 20.11 Changes in plant diversity within habitat types and linear features, Great Britain, 1978 and 1990

	Mean species number									
	Landscape type									
Type of plots	*Arable*		*Pastural*		*Marginal upland*		*Upland*		*GB*	
	1978	*1990*	*1978*	*1990*	*1978*	*1990*	*1978*	*1990*	*1978*	*1990*
Main plots (200 m^2)										
Arable fields	6.7	4.1	6.8	6.1	—	—	—	—	6.7	4.8
Improved grass	9.8	8.7	10.2	10.7	13.9	13.3	7.3	8.7	10.3	10.3
Semi-improved grass	21.5	20.3	22.2	19.2	22.2	23.8	19.8	21.2	21.9	20.6
Woodland	14.9	16.0	14.9	11.7	19.0	10.6	19.8	12.5	16.1	12.9
Upland grass	—	—	25.9	22.1	18.4	19.0	25.1	26.8	23.2	23.4
Moorland	—	—	15.6	12.1	12.4	16.3	18.9	20.2	17.7	19.0
Linear features (10 · 1 m)										
Stream-sides	16.1	14.6	18.1	15.0	20.7	19.5	23.9	20.7	19.5	17.1
Hedges	11.0	10.2	14.4	12.4	17.9	17.5	—	—	13.1	11.8

The evidence (UK Government 1995) suggests that while many native species are still relatively common across the United Kingdom, between about 10 and 20 per cent of native species are threatened. More than one-third of the 2,700 native species of mosses, liverworts, and lichens are threatened or nationally scarce, as are about a quarter of the seed plants, ferns, and related plants (about 2,300 species in total) and invertebrates (about 15,000 species).

Many of these reductions in biodiversity are linked directly or indirectly with changes in land cover, which inevitably reduces, damages or removes habitats. Comparison of results from the 1978 and 1990 Countryside Surveys (Table 20.11) gives an indication of the scale and pattern of change in plant species. For most landscape types there were fewer species recorded in 1990 than in 1978, with declines often in the order of a quarter or more species. High relative decreases were observed in arable fields, semi-improved grass pasture, upland grass, moorland pasture. The number of species recorded in the sample plots of marginal upland woodland fell by nearly a half between 1978 and 1990, and there was major decline in upland woodland. It was not all bad news, however, because biodiversity in some habitat types increased over that period. Relatively small increases were recorded in arable woodland, semi-improved grass in marginal upland, upland grass, and moorland in marginal upland. Most habitat types in the upland landscape type contained more plant species in 1990 than in 1978. Plant species diversity in both stream-sides and hedgerows decreased, and decline was recorded in every landscape type (Table 20.11).

Other dimensions of biodiversity have also changed in the United Kingdom in recent years, for much the same reasons as the changes in plant diversity. Records show that the geographical distributions of just over a half of the species of dragonflies and butterflies within Great Britain have reduced since the 1970s. Bird populations and distributions (environmental indicator r2, Table 20.4) have also been affected by cover change, habitat

TABLE 20.12 Changes in diversity of breeding birds in Great Britain between 1968–72 and 1988–91

	Declining	*Little change*	*Increasing*
Number of species changing in population			
Farmland	22	9	15
Woodland	14	2	19
Coastal	3	5	13
Lowland wet	2	7	20
Number of species changing in distribution			
Farmland	10	14	2
Woodland	11	24	12
Coastal	8	9	7
Lowland wet	5	11	13

loss, pollution and natural factors (Andrews and Carter 1993). Population size of twenty-two species of breeding birds in Britain fell between the 1970s and 1990s, and ten species experienced a decline in geographical distribution (UK government 1996b). The fate of bird populations in woodland, coastal and wetland habitats was more mixed, because more species increased in numbers than decreased (Table 20.12). Almost a half (twenty-two species) of farmland species declined in population size and around 40 per cent (ten species) decreased in their geographical extent. Species reductions were relatively much higher on cultivated land than on grazing land, although populations of some farmland species (such as jackdaw and magpie) more than doubled. All bird species naturally occurring in the wild are protected under the Wildlife and Countryside Act 1981.

Habitat change and loss have also affected mammal populations (environmental indicator r9, Table 20.4). By the late 1990s there were sixty-one species of land-based mammal breeding in Great Britain, comprising thirty-nine native species, twenty-one introduced and feral species, and one migrant species (*Nathusius' pipistrelle* bat). More than a third of British species of mammals appear to have declined in population size between the 1969s and the 1990s, mostly smaller mammals such as rodents and bats. About a quarter – mainly larger mammals such as carnivores or deer, and a number of non-native species – appear to have increased.

Solutions

There is abundant evidence, from the Countryside Surveys and other sources, that the natural capital of the United Kingdom is declining and has been doing so throughout most of the post-war period. The government faces many challenges in turning this situation around, given its desire to manage the nation's environmental estate sustainably and its commitment to international initiatives designed to conserve nature (most particularly the Rio Biodiversity Convention, but also European Union directives and aspirations). It is fighting on many fronts at the same time to try to stem the decline in species diversity, minimise the loss of species, and limit the loss and fragmentation of and damage to habitats, whilst at the same time optimising productive use of the countryside. Conserving the biodiversity of the United Kingdom is an important part of government policy (Palmer 1996).

A key priority is to conserve the habitats of species most under threat. In January 1994 the government published *Biodiversity: the UK Action Plan* in response to Article 6 of the Rio Convention on Biodiversity (UK government 1994b). Amongst other things, the action plan established a Biodiversity Steering Group to formulate and oversee national initiatives designed to conserve species and habitats. The Steering Group recognise that the established network of designated areas – including SSSIs and schemes such as the Environmentally Sensitive Areas – is a good foundation on which to build, and which has already achieved some important successes in preserving the nation's ecological assets. But the existing network is too broad-brush to offer the sort of protection that is essential for the survival of a number of particularly threatened species and habitats.

In December 1995 the Biodiversity Steering Group published a set of proposed specific, costed targets and action plans for 116 priority species and fourteen key habitats of conservation importance (UK government 1996a). The habitat plans cover about 2 per cent of the land area of the United Kingdom. Between 1996 and 1999 the group committed itself to preparing plans for an additional 286 priority species and twenty-four key habitats. The targets and action plans, once agreed by government (UK government 1996c), are intended to form the basis for conservation action in the UK for the foreseeable future.

Set within this broader strategy aimed at conserving the nation's biodiversity is a number of more specific initiatives. One of the most prominent and most important of these is the Species Recovery Programme mounted by English Nature. The programme is designed to maintain or enhance populations of wildlife species that are in decline or threatened with extinction, by directing action to halt or reverse the reduction in their range and number. Recovery objectives are defined for specific species, and between 1991 and 1997 initial recovery objectives were met for twenty-one of the species most at risk. By 1997 work was being carried out on seventy-five species, most of them defined within the UK Biodiversity Action Plan, through a series of partnership projects with a range of organisations and individuals.

Partnership is central to both the spirit and purpose of the government's approach to nature conservation. Thus, for example, Species Action Plans for rare and threatened bird species have been produced by the Joint Nature Conservation Committee, The Royal Society for the Protection of Birds and The Wildfowl & Wetlands Trust (Williams *et al.* 1995). English Nature is also keenly aware of the importance of empowering local groups to take responsibility for their own environment (under Agenda 21), and the educational values of partnership schemes in helping to change people's behaviour and attitudes. English Nature operate a variety of grant support schemes (Table 20.13) designed to engage the assistance of other agencies, organisations and individuals in achieving the government's nature conservation objectives.

Some species of wildlife are already protected under national or international legislation. There are strict controls, for example, on the possession, sale and display of native wild birds, and those that are taken into captivity must be registered under the Wildlife and Countryside Act 1981. Article 6 of EC Regulation 3626/82 also regulates the sale and display to the public for commercial purposes of certain endangered birds. Most

TABLE 20.13 National conservation grant schemes operated by English Nature, 1997

Grant scheme	*Details*
National Conservation Development Grants Scheme	Grants of around £10,000 are available for major nature conservation projects related to England as a whole and GB projects in co-operation with the Countryside Council for Wales and Scottish Natural Heritage
Species Action Grants Scheme	Grants are being made available for work on those priority species in the Biodiversity Action Plan which occur in England and for which Species Action Plans have been prepared
Species Recovery Programme Grants Scheme	Grants are available to organisations and individuals who contribute to a range of partnership projects which aim to restore, maintain or enhance populations of plants and animals that are in severe decline or currently under threat of extinction
Reserves Enhancement Grants Scheme	Funds are available for the comprehensive management of SSSIs by Trusts within the Wildlife Trust Partnership and other voluntary conservation organisations
Lowland Heathland Grants Scheme	Grants in recognition of the international importance of England's remaining heathlands, for heathland management work and for interpretation of heathland sites for the public. Final year 1997–8
Voluntary Marine Nature Reserves (VMNRs) Grants Scheme	Grants were available to support individuals or organisations involved in work on Voluntary Marine Nature Reserves. In April 1998 this scheme became the Voluntary Marine Initiative Grants, providing grants of up to £7,500 for marine projects

UK mammals are also protected under the Wildlife and Countryside Act 1981 and the EC Habitats and Species Directive 1992. The UK government is also a signatory to CITES (the 'Washington' Convention on International Trade in Endangered Species of Wild Fauna and Flora), and takes seriously its responsibility to monitor and control trade in wildlife (such as the trade in tiger and rhino products). The UK also ratified the 'Bonn' Convention on the Conservation of Migratory Species of Wild Animals. Through this it has signed the Agreement on the Conservation of Bats in Europe which came into force in January 1994 and is designed to encourage co-operation within Europe to conserve all its species of bats.

Conclusion

At the close of the millennium the United Kingdom still has an abundance of high quality natural and semi-natural landscapes, and an impressive diversity of wildlife. Despite many centuries of land cover change most of the countryside remains in good condition whilst continuing to support the nation's population by providing food, timber, water and other natural resources. As we have seen, the context within which landscape preservation and wildlife conservation operate has changed during the 1990s in five important ways. First, the natural environment enjoys a much higher profile in national politics than it has done previously, and this is reflected in the institutional changes introduced during the 1990s designed to produce better integration between different sectors of government decision-making and policy formulation. Second, efforts are being made to collect, analyse and make widely available appropriate information on the state of the UK environment and the quantity and quality of its natural capital (including both landscape and wildlife). This is evident in the range of government publications (and their on-line versions), including the annual digests of environmental statistics and the ongoing work on development of environmental indicators. Third, sustainable development has been embraced as an important framework within which to make decisions about all aspects of national resource use, and this has both elevated the status of environmental protection and raised awareness of the interrelatedness and co-dependency of all environmental systems. Fourth, and partly arising from this new focus on sustainable development, national and local interest within nature conservation has been galvanised around the need to preserve biodiversity, for both national and international benefit. This has broadened the perspective within nature conservation away from simply seeking to preserve key and characteristic species, and characteristic or rare habitats, towards a broader appreciation of the need to preserve as many species and habitats as possible. Fifth, whilst the government has an obvious responsibility to preserve the country's natural capital for our own domestic benefits, during the 1990s this responsibility has taken on an increasingly international dimension as well through the UK's commitment to initiatives such as the Rio Biodiversity Convention and European Union environmental directives. In the final analysis, decision-makers within and beyond the UK government recognise the need to preserve an environmental estate that is as natural and diverse as possible, provides as many benefits to the current population as possible, and can be passed on to future generations.

References

Adams, W.M. (1996) *Future Nature: a Vision for Conservation*, London: Earthscan, for British Association of Nature Conservationists.

Adams, W.M., Hodge, I.D. and Bourn, N.A.D. (1994) 'Nature conservation and the management of the wider countryside in eastern England', *Journal of Rural Studies* 10(2): 147–57.

Andrews, J. and Carter, S. (1993) *Britain's Birds in 1990–91: the Conservation and Monitoring Review*. Tring: British Trust for Ornithology/Joint Nature Conservation Committee.

Baldock, D. (1994) 'European environmental policy – trends and issues', *Ecos: a Review of Conservation* 15(1): 11–17.

Barkham, J. (1994) 'Wildlife Trusts – or mistrust?', *Ecos: a Review of Conservation* 15(2): 23–8.

Barr, C.J., Bunce, R.G.H., Clarke, R.T., Fuller, R.M., Furse, M.T., Gillespie, M.K., Groom, G.B., Hallam, C.J., Hornung, M., Howard, D.C. and Ness, M.J. (1993) *Countryside Survey 1990 Main Report*, London: Department of Environment.

Bignal, E.M. and McCracken, D.I. (1996) 'Low-intensity farming systems in the conservation of the countryside', *Journal of Applied Ecology* 33(3): 413–24.

Box, J. (1994) 'Changing the conservation culture', *Ecos: a Review of Conservation* 15(2): 17–22.

Box, J., Douse, A. and Kohler, T. (1994) 'Non-statutory sites of importance for nature conservation in the West Midlands', *Journal of Environmental Planning and Management* 37(3): 361–7.

Curtis, L.F. (1991) 'Conservation and protection', in R.J. Johnston and V. Gardiner (eds) *The Changing Geography of the United Kingdom* (2nd edn), London: Routledge.

Curtis, L.F. and Walker, A.J. (1982) 'Conservation and protection', in R.J. Johnston and J.C. Dornkamp (eds) *The Changing Geography of the United Kingdom* (1st edn), London: Methuen.

Elmes, G.W. and Free, A. (1994) *Climate Change and Rare Species in Britain*, ITE Research Publication 8, London: HMSO, for Natural Environment Research Council/Institute of Terrestrial Ecology.

Goldsmith, F.B. and Warren, A. (eds) (1993) *Conservation in Progress*, London: Wiley.

Hearn, K. (1994) 'The "natural aspect" of the National Trust', *British Wildlife* 5(6): 367–78.

Idle, E.T. (1995) 'Conflicting priorities in site management in England', *Biodiversity and Conservation* 4(8): 929–37.

Jewell, T. and Steele, J. (1996) 'UK regulatory reform and the pursuit of "sustainable development": the Environment Act 1995', *Journal of Environmental Law* 8(2): 283–300.

Organisation for Economic Cooperation and Development (1994) *Environmental Indicators: OECD Core Set*, Paris: OECD.

Owens, S. (1993) 'Planning and nature conservation – the role of sustainability', *Ecos: a Review of Conservation* 14(3–4): 15–22.

Palmer, M.A. (1996) 'A strategic approach to the conservation of plants in the United Kingdom', *Journal of Applied Ecology* 33(6): 1231–40.

Park, C.C. (1997) *The Environment: Principles and Applications*, London: Routledge.

Perkins, P. (1996) 'The North Peak: an environmentally sensitive area in the UK', *Environmentalist* 16(4): 263–8.

Price, M.F. (1996) 'People in biosphere reserves: an evolving concept', *Society and Natural Resources* 9(6): 645–54.

Raustiala, K. and Victor, D.G. (1996) 'Biodiversity since Rio: the future of the Convention on Biological Diversity', *Environment* 38(4): 17–20, 37–45.

Sheail, J. (1995) 'Nature protection, ecologists and the farming context: a UK historical context', *Journal of Rural Studies* 11(1): 79–88.

UK Government (1990) *This Common Inheritance, Britain's Environmental Strategy*, Cm 1200, London: HMSO.

UK Government (1994a) *Sustainable Development: The UK Strategy*, Cm 2426, London: HMSO.

UK Government (1994b) *Biodiversity: the UK Action Plan*, London: HMSO.

UK Government (1995) *Digest of Environmental Statistics*, Number 17, London: HMSO.

UK Government (1996a) *Indicators of Sustainable Development*, London: HMSO.

UK Government (1996b) *Digest of Environmental Statistics*, Number 18, London: HMSO.

UK Government (1996c) *Government Response to the UK Steering Group Report on Biodiversity*, Cm 3260, London: HMSO.

Whitby, M. and Adger, W.N. (1996) 'Natural and reproducible capital and the sustainability of land use in the UK', *Journal of Agricultural Economics* 47(1): 50–65.

Williams, G., Holmes, J. and Kirby, J. (1995) 'Action plans for United Kingdom and European rare, threatened and internationally important birds', *Ibis*, 137 (Supplement 1): 209–13.

World Bank (1995) *Monitoring Environmental Progress: A Report on Work in Progress*, Washington, DC: World Bank.

Further reading

There is a large and rapidly-growing literature on the theme of biodiversity and conservation, both in general and with specific reference to the United Kingdom. A broad introduction to the general field is Richard Huggett, (1998) *Fundamentals of Biogeography* (London: Routledge). Other recent overviews include Mike Jeffries (1997) *Biodiversity and Conservation* (London: Routledge), Gordon Dickinson and Kevin Murphy (1998) *Ecosystems* (London: Routledge), Ian Spellerberg (ed.) (1996) *Conservation Biology* (London: Addison Wesley Longman), and Ian Bradbury (1998) *The Biosphere* (second edition) (London: Wiley). Sound treatments of particular habitats in the UK include Alan Fielding and Paul Haworth (1999) *Upland Habitats* (London: Routledge), and Helen Read and Mark Frater (1999) *Woodland Habitats* (London: Routledge). David Evans (1996) *A History of Nature Conservation in Britain* (second edition) (London: Routledge), offers a detailed review of the subject, whilst a longer-term and broader perspective is given in B. Clapp (1994) *An Environmental History of Britain since the Industrial Revolution* (London: Addison Wesley Longman). The place of nature conservation within environmental policy is outlined in Phillip Lowe and Stephen Ward (eds) (1998) *British Environmental Policy and Europe* (London: Routledge), and in Andrew Gilg (1996) *Countryside Planning* (second edition) (London: Routledge). The whole realm of ethics and rights within nature is also very topical, and this is well illustrated in Robert Elliott (1997) *Faking Nature: The Ethics of Environmental Restoration* (London: Routledge). An important emerging theme is the social construction of nature, which is introduced in Bruce Braun and Noel Castree (1998) *Remaking Reality: Nature at the Millennium* (London: Routledge).

Chapter 21

Growing into the twenty-first century: social futures

Hugh Matthews

Introduction

Like almost every country in the world (the exceptions being Somalia and the USA) the UK government ratified in 1991 the United Nations Convention of the Rights of the Child (UNCRC). This Convention provides a comprehensive framework of human rights for all people under the age of 18. In essence, children have the right to life, which is universal and not contingent on any corresponding responsibilities by themselves (Lansdown 1995). All responsibility rests with adults (either parents or the state) who should ensure that all children are offered the basic rights of protection, provision and participation. As such the Convention establishes an international framework which in many parts challenges the traditional assumptions of the status of children and the meaning of childhood (James *et al.* 1998). In the first of two concluding chapters to this book, I consider what it is like to be a child growing into the twenty-first century within the UK and whether the position of children has been in any way changed or transformed in the decade since the UNCRC's ratification. This is an immense subject and so in order to provide a focus to this review my emphasis will be upon aspects of the impact of only one of the fifty-four Articles outlined in the Convention; that is, Article 12. Under this Article, the government of the UK is required to:

- enable all children to express their views freely about all matters which affect them;
- ensure that their views are given due weight, in keeping with their age and maturity;
- give children the opportunity to be heard in any administrative or legal proceeding which affects them.

Article 12 therefore provides 'a powerful assertion of children's right to be actors in their own lives and not merely passive recipients of adult decision-making' (Lansdown 1995: 2). From a geographical perspective the implications of Article 12 are far-reaching. In effect, this Article establishes the fundamental right for under-18s to be involved in decisions about provision and protection of all kinds, and ratification is seen to be a commitment to the principle of participation. Yet, despite the clamour and fanfare which followed the UK government's signing of the Convention, a culture of non-participation by young people is endemic in the UK, and no more so than in the context of decisions surrounding young people's social, economic and environmental futures. For the most part, young people are seemingly invisible in 'political' decision-making and seldom given opportunities to express their preferences (Matthews 1995; Matthews and Limb 1998). For example, routinely and regularly local councils, health and education authorities and government departments make proposals for development and planning. These cover a broad range of issues from transport, housing and wastes management to leisure, environmental design and landscape conservation. Planning decisions of this sort affect everyone, including children. However, as many commentators point out – for a review see Adams and Ingham (1998) – young people do not feature prominently on these agenda and the way in which they are included in the practices of consultation varies considerably.

The case of structure planning further illustrates young people's general lack of involvement in local and community affairs. Each county, metropolitan and unitary authority is required to produce a structure plan in order to address wider strategic and local policy issues. During the 1990s every borough in London began to prepare a unitary development plan (UDP) along these lines. Plans of this kind provide a framework for development, provision, amenity and access. They are intended to be comprehensive and to be responsive to the disparate needs of local populations. In 1994 a comprehensive review of the policies contained in the UDPs of each of the thirty-three London Boroughs was carried out (Planning Aid for London 1994), with a view to determining whether young people had been consulted or involved in this process and the extent to which local policies were targeted at their needs. The report concluded that few UDPs considered young people's issues seriously and that there was little practice of consultation. There is little evidence to suggest that London differs from elsewhere in the UK with regard to the participation of young people in local planning. For these sorts of reasons the Children's Rights Office (Lansdown 1995) declares that children in the UK – that is, those aged under 18 – constitute a disenfranchised group of 13 million citizens with no public voice. In this chapter I review young people's current involvement in participatory processes, consider discourses about young people's competence as social actors, examine the development of a variety of forms of local participation especially the role of youth councils, and consider what the future may hold for young people with respect to their chances for taking part in society.

Taking part and the right to say

On face value it would seem that the lack of participation by young people in many aspects of UK society is an outcome of their own actions. For example, when young people's participation rates in the general political processes are considered there is considerable evidence for low levels of interest and involvement (Bynner and Ashford 1994; Park 1995; Furlong and Cartmel 1997). People under 25 are more likely than any other group not to be registered to vote, with only 43 per cent voting in the 1992 General Election. Explanations for political apathy of this kind hinge on two competing interpretations (Park 1995). On the one hand, there is a view that non-involvement is something which has always been a universal characteristic of young people and that with age and growing responsibilities political interest will develop. For example, a recent survey in the UK (Bynner and Ashford 1994) has shown that the majority of 15- to 16-year-olds (72 per cent) are not at all interested in politics, an attitude confirmed in parts of Europe, USA, Canada and Australia (Wilkinson and Mulgan 1995), and one which is in stark contrast to overall turn-out rates in national elections. On the other hand, another explanation is that disillusionment and apathy is a recent phenomenon and symptomatic of a trend that will become more apparent as the present 'new' electorate grows older. Evidence in support of this view includes the declining figures of party membership by young people across the whole political spectrum (Cole 1997). The plummeting of Young Conservative membership from 50,000 in 1970 to 10,000 by the early 1990s is typical of this trend. Also, in a report produced by the Industrial Society (1997) prior to the General Election of 1997, only 5 per cent of 12- to 25-year-olds claimed interest in national politics of any sort, and a substantial majority, 80 per cent, of 16- to 25-year-olds felt they were not part of any political party. Possible reasons for this growing sense of 'political disconnection' are that young people are now too 'busy', given the developing range of leisure opportunities, or are more 'satisfied' due to increased material affluence compared to their parents and grandparents. Another suggestion is that political disaffection is strongly associated with a growing cynicism about politics, grounded in accusations of sleaze and corruption which do nothing to inspire the interest of young people (Bynner and Ashford 1994).

An alternative explanation, however, is that young people's lack of participation and interest in politics is a product of their strong sense of marginalisation, an outcome of the ways in which they are treated by adults prior to the voting age. For the most part, despite a governmental commitment to the UNCRC, young people, especially those under 18, are provided with few opportunities to engage in discussions about their economic, social and environmental futures and seldom given chances to express their preferences outside of adult-dominated institutions (Hart 1997; Matthews 1992, 1995; Matthews and Limb 1998). It would seem that despite some changes in recent years, which have begun to shift the relationship between those who hold power and those on the receiving end of that exercise of power, the practice of consultation with young people is not well established. In essence, participation within the UK is conceived to be an adult activity (Oakley 1994). A mid-decade review carried out by the Article 12 Network, an organisation launched in 1996 and run by young people with the aim to ensure that all under-18s know they have a right to express their opinion, draws attention to what is and what is not happening within UK law, planning, policy and practice about the involvement of young people (Table 21.1).

There is ample evidence to suggest, however, that if young people are given more responsibilities and more chances to participate in the running of society, then they will be more willing to engage in the processes of democracy (Hodgkin and Newell 1996). For example, in single issue organisations where young people are encouraged to take part, membership statistics confirm a growing participation rate. Amnesty International's youth section increased from 1,300 in 1988 to 15,000 in 1995; Greenpeace's youth membership rose from 80,000 in 1987 to 420,000 in 1995; and Friends of the Earth report a growth of 125,000 new young members over the same period (British Youth Council 1996).

TABLE 21.1 The right to say: young people's involvement in policy and practice

Policy area	*Positive involvement*	*Limited or no involvement*
Central government		No vote until 18; not allowed to stand as MP until 21. No statutory obligation on central government to consult children about anything. No government office to represent children's views; no independent children's ombudsperson or commissioner for children (unlike many European countries).
Local government	New guidance on planning for children's services encourages providers to find out what children think and to take these opinions into account in the planning process.	No legal obligation to consult children about local planning, housing, transport policy, leisure services, crime prevention or local environmental policies. Not allowed to be a local councillor until 21.
	Agenda 21 guidelines encourage consultation with young people on environmental issues.	There has been no consistent response and many local authorities have not put in place appropriate structures.
	Children Act (1989) for England and Wales requires local authorities (and courts) to consider the wishes and feelings of children when making decisions concerning their welfare. An outcome has been that some local authorities have appointed children's rights officers.	The scope of the Children Act is extremely limited. The obligation to take account of children's wishes and feelings only has application once conflict arises within a family leading to divorce or when the local authority has responsibility for the child's care. There is no general presumption in the Act that children have a right to participate in decision-making or that they should be encouraged to articulate their views.
	In some Scottish authorities 16-year-olds can stand and vote in elections for community councils.	
Health services	The 'Gillick' case (1986) established the principle of the competent child. It recognised that children have the right to consent to medical and dental treatment once judged to have 'sufficient understanding'. This principle has been included only in Scottish law (The Children (Scotland) Act 1995).	The 'Gillick' principle has not been incorporated into primary legislation in England and Wales. Other more recent cases have undermined the ruling. In 1992 the Appeal Court ruled that until 18, if a child refused to give consent the parent could intervene and give consent on their behalf irrespective of the competence of the child. The notion of the competent child has largely been ignored.
	Department of Health guidelines on children in hospital and child health in the community does promote the intentions of the UNCRC.	

TABLE 21.1 continued

Policy area	*Positive involvement*	*Limited or no involvement*
Education	Some schools, mostly secondary schools, have school councils, but their powers are usually very limited.	The right of under-18-year-olds to be school governors removed by the Education Act (1986)
	Code of Practice on special educational needs encourages respect for children's views.	Children have no formal right to participate in matters concerning their education. It is parents not children who are defined as the 'consumers' of education: children are seen as the 'product'.
	Department for Education and Employment 'advice' suggests that behaviour policies of schools should be discussed with pupils.	Children do not have the right to participate in matters such as school choice, curricula, appeals over exclusions, school policy or administration. No requirement to involve children in decisions on, for example, school uniforms, arrangements for school meals, supervision in the playground, tackling bullying or discipline. Schools are not required to introduce complaints procedures. Parents can take their children out of religious education and sex education without the need to consult children.
Children in care	Care authorities have a legal duty since 1975 to consult children and take their views seriously.	A number of recent inquiries into children's homes report that there was a failure to consider the views of children and that there was a general predisposition not to believe children.
	Local authorities are required to establish complaints procedures for children in need.	
Family	In Scotland parents must consult children about all major decisions.	In England and Wales and Northern Ireland there is no obligation on parents to consult with children; there are no governmental guidelines encouraging parents to take children's views seriously.
	When parents divorce or separate courts must consider the views of children.	If parents agree about the arrangements to be made for children following divorce or separation, there is no opportunity for the child's view to be considered. Unlike parents and other family members, children have no right to apply to court for an order about such matters as where they should live and who they should have access to.

Source: Based on Article 12 Network (1996) and Lansdown (1995).

A culture of non-participation

Three factors appear central to this culture of non-participation within the UK. First, there remain discourses within society which question the appropriateness of children's political involvement. Second, there are those who doubt the capability of children to participate. Third, even amongst those who believe in the principle of children's right to say, there are uncertainties about the form that participation should take and the outcomes which might result (Matthews *et al.* 1999).

In spite of a growing lobby in favour of children's rights to participate, particularly fuelled by Article 12, there remains an intransigence in some quarters about whether such involvement is appropriate. Lansdown (1995: 20) identifies three reasons why some adults are reluctant for children to take part in decision-making that will impact on their own life and the lives of others. First, giving children the right to say threatens the basis of family life by calling into question parents' 'natural' authority to decide what is in the best interests of a child. Yet, as Qvortrup *et al.* (1994) suggest, to sustain such an argument, it must be beyond reasonable doubt that adults behave with children's best interests in mind. In practice, this is not always the case. Second, imposing responsibilities on children detracts from their right to childhood, a period in life which is supposed to be characterised by freedom from concern. Such a perspective ignores the fact that many children's lives are full of legitimate concerns which are products of the same social and economic forces that affect adults. A third strand to the argument is that children cannot have rights until they are capable of taking responsibility. This view is based on an idealised view of childhood, yet few children live without responsibilities. Alanen (1994) points out that children's labour and duties within the home are underestimated, whilst the reality of school work and its associated responsibilities are rendered invisible by the label 'education'.

A second, though related, argument against children's participation is based on a conviction that children are incapable of reasonable and rational decision-making, an incompetence confounded by their lack of experience and a likelihood that they will make mistakes. Furthermore, if children are left to the freedom of their own inabilities the results are likely to be harmful (Scarre 1989). Franklin and Franklin (1996) draw attention to a range of libertarian criticisms of these two viewpoints. As a starting point, children are constantly making rational decisions affecting many parts of their daily lives (some trivial, some less so) without which their lives would have little meaning, order or purpose. In addition, adults are often not good decision-makers and history bears this out. Indeed, this observation provides an incentive to allow children to make decisions so that they may learn from their mistakes and so develop good decision-making skills. More radically, it has been argued that the probability of making mistakes should not debar involvement, as such an assumption 'confuses the right to do something with doing the right thing' (Franklin and Franklin 1996: 101). Critics also draw attention to the existing allocation of rights according to age, which is flawed by arbitrariness and inconsistency. For example, within the UK; a young person is deemed criminally responsible at the age of 10, sexually competent at the age of 16, but not politically responsible until the age of 18, when suddenly, without training or rehearsal, young people enjoy the right to suffrage. Last, denying rights of participation to everyone under the age of 18 assumes a homogeneity of emotional and intellectual needs, skills and competences. Both of these positions are imbued with an adultist assumption that children are not social actors in their own right, but are adults-in-waiting or human becomings.

Denigrating children in this way not only fails to acknowledge that children are the citizens of today (not tomorrow), but also undervalues their true potential within society and obfuscates many issues which challenge and threaten children in their 'here and now' (Matthews and Limb 1999).

Third, the debate about children's right to participate is compounded by a divergence of views on the nature, purpose and form that participation should take. For some (Hart 1992, 1997; Lansdown 1995), democratic responsibility is something which does not suddenly arise in adulthood but is a condition which has to be nurtured and experienced at different stages along a transition and so should be a feature of all democratic education. 'It is unrealistic to expect them [children] to become responsible, participating adults at the age of 16, 18 or 21 without prior exposure to the skills and responsibilities involved' (Hart 1992: 5). In addition, there is ample evidence to suggest that the involvement of children in local decision-making acts as a catalyst for participation amongst the community as a whole (Hart 1997). Others (Council of Europe 1993; Storrie 1997) argue, however, that education of this kind is disempowering in that it is designed primarily to integrate young people into existing social and institutional structures on which they are unable to exert any real influence. Instead, if participation is to be truly effective it should be carried out in such a way that the material influence of young people becomes progressively enlarged. Participation here is more broadly conceived to be the right to influence, in a democratic manner, processes bearing upon one's own life and the development of local youth policy. This debate relates closely to notions of education versus empowerment and training versus emancipation (de Winter 1997).

Local places and community involvement

From the work of geographers, there is considerable empirical support for the view that young people are competent social (environmental) actors, with the capability and adeptness to take part in decision-making which affects their everyday lives (Matthews and Limb 1999). For example, both Matthews (1992, 1995) and Hart (1997) have shown that children, from the age of 6 years, have the capacity, ingenuity and motivation to become keenly involved in determining the development and management of local places. Initially, children's horizons are set within a domestic context of care (e.g. care of animals and plants, gardening at home). As they become older so their interests and involvement can be broadened and diversified from taking part in local environmental management schemes (e.g. recycling, weather surveys, wildlife surveys, waste audits) through to a growing range of community-based projects (e.g. school councils, youth club committees, young people's forums). As a result of their involvement children will be drawn into increasingly complex social and political milieux and gain a sense of moral responsibility. Furthermore, participation of this kind engenders feelings of belongingness and rootedness, which are important dimensions of citizenship (Matthews *et al.* 1999).

A number of national organisations have taken a lead in attempting to integrate young people into community planning of this kind. The Council for Environmental Education (CEE), an umbrella body for a broad range of environmental and educational organisations, secured funding in 1996 from the Department for Education and Employment to establish the National Young People's Environment Network (NYPEN). Targeted at 13- to 19-year-olds, and organised into six regional groups, NYPEN aims to raise awareness of environmental issues and to encourage action for local environmental improvement (CEE 1996). For some time, The Tidy Britain Group has encouraged young people to take an interest in their local environments. As part of their 'Going for Green' programme, the Eco-Schools Award Scheme aims to increase environmental awareness and involve young people in decision-making and action in order to improve their school environs. The World Wide Fund for Nature (WWF UK) has a wide-ranging educational programme designed to encourage change in young people's attitudes and behaviour with regard to their environmental lifestyles and to promote their involvement in local planning. The Countryside Commission, English Nature and English Heritage have worked to encourage young people to become involved in their communities and have set in place various opportunities to facilitate participation. Organisations such as the British Trust for Conservation Volunteers and the Groundwork Trust (GT) work in partnership with all sections of the community to create opportunities for young people to engage in community affairs. GT reports that it has developed more than 4,000 regeneration schemes in 120 towns and cities, involving over 46,000 volunteers and 117,000 schoolchildren. None the less, the take-up of all these initiatives is both sporadic and uneven and there is no clear geography. What is apparent, however, is that young people from all social backgrounds and from every kind of geographical locale, whether from a decaying inner city or a leafy suburb, have been involved. Yet, whilst the engagement of young people in environmental project work has great value, there is also a danger that they act as little more than a volunteer labour force, with little chance to develop their own ideas.

Youth councils and the participation of young people

There are some encouraging signs, too, that at the local level attitudes are changing with regard to the involvement of young people in decision-making. There are a number of associated reasons for such a development. First, the momentum given to young people's rights in general by the UNCRC has been added to by the principles set by Local Agenda 21. Amongst its many declarations for a sustainable future is the view that dialogue should be established between the youth community and government at all levels, which enables young people's perspectives and visions to be incorporated as a matter of course into future environmental policy (Freeman 1996). Second, local government reorganisation in the mid-1990s has provided a stimulus for youth issues to be addressed in a strategic manner, partly through a need to demonstrate community consultation and partly to tackle what is perceived to be 'the youth problem' (Griffin 1993; Wynn and White 1997). Third, there is the 'millennium factor'; as we move towards the turn of the century there seems to be an emerging sense that the future is for our children (Hackett 1997; Storrie 1997), and local decision-making is critical to young people's well-being. As part of this movement towards giving young people a *say* has been the development of youth councils. The term 'council' is used to describe the range of ways in which congregations of young people come together, usually, but not exclusively, in committee, to voice their views about their needs and aspirations on their social and physical worlds.

Youth councils have been around for some time. There have been two surges of interest prior to the present day. During the late 1940s and 1950s a considerable number of youth parliaments were set up throughout the UK as a means for supplementing the adult-run Youth Service. In 1949 there were as many as 240 youth councils, based largely on 'rotarian' lines (Joseph 1984). Butters and Newell (1978) identify three ideological pulses behind these developments: *character building*, which aimed to integrate young people into society and so produce mature citizens capable of rebuilding the country; *social education*, which sought to move young people into positions where they could work for institutional reform; and, more radically, *self-emancipation*, conceived as a means to equip young people with the skills and capabilities to challenge and to take control of those organisations (and structures) which effectively disenfranchised them. These early attempts failed, however, partly because of a lack of common purpose, for there was little cohesion between these three strands, and partly because the councils were fundamentally flawed, in that they had been set up by adults with political agendas divorced from the priorities and sensibilities of young people (Crossley 1984). A second wave of youth councils developed during the mid-1980s. The Thompson Report (1982) on the Youth Service laid great stress on the idea that young people should participate in decision-making and that the best way forward was through youth councils (Paraskeva 1992). At the time a number of county youth services sought to establish youth councils in each of their major towns. However, few of these councils lasted more than a few years. Like those established in the earlier round, the driving force behind young people's participation was grounded not upon convictions of desirability and basic rights, but on political expediency. Unfortunately, in their rush to form youth councils many youth services made the fatal mistake of creating makeshift structures and constitutions.

The youth councils of today represent a new wave of interest in this form of political participation. A recent survey (Matthews and Limb 1998) has revealed that there are over two hundred youth councils within the UK. A number of national organisations have played important yet differing roles in their development. A consequence of their varying approaches is an unevenness of provision within the four home countries. In England, the National Youth Agency (NYA) and the British Youth Council (BYC) provide advice and information on request about youth councils. The Wales Youth Agency (WYA) has a similar remit. However, these agencies, although committed proponents of young people's participation, have limited capacity to support development. Because of this, the development of youth councils in England and Wales has largely been a haphazard one. Their form and character thus depend partly on such factors as the demography, political make-up and traditions of a locality, and partly on existing institutional and organisational structures and charismatic individuals.

In Scotland there is a more coherent strategy. Here a partnership between the Scottish Community Education Council (SCEC), Youth Link Scotland and the Principal Community Education Officers Group, which followed four years of research and consultation, gave rise to the 'Connect Youth' programme, launched in 1995. Targeted at 14- to 25-year-olds, this programme seeks to promote effective involvement of young people in the decision-making processes which affect their lives and to engage young people in determining their views on services and the development of opportunities for enhanced community involvement (SCEC 1996). However, these are guiding principles and it is up to individual voluntary and statutory agencies how these ideas are translated into practice. Inevitably, there has been a diversity of outcome. Of major significance, none the less, is the development of a network of youth forums throughout Scotland (located in Ayrshire, Clackmannanshire, Dumfries and Galloway, Dunbarton, Dundee, Falkirk, Fife,

Lanarkshire, Mid-Argyll, Shetland, and Stirling). To help support the transition of this programme into the new single-tier authorities, a number of national initiatives have been developed. These include the creation of a Youth Training Scheme to recruit a hundred young people to support the work of Connect Youth at a local level and the establishment of a Youth Issues Unit to provide a focal point to collect, collate and disseminate information on issues facing young people in Scotland.

By far the strongest tradition of youth councils in the UK is within Northern Ireland. In 1979 the Department of Education established the Northern Ireland Youth Forum (NIYF), with a brief to encourage the development of a network of Local Youth Councils (LYC). Members of the LYCs were recruited from local youth groups, including statutory and voluntary agencies, both uniformed and non-uniformed. Each youth group was eligible to send two representatives aged between 16 and 25 to a LYC. In turn, each LYC elected two young people to the NIYF. In the first ten years of the project between sixteen and twenty LYCs were operational out of an initial target of twenty-nine, and these were supported financially by five Education and Library Boards (Youth Service). The purpose of the LYCs was to get young people involved in tackling local issues and to ensure that their voices were heard by local district councils. The NIYF, on the other hand, took on a broader role and attempted to provide a national platform for young people's issues. In a review of its achievements, the Northern Ireland Youth Forum draws attention to a variation of outcomes. These arose for a number of reasons, including differences in funding between each of the five Boards, a structure which was perceived to be too 'top down' in its approach and emphasis, lack of a clear agenda, and no formal methods of monitoring and evaluating effectiveness. Since then significant changes have taken place. The NIYF now co-ordinates the activities of more than fifty groups and is proactive in campaigning for young people's rights across four major domains: policing, accommodation, employment, and education (NIYF 1996). As a result of high profiling in the media, young people's views are increasingly valued by statutory providers such as the Training and Employment Agency, police authorities, health trusts and Education and Library Boards. Currently being discussed are proposals to get youth representatives on each district council and the formation of a Northern Ireland youth parliament. Amongst the assurances of the new Labour administration is to send a Northern Ireland minister to the Youth Forum every year. None the less, given the geographical spread of constituent groups, some difficulties remain. Notably there is a diversity of infrastructure, inequities in support funding and problems in co-ordinating the activities of groups. Furthermore, policy differences between statutory agencies complicate the ways in which young people's suggestions are taken up.

The survey by Matthews and Limb (1998) suggests that the development of youth councils represents an organic rather than a structured movement towards incorporating the views of young people. Herein lies a major problem. At present, unlike many other European countries (see Matthews *et al.* 1999), there is no single organisation responsible for their inception. Even when national agencies are involved decisions are largely left to individual statutory and voluntary organisations. Given these circumstances there is still a strong sense that these are novel and slightly 'risky' experiments, which reinforces their position outside of the mainstream. Symptomatic of this general lack of organisation is that there is no comprehensive listing of youth councils and only recently has there been any attempt to compile a directory (an initiative launched by the National Youth Agency and the British Youth Council in 1997). However, in the absence of any governmental lead, there is hope that the spread of youth councils will generate a momentum that will see young people's involvement in their local community as something which is normal and natural.

Conclusion

As we move into the millennium, despite an all-party commitment to the UNCRC, we still do not have a culture of listening to children within the UK. Young people's lack of involvement in the formal political process after the age of suffrage within the UK is both a product of their marginalisation from local decision-making when growing up and an outcome of a strong sense of disenfranchisement and powerlessness during childhood. Up until the age of 18 years, young people have little opportunity for 'taking part' and are given little chance to make their views heard. In this process children are denigrated to little more than 'citizens-in-waiting', with little recognition afforded to their developing skills and competences. Indeed, the UN Committee on the Rights of the Child, the international body which was set up to monitor the implementation of the UNCRC, expressed concern in its meeting in January 1995 about the lack of progress made by the UK government in complying with its principles and standards. In particular, attention was drawn to the insufficiency of measures relating to the operationalisation of Article 12. It recommended that:

> greater priority be given to . . .Article 12, concerning the child's right to make their views known and to have those views given due weight, in the legislative and administrative measures and in policies undertaken to implement the rights of the child.
>
> (United Nations 1995: 15)

and went on to suggest that:

> the State party consider the possibility of establishing further mechanisms to facilitate the participation of children in decisions affecting them, including within the family and the community.
>
> (United Nations 1995: 15)

Because of these shortcomings, the Children's Rights Office argues that immediate and radical action is needed by the UK government if the culture of non-participation is to be broken down in the first decade of the millennium. This should involve action both to promote greater social awareness of children's right to participation and to create the necessary framework for participation. These ideas are summarised in Table 21.2.

Further criticism of the UK with regard to the status and position afforded to its young people emanated from the Habitat 2 seminar held in New York in 1996. The focus of this world meeting organised by UNICEF was on the the links between environment participation and children's rights. An outcome of the seminar was a set of recommendations which unequivocally places young people on local environmental planning agenda (Table 21.3). It was felt that the UK rarely matched up to these standards and in general there was insufficient dialogue between adults and young people in local decision-making. In essence,

TABLE 21.2 An action agenda: recommendations for improving children's right to participation

Establishing a social framework

- Ensure that all guidance provided by government departments on services for children reflects the principles of respect for the incorporation of the child's points of view and commitment to the child's right to participate in decision-making, in accordance with age and maturity.
- Promote the provision of public education campaigns in order to raise general awareness about the principles of the UNCRC, especially the need to listen to children and to take their views seriously.
- Establish the provision of parent education in order to fulfil the principles of children's rights to protection, provision and participation.
- Guarantee that the training of all professionals working with children is founded on the principles of the UNCRC, including the skills of communicating with children.

Establishing a legal framework

- Amend family law in England, Wales and Northern Ireland (thus bringing the law into line with that in Scotland) to require that in reaching any major decision relating to a child all those with parental responsibility and those in care and control of children must show regard for the views of the child and give them due consideration according to the child's age and maturity.
- Incorporate the 'Gillick' principle (that is, once a child has 'sufficient understanding', he/she should be able to make decisions for himself/herself) into all primary law relating to jurisdiction, unless there are specific legal restrictions (for example, recognition that the age of sexual consent for girls is 16).
- Provide new legislation which will require schools and local education authorities to put in place structures, such as school councils, designed to ensure that all children are given opportunities to express their views on matters which concern them about school administration and that these views are listened to and taken seriously, and to establish procedures for determining and giving due weight to the views of individual children on matters affecting them, such as school choice and needs assessment.

Source: Based on Children's Rights Office (Lansdown 1995).

these recommendations provide an environmental framework which can be added to the action agenda articulated in Table 21.2.

The burgeoning work of various national agencies and the development of local youth councils provide a range of different ways forward within the UK, both to integrate young people into their local communities and to encourage feelings of political worth and engagement. However, in the absence of a coherent national strategy, young people are likely to remain as 'outsiders' within society and continue to be largely disconnected from all decision-making processes. In this case, the UK has much to learn from the experiences of many parts of mainland Europe. Here, there is ample evidence of effective and well-established participatory structures which operate at a grass-roots level (see Matthews *et al.* 1999). Until co-ordinated policies are put in place which truly empower young people, the majority will continue to remain largely invisible on the social, economic and environmental landscapes of the millennium.

TABLE 21.3 Habitat 2: recommendations for young people's involvement in planning

Establishing an environmental framework

- ◆ In recognition that democratic behaviour in a civil society must be learned through experience, children should be given a voice in their communities, according to their abilities. This will serve as a preparation for their full participation in civil society as adults and will be a means of better meeting their needs as children.
- ◆ Basic education for children should include investigations and dialogue on local development and the local environment in order to facilitate participation for sustainable development.
- ◆ In recognition of the marginalisation of women in decision-making, attention should be given to preparing girls as well as boys with the confidence and skills to be involved as equal participants with their peers.
- ◆ Children should be involved, according to their capacities, in the design of environments intended explicitly for them, such as play places, schools and children's hospitals.
- ◆ Formal democratic mechanisms should be established for giving all citizens, including children, according to their capacities, a voice at the community and municipal level, both as a way of preparing for participation in civil society and as a way of improving the appropriateness and effectiveness of decision-making.
- ◆ Children's participation works best in a society which also encourages adult participation; the participation of adults and children must be complementary and mutually reinforcing.
- ◆ Local authorities should initiate the establishment of innovative partnerships between children, parents, schools, private sector and community-based organisations and NGOs to optimise the effectiveness of the existing structures by involving children in local community services provision. This will strengthen children's awareness and sense of belonging in the community.
- ◆ Children should participate according to their abilities in the management of all institutions and facilities that they use, including schools, recreation facilities, children's organisations and community organisations.
- ◆ Local government authorities should involve children, according to their capabilities, in local governance processes.

Source: UNICEF (1996).

References

Adams, E. and Ingham, S. (1998) *Changing Places. Children's Participation in Environmental Planning*, London: The Children's Society.

Alanen, L. (1994) 'Gender and generation: feminism and the child question', in J. Qvortrup, M. Bardy, G. Sgritta and H. Wintersberger (eds) *Childhood Matters. Social Theory, Practice and Politics*, Aldershot: Avebury Press.

Article 12 Network (1996) 'Taking children's views seriously', *Report Card*, November: 1–4.

British Youth Council (1996) *Young People, Politics and Voting*, London: British Youth Council.

Butters, S. and Newell, S. (1978) *Realities of Training: A Review of the Training of Adults who Volunteered to Work with Young People in the Youth and Community Service*, London: HMSO.

Bynner, J. and Ashford, S. (1994) 'Politics and participation. Some antecedents of young people's attitudes to the political system and political activity', *European Journal of Social Psychology* 24: 223–36.

Cole, M. (1997) 'Politics and youth', *Politics Review* 6(3): 5–9.

Council for Environmental Education (1996) *Earthlines*, Reading: CEE.

Council of Europe (1993) *The Development of an Integrated Approach to Youth Planning at a Local Level*, Strasbourg: European Steering Committee for Intergovernmental Cooperation in the Youth Field.

Crossley, C. (1984) 'The rise (and fall?) of local youth councils', *Youth and Society*, March: 24–5.

de Winter, M. (1997) *Children as Fellow Citizens: Participation and Commitment*, Oxford: Radcliffe Medical Press.

Franklin, A. and Franklin, B. (1996) 'Growing pains: the developing children's right movement in the UK', in J. Pilcher and S. Wagg (eds) *Thatcher's Children: Politics, Childhood and Society in the 1980s and 1990s*, London: Falmer Press.

Freeman, C. (1996) 'Local Agenda 21 as a vehicle for encouraging children's participation in environmental planning', *Local Government Policy Making* 23: 43–51.

Furlong, A. and Cartmel, F. (1997) *Young People and Social Change*, Buckingham: Open University Press.

Griffin, C. (1993) *Representations of Youth*, Cambridge: Polity Press.

Hackett, C. (1997) 'Young people and political participation', in J. Roche and S. Tucker (eds) *Youth and Society*, London: Sage/Open University.

Hart, R. (1992) *Children 's Participation: from Tokenism to Participation*, Florence: International Child Development Centre/UNICEF.

Hart, R. (1997) *Children's Participation: the Theory and Practice of Involving Young Citizens in Community Development and Environmental Care*, London: Earthscan/UNICEF.

Hodgkin, R. and Newell, P. (1996) *Effective Government Structures for Children*, London: Calouste Gulbenkian Foundation.

Industrial Society (1997) *Speaking up, Speaking out: the 2020 Vision Programme, Summary Report*, London: The Industrial Society.

James, A., Jenks, C. and Prout, A. (1998) *Theorizing Childhood*, Cambridge: Polity Press.

Joseph, S. (1984) 'Experience and what? Participation, local youth councils, and the Youth Service Review', *Youth and Society*, August: 13–15.

Lansdown, G. (1995) *Taking Part: Children's Participation in Decision Making*, London: IPPR.

Matthews, H. (1992) *Making Sense of Place: Children's Understanding of Large-scale Environments*, Hemel Hempstead: Harvester Wheatsheaf.

Matthews, H. (1995) 'Living on the edge: children as outsiders', in *Tijdschrift voor Economische en Sociale Geografie* 86(5): 456–66.

Matthews, H. and Limb, M. (1998) 'The right to say: the development of youth councils/forums in the UK', *Area* 30(1): 66–78.

Matthews, H. and Limb, M. (1999) 'Defining an agenda for the geography of children', *Progress in Human Geography* 23(1): 61–90.

Matthews, H., Limb, M. and Taylor, M. (1999) 'Young people's participation and representation in society', *Geoforum* 30(2): 1–10.

National Youth Agency (1996) *Youth Forums*, Leicester: NYA.

Northern Ireland Youth Forum (1996) *Can Young People Break the Glass Ceiling? Young People's Participation in Decision Making*, Belfast: Northern Ireland Youth Forum.

Oakley, A. (1994) 'Women and children first and last: parallels and differences between children's and women's studies', in B. Mayall (ed.) *Children's Childhoods: Observed and Experienced*, London: Falmer Press.

Paraskeva, J. (1992) 'Youth work and informal education', in J. Coleman, and C. Warren-Adamson (eds) *Youth Policy in the 1990s, the Way Forward*, London: Routledge.

Park, A. (1995) 'Teenagers and their politics', in *British Social Attitudes: 12th Report*, Devon: Dartmouth Press.

Planning Aid for London (1994) *Planning for the Future. Children Should be Seen and Not Heard?*, London: PAL.

Qvortrup, J., Bardy, M., Sgritta, G. and Wintersberger, H. (eds) (1994) *Childhood Matters: Social Theory, Practice and Politics*, Aldershot: Avebury Press.

Scarre, G. (1989) *Children, Parents and Politics*, Cambridge: Cambridge University Press.

Scottish Community Education Council (1996) 'Connect Youth: a national initiative to promote greater involvement of young people', *Progress Report*, September.

Storrie, T. (1997) 'Citizens or what?', in J. Roche and S. Tucker (eds) *Youth and Society*, London: Sage/Open University.

United Nations (1995) *Concluding Observations of the Committee on the Rights of the Child: United Kingdom of Great Britain and Northern Ireland*, CRC/C/15/Add.34.

UNICEF (1996) *Children's Rights and Habitat*, Report of the expert seminar convened by UNICEF and UNCHS Habitat, New York: UNICEF.

Wales Youth Agency (1996) 'Involving young people', *Newsline Occasional Paper*, March.

Wilkinson, H. and Mulgan, G. (1995) *Freedom's Children*, London: Demos.

Wynn, J. and White, R. (1997) *Rethinking Youth*, London: Sage.

Further reading

There is little work by geographers on children's rights. For an introduction into this topic read H. Matthews and M. Limb (1998) 'The right to say: the development of youth councils/forums in the UK', *Area* 30(1): 66–78; H. Matthews and M. Limb (1999) 'Defining an agenda for the geography of children', *Progress in Human Geography* (23(1): 61–90); and H. Matthews, M. Limb and M. Taylor (1999) 'Young people's participation and representation in society', *Geoforum* (30(2): 1–10). An excellent essay on children's participation in decision-making is provided by G. Lansdown (1995) *Taking Part: Children's Participation in Decision Making* (London: IPPR). A powerful case for increasing the rights of children is found in R. Hodgkin and P. Newell (1996) *Effective Government Structures for Children* (London: Calouste Gulbenkian Foundation). For an excellent introduction into what is meant by the social construction of children and childhood read A. James, C. Jenks and A. Prout (1998) *Theorizing Childhood* (Cambridge: Polity Press). A general review of growing up in the 1990s is found in J. Pilcher and S. Wagg (eds) *Thatcher's Children: Politics, Childhood and Society in the 1980s and 1990s* (London: Falmer Press). The many works of Roger Hart provide an invaluable guide to what can be achieved by children. In particular, R. Hart (1997) *Children's Participation: the Theory and Practice of Involving Young Citizens in Community Development and Environmental Care* (London: Earthscan/UNICEF).

Chapter 22

Growing into the twenty-first century: environmental futures

Vince Gardiner

This book is about the changing geography of the United Kingdom at the end of the twentieth century. Whatever geography is about (and even professional geographers cannot always agree amongst themselves on this) it does include, as at least part of the heart of the discipline, some consideration of the relationship between people and their environment. We hope that the preceding chapters in this book have gone some way towards describing the major geographical patterns in both the human and physical geography of the UK, and giving some understanding of the powerful social, economic, political and physical processes bringing about these patterns. However, some questions about people living in the physical environment which is the UK today remain largely unanswered, perhaps because the very organisation of the book, with separate chapters addressing individual aspects of the human and physical geography of the country, has marginalised them. These questions concern the real nature of the way in which individuals living in the United Kingdom today experience the complexity of the physical environment in which they live, and the sustainability of present relationships between people and their environment. Are there such things as a 'natural Britain', in which people enjoy a genuine communion with nature 'red in tooth and claw', or a 'technological Britain', with people insulated by technological ingenuity from nature? Does the physical environment have any real influence on the day-to-day lives of the modern technologically dependent and materially insulated Briton? In Britain today, is nature natural? Are the day-to-day lives of Britons becoming, perhaps paradoxically, more differentiated from one another because of increasing social and economic stratification, and spatial marginalisation, despite the impact of technology in removing the reliance of individuals on a highly differentiated physical environment? Is there a single 'correct' interpretation of British landscapes, perhaps as defined in the preceding chapters, or is it more complex?

Consider the case studies in Boxes 22.1 and 22.2, for differing views of parts of the UK.

BOX 22.1 Hilbre island: a natural environment, an experience of nature – or not?

Hilbre – a small, remote island in the centre of the Dee Estuary, formed of resistant red sandstones. Sitting on the grass-covered cliffs at the northern end of the island one looks out across the Irish Sea, with waves breaking below. In the distance to the west are the remote mountains of Wales, whilst to the east the red sandstone of Hilbre and the smaller islands to the south is seen to be a small-scale version of the red sandstone ridges which form the physical framework of the Wirral peninsula. The sand and mud banks of the Dee Estuary stretch out on either side of the island, with abundant seabirds feeding at the water's edge as the tides ebb and flow their daily cycles. Seals bask on the nearby West Hoyle Bank, their plaintive cries eerily sounding across the deep-water channel separating them from the island. There is a real sense of isolation, and no matter which direction one looks, one gazes into distance – across the sea, along the full length of the Dee Estuary, to the Sefton coast, or to the Llandudno Ormes. Alone in this location, one can be at peace with nature as one comes to grips with a geographical understanding of this part of Britain today.

Yet there is another side to Hilbre (or is it another interpretation of the same side?).

To the east beyond Hoylake the buildings of Liverpool are clearly visible – with the Liver and Cunard Buildings a constant reminder of when the United Kingdom faced outwards to an empire, and of the way in which its present spatial marginalisation at the periphery of the UK's European involvement has been sadly reflected in its economic decline since last century. The adjoining settlements of Caldy and West Kirby, from whence one sets off to walk the 2 kilometres to Hilbre at low tide, formerly housed the richest entrepreneurs from Liverpool but now constitute a commuter town for Merseyside and the Wirral, with bright yellow electric trains busying to and fro and the Wirral's roads funnelling commuters through the tunnels into Liverpool. At West Kirby, the Marine Lake serves as a focus for water sports, with the colourful sails of yachts and sailboards clearly visible from Hilbre. On Sundays, the island teems with people who come to watch birds, to walk their dogs, to do the multitude of things they do on the beach, or who simply come in order to say they've been, curiosity fulfilled, with nature communed. Some even come by car, or more usually four-wheel drive vehicle, following the recommended path across the flats. To the west across the Dee is Wales, with individual cars and buildings clearly visible, although to get there involves a 40-minute drive around the Dee Estuary, with historic Chester at its head – the nearness-combined-with-distance serving to undermine the duality of political identity within the countries of the United Kingdom. Arguably, little of even Hilbre itself is natural, with the island managed as part of a nature reserve, several houses and other buildings on its tiny surface, and at night the island dominated by a beacon flashing out its red warning. Indeed, the very existence of Hilbre is only guaranteed because the rocks which form it were protected from the attack of the sea by extensive revettments, constructed by the Liverpool Mersey Docks and Harbour Board, who bought the island in 1856 and erected a telegraph station and tide-gauge there.

BOX 22.2 Suburban London

A flat in Wandsworth. Sitting at the window, the view is of other blocks of flats, both Council and private, TV aerials on the top, satellite dishes facing south-east, like technological sunflowers following the sun. Above, planes roar overhead as, with monotonoous regularity, they approach Heathrow Airport, the busiest international airport in the world – a constant reminder of the capital's role in the UK, as a major international city, and a hub of the global economy. Trains can be heard on the privatised South Western Trains route, and on the Underground, as can traffic on the A3, with commuters ebbing and flowing their daily cycle, feeding and sustaining the economic vitality of the capital. Separately numbered car park spaces house cars, some used only at weekends as their owners use public transport to commute, and others used to transport their owners (usually individually) to work – or to the station only a kilometre away, with their cars never even reaching a pollution-minimising operating temperature. Many carry their lap-top computers ostentatiously during the week, and wear their designer clothes equally ostentatiously at weekends. A few kilometres away is the heart of London – the City, with its temples to the generation of wealth; the West End, with its temples to entertainment and consumerism; the dispersed centres of culture and sport, and globally significant centres of learning.

Surely, this is the heart of modern Britain, where technology and consumerism have triumphed, and nature has no role to play?

Yet there is another side to suburban London (or is it another interpretation of the same side?).

The very existence of the country's largest conurbation is dictated by the physical environment, with the fluvial history of the Thames dictating the early development of a crossing place which has endured and developed throughout the centuries, with even today archaeological evidence of the long occupance of the site emerging from beneath the urban fabric to be rescued for posterity. The very shape of the urban area is determined at some levels by the physical environment, with river terraces introducing interesting topographic variety to the city centre, and areas of poorer sandy soils in the suburbs being conserved as commons. Indeed, nature has never been stronger than within this leafy enclave of the urban area. Wimbledon Common is a hundred metres away, with its historic status as a common land protected by legislation, and the Commoners managing it to conserve its wildlife and diversity. Urban foxes regularly cross the car park at night, and the regular sightings of magpies testify to all being well in the ecological food pyramid. Magnificent pine trees around the car park enjoy protected status, and teem with grey squirrels. The very fabric of the urban area modifies the climate – funnelling winds, increasing convection and hence rainfall – and interactions with air pollution intensify these effects.

What is the geographical reality of these case studies? We are not going to answer this question, although it might dismay student readers of this book who are looking for 'the answers'. For one editor of this book (myself) both case studies, and both interpretations of both case studies, have elements of reality; this book was edited whilst employed in London and Liverpool, from houses in Wandsworth and West Kirby. But how this editor saw and experienced those landscapes is unique to that editor. If there is an answer to the question it is that each landscape of Britain is not only a complex product of physical and human processes but is also the landscape perceived both through the vision of each individual's eyes and also through the filters of their attitudes, knowledge, experience and even mood. The physical environment has not one definite place in the nation's consciousness, but an infinite number of different places.

A striking example of how physical 'reality' and people's consciousness interact in the context of one part of the physical environment is Tunstall and Penning-Rowsell's (1998) study of the English beach. They point out that the beach has a special place in the nation's consciousness, with the typically British experience of the day at the beach being one rich in ritual, symbolism, nostalgia and myths. Perhaps one of the most important elements it gives is the opportunity for tactile physical contact with the physical world, where children and even adults are allowed or even expected to pick up, touch, shape and play with its physical material – for example, sand and water, and the associated wildlife such as crabs and shellfish – without restrictive rules concerned with protecting the environment, or health and safety. The beach visit therefore reconnects people with nature in a very traditional way, as well as having definite memories associated with family, childhood and romance for many people. People seem in general to have the view that the seaside should always be like the seaside always was. However, a problem occurs when this has to be reconciled with the physical reality of a changing coastline. People seem to prefer to see the familiar beach, promenade or whatever, as they had always known them – no matter that this might be a long way from any 'natural' state in which physical processes could establish any degree of equilibrium between erosion and rock types. People are resistant to any notion of allowing natural processes to do their work at the coast, with a clear conflict between notions of 'nature' and 'control', and holding dear the notion of 'naturalness' – in Tunstall and Penning-Rowsell's words' 'taking in the view of the incoming tide while standing by the amusement arcades' (1998: 331). They see these attitudes as being a throwback to the Victorian image of coast as adventure and freedom, in a controlled environment where adventure and freedom are rendered risk-free by people's intervention, with the sea and its power being comfortably viewed from safe refuge. Such notions are of immense importance in devising ways for the management of coastlines in ways that are publicly acceptable as well as being physically sustainable.

How natural are the landscapes of the UK? If it were possible to remove the direct evidence of the impact of people (the structures such as buildings, roads, railways and so on, as well as crops and other planted and managed vegetation), would the landscapes remaining be in any sense 'natural'? Whilst the answer in terms of the gross relief (the existence of major valleys, hills and so on) must be 'yes', the answer in terms of much of the detail of the relief must be 'no'. Research has shown that people's activity, from the Neolithic onwards, has had major effects upon land use, and hence upon geomorphological processes such as sediment yield and sediment transport. Thus river patterns have changed, producing fluvial landscapes very different from what might have existed had people not intervened. Palaeoecological reconstruction based on evidence such as that of the pollen preserved in upland peat shows that people's impact upon vegetation has been dramatic. Whilst the upland landscapes of areas such as Dartmoor and the Highlands of Scotland might seem wild and unspoilt, they are in some senses every bit as unnatural as our urban areas. The climatic climax vegetation for most of the UK would be some form of forest, but virtually no natural forest now remains, with most of even our oldest woodlands being planted or managed to some degree, or at least modified in detail, and our 'wild' upland areas being produced by early anthropogenic activity, clearing the natural vegetation for agriculture.

Just how real is the physical environment to the average Briton? Two hundred years ago almost everyone had direct and intimate contact with the environment – the walk to work; work itself, often outdoors; recreation, if any, often in the outdoors. Even buildings arguably offered a far lesser degree of protection from the vagaries of the weather, although a case can be made for 'traditional' building materials and methods being much more effective in some ways than our theoretically more advanced methods. It could now at least in theory be possible to live life without any explicit contact with a natural environment – transported to and fro by air-conditioned car or train; working in a completely artificially controlled environment; recreation in completely artificial environments within leisure centres, or at home, with the workplace's workstation being echoed by the home's playstation, providing arcade-type games, and the Internet providing, potentially, access to every computer in the world. Yet in many ways the technological sophistication of people's activity has, at least through the rose-tinted spectacles of retrospect, perhaps paradoxically made life more rather than less dependent upon the physical environment – steam trains usually made their destinations, despite wet leaves on the line or the wrong kind of snow to which electric and diesel trains are prone to fall victim; buildings and other structures were less vulnerable to storms – or were able to be replaced without the need for substantial investment; and less-intensively bred crops were less vulnerable to changes in climate and crop pests. At one level, people in Britain have never been in more contact with their environment, as membership of organisations with an environmental or wildlife concern, such as the National Trust, Greenpeace, and Royal Society for the Protection of Birds continues to rise, and growing numbers of people continue to participate in outdoor recreation of various kinds. However, at another level, that of genuine oneness with the physical environment, contact has never been less than in the past. Fewer and fewer people work regularly outdoors, with the number of people employed in activities such as agriculture, extractive industry and fishing declining. Fitness levels of children in schools are a serious concern of government, with the decline of organised competitive outdoor sport, the rise of solitary and sedentary activities such as computer gaming, and undesirable habits such as smoking, particularly amongst females, having unwelcome effects amongst young people. For many, the environment, and all that it stands for, is packaged as a commodity, to be experienced in a tamed and moderated way, rather as Tunstall and Penning-Rowsell describe for the beach visit. A day out in the country might consist for many of a drive to a country park, a leisurely stroll around a nature trail, and an ice-cream from the convenient vendor. The wild osprey at Symond's Yat is observed nesting through the telescope from the viewpoint, with RSPB guide at hand – certainly it's better than no osprey at all, but is it really a genuine contact with nature?

Questions such as those raised above are of great relevance to the question of whether the present existence of people in the physical environment of the UK is a sustainable one; if people do not experience the environment, can they value it, cherish it and safeguard it in any meaningful way? Apart from its people, Britain's finest resource is its rich diversity of landscapes, yielding many tangible physical resources, such as water, minerals and soil, as well as less tangible resources such as landscape and a topography suitable for human activities of many kinds. The sustainable use of this resource, against a background of changing economic, social, political and physical processes, is perhaps the biggest challenge faced by Britons today. By sustainable is meant the present use without adversely affecting the resource for future generations. Too often in the past attempts have been made to separate the activity of people from environmental hazard, and to mitigate the impact of people on the environment, by what might be termed hard engineering, based on a rigorous understanding of environmental processes and relevant scientific principles. However there is increasingly a realisation that many physical processes are inherently unpredictable, at least at the detailed level which is relevant to the activity of people, being often dependent in a very sensitive way upon the sequence of historical events, as well as sometimes upon their spatial distribution. Thus policies for environmental management should be flexible, and should recognise this inherent unpredictability of many physical processes. This argument is developed in detail for coastal management by Turner *et al.* (1998), who in addition point out that for many ecological economists policy analysis must also incorporate both ethical and philosophical dimensions, and those related to considerations of social equity. For example, in considering pollution, it can be argued that those who suffer from the actions of others should be given full compensation; more generally, those who benefit from actions which adversely affect the environment should compensate those who suffer. Whilst monetary compensation does not necessarily ensure sustainability, it does provide an incentive towards it.

The future will hold many challenges for the ways in which people interact with their environment in the UK, with the complexly linked effects of climatic change, the Greenhouse effect, stratospheric ozone depletion and sea-level change being obvious contenders for the most significant. However, other developing issues and potential problems identified by some chapters in this book, such as increasing urbanisation, decreasing biodiversity, diminishing stocks of primary resources, increasing subsidiarity of the UK's political system to that of the EU, increasing globalisation of economic activity, and increasing commodification of nature might in the final analysis be more important. This chapter set out to raise questions, not to provide answers. If it has stimulated thinking which will in any way contribute to an increased effectiveness of the custodianship of the land of the UK by its citizens of the twenty-first century it will have served its purpose, as will the rest of this book.

References

Tunstall, S. and Penning-Rowsell, E.C. (1998) 'The English beach: experience and values', *Geographical Journal* 164(3): 319–32.

Turner, R.K., Lorenzoni, I., Beaumont, N., Bateman, I.J., Langford, I.H. and McDonald, A.L. (1998) 'Coastal management for sustainable development: analysing environmental and socio-economic changes on the UK coast', *Geographical Journal* 164(3): 269–81.

Index

Note: **Emboldened** page numbers indicate chapters and *italicised* page numbers indicate maps and boxes. People mentioned only once and places and organisations mentioned only once or twice have generally been omitted.